CLASSICAL MECHANICS

CLASSICAL
MECHANICS

John Michael Finn

JONES AND BARTLETT PUBLISHERS

Sudbury, Massachusetts

BOSTON   TORONTO   LONDON   SINGAPORE

World Headquarters

Jones and Bartlett Publishers
40 Tall Pine Drive
Sudbury, MA 01776
978-443-5000
info@jbpub.com
www.jbpub.com

Jones and Bartlett Publishers
Canada
6339 Ormindale Way
Mississauga, ON L5V 1J2
Canada

Jones and Bartlett Publishers
International
Barb House, Barb Mews
London W6 7PA
United Kingdom

Jones and Bartlett's books and products are available through most bookstores and online booksellers. To contact Jones and Bartlett Publishers directly, call 800-832-0034, fax 978-443-8000, or visit our website at www.jbpub.com.

Substantial discounts on bulk quantities of Jones and Bartlett's publications are available to corporations, professional associations, and other qualified organizations. For details and specific discount information, contact the special sales department at Jones and Bartlett via the above contact information or send an email to specialsales@jbpub.com.

ISBN 978-0-7637-7960-3

Production Credits
Publisher: David Pallai
Cover Design: Tyler Creative
Printing and Binding: Malloy, Inc.
Cover Printing: Malloy, Inc.

Library of Congress Cataloging-in-Publication Data
Finn, J. Michael.
  Classical mechanics / J. Michael Finn.
    p. cm.
  ISBN 978-1-934015-32-2 (hardcover)
  1. Mechanics. I. Title.
  QC125.2.F56 2008
  531—dc22

                              2008015455

6048
Printed in the United States of America
13 12 11 10 09   10 9 8 7 6 5 4 3 2 1

In Memoriam: Galileo Galilei (1564–1642)

Once again, the Inquisition called me in for questioning,
    "Are you quite certain
    Of what you have claimed to see?"
    I tell you, I sweated, despite the chill in the air.

Misunderstanding their intent, I persisted,
    "Through my lenses, I saw what I saw: spots on the sun;
    Valleys, mountains, and plains on the moon;
    More moons circling Jupiter; the phases of Venus.

"The Heavens are a sight to behold.
    God's creation is grander than we dared imagine.
    Would anyone here care to see for himself?
    Perhaps make use of the telescope lying before you on the table?"

Their response, when it finally came, was solemn and sure,
    "This Council will not be lectured
    On that, of which
    We have already decided.

"Let us speak plainly:
    You are disturbing the people.
    The truth will be what we say it is to be,
    Neither one word more, nor one word less."

Defeated at last,
    Blinded by age,
    Broken in spirit,
    I consented to their demands.

"Your will be done, but, nevertheless,
    Earth turns round the sun still,
    No matter what you choose to believe,
    Or what I am forced to confess."

—John Michael Finn, adapted from The Butterfly Girl ©2002

CONTENTS

PREFACE

This textbook presents an updated treatment of the classical dynamics of particles and their interaction fields, suitable for students preparing for advanced study of physics and closely related fields, such as astronomy and many of the applied engineering sciences. The material is based on lectures notes that I have developed over a number of years for use in Physics 601, Classical Mechanics, taught at The College of William and Mary in Virginia. Physics 601 is required for all entry-level graduate physics majors at our University.

The text is designed for a one or two semester course of study at the entering graduate student level, or for use by well-prepared students at the advanced undergraduate level. All twelve chapters can be covered in a two-semester sequence. With a judicious choice of material, up to eight to ten chapters can be covered in an accelerated, one-semester, course of study. Optional, advanced, sections are marked with an asterisk as guidance to the instructor. A caveat, the text contains more explanatory material, examples, and historical references than is intended to be covered in the lectures. Instead of "teaching from the book," the lecturer should focus on presenting the key concepts, proofs, and critical examples. Introductory material can be quickly summarized for more advanced students, or amplified as needed for novices.

In deciding on the level of presentation of this material, I tried to follow Einstein's dictum, "Everything should be made as simple as possible, but not one bit simpler." There are concepts and techniques, which future practitioners of any field must acquire, for which it serves no useful purpose to evade or postpone. Many of the older traditional treatments of mechanics, for example, attempt to avoid the complexities of tensor analysis by using orthogonal coordinates, avoiding the metric tensor, and obscuring the differences between covariant and contravariant tensors. This works fine until one gets to special relativity, where the student suddenly finds himself having to unlearn things he thought he already knew. Moreover, the notation he has learned is dated and has little in common with that being taught in other advanced courses. The concepts of invariance and covariance are too important to bypass.

On the other hand, most entering graduate students in American universities have a limited mathematical background. One cannot assume that they are already fluent in the latest mathematical techniques of differential geometry. I read somewhere once that one can only teach others that which they already know and are prepared to accept. One should use the familiar to illuminate the unknown. The approach that I have chosen to follow is to start with the standard vector calculus notation, commonly used in undergraduate programs, and build from this foundation.

Compared to older textbooks on this subject, the mathematical treatment has been updated to better prepare the student for the study of advanced topics in quantum, statistical, nonlinear, and orbital mechanics. For example, the concept of phase space is introduced in the first chapter; the metric structure of space and curvilinear coordinates are introduced in the second chapter; and generalized coordinates are introduced in the third chapter. In the section on rotational motion, geometric algebra concepts are used to develop the spin rotation group and the results compared to the standard vector analysis formulations. Dirac bra-ket notation is introduced as an alternative to the older tensor formulation of dyads and dyadics, and may be used interchangeably.

The text begins with a review of the principles of classical Newtonian dynamics of particles and particle systems. It proceeds to show how these principles have been modified and extended by subsequent developments in the field. The text ends with the unification of space and time given by the special theory of relativity in Chapter 11 and with an introduction to the theory of continuous fields in Chapter 12.

Hamiltonian dynamics and phase space concepts are introduced early on. This allows integration of chaotic behavior and other nonlinear effects into the main flow of the text, rather than as a separate chapter, tacked on as an afterthought. For example, Poincaré sections are used in Chapter 5, "Central Force Motion," to illustrate the onset of chaos in the nonrelativistic three-body problem. The symplectic structure of Hamiltonian dynamics is used to demonstrate the conservation of phase space and to introduce the discussion on Liouville's theorem.

The role of symmetries and the underlying geometric structure of space-time are key themes. Particular emphasis is placed on the role played by Galilean and Poincaré invariance of isolated physical systems. Noether's theorem, relating the continuous symmetries of the Lagrangian to conservation laws, is developed early in the text. The relationship between physical observables and generators of the motion are explored.

In the latter chapters, the connection between classical and Quantum Mechanics is examined. For example, the close connection between the Poisson bracket formulation of Classical Mechanics and the Heisenberg equations of motion of Quantum Mechanics is explored in Chapter 9, "Canonical Transformation Theory". Likewise, the relationship between Hamilton's principal function and the geometric limit of the Schrödinger wave equation is explored in the following chapter on Hamilton-Jacobi theory. Chapter 11, covering special relativity, prepares the student for the breakdown of the classical particle picture. Possible solutions are explored in the following chapter on the continuous theory of fields.

The text attempts to strike a balance between analytic and computational approaches to the subject matter. Analytic techniques provide great insight into the systematic behavior of dynamical processes, but are limited to integrable systems. Numerical methods allow one to generate numerically accurate results and greatly assist in the visualization of complicated physical systems. My own preference is to allow computer-generated plots and tabular data, but to require that most other results and explanations be translated into a human-readable dialect (such as English) using standard mathematical notation for equations.

There are a number of high-level scientific analysis and visualization software packages available on the commercial market such as Maple™, Mathcad®, Mathematica®, and Matlab®, which can be used in conjunction with this course of study. Many universities provide site licenses for one or more of these products. Matlab was used to assist with the generation of a number of the illustrations presented in the text. However, the text itself makes no assumption as to which, if any, software tools are to be used to assist in completing the exercises. This is a matter that is best left to the discretion and preference of the instructor. The present state of software design is evolving so fast that it would rapidly date the text to rely too heavily on a particular programming solution.

The International System of Units (SI) are used throughout[1]. Conversion factors to other commonly used systems of units are presented in Appendix A on constants and conversion factors. Transformation to dimensionless units is a commonly used technique—mathematical functions are defined over dimensionless fields in any event—a summary of the most common constants is given in Appendix B. Students are encouraged to use primary sources for physical constants by directly accessing the latest Committee on Data for Science and Technology (CODATA) values of physical constants. These are readily available online [Mohr 2007]. The Particle Data Group's compilation of fundamental physics data [Yao 2006] is a useful alternative resource. Likewise, complete tables of integrals and functions are readily available from a number of sources, for example see, for example, [Abramowitz 1965; Beyer 1987]. Students should consider acquiring a good desk copy of one of these reference works. Recommended online resources (as of 2007) include Wolfram's Mathworld[2] and the upcoming NIST Digital Library of Mathematical Functions[3], which is presently under development and scheduled to be completed in 2008. I personally find it easier to enter a formula into a symbolic integrator to find a solution than searching through tables of integrals. After an initial solution is found, it is then

---

[1]See the NIST Reference on Constants, Units, and Uncertainty [WWW: NIST]. Available at *http://physics.nist.gov/cuu/index.html.*
[2]WWW: Woram Mathworld] Available at *http://mathworld.wolfram.com/.*
[3][WWW: DLMF] Book and web site are currently under development, and are expected to become available in 2008 at *http://dlmf.nist.gov.*

useful to resort to a comprehensive reference to study the properties of the function, to verify and optimize the result, and to gather greater insight into the systematic behavior of the system being studied.

The concept and organization of this work is a product from my years of experience in teaching classical mechanics, initially following the outlines presented by Goldstein [Goldstein 2002] and Fetter and Walecka [Fetter 1980], who drew on older classics such as Landau and Lifshitz [Landau 1976]. My approach to the material has been strongly influenced by recent pedagogical advances made by a number of present day instructors such as Professors Morii of Harvard [Morii 2003] and Professor Tong of Cambridge [Tong 2005], who have been kind enough to place their lecture notes on the World Wide Web, and by more recent textbooks significantly updating the mathematical formalism [Doran 2003; Hestenes 2002; José 1998] or reexamining its philosophical foundations [de Gossen 2001]. My own research into subatomic structure via electroweak interactions has certainly influenced the choice of topics, as has my interest in understanding physics concepts in their historical context. (In my opinion, the significance of an idea can be more fully appreciated by examining it from the context of the cultural milieu in which the idea was first framed.) The bibliographic listing, presented in the reference section, presents a more complete acknowledgement than is possible here. I would be remiss in not acknowledging my students for their patience and feedback. My daughter, Danette Zeh, contributed her efforts to some of the line art.

The initial reviewers of this book made many helpful suggestions that I attempted to incorporate and for which I am grateful. Finally, I want to thank my publisher, David Pallai, for his help and guidance with this project.

Classical Mechanics, the study of the dynamical behavior of physical objects, is not some musty relic, abandoned to time. From the moment that the proverbial apple fell on Newton's head to the very present, it has continued to evolve. Old issues have been resolved, while new issues have arisen; and, in the natural course of evolution, the old can sometimes become the new again. In the end, all physics comes down to the observation and measurement of positions, angles, and intervals of time: the very stuff of geometry. When all ideas have been exhausted, the wonder of geometry remains.

Although the center of fundamental physics research has moved on to the edges of the known universe—to the very small, the very large, and the exotic—a solid grounding in the principles of mechanics is still considered to be foundational for the advanced study of physics. The very concepts of space, time, matter, and energy flow out of classical principles of dynamics. And, like it or not, measurement remains essentially a classical construct.

That which is "real" or "true" can only be extracted using the filter of human experience. The only true constraint imposed by the scientific method is to be willing to abide by the results of experimental reality. Too much is made about the theoretical-experimental split in the field of physics, which, to my mind, is mostly a reflection of temperament and interest. To attempt to separate the concrete from the abstract is to create a false division. Experiment without theoretical insight is clueless, while theory without experimental guidance is sterile.

—John Michael Finn
Williamsburg, VA
2008

REVIEW OF NEWTONIAN PARTICLE MECHANICS

Chapter **1**

In this chapter, the Newtonian principles underpinning the study of the classical dynamics of particle systems will be reviewed in their historical context. The most important results of Newton's *Principia* will then be summarized. Elementary analytic and numerical methods of solving the initial value problem will be examined next, and the concept of phase space introduced. Solutions to selected one-dimensional systems will then be explored.

1.1 INTRODUCTION TO CLASSICAL MECHANICS

In his *Philosophiae Naturalis Principia Mathematica*[1] [Newton 1687], or the *Principia* for short, Isaac Newton (1643–1727) introduced the abstraction of point-like *particles* as a basis for understanding the time evolution of the material universe. Extended objects, which, in the continuum limit, may include liquids and solids, are treated as aggregates of material particles having a property called *inertia*. Inertia, in turn, describes the resistance of particles to a change in their state of motion. *Mass* is the proper measure of inertia. Likewise, *linear momentum* **p** is the proper measure of uniform motion. A change in the motion of a particle is attributed to the action of an *applied force*.

The physical particle of Newton's conceit is an idealization for a closed physical system having constant mass and negligible dimensions. It allows one to ignore the complicated internal forces that must be carefully balanced for such structures to persist. The Newtonian particle is the prototypical example of the simplifying hypothesis, which is often essential for scientific progress.

Classical Mechanics deals with the description (*kinematics*) and origin (*dynamics*) of the motion of particles and their interaction fields. It is predicated on the *Aristotelian premise* that every effect (a change of motion)

---

[1] Latin for the "mathematical principles of natural philosophy."

must have a sufficient cause (a net applied force). Since force originates from the mutual interaction of particles, cause and effect, like yin and yang, are intractably intertwined. Response to a cause gives rise to new causes.

Dynamics and kinematics are connected through Newton's *Second Law of Motion*, written as

$$\mathbf{F}(\mathbf{r}, \mathbf{p}) = d\mathbf{p}/dt, \tag{1.1}$$

where the force $\mathbf{F}$ on a particle of mass m is understood to be a function of its coordinate points $\mathbf{r}$ and linear momenta $\mathbf{p}$. This leads to a coupled set of deterministic second-order differential equations to be solved. The equations can also be expressed as coupled first-order differentials in phase space by defining the velocity in terms of the momentum.

$$\dot{\mathbf{p}} = \mathbf{F}(\mathbf{r}, \mathbf{p}), \quad \dot{\mathbf{r}} = \mathbf{v}(r, p) = \mathbf{p}/m. \tag{1.2}$$

The motion of the particle is then interpreted as a *flow* in phase space.

The goal of dynamics is to predict the future evolution of a system of particles, given a complete set of initial coordinates and particle momenta. Analysis of this motion is complicated by the interaction of the particles with their environment. The concept of the *physical field*, as a first order property of the continuum, would develop slowly over the two centuries following Newton, culminating in the work of James Clerk Maxwell (1831–1879) on *Electromagnetic Theory* [Maxwell 1864].

The description of the motion of a system of particles can be represented in a variety of ways. One can specify the trajectories of every particle in an ensemble by individual curves $\mathbf{r}_i(t)$ in a 3-dimensional Euclidean space as shown in the left side of Figure 1.1, or one can consider the trajectory of a single point $\mathbf{R}(t)$ in a 3N-dimensional, multiparticle, *configuration space*

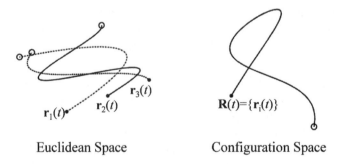

Euclidean Space   Configuration Space

FIGURE 1.1   Trajectories of a system of particles in 3-dimensional Euclidean space and in 3N-dimensional configuration space. The closed circles denote the initial state and the open circles the final state of the system.

as shown in right side of the figure. Here, $\mathbf{R}(t) = \{\mathbf{r}_i(t)\}$ denotes an n-tuple array of the vectors $\mathbf{r}_i(t)$. The two pictures are equivalent formulations. A complete description of the system at any time requires specification of the both positions and velocities of the particles at that time.

Alternatively, both the initial coordinates and the momenta of a system of particles can be specified in a $2n$-dimensional *phase space*. First-order differential equations of motion can then be used to evolve this system in time. This is the preferred method when using numerical analysis techniques.

Classical dynamics as a fundamental endeavor was largely superseded in the twentieth century by the development of *Quantum Mechanics*. Nevertheless, the classical approach has its virtues. In particular, it provides a simple, direct way of thinking about the motion of material objects that facilitates development and articulation of basic physical concepts. Therefore, the study of Classical Mechanics continues to be considered foundational for more advanced study of physics. Moreover, the development of *chaos theory* in the latter half of the twentieth century shows that the classical approach is still capable of surprising insights. It well may be that the future development of grand unified theories may yet revert to a dynamical interpretation of the motion of matter, combining nonlinear soliton models, chaotic behavior, and hidden degrees of freedom to bridge the gap between the quantum mechanical and classical world views.

Before examining the *Principia*, this review of Newtonian Mechanics will present a brief history of earlier western science, so that Newton's achievement may be considered in its historical context. The most important themes of the *Principia* will then be summarized. The reader should already be aware of the major results from previous study. Elementary methods of solving the equations of motion will then be presented. The chapter ends with an analysis of some representative one-dimensional systems. The remaining chapters of the book will develop deeper insights into the nature of classical dynamics.

1.2   HISTORICAL CONTEXT

The Greco-Roman[2] world had an excellent grasp of statics and material properties as witnessed by their temples, aqueducts, coliseums, and other artifacts they left for the ages to admire. In the third century before the Common Era, Archimedes of Syracuse (ca. 287 BC–ca. 212 BC) appears to be among the first to formally expound on the nature of force, developing the foundations of hydrostatics and explaining the principle of the lever. An unbalanced force

---

[2]For an overview of ancient Greek science see [Farrington 1981].

was recognized to be a sufficient cause for motion. The essential vector character of force, its magnitude and direction, was understood very early on, and static systems could be analyzed using the methods of Euclidean[3] geometry and plane trigonometry.

Static equilibrium of a rigid body requires a neutral balance of forces $\mathbf{F}_i$ and their *torques* $\mathbf{N}_i$, where the torque is the first moment of the force with respect to a rotation about an axis. In modern mathematical notation, one requires that the vector sums of the forces and of the torques acting on all parts of the system vanish

$$\sum_i \mathbf{F}_i = 0, \tag{1.3}$$

$$\sum_i \mathbf{N}_i = \sum_i \mathbf{r}_i \times \mathbf{F}_i = 0. \tag{1.4}$$

For example, Figure 1.2 shows a two-dimensional diagram of a ball of mass m that is suspended by two strings, one of which happens to be parallel to the ground. The force diagram (see inset), given by $\mathbf{T_1} + \mathbf{W} + \mathbf{T_2} = 0$, forms a right triangle for this particular example. Measurements of two angles and the magnitude of the weight allow one to solve for the string tensions

$$W_y = mg = T_1 \cos\theta, \quad T_2 = T_1 \sin\theta. \tag{1.5}$$

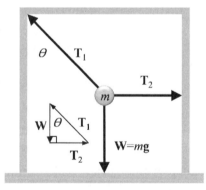

FIGURE 1.2 Equilibrium diagram of a ball of weight suspended from two ropes. The result is a triangle that can be analyzed using the methods of Euclidean geometry and trigonometric functions.

---

[3]Euclid of Alexandria (ca. 325 BC–ca. 265 BC).

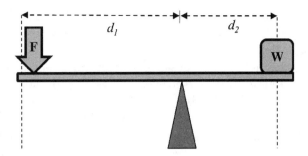

FIGURE 1.3   A lever applies mechanical advantage to lift a heavy weight with a smaller applied force.

The ancients also developed a number of simple *machines*, such as levers, pulleys, screws, and inclined planes. A machine, in its classical sense, describes a device for doing work. For example, the lever achieves its purpose by balancing the torques exerted on either side of a fulcrum, as is illustrated in Figure 1.3. By adjusting the relative distances from the fulcrum of an applied force **F** and a counterbalancing weight **W** to be lifted, one can gain *mechanical advantage*. That is, a smaller force can be used to balance a larger force. In this diagram, the mechanical advantage is given by the proportionality of the ratio of forces to the inverse ratio of their transverse displacements from the fulcrum

$$W : F :: d_1 : d_2. \tag{1.6}$$

When moving the lever *adiabatically*, the work done by the applied force is equal to the work extracted in lifting the weight through an infinitesimal displacement

$$\mathbf{F} \cdot d\mathbf{r}_1 = -\mathbf{W} \cdot d\mathbf{r}_2. \tag{1.7}$$

1.2.1 Ancient Cosmology

Unlike statics, the ancient world's understanding of the science of motion was deeply flawed. They failed, for example, to appreciate the relative nature of motion. Most viewed the Earth as forming an absolute frame of reference, the so-called "fixed foundation for the firmament." The prevailing view was that it required a force to generate and maintain a state of motion on the surface of the planet—force was somehow related to absolute velocity— the contrary evidence that the stars and planets appeared to move without resistance on the celestial sphere was explained away by adopting a dualistic cosmology: the heavens were perfect, and the Earth was imperfect.

It was as natural for the stars to revolve continuously about the pole of the Earth's axis as it was for a ball to come to rest at its lowest point on terra firma.

Nevertheless, ancient astronomy was surprisingly sophisticated in its predictive abilities. In some civilizations, advanced calendars and observatories were developed long before writing systems became generally available. The circle being the most perfect of shapes, the *Ptolemaic system*[4] of the Alexandrian Greeks successfully explained the motion of the planets, sun, and moon as the friction-free motion of *epicycles* built upon compounded circular motion. This is an early example of the application of a series solution. Given enough epicycles (circles within circles), complicated motions, such as the regression of the planets, could be explained and the future motion of heavenly bodies could be accurately predicted. The Ptolemaic system was a geocentric theory, wherein the planets and the stars of the celestial sphere circled a stationary Earth at the center of the universe.

True, there were those with differing worldviews. Aristarchus (ca. 310 BC–ca. 230 BC), for one, showed that the sun was much larger than the Earth, which led him to propose that the Earth circled the sun, but these voices had little lasting influence on the development of western thought.

Noteworthy among the earliest Greek contributions, Pythagoras (ca. 569 BC–ca. 475 BC) and his school made a number of advances in astronomy. Pythagoras recognized that the orbit of the moon was inclined to the equator of the Earth. He was one of the first to realize that Venus as an evening star was the same planet as Venus as a morning star. However, although there is an appeal to observational evidence in these discoveries, Pythagoras's school of thought was based fundamentally on the supremacy of mathematical perfection, which is, by its nature, in opposition to the scientific approach. However, Pythagorean philosophy does contain at least one essential concept that has had a lasting impact: namely, the idea that complex phenomena can be understood in terms of simpler ones. This has proved a powerful paradigm throughout the history of science. It played an important role in the development of the genius of Newton and Einstein.

The Greeks deserve much credit for developing the *scientific method*, that is, the use of observation to validate and refine hypotheses. Aristotle (384 BC–322 BC), for one, used the circular shapes of the phases of the moon to argue that the Earth must be a sphere. But the Greek thinkers were primarily philosophers, not scientists, and tended to rely excessively on pure reason.

---

[4]Claudius Ptolemaeus (ca. 90–ca. 168 AD) known in English as Ptolemy.

After the collapse of the Western Roman Empire, the center for scientific discovery moved eastward. The Arabs made important contributions to algebra, optics, and astronomy that would have a powerful impact on western culture at the dawn of the Renaissance and the beginning of the Age of Discovery. Arabic numerals are a product of their interaction with the culture of the Indian subcontinent. China was a center for technical innovation, producing machines that would be unrivalled until the modern age.

1.2.2 Copernican Revolution

The *Copernican Revolution* challenged the prevailing medieval European worldview by placing the sun at the center of the universe and putting the Earth into motion about it. In this, the Catholic bishop and worldly diplomat, Copernicus (1473–1543), simply re-created what many ancient Greeks already knew or had suspected. The major difference was the existence of the Gutenberg (c. 1400–1468) printing press, which allowed the wide spread distribution of his epoch-shattering book, *De revolutionibus orbium coelestium (On the Revolutions of the Heavenly Spheres)*. Copernicus's version of the *heliocentric theory* maintained the use of epicycles to correct for the details of the motion of the planets about the sun. Since the orbits of the major planets are nearly circular in their motion about the sun, a system of epicycles about the sun converges faster than those built on epicycles about the Earth. Nevertheless, the new scheme was not sufficiently superior to the Ptolemaic system to supplant it. In the years immediately preceding Newton, further developments of the heliocentric theory, based on the works of Kepler and Galileo, would prove to be particularly influential.

1.2.3 Kepler's Laws of Planetary Motion

Johannes Kepler (1571–1631), serves as an appropriate archetype for the computational physicist. Kepler analyzed the detailed planetary observations of the astronomer Tycho Brahe (1546–1601) and developed accurate, although empirical, laws of planetary motion based on the heliocentric model of the universe. Kepler's three laws, which will be examined in detail in Chapter 5, "Central Force Motion," are as follows:

1. The *law of ellipses* asserts that the orbits of planets are ellipses with the sun at one of the foci.
2. The *law of areas* states that the planets sweep out equal areas in equal times.
3. The *law of periods* states that the square of the periods of the planets are proportional to the cube of the semimajor axes of the orbits.

1.2.4 Galileo and the Principle of Relativity

Kepler corresponded with Galileo Galilei (1564–1642) [Gamov 1961; Copper 2001] who further popularized the Copernican system.[5] Hearing of the invention of the telescope, Galileo constructed one of his own. What he found, when he pointed his instrument at the heavens, would change man's place in his universe forever.

The Milky Way consisted of an uncountable number of stars. The moon, far from being a perfect sphere, had surface features like the Earth. The planets were spheres, like the Earth and moon, some with moons of their own revolving about them. The sun had sunspots. The heavens and the Earth were apparently more similar than dissimilar in composition.

The stars were too distant to be resolved, but by the time of Newton, Edmond Halley (1656–1742), was able to discern the proper motions of the closest stars. It did not take much imagination to suppose, at their calculated distances, that they must be powerful beacons of light—suns like our sun. The universe was much larger than humankind had ever suspected.

If Kepler could be called a computational physicist, then Galileo was the model of the modern experimental physicist. He performed numerous experiments to validate his hypotheses. Going beyond mere observation, he was a true student of the scientific method. For example, he measured the acceleration of gravity, using his pulse as a timepiece, since accurate clocks would be a latter invention. He discovered that all bodies tended to fall with a constant acceleration towards the center of the Earth. From these observations, which Galileo published and which were widely read, Newton was able to deduce that the net force acting on a mass point must be proportional to its acceleration and not to its velocity.

Galileo also emphasized the need to account for all the forces acting on the system. In his writings, he surmised that a feather and a cannon ball falling in a vacuum, thereby eliminating the greater effect of air resistance on the feather, would necessarily fall at the same rate. This early thought experiment would be dramatically demonstrated in the vacuum of the moon almost four centuries later, when Apollo 15 astronaut Dave Scott simultaneously dropped

---

[5]By the time of Kepler and Galileo, specific identification of the country of origin for what had already become a European-wide Renaissance in thought is largely superfluous. The German Kepler and the Italian Galileo were in close communication. Their scientific descendent, the Englishman, Isaac Newton, was well aware of continental discoveries. In the modern age, Physics has become a unified human endeavor embraced by a global community of scientists.

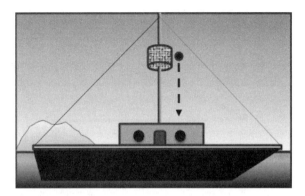

FIGURE 1.4 Ignoring air resistance, a cannon ball dropped from a mast falls straight down to the deck of a smoothly moving ship irrespective of the ship's relative motion to fixed land.

a feather and a hammer to determine which would fall faster in the Moon's gravitational field.[6]

Galileo also expounded on the relative character of motion, which he propounded in a number of discourses. In his *Dialogue Concerning the Two Chief World Systems* [Galileo 1632], Galileo has his protagonist observe that a cannon ball, released at rest from the crow's nest on the mast of a ship moving uniformly on smooth seas, would drop straight to the base of the mast. The result, as illustrated in Figure 1.4, was independent of the motion of the ship relative to fixed land. In like manner, the uniform motion of the Earth in the heavens would have no effect on the relative motion of objects on the Earth.

The *Principle of Galilean Relativity* requires that objects moving in an isolated room behave in the same manner, regardless of whether the room is moving uniformly, or remains at rest, with respect to a fixed frame. Velocities in different frames are related by the vector addition of the relative velocity of the different reference frames. Let $\mathbf{v}_A$ and $\mathbf{v}_B$ denote a particle's velocity in frames A and B respectively, and let $\mathbf{v}_{rel}$ be the relative velocity of the two frames; then the law of vector addition gives

$$\mathbf{v}_B = \mathbf{v}_A + \mathbf{v}_{rel}. \tag{1.8}$$

The sum of vectors forms a triangle as shown in Figure 1.5. The transformation implicitly assumes that time flows uniformly for all observers.

---

[6]A video of the event, originally televised live to a worldwide audience, is available at the *NASA Lunar Feather Drop Home Page*: *http://www1.jsc.nasa.gov/er/seh/feather.html.*

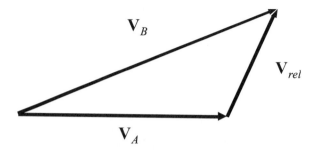

FIGURE 1.5 Vector addition of velocities using the Galilean velocity transformation. The velocity measured in frame *B* is composed by the vector addition of the relative velocity of the two frames to the velocity measured in frame *A*.

1.2.5 Descartes and the Development of Analytic Geometry

A contemporary of Galileo, Rene Descartes (1596–1650) succeeded in reducing geometry to algebraic methods by developing analytic geometry. Analytic geometry relies on using coordinates to describe the trajectory of points. His Cartesian coordinates use a rectangular grid to position points with respect to a fixed frame of reference. Reportedly, Descartes was inspired in development of his coordinate system by the rectangular pattern of tiles in his bathroom. Descartes' use of infinitesimal measures to calculate tangent lines laid the foundations for the later development of the calculus by Isaac Newton and Gottfried Leibniz.

Descartes believed that the universe could be understood in terms of the motion of corpuscles of matter. In his view, nature abhors a vacuum; therefore, the motion of bodies causes other bodies to fill the void, giving rise to the circular motion of the planets. In Chapter 7 of his *Treatise on the World*, dealing with the laws of nature, he develops three principles of motion, which anticipate the law of conservation of momentum. According to Descartes [Mahoney 2000]:

1. Each individual part of matter always continues to remain in the same state unless collision with others constrains it to change that state.
2. When one of these bodies pushes another, it cannot give the other any motion except by losing as much of its own at the same time.
3. When a body is moving, each of its individual parts tends always to continue its motion along a straight line.

Kepler, Galileo, Descartes, and a whole host of other luminaries created the milieu in which Newton worked. Newton would acknowledge as much, stating in a letter to Robert Hooke (1635–1703) dated February 5, 1675: "If I have seen further, it is by standing on the shoulders of giants."

1.3   NEWTON'S *PRINCIPIA*

Newton, the quintessential prototype for the theoretical physicist, synthesized the evidence gathered by his colleagues of the Age of Reason. As Albert Einstein (1879–1955) would later point out, Newton's *Principia* was the first successful attempt to construct a uniform theoretical foundation for the science of motion [Einstein 1940]. According to Einstein, in Newton's system, everything could be reduced to the following fundamental concepts [Einstein 1931; 1940]:

1. The concepts of absolute space and time
2. The existence of mass points with invariable mass
3. Force, representing the reciprocal action of material points on each other, via action at a distance
4. A set of laws describing the motion of mass points

The concept of a particle, which in common usage is roughly equivalent to the Greek *atom*, is one that bears analyzing in the context of Newton's philosophy. The Greek atom is the idealized building block, the irreducible element, devoid of internal structure, identical to all its brethren, which, when combined in a myriad of ways, gives rise to the complexity of matter.

The *atomic theory*, however, was not generally accepted as proven until Einstein's paper on *Brownian motion* appeared in 1905. The concept of particle creation and annihilation are even more recently discovered phenomena. A Newtonian particle is both simpler and more ambiguous than either the Greek atom or the modern concept of the *elementary particle*. It is an abstraction of an object with insignificant size, having constant mass and ignorable internal degrees of freedom. In context, even extended bodies may be properly considered as particles. Some examples of objects that can be treated as particles on some scale of behavior include:

−   Electrons moving in a vacuum tube
−   A baseball in flight
−   The planets orbiting the sun

Newton's implicit assumption of mass conservation for *closed systems* would be superseded in the *Theory of Special Relativity* by the concept of total energy conservation, but, at nonrelativistic speeds, mass and total internal energy are essentially the same, as expressed by Einstein's famous formula $E = mc^2$ relating mass and energy.

A classical Newtonian particle has a well-defined position $\mathbf{r}(t)$, a velocity $\mathbf{v}(t) = d\mathbf{r}/dt$, a mass m, and a linear momentum $\mathbf{p} = m\mathbf{v}$. Its trajectory can be analyzed using the methods of the differential calculus, which Newton co-invented, independently of Gottfried Wilhelm Leibniz (1646–1716).

1.3.1 Absolute Time

In addition to his three laws of motion, Newton introduces two underpinning hypotheses, the concepts of absolute time and of absolute space. Newton is clear about the hypothetical character of both. About *absolute time*, Newton, in his *Principia*, writes:

> "Absolute, true, and mathematical time, in and of itself, and of its own nature, without reference to anything external, flows uniformly."   [Newton 1687]

In other words, time is a scalar quantity that flows uniformly for all observers and for all times. It should be emphasized that, at the time of Newton, accurate means of telling time did not exist. The only reliable clock was the motion of the stars about the polar axis and that was only available at night. Other clocks were variable by design. The Roman day was twelve hours from sunrise to sunset, so that the hours of the day in the Roman Empire varied both with the season and with one's latitude. Newton elaborates

> "For the natural days are truly unequable, though they are commonly considered as equal and used for a measure of time. Astronomers correct this inequality for their more accurate deducing of the celestial motions. It may be that there is no such thing as an equable motion, whereby time may be accurately measured. All motions may be accelerated and retarded, but the true or equable progress of absolute time is liable to no change."   [Newton 1687].

Nevertheless, absolute time was understood by Newton to be a fiction. It is not measurable. Only relative duration, or time difference, is physically important. Newton writes further:

> "Relative, apparent, and common time is any sensible and external measure (precise or imprecise) of duration by means of [cyclic] motion. Such a measure—for example, an hour, a day, a month, a year—is commonly used instead of true time." [Newton 1687].

That is, the absolute start time of a clock is irrelevant to measurements of elapsed time. Since time flows uniformly, and only time intervals are measurable, universal time is *translationally invariant*. This symmetry of nature eventually leads to the concept of *energy conservation*.

1.3.2 Absolute Space

Newton simultaneously introduced a second fiction, that of *absolute space*. At first, this seems to be a surprising regression, since Galileo had already demonstrated that only relative motion was physically relevant. For how else can the motion of a body be measured, except by reference to a second body? Should this not imply that only relative displacements are significant as well? However, absolute space is introduced for a specific purpose, created out of the fabric of shear necessity, in order to explain the existence of special, preferred, *inertial frames of reference*. This was an essential insight according to Einstein, who asserted that Newton found the only possible way to proceed in his day and age [Cohen 1955]. Like the fiction of absolute time, absolute space is not directly measurable:

> Absolute space, in its own nature, without regard to anything external, remains always similar and immovable. Relative space is some moveable dimension or measure of the absolute spaces, which our senses determine by its position to bodies, and which is vulgarly taken for immovable space. ... and so instead of absolute places and motions, we use relative ones ... but in philosophical disquisitions, we ought to abstract from our senses, and consider things themselves, distinct from what are only sensible measures of them. For it may be, that there is no body really at rest, to which the places and motions of others may be referred. [Newton 1687].

To Newton, absolute space means three-dimensional *Euclidean space*, devoid of matter and extending uniformly in all directions.[7] Absolute space is rotationally and translationally invariant. Coordinate systems have an origin, but the absolute origin is not discernable, since physical forces depend only on the relative coordinates and relative velocities of the interacting particles from which they originate. Mathematically, such spaces are called *affine spaces*.

Relative position, or displacement, in absolute space can be determined by specifying the *Cartesian coordinates* of a point relative to some origin

$$\mathbf{r}(x, y, z) = x\mathbf{e}_1 + y\mathbf{e}_2 + z\mathbf{e}_3, \tag{1.9}$$

where $\mathbf{e}_1, \mathbf{e}_2, \mathbf{e}_3$ are three constant, orthogonal, and normalized Euclidean basis vector operators that span the three independent dimensions of space. The length of a *vector displacement* in this space is defined by the use of the *Pythagorean theorem*. The absolute origin of Euclidean space disappears when measuring relative displacements.

---

[7]There is an apparent paradox here. The empty universe has no clocks and no rulers, and, therefore, no way to measure relative motion. The laws of mechanics may be predicated on a not so empty universe.

1.3.3 First Law of Motion

Newton's *first law of motion* states that a physical particle remains in a state of rest or of uniform linear motion unless acted upon by a net externally applied force. Newton's first law has various interpretations, one being that it defines an *inertial frame*, another that it defines motion as linear momentum, and yet another, that it defines a uniform measure for the flow of time. By *uniform linear motion*, Newton essentially means the linear momentum of the particle, which is a product of its scalar mass and its vector velocity

$$\mathbf{p} = m\mathbf{v} = m(d\mathbf{r}/dt). \tag{1.10}$$

Newtonian particles have mass, which is a measure of resistance to a change of motion. Mass is an additive scalar quantity, representing the amount of matter in a system. The more massive an object, the greater quantity of matter present.

A Newtonian particle is stable, which means that its internal forces must be balanced. Indeed, the internal forces acting on a particle must identically cancel or a particle could accelerate on its own volition. This would certainly defy common sense expectations.

Newton's laws of motion, however, are not valid in all frames, but only in special frames called *inertial frames*. This limitation of the laws of physics to special frames of reference would not be remedied until the advent of *general relativity*. The first law provides an experimental mechanism for determining whether or not a frame is inertial. Simply isolate a particle, and observe whether it appears to move with constant velocity. If it accelerates, either there remains some as yet unaccounted for force, or the reference frame is noninertial. However, due to the action of long-range forces, such as gravity, it may be that no particle can ever be truly isolated.[8]

Suitable approximations for inertial frames depend on the scale of the phenomena being studied. A measurement occurring in the confines of a laboratory on the surface of the Earth, requiring only a matter of minutes to complete, can often consider the surface of the Earth to be an approximate inertial frame. Phenomena occurring within the confines of the solar system can often use the distant stars to define a suitable inertial frame.

1.3.4 Second Law of Motion

Newton defines force as the originating cause for the change of motion of a particle. Newton's *second law of motion* states that the net applied force

---

[8]Another possibility is that both particle and observer are in freefall with respect to some local gravitational field.

acting on a particle is equal to its time rate of change of its motion. The application of force results in a time rate of change of the particle's linear momentum. Force is therefore a vector, since velocity is a vector and mass and time are scalars. The vector sum of all the physical forces acting on an object is the net applied force acting on that object

$$\mathbf{F}_{applied} = \sum_i \mathbf{F}_{(i)} = \frac{d\mathbf{p}}{dt} = \frac{d(m\mathbf{v})}{dt}. \tag{1.11}$$

For a particle with fixed mass this reduces to

$$\mathbf{F} = m\frac{d\mathbf{v}}{dt} = m\mathbf{a}. \tag{1.12}$$

The effect (kinematical acceleration) of a given force (dynamical principle) is inversely proportional to the inertial mass of the system to which it is applied. Thus, the inertia of the body is a measure of its resistance to an applied force. Consequently, forces must have a target. They can only be applied to physical objects that can carry and exchange energy and momentum. Note that the physical applied force acting on a particle is the same in all inertial frames, since inertial frames differ by at most a *constant relative velocity*, which does not affect the acceleration of the particle

$$\mathbf{F} = m\frac{d\mathbf{v}}{dt} = m\frac{d(\mathbf{v}' - \mathbf{v}_{rel})}{dt} = m\frac{d\mathbf{v}'}{dt} = \mathbf{F}'. \tag{1.13}$$

1.3.5 Third Law of Motion

Newton's *third law of motion* states that for every action there is an equal and opposite reaction. That is, forces (actions and reactions) arise out of mutual interactions of particles. This postulate was used by Newton to account for the conservation of momentum in two-body interactions. However, the assumption of two-body forces underlying this statement is not always valid. There may be intermediary fields present that can carry off momenta. Directly postulating the *conservation of momentum* for all internal processes of systems of particles (and their fields) would have provided a surer foundation for Newton's dynamics.

To derive the law of *action and reaction* from the law of conservation of momentum, consider the collision of two bodies labeled A and B. External, long-range forces can be ignored compared to the momentum transferred during the impulsive collision shown in Figure 1.6. The system is effectively isolated over a short enough time interval. The total momentum of an isolated system is conserved, giving

$$(\mathbf{p}_A + \mathbf{p}_B)|_{t=t_0-\varepsilon} = (\mathbf{p}'_A + \mathbf{p}'_B|_{t=t_0+\varepsilon}, \tag{1.14}$$

where A and B denote the initial particle states and A' and B', the final particle states. Rearranging terms gives

$$\begin{aligned}
(\mathbf{p}'_A - \mathbf{p}_A) &= -(\mathbf{p}'_B - \mathbf{p}_B), \\
\Delta\mathbf{p}_A &= -\Delta\mathbf{p}_B, \\
\mathbf{F}_A \Delta t &= -\mathbf{F}_B \,\Delta t, \\
\mathbf{F}_A &= -\mathbf{F}_B,
\end{aligned} \tag{1.15}$$

where $\mathbf{F}_A$ is the force applied by particle B on particle A, and $\mathbf{F}_B = -\mathbf{F}_A$ is the reaction force applied to particle B by particle A. Therefore, the third law can be thought of as a restricted statement of the principle of momentum conservation, applicable when the forces involved are two-body in character.

Newton assumed that when the two interacting particles are separated by distance, the action-reaction forces act *instantaneously* on both particles. This is referred to as *action-at-a-distance*. The Newtonian concept of action-at-a-distance, illustrated in Figure 1.6, was problematic from the beginning. It survived, for as long as it did, only because it worked better than anyone could have anticipated.

As an alternative to action-at-a-distance, a mechanical *aether* (a medium transferring energy and momentum) was felt to be necessary to convey force.

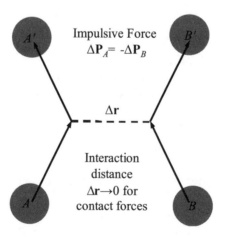

FIGURE 1.6 In an impulsive two-body interaction, the action-reaction forces transfer momentum from one part of the system to another in a nearly instantaneous manner. The total momentum of the two-body system as a whole is unchanged.

It was immediately understood that the aether would need to have great rigidity to account to account for the high value of the speed of light and for its transverse polarization. Newton, therefore, rejected the concept of the aether. For similar reasons, he also rejected the wave theory of light in favor of a corpuscular one. Einstein's analysis of the photoelectric effect in 1905 showed that such corpuscles do indeed exist in the form of light quanta [Einstein 2005].

Despite many experimental attempts, no evidence for the mechanical aether's existence has ever been found. The Michelson-Morley Interference Experiment [Michelson 1887] should have conclusively demonstrated its existence, based on the prevailing physical theories of the time. Instead, it produced a null result. Eventually, both the concepts of action-at-a-distance and of the mechanical aether were supplanted by the modern concept of a field. A *field* is an intrinsic property of the space-time continuum that requires no underlying medium for its support or propagation.

The third law provides a means to determine mass ratios. For bodies undergoing two-body interactions, where external influences can be ignored, one obtains

$$\frac{m_2}{m_1} = \frac{|\mathbf{a}_1|}{|\mathbf{a}_2|}. \tag{1.16}$$

This result is independent of the specific character of the force law. Given a test particle of unit mass, the inertial masses of other particles can be determined experimentally.

The *weak statement of the third law* requires only that action-reaction forces cancel in pairs, as shown in the right-hand-side of Figure 1.7. There may be a net torque arising from the force couple. The *strong statement of the third law*, in addition to requiring that the interaction forces between particles be equal and opposite, also requires that they lie along the line joining their centers. In this limit, there is no net torque applied to the

Action-Reaction Pairs

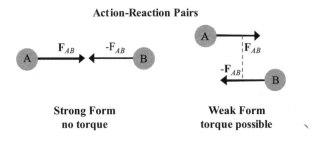

Strong Form
no torque

Weak Form
torque possible

FIGURE 1.7 **Weak and strong forms of the Law of Action and Reaction.**

But, if the cart exerts and equal and opposite
force on the horse (poor horse) how can they
possibly move?

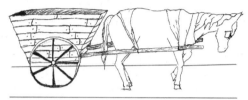

FIGURE 1.8   In this force diagram, the action-reaction pairs acting between the horse
and cart are internal to the coupled system. They identically cancel. (Drawing of horse
and cart by Danette Zeh.)

system. The strong form of the Third Law applies whenever the forces are
conservative in nature.

Action and reaction pairs act on different objects. This will result in
a net exchange of momentum between parts of the system, if the objects
in question are not subject to additional *constraints,* such as requiring the
components to move together as a single object. The hitch connecting the
horse and cart shown in Figure 1.8 is an example of such a constraint. Any
change in the motion of the horse and cart system is due to the unbalanced
force applied by the horse on the ground, which, in turn, results in an equal,
but opposite, applied reaction force on the horse, propelling the horse-cart
system forward.

1.3.6 Law of Universal Gravitation

Newton also postulated that the heavenly motions of the planets and the
earthly motion of falling objects originate from a common force of attraction
between all material bodies called the *Law of Universal Gravitation. Gravity*
between two mass points obeys a central force law, proportional to the product of masses, and inversely proportional to the squares of their separations,
given by the formula

$$\mathbf{F}_{12} = -G\frac{m_1 m_2}{r^2}\hat{\mathbf{r}}_{12}, \tag{1.17}$$

where $\hat{\mathbf{r}}_{12}$ is a unit vector directed along the line connecting the two particles,
G is Newton's gravitation constant, and r is their separation distance. According to Newton, gravity acts instantaneously and simultaneously on both
bodies. The force of gravity from all sources on a material point is given by

$$\mathbf{F}_G = m\mathbf{g}(\mathbf{r}), \tag{1.18}$$

where $\mathbf{g}(\mathbf{r})$ is the *gravitational acceleration field* at the position $\mathbf{r}$ of a particle of mass m due to all the other particles in the system. The modern viewpoint is that gravity is due to an all-permeating *gravitational field*, propagating at the speed of light, whose source arises from the presence of matter, and whose existence affects the motion of matter.

1.3.6.1 Unifying Projectile and Orbital Motion

Using his laws of motion, Newton was able to rectify projectile motion on the surface of the Earth with the motion of the planets in the sky. Figure 1.9 shows a sketch of a cannon ball fired from a mountain peak, an updated version of an early sketch by Newton. It illustrates the motion of a projectile in free fall towards the Earth under the influence of universal gravitation. As one increases the muzzle velocity, the cannon ball hits the surface of the Earth further from the peak. At a sufficiently high velocity, it enters into an orbit, similar to the orbital motion of the moon about the Earth. The dream of men creating artificial satellites of their own was at least a hypothetical possibility from this time forward.

1.3.6.2 Kepler's Third Law for Circular Orbits

An example of motion under a central force is given by a ball attached to a string of length r (Figure 1.10), rotating with constant angular velocity $\dot{\phi} = v/r$ about its pivot point. The circular motion results in a *centripetal* (center-pointing) acceleration $\mathbf{a} = -(v^2/r)\hat{\mathbf{r}}$. The mass times acceleration

FIGURE 1.9 **A conceptual drawing of projectiles shot from an elevated cannon on the surface of the Earth, based on a sketch from Newton's Principia. As the muzzle velocity increases, the projectiles fall further from the base before impact with the planet. At sufficiently high muzzle velocity, they enter into orbital motion. (Drawing of cannon by Danette Zeh.)**

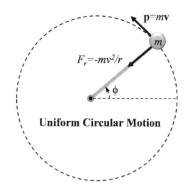

FIGURE 1.10   **A rotating ball attached to a rope illustrates uniform circular motion.**

matches the tension in the string, giving the force equation $T_r = ma_r = -mv^2/r$. If the string is cut, the force constraint is removed, and the ball resumes uniform linear motion with momentum $\mathbf{p} = m\mathbf{v}$, where $\mathbf{v}$ is its velocity at the moment of release.

Since the orbits of most planets are nearly circular, one can approximately derive Kepler's Law of Periods by assuming uniform circular motion and Newton's Law of Gravitation. Replacing the string tension with the force of gravity, the force law can be written as

$$Gm_pm_s/r^2 = m_pv^2/r. \tag{1.19}$$

where m_s is the mass of the sun and m_p is the mass of a planet. Substituting a constant magnitude for the velocity gives $v = 2\pi a/T$, where a is the mean radius of the orbit, $2\pi a$ is the circumference, and T is the period. Letting $r = a$ gives

$$T^2 = \frac{(2\pi)^2}{Gm_s}a^3. \tag{1.20}$$

The constant radius of a circular orbit in this simple analysis replaces the semimajor axis of the ellipse required for a more rigorous proof of Kepler's third law.

1.3.7 Torque and Angular Momentum

Torque $\mathbf{N}$ is the first moment of a force relative to a pivot point. It is equal to the time rate of change of *angular momentum* $\mathbf{L} = \mathbf{r} \times \mathbf{p}$

$$\mathbf{N} = \mathbf{r} \times \mathbf{F} = \mathbf{r} \times \frac{d\mathbf{p}}{dt} = \frac{d(\mathbf{r} \times \mathbf{p})}{dt} - \cancel{\mathbf{v} \times \mathbf{p}} = \frac{d\mathbf{L}}{dt}. \tag{1.21}$$

In the absence of torque, the angular momentum of a particle is conserved. Newton showed that angular momentum is always conserved for central force motion. Kepler's law of areas is, in fact, a direct result of the conservation of angular momentum for planetary orbits. Central force motion lies in a plane, and its angular momentum is proportional to the *areal velocity* $dA/dt = 1/2r^2\dot{\phi}$ giving

$$L_{plane} = (\mathbf{r} \times \mathbf{p})|_{\text{plane}} = p_\phi = mr^2\dot{\phi} = 2m\frac{dA}{dt}. \tag{1.22}$$

1.3.8 Impulse Approximation

The first integral of force with respect to time is its *impulse*

$$\Im = \int_1^2 \mathbf{F}_{\text{net}}dt = \Delta\mathbf{p} = \mathbf{p}_2 - \mathbf{p}_1. \tag{1.23}$$

The impulse of an applied force results in a change in the momentum of its object; see Figure 1.11. An *impulsive force* is one that acts for a short time (nearly instantaneously), allowing one to ignore the change of positions of the particle over the time interval of the impulse

$$\Delta\mathbf{r} = \mathbf{r}_2 - \mathbf{r}_1 = 0, \quad \Delta\mathbf{p} = m(\mathbf{v}_2 - \mathbf{v}_1) = \Im. \tag{1.24}$$

This is an example of applying *boundary conditions* to incorporate the essential dynamics of complex interactions, rather than by providing a detailed analysis of the forces involved.

1.3.9 Rocket Motion in a Gravitational Field

The *impulse approximation* can be used to analyze the motion of a rocket in a gravitational field $\mathbf{g}(\mathbf{r})$. This is a case where the mass of the rocket m_R is

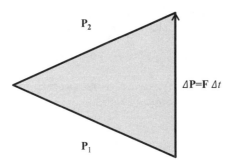

FIGURE 1.11   An impulsive force changes the momentum of a particle.

not constant, but decreases due to the ejection of spent fuel with mass m_F at a rate $dm_F/dt = -dm_R/dt \geq 0$. The expelled fuel has a velocity $\mathbf{v}_{rel}$ relative to the rocket.

The starting point for the analysis assumes that the rate of change of momentum for the system as a whole is given by the external force, and that mass is conserved for a closed system. Gas is emitted from the rear of the rocket, causing a change of momentum of the rocket, propelling it forward. At a given instant of time, define the initial system to be the rocket alone, with an instantaneous net mass m_R, and let its velocity be $\mathbf{v}$. A short time later, the rocket has a reduced mass $(m - \Delta m_{\text{fuel}})$ and increased velocity $(\mathbf{v} + \Delta\mathbf{v})$. However, the closed system must now include an element of exhaust gas Δm_{fuel} with velocity $(\mathbf{v} + \mathbf{v}_{rel})$. The initial and final momenta of the system are

$$\mathbf{p}_i = m_R\mathbf{v}, \quad \mathbf{p}_f = (m_R - \Delta m_{\text{fuel}})(\mathbf{v} + \Delta\mathbf{v}) + \Delta m_{\text{fuel}}(\mathbf{v} + \mathbf{v}_{rel}), \qquad (1.25)$$

and the net impulse on the system is given by

$$\Delta\mathbf{p} = \mathbf{p}_f - \mathbf{p}_i = \mathbf{F}_{\text{external}}\Delta t, \qquad (1.26)$$

giving

$$\Delta\mathbf{p} = m_R\Delta\mathbf{v} + \Delta m_{\text{fuel}}(-\Delta\mathbf{v} + \mathbf{v}_{rel}) = m_R\mathbf{g}(\mathbf{r})\Delta t, \qquad (1.27)$$

where $\mathbf{v}_{rel}$ is the velocity of the gas relative to the rocket. Dividing by Δt and taking the limit as Δt goes to zero gives the rocket's equation of motion

$$m_R\frac{d\mathbf{v}}{dt} = \frac{dm_R}{dt}\mathbf{v}_{rel} + m\mathbf{g}(\mathbf{r}). \qquad (1.28)$$

Second-order terms $-\Delta m_{\text{fuel}}\Delta\mathbf{v}$ in Equation (1.27) are treated as negligible in the limit as $\Delta t \to 0$. The mass rate of change term $\dot{m}_R\mathbf{v}_{rel}$ is called the *thrust* on the rocket.

In Newtonian mechanics, closed systems have constant mass. The change in mass of a system of particles comes from the addition or removal of particulate matter into or out of the system. *Open systems*, like the rocket example discussed previously, allow matter to escape the system.

1.3.10   Work and Energy

Work is the energy transferred by force acting through a distance and is defined by the path integral

$$W = \int_a^b \mathbf{F} \cdot d\mathbf{r}. \qquad (1.29)$$

The time rate of change of work is its *power*. It is straightforward to show that the time rate of change of the *kinetic energy* is equal to the net power transferred by all the applied forces. First, multiply the force equation by applying the dot product of the velocity to both sides of the equation

$$\mathbf{F}_{net} \cdot \mathbf{v} = m\mathbf{a} \cdot \mathbf{v}, \tag{1.30}$$

then, simple manipulation gives the desired result

$$\mathbf{F}_{net} \cdot \mathbf{v} = \left(m\frac{d\mathbf{v}}{dt} \right) \cdot \mathbf{v} = \frac{d}{dt}\left(\frac{1}{2}m\mathbf{v} \cdot \mathbf{v} \right) = \frac{dT}{dt}, \tag{1.31}$$

where

$$T = \frac{1}{2}mv^2 = \frac{1}{2}\mathbf{v} \cdot \mathbf{p} = \frac{1}{2m}p^2 \tag{1.32}$$

is the kinetic energy of a point particle. The time integral of the power is the work done by a force. The net work done on a particle by the combined effect of all external forces acting on it is equal to its change in kinetic energy T

$$\Delta T = \int_1^2 (\mathbf{F}_{net} \cdot \mathbf{v})dt = \int_1^2 \mathbf{F}_{net} \cdot d\mathbf{r}, \quad \Delta T = W_{net}|_1^2. \tag{1.33}$$

The *work-energy theorem* requires that the change in the kinetic energy of a particle is equal to the work done by the net applied force acting on the particle.

1.3.10.1 Conservative Forces and Potential Energy

If the work done on a particle is the same for all possible paths between any two points (see Figure 1.12), then the work done about any closed loop is zero, and the force is *conservative*. A conservative force can be derived

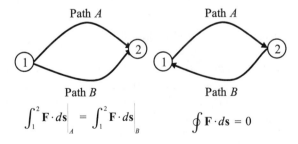

FIGURE 1.12 If the work is the same for all paths between points 1 and 2, the force is conservative; equivalently, if one constructs a closed loop, the net work integrated over the closed path is zero for conservative forces.

as the gradient of a *scalar potential* function by invoking Stokes' theorem $\oint_{\partial A} \mathbf{F} \cdot d\mathbf{r} = \int_A \nabla \times \mathbf{F} \cdot d\mathbf{A}$. If the integral over the closed path is path-independent, it follows that

$$\oint \mathbf{F}_C \cdot d\mathbf{r}\bigg|_{\text{any path}} = 0 \rightarrow \nabla \times \mathbf{F}_C = 0 \rightarrow \mathbf{F}_C = -\nabla V(r). \qquad (1.34)$$

The work done by a conservative force is independent of the path taken by the particle, which can be parameterized by the scalar quantity s ($d\mathbf{r} = d\mathbf{s} = \hat{\mathbf{s}}\, ds$) where $\hat{\mathbf{s}}$ is the unit tangent vector to the path). It depends only on the values of the potential at the end points of the path. The chain rule gives $dV(\mathbf{r}) = (\partial V/\partial x^i)dx^i = (\partial V/\partial s)ds$, where $\partial V/\partial s$ is the directional derivative along the tangent, therefore

$$\int_1^2 \mathbf{F}_C \cdot d\mathbf{r} = \int_1^2 -\nabla V \cdot d\mathbf{s} = -\int_1^2 \frac{\partial V}{\partial s}ds = -\int_1^2 dV = -(V_2 - V_1). \qquad (1.35)$$

1.3.10.2 Work-Energy Theorem

The work-energy theorem also relates a change in the *mechanical energy* of a particle to the work done by nonconservative forces. Partition the force into conservative $\mathbf{F}_c$ and nonconservative $\mathbf{F}_{nc}$ parts

$$\mathbf{F} = \mathbf{F}_c + \mathbf{F}_{nc} = -\nabla V + \mathbf{F}_{nc}, \qquad (1.36)$$

and define the mechanical energy as the sum of the kinetic and potential energies

$$E = T + V. \qquad (1.37)$$

Then

$$\Delta E = \Delta(T + V) = \int_1^2 \mathbf{F}_{nc} \cdot d\mathbf{s}, \qquad (1.38)$$

leading to the following alternate forms of the work-energy theorem:

1. The change of the mechanical energy of a particle is the work done by the nonconservative forces acting on a particle.
2. If the forces acting on a particle are all conservative, its mechanical energy is conserved.

1.3.11 Galilean Group

The geometry of space and time as described in Newton's *Principia* has a number of *symmetries* that can be summarized by the ten-parameter

Galilean Group. The Galilean Group transforms inertial frames to other inertial frames. Let $\mathbf{r} = \mathbf{P} - \mathbf{O}$ denote a displacement vector of a point from an origin in space, then in Newtonian mechanics, the invariance group allows the following point transformations [Doran 2005]:

- *Space is isotropic*, the same in all directions, leading to the three-parameter special orthogonal rotation group

$$\mathbf{r}' = \mathbf{A} \cdot \mathbf{r}. \tag{1.39}$$

- *Space is homogeneous*, the same at all points, leading to a three-parameter translation group

$$\mathbf{r}' = \mathbf{r} + \Delta\mathbf{r}_0. \tag{1.40}$$

- *Motion is relative.* There are three possible relative velocity transforms

$$\mathbf{r}' = \mathbf{r} + \mathbf{v}_{\text{rel}}(t - t_0). \tag{1.41}$$

- *Time is homogeneous* under a transformation of origin

$$t' = t + \Delta t, \quad t_0' = t_0 + \Delta t \rightarrow \Delta t' = \Delta t. \tag{1.42}$$

Therefore, points in any inertial frame can be related to points in another inertial frame by the combined *Galilean transformation*

$$\mathbf{r}' = \mathbf{A} \cdot \mathbf{r} + \Delta\mathbf{r}_0 + \mathbf{v}_{\text{rel}}(t - t_0). \tag{1.43}$$

These symmetries are required for the equations of motion. The best way to ensure this is to derive the equations of motion from a scalar *Lagrangian function* (to be developed in Chapter 3) that manifestly incorporates these symmetries. In effect, homogeneity of space requires that the forces $\{\mathbf{F}_n\}$ acting on an isolated system of particles depend only on their relative coordinates, while Galilean invariance requires that the applied forces depend on only relative velocities

$$\mathbf{F}_n = \mathbf{F}_n(\mathbf{r}_i - \mathbf{r}_j, \dot{\mathbf{r}}_i - \dot{\mathbf{r}}_j). \tag{1.44}$$

Conservation laws imply the existence of symmetries. The converse is also true. The previous symmetries are responsible for a number of conservation laws including total energy, linear momentum, and angular momentum conservation.

In Chapter 12, introducing the classical theory of fields, the continuous symmetries of nature will be related to the existence of *gauge fields* that generate the fundamental laws of motion in a manner consistent with the prescribed symmetries. In moving to the four-dimensional space-time of the theory of relativity, the ten-parameter Galilean group is replaced by the ten-parameter *Poincaré group*. In *gravitational gauge theories*, gravity is the gauge field associated with the *Poincaré invariance* of space-time.

1.4 MOTION OF A SYSTEM OF PARTICLES

Adding particles to a system is easy: just add particle label indices. The position vector of the jth particle in an N-particle system is given by

$$\mathbf{r}_j = x_j \mathbf{e}_1 + y_j \mathbf{e}_2 + z_j \mathbf{e}_3. \tag{1.45}$$

Newton's second law for a multiparticle system becomes a system of coupled-differential equations

$$\mathbf{F}_j(\mathbf{r}, \dot{\mathbf{r}}, t) = m_j \ddot{\mathbf{r}}_j. \tag{1.46}$$

The applied force on a particle has an external component and an internal component, the latter coming from its interaction with other particles in the system

$$\mathbf{F}_i = \mathbf{F}_i^{\text{ext}} + \sum_j \mathbf{F}_{ij}^{\text{int}}. \tag{1.47}$$

Self-consistency requires that the force of a particle acting on itself vanishes

$$\mathbf{F}_{ii}^{\text{int}} = 0. \tag{1.48}$$

1.4.1 Center of Mass Motion

The *center of mass point* of a system of particles $\mathbf{r}_{\text{cm}}$ is defined as the mass-weighted centroid of the particle positions

$$m_\Sigma \mathbf{r}_{\text{cm}} = \sum_i m_i \mathbf{r}_i, \tag{1.49}$$

where the total mass of the system is denoted by it sum $m_\Sigma = \sum_i m_i$, and $\sum_i = \sum_{i=1}^N$ denotes the sum over all particles. The total momentum $\mathbf{p}$ of the system is the vector sum of the individual momenta

$$\mathbf{p} = \sum_i \mathbf{p}_i = \sum_i m_i \dot{\mathbf{r}}_i = m_\Sigma \dot{\mathbf{r}}_{\text{cm}}. \tag{1.50}$$

The net (total) force acting on a system is therefore given by

$$\mathbf{F} = \sum_i \mathbf{F}_i = \sum_i \dot{\mathbf{p}}_i = \dot{\mathbf{p}} = \sum_i m_i \ddot{\mathbf{r}}_i = m_\Sigma \ddot{\mathbf{r}}_{cm}. \qquad (1.51)$$

This looks exactly like Newton's second law for a point particle of mass m_Σ located at the center of mass point $\mathbf{r}_{cm}$. However, there are now internal forces that have to be eliminated. Expanding these as pairwise forces in an ordered sum gives

$$\mathbf{F} = \sum_i \mathbf{F}_i^{ext} + \sum_{i \neq j} \mathbf{F}_{ij}^{int} = \sum_i \mathbf{F}_i^{ext} + \sum_{i < j} (\mathbf{F}_{ij}^{int} + \mathbf{F}_{ji}^{int}). \qquad (1.52)$$

Assuming the weak form of Newton's third law, which requires only that the action and reaction pairs cancel, then $\mathbf{F}_{ij}^{int} = -\mathbf{F}_{ji}^{int}$. Therefore, the internal forces cancel pairwise, and the net applied force is due solely to the external forces acting on the system $\mathbf{F} = \sum_i \mathbf{F}_i^{ext}$. Provided that the weak form of the laws of action and reaction holds, the global motion of a system of particles can be treated as if it were an effective point of mass, located at the center of mass, moving under the influence of the net externally applied force $\mathbf{F} = \dot{\mathbf{p}} = m_\Sigma \ddot{\mathbf{r}}_{cm}$. The *center of mass motion* of a particle system acts as if it were a particle, carrying the total mass of the system, acting under the influence of the net externally applied force.

For example, in Scene 8 of the farcical movie, *Monte Python and the Holy Grail*, the French defenders of a castle use a catapult to propel a cow over the castle wall. The center of mass of the cow moves as a point particle subject to the force of gravity, as shown in the cartoon of Figure 1.13.

1.4.2 Conservation of Linear Momentum

Just as the linear momentum of an isolated particle is conserved, the linear momentum of an isolated particle system is conserved. The *Law of Conservation of Linear Momentum* states that the linear momentum of a closed particle system is conserved in the absence of an applied external force. Momentum is conserved when two particles collide with each other. If the collision is *elastic*, then kinetic energy is conserved as well. Assuming the interaction time is short, the conservation laws require

$$m_1 \mathbf{v}_1 + m_2 \mathbf{v}_2 = m_1 \mathbf{v}_1' + m_2 \mathbf{v}_2' \rightarrow \Delta \mathbf{p}_1 = -\Delta \mathbf{p}_2,$$
$$\frac{1}{2} m_1 \mathbf{v}_1^2 + m_2 \mathbf{v}_2^2 = \frac{1}{2} m_1 \mathbf{v}_1'^2 + m_2 \mathbf{v}_2'^2 \rightarrow \Delta T_1 = -\Delta T_2. \qquad (1.53)$$

where the unprimed coordinates represent the particle system before collision and the primed coordinates represent the particle system after scattering.

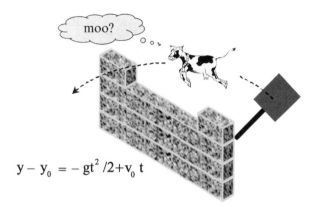

$$y - y_0 = -gt^2/2 + v_0 t$$

FIGURE 1.13 Stripped to its essentials, from the singular viewpoint of a ballistics analyst, a flying cow is little more than a point mass moving in a well-defined gravitational trajectory; hence the phrase, "consider a spherical cow." (Drawing of cow by Danette Zeh.)

These equations can be rewritten as

$$m_1(\mathbf{v_1} - \mathbf{v'_1}) = -m_2(\mathbf{v_2} - \mathbf{v'_2})$$
$$\frac{1}{2}m_1(\mathbf{v_1} - \mathbf{v'_1}) \cdot (\mathbf{v_1} + \mathbf{v'_1}) = -\frac{1}{2}m_2(\mathbf{v_2} - \mathbf{v'_2}) \cdot (\mathbf{v_2} + \mathbf{v'_2}). \tag{1.54}$$

For one-dimensional scattering problems, the boldface vector notation can be dropped. Dividing the second equation by the first and rearranging terms then gives

$$v_1 - v_2 = v'_2 - v'_1, \quad v_{\text{rel}} = -v'_{\text{rel}}. \tag{1.55}$$

The effect of the elastic collision is to reverse the relative velocities of the two particles.

For *inelastic* collisions, a useful parameter is the *coefficient of restitution*, which is defined as by the ratio of the relative velocities before and after scattering

$$e = \frac{|v'_{\text{rel}}|}{|v_{\text{rel}}|}. \tag{1.56}$$

Solving for the final velocities and allowing the second particle to be initially at rest in the lab ($v_2 = 0$) frame gives the one-dimensional scattering formulas

$$v'_1 = \frac{m_1 - em_2}{m_1 + m_2}v_1, \quad v'_2 = \frac{(1 + e)m_1}{m_1 + m_2}v_1. \tag{1.57}$$

For $e = 1$, one gets the elastic scattering limit. If $e = 0$, then $v_1' = v_2'$ and one gets the *totally inelastic limit*, where the two particles stick together after collision. Other cases worth examining are the limits when $m_1 \gg m_2$, $m_1 \ll m_2$, and $m_1 = m_2$:

$$
\begin{aligned}
m_1 \gg m_2 : \quad & v_1' = v_1, & v_2' = (1+e)v_1, \\
m_1 \ll m_2 : \quad & v_1' = -ev_1, & v_2' = \frac{(1+e)m_1}{m_2}v_1 \simeq 0, \\
m_1 = m_2 : \quad & v_1' = \frac{1-e}{2}v_1, & v_2' = \frac{(1+e)}{2}v_1.
\end{aligned}
\tag{1.58}
$$

1.4.3 Work Done on a System of Particles

The total work done on a system of particles is given by summing over all elements of work

$$
dW = \sum_i \mathbf{F}_i \cdot d\mathbf{r}_i = \sum_i m_i \ddot{\mathbf{r}}_i \cdot d\mathbf{r}_i = \sum_i d\left(\frac{1}{2}m_i \dot{\mathbf{r}}_i \cdot \dot{\mathbf{r}}_i\right) = dT. \tag{1.59}
$$

This implies that the time rate of change of the total kinetic energy is given by the total work done on the system, extending the work-energy theorem to a system of particles

$$
\Delta T = W|_1^2 = \sum_i \int_1^2 \mathbf{F}_i \cdot d\mathbf{r}_i, \tag{1.60}
$$

where the total kinetic energy of a particle system is given by

$$
T = \sum_i \frac{1}{2}m_i \dot{\mathbf{r}}_i \cdot \dot{\mathbf{r}}_i = \sum_i \frac{1}{2}\mathbf{v}_i \cdot \mathbf{p}_i = \sum_i \frac{1}{2m_i}\mathbf{p}_i \cdot \mathbf{p}_i. \tag{1.61}
$$

1.4.4 Energy

If all the external forces are conservative in nature, they can be derived as the gradient of a potential energy function

$$
\mathbf{F}_i^{\mathrm{ext}} = -\nabla_i V^{\mathrm{ext}}. \tag{1.62}
$$

The work done by the external forces can be expressed in terms of a potential difference

$$
\begin{aligned}
W_{12}^{\mathrm{ext}} &= \sum_i \int_1^2 \mathbf{F}_i^{\mathrm{ext}} \cdot d\mathbf{r}_i = -\sum_i \int_1^2 \nabla_i V^{\mathrm{ext}} \cdot d\mathbf{r}_i \\
&= -\int_1^2 dV^{\mathrm{ext}} = -\Delta V^{\mathrm{ext}}.
\end{aligned}
\tag{1.63}
$$

If one assumes that the internal forces are conservative as well, the force couples due to central potentials depend only on the magnitudes of the relative displacements

$$\mathbf{F}_{ij}(|\mathbf{r}_i - \mathbf{r}_j|) = -\nabla_i V_{ij}, \tag{1.64}$$

$$W_{12}^{\text{int}} = \sum_{i \neq j} \int_1^2 \mathbf{F}_{ij} \cdot d\mathbf{r}_i$$

$$= \frac{1}{2} \sum_{i \neq j} \int_1^2 (\mathbf{F}_{ij}) \cdot (d\mathbf{r}_i - d\mathbf{r}_j) = -\frac{1}{2} \sum_{i \neq j} \int_1^2 dV_{ij}, \tag{1.65}$$

where $\mathbf{F}_{ij}$ is the force on particle i due to particle j. The two-body potential $V_{ij} = V_{ij}(|\mathbf{r}_i - \mathbf{r}_j|)$ is an even function under exchange of displacements $\mathbf{r}_i \leftrightarrow \mathbf{r}_j$. Therefore, action-reaction pairs are odd under exchange of indices

$$\mathbf{F}_{ij} = -\nabla_i V_{ij} = +\nabla_j V_{ij} = -\mathbf{F}_{ji}. \tag{1.66}$$

The total potential energy function of a system is the sum of the external and internal contributions

$$V = \sum_i V_i^{\text{ext}} + \frac{1}{2} \sum_{i \neq j} V_{ij}. \tag{1.67}$$

The second term in the potential energy in the above formula is the internal potential energy contribution. It depends only on the relative separation distances between pairs of particles. The internal potential energy is constant (and therefore ignorable) if relative configuration of particles in the system is fixed, as, for example, would be the case for *rigid body motion*. From the above considerations, one can deduce the *Law of Conservation of Mechanical Energy,* which states that the total mechanical energy of the system is conserved if the forces acting on a particle system are all conservative in character.

1.4.5 Total Angular Momentum of a Particle System

The total angular momentum $\mathbf{L}$ of a system is given by the sum of its component angular momenta

$$\mathbf{L} = \sum_i \mathbf{r}_i \times \mathbf{p}_i. \tag{1.68}$$

Taking the time derivative and using Equation (1.52) gives the torque equation

$$\mathbf{N} = \frac{d\mathbf{L}}{dt} = \sum_i \mathbf{r}_i \times \mathbf{F}_i^{\text{ext}} + \sum_{i<j} (\mathbf{r}_i - \mathbf{r}_j) \times \mathbf{F}_{ij}. \tag{1.69}$$

The first term on the right-hand side is the net external torque $\mathbf{N}^{\text{ext}}$. The second term is due to torque couples coming from the internal forces. A torque couple vanishes if $\mathbf{F}_{ij}$ satisfies the strong statement of the Third Law. Therefore, the internal torque cancels whenever the internal forces are conservative in nature. Assuming the strong form of action and reaction, the torque equation can be written as

$$\frac{d\mathbf{L}}{dt} = \mathbf{N} = \sum_i \mathbf{r}_i \times \mathbf{F}_i^{\text{ext}}, \tag{1.70}$$

and, therefore, one can deduce the *Law of Conservation of Angular Momentum,* which states that the total angular momentum of a particle system is conserved in the absence of an externally applied torque, provided that the strong form of the third law applies.

1.4.6 Center of Mass Coordinates

Defining a particle's position in terms its displacement from the center of mass of the system gives the *center of mass coordinates*

$$\mathbf{r}_i' = \mathbf{r}_i - \mathbf{r}_{\text{cm}}. \tag{1.71}$$

From the definition of the center of mass, one gets the identity

$$\sum_i m_i \mathbf{r}_i' = \sum_i m_i(\mathbf{r}_i - \mathbf{r}_{\text{cm}}) = \sum_i m_i \mathbf{r}_i - m_\Sigma \mathbf{r}_{\text{cm}} = 0. \tag{1.72}$$

The velocities in the center of mass frame are given by

$$\mathbf{v}_i' = \dot{\mathbf{r}}_i' = \mathbf{v}_i - \mathbf{v}_{\text{cm}}, \tag{1.73}$$

where $\mathbf{v}_{\text{cm}} = \dot{\mathbf{r}}_{\text{cm}}$ is the velocity of the center of mass point. The system has no net momentum in its center of mass frame

$$\mathbf{p}' = \sum_i \mathbf{p}_i' = \sum_i m_i \mathbf{v}_i' = \frac{d}{dt} \sum_i m_i \mathbf{r}_i' = 0. \tag{1.74}$$

The total angular momentum of the system can be expressed using center of mass coordinates by

$$\mathbf{L} = \sum_i \mathbf{r}_i \times \mathbf{p}_i = \sum_i (\mathbf{r}_i' + \mathbf{r}_{\text{cm}}) \times m_i(\mathbf{v}_i' + \mathbf{v}_{\text{cm}}). \tag{1.75}$$

One can use $\sum_i m_i \mathbf{r}_i' = 0$ and $\sum_i m_i \mathbf{v}_i' = 0$ to show that the cross terms in the above products cancel. Therefore, the angular momentum in any frame reduces to two terms, representing external and internal contributions

$$\mathbf{L} = \mathbf{L}_{cm} + \mathbf{L}' = \mathbf{r}_{cm} \times \mathbf{p} + \sum_i \mathbf{r}'_i \times \mathbf{p}'_i. \tag{1.76}$$

The first term on the right-hand side of this equation represents the contribution $\mathbf{L}_{cm}$ to the angular momentum from an effective point particle of mass concentrated at the center of mass point, and carrying the total momentum of the system. The second term represents internal contributions $\mathbf{L}'$ to the angular momentum of the system evaluated in the center of mass frame.

Similarly, the kinetic energy in center of mass coordinates is the sum of the particle kinetic energies

$$T = \sum_i \frac{1}{2} m_i (\mathbf{v}'_i + \mathbf{v}_{cm}) \cdot (\mathbf{v}'_i + \mathbf{v}_{cm}). \tag{1.77}$$

The cross terms in the products cancel, therefore the kinetic energy reduces to

$$T = T_{cm} + T' = \frac{1}{2} \mathbf{v}_{cm} \cdot \mathbf{p} + \sum_i \frac{1}{2} \mathbf{v}'_i \cdot \mathbf{p}'_i, \tag{1.78}$$

where the first term on the right-hand side of the previous equation is the kinetic energy contribution T_{cm} from the effective center of mass point. Likewise, the second term is the internal kinetic energy T' of the system defined with respect to the center of mass frame.

1.4.7 Reprise of the Conservation Laws

From the weak law of action-reaction, one can infer conservation of linear momentum for isolated particle systems. From the strong law of action-reaction one can additionally infer angular momentum conservation for isolated particle systems. Actually, the cases for these conservation laws are much stronger than the case that can be made for Newton's third law of motion.

In Chapter 4, "Hamilton's Principle," it will be shown that the total linear momentum $\mathbf{p}$ and angular momentum $\mathbf{L}$ of an isolated system of particles must be conserved if empty space is homogeneous and isotropic. Homogeneity implies that there is no special origin to absolute space, and isotropy implies that there is no special orientation to absolute space. If one accepts these statements as fundamental principles, then all fundamental forces must be generated in a manner consistent with the laws of conservation of total linear and angular momentum.

In some cases, particle interactions can transfer energy and momentum to or from the interaction fields, violating the third law. By accounting for the field contributions to the energy and momentum, the conservation laws can be maintained.

1.5 SOLUTION OF THE INITIAL VALUE PROBLEM

Newton's equations of motion in Cartesian coordinates are $3N$-second-order differential equations with respect to the time. By solving for the accelerations, they can be written in the following form

$$\ddot{\mathbf{x}}_{(i)} = \frac{\mathbf{F}_{(i)}(x, \dot{x}, t)}{m_{(i)}}, \tag{1.79}$$

where (i) denotes a particle numbering index ranging over the N particles of the system. Working in Cartesian coordinates for now, the equations can be treated as a single second-order differential equation for an n-dimensional vector in configuration space $\ddot{x}^j = a^j(x, \dot{x}, t)$, where $x = [x^1, \ldots, x^n]$ denotes a system vector in an n-dimensional configuration space, and $n = 3N$ for motion in three space dimensions. The solution $x(t)$ represents the time evolution of a system point in this space. Note that knowing $x(t_0)$ at some initial time is not sufficient to completely describe the system. One needs to specify the initial velocities $\dot{x}(t_0)$ as well.

1.5.1 Numerical Integration

For both visualization and numerical programming purposes, it is better to recast Newton's equations as a system of $2n$ first-order differential equations. This is achieved by treating the momentum and coordinates as independent observables. Newton's equations become

$$\dot{\mathbf{p}}_{(i)} = \mathbf{F}_{(i)}(x, p, t), \quad \dot{\mathbf{x}}_{(i)} = \frac{\mathbf{p}_{(i)}}{m_{(i)}}, \tag{1.80}$$

which can be rewritten in configuration space coordinates as

$$\dot{p}_i = F_i(x, p, t), \quad \dot{x}^i = v^i(x, p, t). \tag{1.81}$$

If the right-hand side of the equation is not explicitly dependent on the time, the equation is said to be *autonomous* in the independent variable. Note that raised indices are used to denote position coordinates and lowered indices are used to denote momentum coordinates. To numerically integrate these equations, the force law has to be rewritten as a function of the positions and momenta instead of as a function of the positions and velocities. The equations require $2n$ initial conditions for their solution.

A large number of ordinary *differential equation* (ODE) *solvers* exist for solving the *initial value problem* for coupled first-order differential equations of the form

$$\frac{dy^i}{dt} = f^i(t, y), \qquad (1.82)$$

where the index ranges over the $2n$ coordinates defined by the column vector

$$y = \begin{bmatrix} x \\ p \end{bmatrix}. \qquad (1.83)$$

The n-tuple $y = \{y^i\}$ defines a $2n$-dimensional *phase space* vector, where $2n = 6N$ for N particles in three-dimensional space, subject to the initial conditions $y(t_0) = y_0$. One advantage of using a phase space formulation of the dynamical problem is that the *lines-of-flow* for the first-order differential equations can never cross, since the instantaneous direction of the *flow* $\dot{y} = f(t, y)$ must be a single-valued function of the time. A point in phase space specified by giving the initial values of position and momentum for all the particles therefore uniquely defines the evolution of the system for all future times.

1.5.1.1 Runge-Kutta Methods

Euler's method for solving first-order differential equations is to convert the first-order differential equations into a set of first-order *difference equations*, and to numerically solve these by stepwise iteration

$$y(t + h) = y(t) + h f(t, y(t)), \qquad (1.84)$$

where $h = \Delta t$ is the step size of the interval. Euler's difference equation solver is an example of a first-order *Runge-Kutta method* with the error proportional to the step size of the time interval.

Greater accuracy for a given step size is given by using higher-order Runge-Kutta integrators. For example, the explicit Runge-Kutta equation solver *RK4* is given by[9]

$$y(t + h) = y(t) + \frac{h}{6}(k_1 + 2k_2 + 2k_3 + k_4), \qquad (1.85)$$

---

[9]A good discussion of Runge-Kutta techniques can be found at *Wolfram Mathworld*, available at *http://mathworld.wolfram.com*.

where

$$k_1 = f(t, y(t)), \quad k_3 = f\left(t + \frac{h}{2}, y(t) + \frac{h}{2}k_2\right),$$

$$k_2 = f\left(t + \frac{h}{2}, y(t) + \frac{h}{2}k_1\right), \quad k_4 = f(t + h, y(t) + hk_3).$$

(1.86)

The RK4 solver is a fourth-order method, meaning that the error in each step is on the order of h^5, while the accumulated error is of order h^4.

1.5.1.2 Controlling Numerical Error

Numerical methods suffer from two special sources of error:

- *Round-off* error
- *Truncation* error

Round-off error increases with the number of integration steps. Truncation error comes from the finite step size.

Runge-Kutta methods know nothing about the underlying dynamics of the ordinary differential equations (ODE). For example, the algorithms do not automatically preserve constants of the motion. This can lead to artificial computational effects, such as numerical *energy dissipation* due to truncation errors. The error can grow significantly over long integration times. To reduce this effect, one can rewrite the algorithm to enforce constraints on the constants of the motion.

Adaptive *variable step-size methods* have been developed to deal with *stiff differential equations*. These are differential equations where rate of change varies dramatically over the limits of integration. Typically, these algorithms work by automatically adjusting the step size so that the local error per step is kept below some prescribed tolerance limit. Such integrators are included in higher level Mathematical software programs, such as Maple™, Mathematica™, Mathcad™, and Matlab™, or in advanced mathematical libraries for common programming languages.[10] When using numerical integration, it is useful to perform "sanity checks" to determine if the algorithms are behaving properly. Tests for convergence, error propagation, and boundary conditions are commonly performed.

Although one can write one's own numerical integrator, the differential equation solvers provided by commercial mathematical software packages

---

[10] A good source for numerical routines written for use with Maple, Mathematica, Mathcad, and MATLAB is the *Holistic Numerical Methods Institute* of the University of South Florida web site, available at *http://numericalmethods.eng.usf.edu/index.html*.

such as Matlab or Mathematica makes the task easier and nearly painless to perform. One can often simply specify a tolerance and let the Runge-Kutta solver do its work, adjusting the step size automatically as needed. *Event triggers* can be placed on the integration to allow handing of exceptional conditions or to extract interesting events, such as detection of zero-crossings of a plane in order to generate Poincaré sections. Moreover, the results are provided in a form directly useful for plotting.

1.5.1.3 Analytical Solutions

The alternative to numerical integration is to solve the problem using analytical methods. A set of $2n$ first order differential equations is *exactly integrable* if can be completely reduced to quadrature, yielding $2n$ *constants of integration*. *Quadrature* is defined as the reduction of an ordinary differential equation to the integral of a function of a single variable with respect to that variable. The results can then be integrated, numerically or analytically. Problems reduced to quadrature are considered to be formally solved. The long-term stability of *integrable systems* can easily be determined. The constants of integration can often be related to the symmetries of the problem, yielding important insights into the dynamics of the physical system.

1.6 ONE-DIMENSIONAL MOTION

In the following subsections, several representative one-dimensional, oscillatory systems are examined. These can be solved either numerically or by reduction to quadrature.

1.6.1 Simple Harmonic Oscillator

An example of a one-dimensional potential is the *Simple Harmonic Oscillator* (SHO) of mass m and spring constant k. The SHO has a linear restoring force, obeying the equation of motion

$$F(x) = m\ddot{x} = -kx, \tag{1.87}$$

where $x = 0$ is the equilibrium position. This equation represents the limiting behavior for the solutions for a large class of oscillatory problems involving small displacements from stable equilibrium. The potential energy of the SHO is given by

$$V(x) = -\int F(x)dx = kx^2/2. \tag{1.88}$$

Note that the total mechanical energy of the system is conserved

$$E(x, p) = \frac{p^2}{2m} + \frac{kx^2}{2}.$$

(1.89)

Reduction to quadrature gives

$$\int dt = \pm \int \frac{m\,dx}{\sqrt{2m(E - kx^2/2)}},$$

(1.90)

where the positive and negative roots describe motion forwards and backwards in time, respectively.

The analytic solution provides the time as a multivalued function of the position coordinate $t(x)$. The correct branch has to be chosen in order to invert the solution, thereby obtaining the position as a single-valued function of the time $x(t)$. The well-known analytic solution for the time evolution of the oscillator equation is

$$x(t) = A \cos(\omega_0 t + \phi),$$

(1.91)

where A is the amplitude of the oscillation, $\omega_0 = \sqrt{k/m}$ is the angular frequency of oscillation, and ϕ is a phase angle. The period of a complete cycle, $T = 2\pi/\omega_0$, is independent of the amplitude. Substitution of the solution into the equation of motion is the easiest way to validate the result.

Potential energy plots are useful in helping to characterize the motion and to identify branch points, if any. Figure 1.14 shows the potential energy diagram for the simple harmonic oscillator problem. Allowed values of the

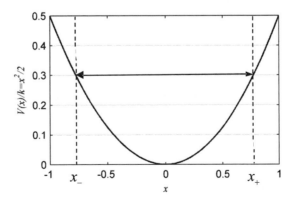

FIGURE 1.14 Potential energy diagram for the simple harmonic oscillator.

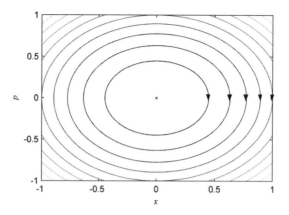

FIGURE 1.15 Phase space diagram for the simple harmonic oscillator. All paths are closed vibrational modes, centered on the stationary point marked with an x symbol. The direction of the flow in time is clockwise along ellipses of constant energy.

energy are $E \geq 0$. The motion is symmetric about the origin $x = 0$. All orbits are bound, with turning points at $x_{\pm} = \pm\sqrt{2E/k}$.

Contour plots of constant energy $E(p, x)$ can be used to generate a more complete two-dimensional *phase space portrait* of the motion. In the SHO case, shown in Figure 1.15, the trajectories are ellipses in phase space centered on the stationary point $p = x = 0$. For increasing time, the direction of flow is clockwise about the stationary point.

1.6.2 Periodic Motion of Bound Systems

The one-dimensional motion of a particle subject to a conserved force is given by the force equation

$$m\ddot{x} = F(x) = -\partial V(x)/\partial x, \tag{1.92}$$

where $V(x)$ is the potential energy function. Numerical solutions to the corresponding first-order differential equations are usually presented in momentum phase space, obtained by substituting the particle momentum $p = mv$ for the particle velocity v. This results in the following coupled first-order ODEs that can be solved numerically

$$\dot{p} = F(x), \quad \dot{x} = p/m. \tag{1.93}$$

In the one-dimensional case, however, the problem is completely integrable and easily reduces to quadrature. Multiplying both sides of Equation (1.92) by the velocity and integrating both sides of the result gives a first integral of the motion

$$m\dot{x}\ddot{x} = \dot{x}F(x) = -\dot{x}\partial V(x)/\partial x \rightarrow m\dot{x}d\dot{x} = -\frac{dV}{dx}dx = -dV$$

$$\rightarrow d\left(\frac{m\dot{x}^2}{2} + V\right) = d(T + V) = 0 \rightarrow T + V = E. \tag{1.94}$$

Here, the constant of integration E represents the conserved mechanical energy of the system, which can be written as a function of the particle position and velocity

$$E(x, \dot{x}) = \frac{1}{2}m\dot{x}^2 + V(x), \tag{1.95}$$

or equivalently as a function of the position and momentum

$$E(x, p) = p^2/2m + V(x). \tag{1.96}$$

The time dependence of the motion can now be solved, by solving for the velocity

$$v = \frac{p}{m} = \pm\sqrt{\frac{2}{m}(E - V)}, \tag{1.97}$$

and then formally integrating the result

$$\int_{t_0}^{t} dt = \pm\int_{x_0}^{x} \frac{mdx}{\sqrt{2m(E - V)}}. \tag{1.98}$$

For the initial value problem, time is monotonically increasing. The two branches in the previous equation correspond to increasing ($x > x_0$) or decreasing ($x < x_0$) values of the space coordinate. The *turning points* in the motion occur when the radical in the denominator of the integral goes to zero.

In this example, reduction to quadrature gives rise to a multivalued function $t(x)$, which can then be inverted to yield the single valued function $x(t)$. In the solution for the momentum, the two branches of the function $t(x)$ correspond to the two roots of the square root function. The branches occur when $E - V(x_\pm) = 0$. A particle trapped in a region of space will oscillate between the two limiting values of the motion $x_- \leq x \leq x_+$ corresponding to its turning points. Such oscillatory motion is referred to as *libration*. The *period T* of the motion is the time required for a bound system to return to its initial state. Since the integral is symmetric about the turning points, one gets

$$T = 2\int_{x_-}^{x_+} \frac{dx}{\sqrt{\frac{2}{m}(E - V(x))}}. \tag{1.99}$$

1.6.3 Small Oscillations about Equilibrium

For small displacements $\delta x = x - x_0$ about a point of stable equilibrium x_0, one can expand the potential in a Taylor's series expansion

$$V(x) = V|_{x=x_0} + \left.\frac{dV}{dx}\right|_{x=x_0} \delta x + \frac{1}{2}\left.\frac{d^2V}{dx^2}\right|_{x=x_0} \delta x^2 \cdots$$

$$= V_0 - F(x_0)\delta x + \frac{1}{2}k\delta x^2 + O(\delta x^3), \tag{1.100}$$

where $F(x_0) = 0$ and V_0 is a physically unimportant constant. The potential is a simple harmonic to the lowest order in the expansion with a spring constant given by

$$k = \left.\frac{d^2V}{dx^2}\right|_{x=x_0} = m\omega_0^2 > 0. \tag{1.101}$$

The equation of motion, to the lowest order in the perturbation, satisfies the SHO equations of motion

$$\frac{d^2}{dt^2}\delta x = -\omega_0^2 \delta x, \tag{1.102}$$

with a period of small oscillations given by

$$T = \frac{2\pi}{\omega_0}. \tag{1.103}$$

Note that the criterion for stability is $k > 0$. If $k < 0$, the motion is unstable. If $k = 0$, the leading order approximation fails, and one needs to carry out the expansion to the first nonvanishing term in the Taylor's series.

1.6.4 Plane Pendulum

For larger oscillations in a potential well, the period is not generally a constant, as can be seen by analyzing the motion of a plane pendulum. The *simple pendulum* moving in a plane subject to a constant gravitational field is illustrated in Figure 1.16. The motion can be written as a solution to the one-dimensional torque equation

$$\tau = I\ddot{\theta} = -mgl\sin\theta, \tag{1.104}$$

where $I = ml^2$ is the moment of inertia of the pendulum, and l is the length of a light, but rigid, rod connecting a plumb-bob of mass m to the pivot point. Note that the constraint force of the tension $\mathbf{F}_T$ in the rod does not

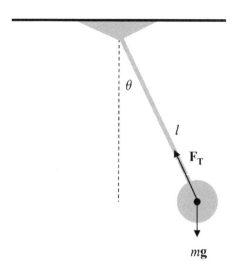

FIGURE 1.16 The plane pendulum.

contribute to the torque since $\mathbf{r} \times \mathbf{F_T} = 0$. The mechanical energy of the system is given by

$$E = \frac{1}{2}I\dot{\theta}^2 + mgl(1 - \cos\theta). \tag{1.105}$$

The amplitude $\theta_{\max}$ of oscillatory motion is related to the energy by

$$E = mgl(1 - \cos\theta_{\max}) < 2mgl. \tag{1.106}$$

The potential energy diagram for the plane pendulum is shown in Figure 1.17. For small oscillations, the motion is similar to the simple harmonic oscillator equations of motion

$$\ddot{\theta} \approx -\omega_0^2\theta, \tag{1.107}$$

with an angular frequency of small oscillations given by

$$\omega_0 = \sqrt{g/l} \tag{1.108}$$

and a period $T_0 = 2\pi/\omega_0$.

However, the period increases as the amplitude increases, becoming infinite when $E = E_S = 2mgl$, where E_S is the separation energy between modes. For $E > E_S$ the motion becomes rotational.

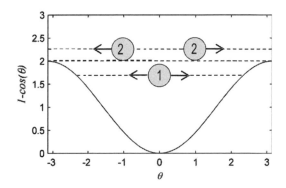

FIGURE 1.17 Potential energy diagram for the plane pendulum (a periodic cosine potential). Region 1 shows a bounded librational mode. Region 2 shows a bifurcation into two cyclic rotational modes, one rotating clockwise and the other counterclockwise in configuration space.

Numerical solution of the equations of motion can be obtained by solving the coupled ordinary differential equations (ODEs). These are given by writing the torque equation in terms of first order phase space equations

$$\dot{p}_\theta = -mgl\sin\theta, \quad \dot{\theta} = p_\theta/ml^2. \tag{1.109}$$

Here, the independent degrees of freedom are chosen to be the rotation angle θ and its conjugate angular momentum, defined as $p_\theta = \partial T/\partial\dot{\theta} = I\dot{\theta}$. Figure 1.18 shows the phase space diagram for the plane pendulum.[11] The phase space geometry is that of an upright cylinder which is periodic on interval $\Delta\theta = 2\pi$. The dashed line indicates the boundary between oscillatory and rotational dynamics. The direction of flow in the diagram is clockwise about the stable attractor at the origin, which is marked with an x. Unstable equilibrium occurs at the saddle points $(p_\theta = 0, \theta = \pm\pi)$.

The plane pendulum has two modes of motion: libration (oscillation) and rotation. The dividing line between the two modes is called the *separatrix*. Solving the energy equation for $\dot{\theta}$ allows the problem to be reduced to quadrature

$$\phi = \omega_0(t - t_0) = \pm\frac{1}{\sqrt{2}}\int_{\theta_0}^{\theta}\frac{d\theta}{\sqrt{\frac{2E}{E_s} + \cos\theta - 1}}. \tag{1.110}$$

---

[11]The term "phase-space" without further qualification will be reserved for the product space of positions and canonical momenta.

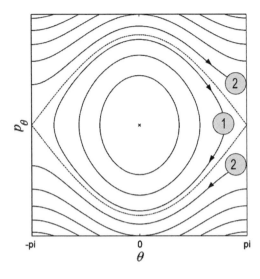

FIGURE 1.18 **Phase-space plot for the plane pendulum. The stable point is marked with an x. The separatrix between oscillatory and rotational dynamics is indicated by the dashed line. The direction of flow is clockwise about the stationary point.**

where $\omega_0 = \sqrt{g/l}$ and ϕ is a dimensionless measure of the time. The solution for the period of a libration cycle can be written a function of the small oscillation period $T_0 = 2\pi/\omega_0$ and the amplitude θ_{max}

$$T = 4\frac{T_0}{2\pi}\frac{1}{\sqrt{2}}\int_0^{\theta_{max}}\frac{d\theta}{\sqrt{\frac{2E}{E_s} + \cos\theta - 1}}, \tag{1.111}$$

where the amplitude can be expressed as function of the energy

$$E(\theta_{max}) = E_s(1 - \cos\theta_{max})/2. \tag{1.112}$$

A particle with energy $0 \le E < E_s$ will oscillate, with an amplitude θ_{max}, about the stable point. Traditionally, such integrals are solved by substitution of variables and/or the use of a good set of integral tables. However, modern symbolic math programs are quicker and easier to use.[12] Making the substitution

$$\sin\theta/2 = (\sin\theta_{max}/2)\sin\phi = k\sin\phi, \tag{1.113}$$

---

[12]Wolfram Mathworld sponsors a free online function integrator that can be to find analytic solutions, available at *http://mathworld.wolfram.com/*. Hand tweaking of the solution, such as simplifying the results and validating that the correct branch has been returned, is often necessary.

gives the exact solution for oscillatory motion in terms of an elliptical integral of the first kind $F(\phi, k^2)$

$$\omega_0 t = F(\phi, k) = \int_0^\phi \left(\sqrt{1 - k^2 \sin^2 \phi}\right)^{-1} d\phi = sn^{-1}(\sin \phi, k^2). \quad (1.114)$$

Inverting the solutions gives the result in terms of a *Jacobi elliptic function* $sn(u, k^2)$

$$\sin \phi = \frac{\sin \theta/2}{\sin \theta_{\max}/2} = sn(\omega_0 t, k^2), \quad (1.115)$$

where $k = \sin \theta_{\max}/2 = \sqrt{E/E_s}$. The period for oscillatory orbits is given by

$$T_{\text{osc}} = \frac{4}{\omega_0} F(\pi/2, k) = \frac{4}{\omega_0} \int_0^{\pi/2} \left(1/\sqrt{1 - k^2 \sin^2 \phi}\right) d\phi$$

$$= \frac{2T_0}{\pi} K(k^2) = \frac{2T_0}{\pi} K(E/E_s). \quad (1.116)$$

Here, $K(k) = F(\pi/2, k^2)$ is the *complete elliptic integral of the first kind*, with $K(0) = \pi/2$ and $K(1) = \infty$. The integral becomes infinite as $E \to E_s$. Therefore, a particle located on the separatrix takes an infinite amount of time to reach the point of unstable equilibrium ($\theta_0 = \pm\pi$). Once there, it stays there indefinitely, until pushed off by a small perturbation.

For $E > E_s$ the motion is rotational. There is a factor-of-two difference in the definition of the rotational period versus the librational period, since the rotational motion never changes direction. (This bifurcation effect should be evident from careful examination of the phase space plots.) The rotational period of the plane pendulum as a function of its energy is given by

$$T_{\text{rot}} = \frac{T_0}{\pi \sqrt{E/E_s}} K\left(\frac{1}{E/E_s}\right). \quad (1.117)$$

Figure 1.19, shows a plot of how the pendulum period behaves as a function of the energy. The period is plotted in dimensionless units, with time in units of the small amplitude period T_0 and energy in units of the branch-point energy E_s. At very high energies, $K \to \pi/2$, the period decreases linearly with the inverse of its angular velocity, $T_{Rot} \sim 2\pi r/v$, as expected when the potential energy term can be ignored.

1.6.5 Dissipative Terms and Driving Forces

More realistic treatment of oscillatory systems require the inclusion of *dissipative* and *driving forces*. Driving forces are time dependent forces, while

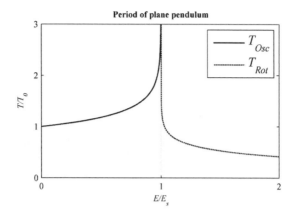

FIGURE 1.19 The period of the plane pendulum is plotted as a function of its energy. At the branch point between oscillatory and rotational motions, the period becomes infinite. There is a factor-of-two difference in the definition of the rotational period versus the oscillatory period, due to an absence of a turning point for the rotational motion case.

dissipative forces dissipate energy through *stochastic processes*. Stochastic processes are random in nature, and their (mean) effects are often determined empirically. Assume, for simplicity, that the damping force is linear in the velocity. Then, the equation of the damped linear harmonic oscillator becomes

$$m(\ddot{x} + \gamma \dot{x} + w_0^2 x) = F(t), \qquad (1.118)$$

where $-m\gamma\dot{x}$ is a velocity dependent damping force, γ is the damping coefficient, and $F(t)$ is an externally applied time-dependent driving force. The equivalent first-order coupled differential equations can be written as functions of position and momentum

$$\dot{p} = -kx - \gamma p + F(t), \quad \dot{x} = p/m. \qquad (1.119)$$

Due to the presence of damping terms and of a driving force, the mechanical energy of the system is no longer conserved. The rate of energy transfer into the system is given by the contributions of the non-conserved terms

$$\frac{dE}{dt} = -m\gamma\dot{x}^2 + F(t)\dot{x} = -2\gamma T(\dot{x}) + P(t, \dot{x}), \qquad (1.120)$$

where $P(t, \dot{x})$ is the power inputted by the time-dependent force.

The total solution to linear equations of motion can be found by superposition of *homogenous* and *inhomogeneous* solutions. The homogeneous solution is a general solution to the source-free homogeneous equation

$$m\left(\frac{d^2}{dt^2} + \gamma\frac{d}{dt} + \omega_0^2\right)x_h = 0, \tag{1.121}$$

while the particular solution is any solution to the inhomogeneous equation which contains the source terms

$$m\left(\frac{d^2}{dt^2} + \gamma\frac{d}{dt} + \omega_0^2\right)x_p = F(t). \tag{1.122}$$

1.6.5.1 Linear Damped Oscillator

For the homogeneous equation, one can find time-dependent normal modes of the form

$$x_h(t) = e^{\lambda t}, \tag{1.123}$$

where the λ's are generally complex solutions to the *secular equation*

$$\lambda^2 + \gamma\lambda + \omega_0^2 = 0. \tag{1.124}$$

There are three domains, corresponding to whether the solutions are *under-damped* (I), *over-damped* (II), or *critically-damped* (III):

$$\lambda = \begin{cases} \dfrac{-\gamma}{2} \pm i\sqrt{\omega_0^2 - \dfrac{\gamma^2}{4}}, & \omega_0^2 > \dfrac{\gamma^2}{4}, & (I) \\[3mm] \dfrac{-\gamma}{2} \pm \sqrt{\dfrac{\gamma^2}{4} - \omega_0^2}, & \omega_0^2 < \dfrac{\gamma^2}{4}, & (II) \\[3mm] \dfrac{-\gamma}{2}, & \omega_0^2 = \dfrac{\gamma^2}{4}. & (III) \end{cases} \tag{1.125}$$

The real-valued solution for the underdamped case, shown in the left-hand-side of Figure 1.20, is a decaying sinusoidal function of the form

$$x_h(t) = Ae^{-\gamma t/2}\sin(w't + \delta), \quad w' = \sqrt{\omega_0^2 - \frac{\gamma^2}{4}}. \tag{1.126}$$

The results are plotted by setting $\gamma = w'/2$ for three different phase angles $\delta = (0, \pm\pi/3)$. The corresponding phase space portraits of these trajectories, shown in the right-hand-side of Figure 1.20, are decaying spirals. Phase space volume is not conserved for dissipative systems, but, nevertheless, lines of flow never cross, although they eventually reach a collection point or an invariant subspace referred to as the *attractor*.

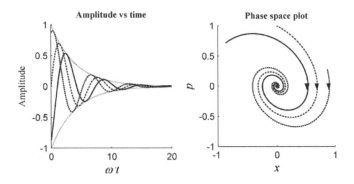

FIGURE 1.20 (Left hand side) the time evolution of the underdamped harmonic oscillator for three different values of the phase angle; (Right hand side) phase space portraits for the three curves shown on the left.

The underdamped solution decays to a *stable attractor* at the point $(x = p = 0)$. It has a $1/e$ time constant of $\tau_D = 2/\gamma$, leaving only the driven inhomogeneous part, if any, to survive at long times.

The critically damped oscillator poses a problem in that the two roots are degenerate. A second independent solution is needed. The general solution can be found by making the guess $x_2(t) = u(t)e^{-\gamma t/2}$, with solution $x_2(t) = c_2 t e^{-\gamma t/2}$ The general solution for the critically damped case is therefore

$$x(t) = \left(x_0 + t\left(\frac{p_0}{m} + \frac{\gamma x_0}{2}\right)\right)e^{-\gamma t/2}. \tag{1.127}$$

1.6.5.2 Resonance Response of the Driven Oscillator

To evaluate a particular solution to the inhomogeneous equation, it is useful to expand the driving force as an integral over frequency components. The *Fourier transform* is an invertible integral transform, which can be written in the symmetrized form

$$F(t) = \frac{1}{\sqrt{2\pi}} \int_{-\infty}^{\infty} F(\omega)e^{i\omega t}d\omega, \quad F(\omega) = \frac{1}{\sqrt{2\pi}} \int_{-\infty}^{\infty} F(t)e^{-i\omega t}dt, \tag{1.128}$$

For a real driving force, Fourier amplitudes come in complex conjugate pairs

$$F(-\omega) = F^*(\omega). \tag{1.129}$$

Solving the problem in the *frequency domain* converts the differential equation to an algebraic one,

$$m \left(\frac{d^2}{dt^2} + \gamma \frac{d}{dt} + \omega_0^2 \right) \int_{-\infty}^{\infty} x_p(\omega) e^{i\omega t} d\omega$$

$$= m \int_{-\infty}^{\infty} (-\omega^2 + i\omega\gamma + \omega_0^2) x_p(\omega) e^{i\omega t} d\omega$$

$$= \int_{-\infty}^{\infty} F(\omega) e^{i\omega t} d\omega, \tag{1.130}$$

with the solution

$$m(-\omega^2 + i\omega\gamma + \omega_0^2) x_p(\omega) = F(\omega)$$

$$x_p(\omega) = \frac{F(\omega)/m}{(-\omega^2 + i\omega\gamma + \omega_0^2)} = R(\omega, \omega_0, \gamma) F(\omega)/m. \tag{1.131}$$

To match the initial conditions, one needs to add a homogeneous part to this particular inhomogeneous solution, yielding the general solution

$$x(t) = A e^{-\gamma t/2} \sin(w't + \delta) + \frac{1}{\sqrt{2\pi}} \int_{-\infty}^{\infty} R(\omega, \omega_0, \gamma) \frac{F(\omega)}{m} d\omega. \tag{1.132}$$

The homogeneous part of the solution decays exponentially, leaving only the particular solution from the driving force to contribute for long time scales. The result of a frequency domain analysis shows that the response to a given driving frequency has a *resonance line shape* $R(\omega)$ given by

$$R(\omega) = \frac{1}{(\omega_0^2 - \omega^2 + i\omega\gamma)} = |R| e^{-i\phi}, \tag{1.133}$$

where ϕ is the *phase lag* of the response, and $|R|$ is it amplitude. The response is resonant when $\omega = \omega_0$, and it has a resonance width γ. The amplitude of the response is given by

$$|R| = \frac{1}{\sqrt{(\omega_0^2 - \omega^2)^2 + (\omega\gamma)^2}}. \tag{1.134}$$

The *quality factor* of the resonance, related to the sharpness of the resonance, is given by

$$Q = \omega_0/\gamma. \tag{1.135}$$

Figure 1.21 shows how the amplitude of the response broadens for smaller values of the quality factor Q. Also shown is a phase lag plot. The phase lag

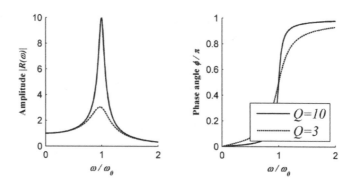

FIGURE 1.21 Resonance response of the damped driven oscillator for two different quality factors.

of the driven response is $0°$ at zero driving frequency, goes through a phase angle of $90°$ as its passes through the resonance frequency, and approaches $180°$ as the driving frequency goes to infinity.

1.7 SUMMARY

In Newtonian dynamics, matter consists of particles of mass m_i with positions $\mathbf{r}_i$ and momenta $\mathbf{p}_i$. In the absence of applied forces, the momenta of particles are conserved. In the presence of force, the particle system is a solution to the coupled first-order differential equations of motion

$$\dot{\mathbf{p}}_i = \mathbf{F}_i(x, p, t), \quad \dot{\mathbf{x}}_i = \frac{\mathbf{p}_i}{m_i}. \tag{1.136}$$

In this form, the equations of motion can be numerically integrated using Runge-Kutta techniques. Alternatively, an analytic solution might be attempted by reduction to quadrature (formal integration). Phase space portraits of the motion can be useful in understanding the dynamics of a system.

According to Newton's third law of motion, forces arise from the interaction of particles giving rise to equal and opposite action-reaction force pairs

$$\mathbf{F}_A + \mathbf{F}_R = 0. \tag{1.137}$$

In the nonrelativistic limit, where the speed of light goes to infinity, noncontact forces due to electromagnetic and gravitational force interactions act instantaneously on each other via action at a distance.

The kinetic energy of a system of particles is a positive definite quantity given by the sum of the kinetic energies of the individual particles

$$T = \sum_i \frac{m_i v_i^2}{2} = \sum_i \frac{1}{2} \mathbf{v}_i \cdot \mathbf{p} = \sum_i \frac{1}{2m_i} \mathbf{p}_i^2. \qquad (1.138)$$

Conservative forces can be derived from a universal scalar potential

$$\mathbf{F}_i^c = -\nabla_i V. \qquad (1.139)$$

If all the forces acting on a system are conservative in character, the mechanical energy

$$E = T + V \qquad (1.140)$$

of the system, defined as the sum of the kinetic and potential energies, is a constant of the motion. The work-energy theorem states that the change in the mechanical energy of the system is equal to the work done by the non-conservative forces. In differential form, the rate of change of the mechanical energy is due to the power transferred by the nonconservative forces given by

$$\frac{dE}{dt} = \sum_i \mathbf{v}_i \cdot \mathbf{F}_i^{nc}. \qquad (1.141)$$

The center of mass position $\mathbf{r}_{cm}$ of a particle system moves as if it were a particle, carrying the total mass m_Σ of the system, acting under the influence of the net externally applied force.

$$\dot{\mathbf{p}} = \sum_i \mathbf{F}_i, \qquad (1.142)$$

where

$$m_\Sigma = \sum_i m_i, \quad m_\Sigma \mathbf{r}_{cm} = \sum_i m_i \mathbf{r}_i, \quad \mathbf{p} = \sum_i m_i \dot{\mathbf{r}}_i = m_\Sigma \dot{\mathbf{r}}_{cm}. \qquad (1.143)$$

In center of mass coordinates, the relative momenta and velocities are defined by

$$\mathbf{v}_i' = \dot{\mathbf{r}}_i - \dot{\mathbf{r}}_{cm} = \mathbf{v}_i - \mathbf{v}_{cm}, \quad \mathbf{p}_i' = m_i(\mathbf{v}_i - \mathbf{v}_{cm}). \qquad (1.144)$$

The mechanical energy of a particle system can be separated into a contribution from the center of mass point and a contribution from the internal kinetic energy of the system

$$T = \frac{1}{2}\mathbf{v}_{cm} \cdot \mathbf{p} + \frac{1}{2}\sum_i \mathbf{v}'_i \cdot \mathbf{p}'_i. \tag{1.145}$$

Likewise, the angular momentum for a system of particles can be written in center of mass coordinates as

$$\mathbf{L} = \mathbf{r}_{cm} \times \mathbf{p} + \sum_i \mathbf{r}'_i \times \mathbf{p}'_i. \tag{1.146}$$

1.8 EXERCISES

1. An object of mass M at rest explodes into three equal fragments of mass. If one fragment moves in the $\hat{\mathbf{x}}$ direction with an initial velocity 20 m/s and a second fragment moves in the $\hat{\mathbf{y}}$ direction with an initial velocity of 21 m/s, in what direction and with what initial speed does the third fragment move?

2. A particle of mass 1 Kg and kinetic energy 1 kJ scatters elastically from a second particle of mass 2 Kg, which is initially at rest. If the first particle is emitted at an angle of 30° with respect its initial direction of motion, find the scattering angle of the second particle. What are the final kinetic energies of the two particles?

3. A particle falling vertically in a uniform gravitational field g sees a velocity dependent force given by $m\ddot{y} = -mg - b\dot{y}$. Solve for $y(t)$, given $y(0) = h, \dot{y}(0) = v_0$. What is the terminal velocity of the system? Find the power loss as a function of the particle velocity. Show that at the terminal velocity, power loss due to drag is equal to power gain from the gravitational potential energy term.

4. A particle in a spherical harmonic potential sees a drag force proportional to its velocity. The force equation is $\mathbf{F} = -k\mathbf{r} - b\dot{\mathbf{r}} = m\ddot{\mathbf{r}}$. Show that the time rate of change of the particle's mechanical energy is proportional to the kinetic energy of the particle.

5. A particle of mass m moving in the x direction sees an attractive potential energy function of the form $V(x) = k|x|$ where $k > 0$. If it is released from rest with amplitude x_0, what is its velocity when it passes the origin? Find the period, by reduction to quadrature, for a single complete cycle of oscillation.

6. A particle of mass m moving in the x-direction sees a potential energy function of the form $kx/(x^2 + x_0^2)$. Draw a sketch of the potential

energy function. Find the position of the potential energy minimum. Calculate the angular frequency of small oscillations about this minimum.

7. A mass of 1 kg is attached to a spring of equilibrium length $l_0 = 0.5$ m. Find the value of the spring constant k if the angular frequency of oscillations is 3 rad/s. If the spring and mass is hung vertically in the Earth's constant gravitational field in the lab, show that the frequency of oscillations remains unchanged, but that the equilibrium length of the spring stretches. Find the offset distance to the new equilibrium position if $g = 9.8$ m/s^2. What is the amplitude of oscillations if the spring is given a vertically upward impulse from rest at time $t = 0$ resulting in an initial velocity of 3 m/s? Find the time dependence of the subsequent motion.

8. A particle of mass m moves in the x-direction in the positive half-plane $x > 0$ under the action of a force $F = -kx + b/x$, where k and b are positive constants. Find the location of its stationary equilibrium point. Expand the force about the equilibrium position, and find the frequency for small oscillations about this point.

9. Consider a particle moving in one dimension subject to the force law $F(x) = -kx + b/x^3$, where k and b are positive constants. Find a potential energy function for this system and sketch it. Where are the equilibrium points? Where are the turning points for a given mechanical energy E? Sketch the velocity phase space curves. Find the period as a function of the mechanical energy by formal reduction to quadrature.

10. For a particle of mass m in a double-hump potential $V(x) = V_0(1 - 2x^2 + x^4)$, find the stationary points, and discuss separately the cases where the energy is greater than, equal to, or less than V_0. What are the frequencies for small amplitude oscillation about the points of stable equilibrium? Sketch the contours of constant energy in a velocity phase space plot $(x, \dot{x})$ using dimensionless units $(m = V_0 = 1)$. Characterize the motion.

11. A rocket is propelled vertically upward in a nearly uniform gravitational field at the surface of the Earth. The rate of mass loss of the fuel is a constant, and the fuel is entirely gone in 60 s. The relative velocity of the exhaust gas is 2.4 km/s. If one assumes that the rocket has just achieved minimal orbital escape velocity from the Earth as the fuel runs out, what is the ratio of the final mass of the rocket to its initial mass at takeoff?

12. A dump truck, coasting with negligible friction on a level road at 50 km/hr, has an empty mass of 5 metric tons. It starts raining, and

the truck begins filling with water at a rate of 2 kg/sec. Plot the velocity of the truck as a function of time, assuming the rain is falling vertically downward with respect to the road.

13. Solve for the tension in the rod of the simple plane pendulum of length l and mass m as function of the pendulum's angle and energy. Assume the rod is massless. If the rod is replaced by a flexible string, what constraint must be put on the tension?

14. An underdamped harmonic oscillator satisfies the equation of motion $m\ddot{x} + b\dot{x} + kx = 0$. Assume that the oscillator, initially at rest, suffers an impulsive force $F(t) = \delta(t - t_0)$ at time $t = t_0$, resulting in a unit change of momentum, $\Delta p = 1$. Show that this yields the Green's function solution given by

$$\left(m\frac{d^2}{dt^2} + b\frac{d}{dt} + k \right) G(t - t_0) = \delta(t - t_0),$$

$$G(t - t_0) = \begin{cases} 0, & t < t_0, \\ \dfrac{1}{m\omega'}e^{-b(t-t_0)/2m} \sin \omega'(t - t_0), & t \ge t_0. \end{cases}$$

where $\omega' = \sqrt{k/m - b^2/4m^2}$.

15. Using the impulse solution to Exercise (1.14), show that a particular solution to a damped oscillator with a driving force $m\ddot{x} + b\dot{x} + kx = F(t)$ is given by

$$x(t) = \int_{-\infty}^{t} G(t - t')F(t')dt'.$$

16. An undamped spring with attached mass m and oscillator frequency ω, initially at rest, is acted on by a constant force F_0 from time $t = 0$ to time $t = \Delta t$. Find the amplitude and phase of the motion given by the equation $A \sin(\omega t + \phi)$ for subsequent times $t > \Delta t$. Use the result from Exercise (1.15).

17. A particle dropped from a height h in a uniform gravitational field has a coefficient of restitution of e, i.e., $|v'_{n+1}| = e|v_n|$ at the point of contact, when bouncing vertically from a level horizontal surface. Find the time required for the particle to come to rest and the total distance that it travels. It is useful to recall that the sum of a geometric series is given by

$$S(a, r) = \sum_{n=0}^{\infty} ar^n = a/(1 - r), \quad 0 \le r < 1.$$

18. Show that the force $\mathbf{F}(x, y) = k(y\,\hat{\mathbf{x}} - x\,\hat{\mathbf{y}})$ is nonconservative by explicitly calculating $\nabla \times \mathbf{F}$, where $\hat{\mathbf{x}}$ is a unit vector in the Cartesian $\hat{\mathbf{x}}$-direction and $\hat{\mathbf{y}}$ is a unit vector in the Cartesian $\hat{\mathbf{y}}$-direction. Find the work $\oint \mathbf{F} \cdot d\mathbf{s}$ done on a particle moving counter-clockwise over a complete cycle on the closed circular path $x^2 + y^2 = r_0^2$. (Hint: parameterize the path by $x = r_0 \cos\theta, y = r_0 \sin\theta$.)

19. A particle moving with velocity v_1 in the positive x direction collides in one dimension with a particle that is initially at rest. Prove that the final velocities of the two particles are given by

$$v_1' = \frac{m_1 - em_2}{m_1 + m_2}v_1, \quad v_2' = \frac{(1+e)m_1}{m_1 + m_2}v_1,$$

where $e = (v_2' - v_1')/(v_1 - v_2)$ is the coefficient of restitution for the collision.

20. Draw a phase portrait diagram for the critically damped oscillator $\omega_0^2 = \gamma^2/4$. Use dimensionless units $m = k = 1$. Pick several different values of $\frac{1}{2}(x_0^2 + p_0^2) = 1$ as initial values for the phase space trajectories (i.e., the particles start out with the same initial energy). Note that it takes an infinite amount of time of the particles to reach the attractor at $x = p = 0$.

   a. Compare the analytic results with the Euler equation ODE solver results for numerical differentiation of the simple harmonic oscillator equation

   $$\frac{d^2x}{d\phi^2} = -x(\phi) : \quad \left[x(0) = 1, \frac{dx}{d\phi}(0) = 0\right],$$

   where $\phi = \omega t$, with the exact solution $x(\phi) = \sin\phi$,

   b. Plot the difference in amplitude $\Delta x(\phi) = x_{\text{Euler}}(\phi) - \sin\phi$,

   c. For $h = \Delta\phi = 0.05$ rad, plot the numerical energy dissipation

   $$\Delta E(\phi) = \frac{1}{2}(\dot{x}^2(\phi)_{\text{exact}} - x^2(\phi)_{\text{Euler}}) - \frac{1}{2},$$

   d. Compare your results with those using a fourth order Runge-Kutta ODE solver with the same step size.

21. A simple pendulum subject to a linear damping term satisfies the first order differential equations

$$\dot{p}_\theta = -\omega_0^2 \sin\theta - \gamma p_\theta, \quad \dot{\theta} = p_\theta/I.$$

Numerically evaluate the phase space plot (θ, p_θ) for the motion, given the initial conditions $(\theta_0 = \pi/3, p_\theta = 0)$. Use dimensionless units $(I = \omega_0 = 1)$, and set the damping coefficient to $\gamma = 1/2$.

22. A damped non-linear oscillator satisfies the differential equation $\ddot{x} + \frac{1}{4}\dot{x}(x^2 - 1) + x = 0$. Note that the stationary point $x = \dot{x} = 0$ is unstable. Numerically integrate this equation and plot several phase space flow curves for this system, illustrating that, independent of the initial starting conditions (with the exception of the unstable stationary point), the trajectories decay to an invariant curve called the limit cycle . The equations to be solved can be linearized using $p = \dot{x}$ in these units to give $\dot{p} = -x - \frac{1}{4}p(x^2 - 1), \dot{x} = p$.

CHAPTER 2

VECTOR SPACES AND COORDINATE SYSTEMS

I n this chapter, the concept of an inner product vector space will be developed. Coordinate maps will be introduced, and common orthogonal curvilinear coordinate systems will be defined. Curves and surfaces will be treated as manifolds embedded in Euclidean space. This chapter provides a basic overview of the elements of vector calculus. Detail properties of the rotational group will be discussed later in Chapter 7.

2.1 POINTS AND VECTORS

Space and time define a continuous differentiable arena or manifold in which particles and fields interact. In nonrelativistic mechanics, space is considered to form a three-dimensional continuum, while time forms a separate one-dimensional continuum. In relativistic mechanics, *spacetime* forms a single four-dimensional continuum of indefinite metric signature. The nonrelativistic picture will be used for now.

A physical particle is a point-like object with a well-defined position and velocity, to which additional physical attributes, such as mass and charge, can be assigned. A point is an entity with a location but no extent. The motion of a particle is given by the trajectory of its point. *Points* are fundamental to the language of mathematics. In mathematics, any form of space can be constructed from elemental points.

Newton's fiction of absolute space, existing independent of matter, is perhaps the easiest way to introduce the concept of a geometric vector space.

Consider a point $\mathbf{P}$ in an absolute three-dimensional Euclidean space, parameterized by three, independent, continuous Cartesian coordinates $(x^1, x^2, x^3) = (x, y, z)$ with respect to a constant, orthogonal Euclidean basis $(\mathbf{e}_1, \mathbf{e}_2, \mathbf{e}_3)$. The origin $\mathbf{O} = (0, 0, 0)$ is embedded in the space. The location of a point defined on E^3 with respect to a Cartesian basis is given by[1]

$$\mathbf{P}(x, y, z) = x^i \mathbf{e}_i = x \, \mathbf{e}_1 + y \, \mathbf{e}_2 + z \, \mathbf{e}_3. \tag{2.1}$$

2.1.1 Displacement

The *displacement* of a point from the origin is given by

$$\mathbf{r} = \mathbf{P} - \mathbf{O}. \tag{2.2}$$

Infinitesimal displacements between points are vectors. With respect to a fixed origin, one gets

$$d\mathbf{r} = dx^i \mathbf{e}_i. \tag{2.3}$$

The Euclidean basis of absolute space is constant in time and position independent; therefore, all of their partial derivatives vanish

$$\partial \mathbf{e}_i / \partial x^j = 0 \quad \partial \mathbf{e}_i / \partial t = 0. \tag{2.4}$$

Euclidean space is translationally and rotationally invariant. Euclidean space is an example of a flat metric space. It has no intrinsic curvature. The length of a relative displacement between two points in this space is given by the *Pythagorean theorem*:

$$(d\mathbf{r})^2 = \delta_{ij} dx^i dx^j = (dx)^2 + (dy)^2 + (dz)^2. \tag{2.5}$$

Infinitesimal displacements $d\mathbf{r}$ are the prototypes for all vectors. A point on a general manifold is not a vector, since a point has a fixed origin of reference, while a vector quantity, like the velocity, only has a magnitude and direction. This distinction becomes important when discussing curved spaces. A common example of a vector is a force—force has a magnitude and an orientation. Multiple forces acting on an object point are summed according to the parallelogram law of vector addition.

The chain rule for partial derivates can be used to map infinitesimal displacements into multiple coordinate systems

$$d\mathbf{r}(\xi(x, y, z)) = d\xi^i \frac{\partial \mathbf{r}}{\partial \xi^i},$$
$$d\mathbf{r} = dx^i \mathbf{e}_i = d\xi^i \boldsymbol{\lambda}_i. \tag{2.6}$$

---

[1]Please note that the *Einstein summation convention* will be used throughout this book: the sum (contraction) over repeated raised and lowered tensor indices is implied in an equation unless otherwise noted by use of the acronym (nsc) denoting "No Summation Convention."

A coordinate map or *patch* is allowable in any region of space where it is differentiable and invertible (non-singular). The chain rule yields a *form invariant* expression for generalized *contravariant* elements of displacement $d\xi^i$ with respect to a *covariant* vector basis $\boldsymbol{\lambda}_i$ defined by

$$\boldsymbol{\lambda}_i = \frac{\partial \mathbf{r}}{\partial \xi^i}. \tag{2.7}$$

A space point can be located by specifying its generalized coordinates ξ^i. The motion of the point is a parameterized curve called its *trajectory*

$$\mathbf{r}(t) = \mathbf{r}(\xi^i(t)), \tag{2.8}$$

where the function $\mathbf{r}(t)$ is twice differentiable with respect to the time. Since a particle has a well-defined location at any instance of time, the equation of a particle trajectory must be a *single-valued function* of the time.

2.1.2 Velocity

The velocity of a particle is the vector tangent to the trajectory obtained by taking the total time derivative of its position

$$\mathbf{v} = \frac{d\mathbf{r}}{dt} = \dot{\xi}^i \boldsymbol{\lambda}_i. \tag{2.9}$$

A *vector* is a special case of a *tensor*, or a form-invariant object. The components of velocity $\dot{\xi}^i$ with respect to a general tensor basis $\boldsymbol{\lambda}_i$ are its *generalized velocities*. The tensor itself is an invariant object, but its components transform by covariance under a coordinate transformation. Vectors are rank-1 tensors and can be used to create tensors of greater rank. Vectors form the foundation for vector fields and vector analysis. Tensors of arbitrary rank can be expressed as multidimensional arrays with respect to some tensor basis, although as objects, in and of themselves, they are invariants, independent of any choice of coordinates. They conform to Einstein's principle of relativity, which asserts that the inherent properties of physical objects should not depend on choice of reference frame used to represent them.

2.2 INNER PRODUCT SPACES

Geometric vector spaces are inner product spaces with a metric, which provide a measure of separation. They are defined over the field of real numbers. If the space has a mixed metric signature, such as Minkowski Space, the subspace of null vectors separates time-like from space-like separations. If the space has definite signature, such as Euclidean Space, the only null vector is the additive identity element zero. The norm-squared of a vector is calculated by the process of scalar contraction and is denoted by the dot ($\cdot$) product. The measure of infinitesimal displacement is defined locally in terms of its metric tensor as

$$ds^2 = d\mathbf{r} \cdot d\mathbf{r} = (\boldsymbol{\lambda}_i \cdot \boldsymbol{\lambda}_j) d\xi^i d\xi^j = g_{ij} d\xi^i d\xi^j. \tag{2.10}$$

2.2.1 Vector Reversion

In general, inner product spaces are dual spaces,[2] where the contraction of a vector $\mathbf{V}$ (sometimes called a ket-vector) with its *vector adjoint*, denoted as $\bar{\mathbf{V}}$ (sometimes called a bra-vector), results in a real-valued measure of its length. The vector adjoint operation, which is referred to as *reversion*, implies a reversal (*transposition*) of the order of all vector operators in a product, plus *field conjugation*.[3,4] The operation of reversion will be denoted by use of an overstrike placed over the object. The real number field is self-conjugate under reversion, as are all scalar fields,

$$\bar{\mathbb{R}} = \mathbb{R}. \tag{2.11}$$

Since geometric vector spaces are defined over the field of real numbers, geometric vectors are self-adjoint under reversion[5]

$$\bar{\mathbf{V}} = \mathbf{V}, \quad V^i \in \mathbb{R}. \tag{2.12}$$

For the purpose of later reference, the adjunct of a complex or hypercomplex (quaternion) vector under reversion will be defined as its complex or

---

[2]A relationship transforming an object to another object is said to represent a dual relationship if repeating the transformation a second time restores the original object (for example, $\bar{\bar{a}} = a$ for the dual operation of reversion.)

[3]If the space has a mixed metric signature, not all the basis vectors are Hermitian. Nevertheless, they are defined to be self-adjoint under reversion. The vector transpose conjugate is, therefore, not necessarily the same as the Hermitian conjugate of its matrix representation. This distinction becomes important when considering Lorentz transformations in Minkowski Space.

[4]From the geometric point of view, the imaginary and hypercomplex numbers represent rotation operators in some internal spin space. Rotation operators are always antiadjoint under vector reversion. Field conjugation results from the consistent application of the concept of vector reversion.

[5]The overstrike denoting the transposed conjugate of a real vector is redundant, but will be displayed, when useful, to emphasize how the Gibbs vector notation is related to the Dirac bra-ket notation defined in Section 2.2.7.

TABLE 2.1   Basic rules of Vector Algebra (**a**, **b**, and **c** denote vector quantities; r and s denote scalars)

Commutativity under vector addition	$\mathbf{a} + \mathbf{b} = \mathbf{b} + \mathbf{a}$
Associativity under vector addition	$(\mathbf{a} + \mathbf{b}) + \mathbf{c} = \mathbf{a} + (\mathbf{b} + \mathbf{c})$
Additive vector identity	$\mathbf{0} + \mathbf{a} = \mathbf{a}$
Additive vector inverse	$\mathbf{a} + (-\mathbf{a}) = 0$
Associativity under scalar multiplication	$r(s\,\mathbf{a}) = (rs)\mathbf{a}$
Scalar multiplicative identity	$1\mathbf{a} = \mathbf{a}$
Distributivity of vector sums	$r(\mathbf{a} + \mathbf{b}) = r\,\mathbf{a} + r\,\mathbf{b}$
Distributivity of scalar sums	$(r + s)\mathbf{a} = r\,\mathbf{a} + s\,\mathbf{a}$

quaternion conjugate

$$\bar{\mathbf{V}} = \mathbf{V}^*, \quad V^i \in \mathbb{C} \text{ or } \mathbb{H}. \tag{2.13}$$

Conjugation is required to get a real-valued result for the norm-squared of a complex or quaternion vector.

The norm-squared of a geometric vector is a real-valued scalar that is defined to be invariant under reversion

$$\overline{\mathbf{V} \cdot \mathbf{V}} = \mathbf{V} \cdot \mathbf{V}. \tag{2.14}$$

The vector dot product of a geometric vector with itself is equivalent to taking its square, which in turn is defined in terms of the metric tensor, giving the definition

$$V^2 = \mathbf{V} \cdot \mathbf{V} = V^i g_{ij} V^j = V_j V^j. \tag{2.15}$$

In the language of differential geometry, the vector components V^i, with raised indices, are defined as the components of a vector field (representing contravariant vectors) and $V_i = g_{ij}V^j$, with lowered indices, are defined as the components of a differential one-form (representing covariant vectors). The contraction of a vector with a one-form is a scalar invariant. *Contraction* involves the summing over products of contravariant and covariant tensors with the same index.

The elementary rules of Vector Algebra are summarized in Table 2.1.

2.2.2 Reciprocal Basis

A complete covariant vector basis $\{\boldsymbol{\lambda}_i\}$ spans the allowed directions of motion in the *tangent vector space*. With respect to such a basis, one can define a reciprocal, contravariant basis $\{\boldsymbol{\lambda}^i\}$ that is everywhere orthogonal to the first basis. The inner product of basis vectors with their reciprocals is defined using the *Kronecker Delta* identity matrix

$$\boldsymbol{\lambda}^i \cdot \boldsymbol{\lambda}_j = \boldsymbol{\lambda}^i \cdot \mathbf{1} \cdot \boldsymbol{\lambda}_j \equiv \delta^i_j. \tag{2.16}$$

The inverse basis exists wherever the determinant of the metric tensor is finite and nonvanishing. The symbol $|g|$ is used to denote the determinant of the metric tensor

$$|g| = \det(g_{ij}). \tag{2.17}$$

The reciprocal basis can be expressed in terms of the covariant basis vectors by defining the reciprocal metric g^{ij}, which raises the index of a covariant tensor, converting it to a contravariant tensor

$$\boldsymbol{\lambda}^i = g^{ik}\boldsymbol{\lambda}_k, \tag{2.18}$$

$$\boldsymbol{\lambda}^i \cdot \boldsymbol{\lambda}_j = g^{ik}\boldsymbol{\lambda}_k \cdot \boldsymbol{\lambda}_j = g^{ik}g_{kj} = \delta^i_j. \tag{2.19}$$

One can always convert to and from contravariant/covariant representations of vector components by using the metric tensor. Vectors can therefore be expressed in either covariant or contravariant form

$$\mathbf{a} = a^i\boldsymbol{\lambda}_i = a_i\boldsymbol{\lambda}^i. \tag{2.20}$$

The dot product of two vectors is formed by contraction of contravariant and covariant components of vectors

$$\mathbf{a} \cdot \mathbf{b} = (a^i\boldsymbol{\lambda}_i) \cdot (b^j\boldsymbol{\lambda}_j) = g_{ij}a^ib^j = a^ib_i = a_jb^j. \tag{2.21}$$

The n-tuple of vector components $\{a^i\}$ can be represented as entries in column matrices with the contravariant representation

$$\{a^i\} = \begin{bmatrix} a^1 \\ a^2 \\ a^3 \end{bmatrix} = \begin{bmatrix} a^1 & a^2 & a^3 \end{bmatrix}^{\mathrm{T}}, \tag{2.22}$$

and their bra-vectors can be represented as entries in row-matrices with the covariant representation

$$\{\bar{b}_i\} = \begin{bmatrix} \bar{b}_1 & \bar{b}_2 & \bar{b}_3 \end{bmatrix}, \tag{2.23}$$

such that one gets

$$\bar{\mathbf{b}} \cdot \mathbf{a} = (\bar{\mathbf{b}} \cdot \boldsymbol{\lambda}_i)(\boldsymbol{\lambda}^i \cdot \mathbf{a}) = \begin{bmatrix} \bar{b}_1 & \bar{b}_2 & \bar{b}_3 \end{bmatrix} \begin{bmatrix} a^1 \\ a^2 \\ a^3 \end{bmatrix} = \bar{b}_ia^i. \tag{2.24}$$

where $\{\boldsymbol{\lambda}_i\}$ defines a covariant projection basis and $\{\boldsymbol{\lambda}^i\}$ defines a covariant projection basis. For real valued vector fields, $\bar{b}_i = b_i$ and therefore $\bar{\mathbf{b}} \cdot \mathbf{a} = \mathbf{b} \cdot \mathbf{a} = b_i a^i$.

2.2.3 Projection of Vector Components

Basis vectors are *projection operators*. They can be used to extract the contravariant or covariant components of a vector. The scalar projection of contravariant and covariant vector components of the vector a are given by real numbers or real fields

$$
\begin{aligned}
\boldsymbol{\lambda}^j \cdot \mathbf{a} &= \boldsymbol{\lambda}^j \cdot (a^i \boldsymbol{\lambda}_i) = a^i \delta_i^j = a^j, \\
\boldsymbol{\lambda}_j \cdot \mathbf{a} &= \boldsymbol{\lambda}_j \cdot (a_i \boldsymbol{\lambda}^i) = a_i \delta_j^i = a_i.
\end{aligned}
\tag{2.25}
$$

Note that scalar projection of the components of a vector is analogous to the process of the measurement of its length along a specific axis. The basis vector plays the role of a laboratory measurement apparatus, defining the geometric orientation and scale parameters of a particular geometric degree of freedom. Scalar projection is defined to return a real-valued result, since the measurement of a classical observable must always return a real valued result.

Some semantics may help clarify the meaning of scalar projection. The statement "the components of vector $\mathbf{a}$ are (a^1, a^2, a^3)" is a covariant statement that depends on the frame of reference; whereas, the statement "the components of vector $\mathbf{a}$ are (a^1, a^2, a^3) as measured with respect to reference frame B" is an invariant statement that all observers should agree with.

2.2.4 Coordinate Transformations

Points transform by invariance under invertible, differential coordinate transformations. That is,

$$
\mathbf{P}(x^1(\xi), x^2(\xi), x^3(\xi)) = \mathbf{P}(\xi^1, \xi^2, \xi^3)
\tag{2.26}
$$

refers to the same point under a coordinate transformation. The set of all invertible coordinate transformations forms a group. Invertiblility is equivalent to requiring that the determinant of the metric tensor be finite and non-zero wherever the coordinate map is applicable.

Infinitesimal displacements $d\xi^i$ are the prototypes for all contravariant vectors. In a flat Euclidean space, they are defined relative to a covariant basis given by

$$\lambda_i = \frac{\partial \mathbf{P}(\xi^1, \xi^2, \xi^3)}{\partial \xi^i} = \frac{\partial x^j(\xi^1, \xi^2, \xi^3)}{\partial \xi^i} \mathbf{e}_j. \tag{2.27}$$

A unit vector in a given direction is often denoted by using a hat caret over the coordinate label

$$\hat{\boldsymbol{\xi}}_i = \lambda_i / |\lambda_i| = \hat{\lambda}_i. \tag{2.28}$$

Using the chain rule for differentiation, one obtains the following transformation rules

$$\lambda'_i = \frac{\partial \xi^j}{\partial \xi'^i} \lambda_j, \tag{2.29}$$

$$d\xi'^i = \frac{\partial \xi'^i}{\partial \xi^j} d\xi^j, \tag{2.30}$$

$$d\mathbf{r}' = \lambda'_j d\xi'^j = \lambda_i d\xi^i = d\mathbf{r}. \tag{2.31}$$

Note that the chain rule is an invariant form consisting of the inner product of covariant and contravariant tensors. Covariant and contravariant vector components transform under reciprocal rules of transformation. That is, the transformation rule for a vector's components is "contravariant" to the "covariant" transformation rule for the basis of the frame of reference. Raised (superscript) and lower (subscript) indices are used to distinguish between the two classes of vectors. The rules of transformation can be summarized as

1. The n-tuple $\{v^i\}$ forms the components of a *contravariant vector* if it transforms in the same manner as the components of an infinitesimal displacement $\{d\xi^i\}$ under an allowed coordinate transformation.
2. The n-tuple $\{v_i\}$ forms the components of a *covariant vector* if it transforms in the same manner as components of the gradient operator $\{\partial/\partial\xi^i\}$ under an allowed coordinate transformation.

Covariant transformation matrices $\partial \xi^j / \partial \xi'^i$ have an inverse relationship to contravariant transformation matrices $\partial \xi'^i / \partial \xi'^j$ yielding

$$\frac{\partial \xi^j}{\partial \xi'^i} \frac{\partial \xi'^i}{\partial \xi^k} = \frac{\partial \xi^j}{\partial \xi^k} = \delta^j_k. \tag{2.32}$$

This reciprocality ensures that the inner product of contravariant and covariant vectors are invariant under all invertible coordinate transformations

$$b'_i a'^i = \frac{\partial \xi^{j'}}{\partial \xi'^i} b_{j'} \frac{\partial \xi'^i}{\partial \xi^k} a^j = \delta^{j'}_j b_{j'} a^j = b_j a^j. \tag{2.33}$$

As another example, the total derivative of a scalar function of the coordinates with respect to the scalar time must clearly be a scalar. By the chain rule, this scalar can be written as the dot product of two vectors

$$\frac{df(\xi)}{dt} = \left(\frac{\partial f}{\partial \xi^i}\right)\left(\frac{d\xi^i}{dt}\right) = (\nabla f) \cdot \dot{\xi}, \tag{2.34}$$

where the *generalized velocities* $d\xi^i/dt$ are the components of a contravariant vector, and *the gradient functions* $\partial f/\partial \xi^i$ are the components of a covariant vector or one-form. Contraction over pairs of covariant and contravariant indices is necessary get a scalar invariant result. The generalized velocities and the gradient functions are prototypes defining the transformation properties of all vectors and one-forms.

As an example of the power of the vector notation, consider that the kinetic energy can be written as a bilinear product of its velocity and Newtonian momentum vectors

$$T = \frac{1}{2}m\mathbf{v} \cdot \mathbf{v} = \frac{1}{2}\mathbf{p} \cdot \mathbf{v} = \frac{1}{2m}\mathbf{p} \cdot \mathbf{p}. \tag{2.35}$$

In component notation, these relations can be expressed in the contracted index form

$$T = \frac{m}{2}g_{ij}\dot{\xi}^j\dot{\xi}^i = \frac{1}{2}p_i\dot{\xi}^i = \frac{1}{2m}g^{ij}p_ip_j, \tag{2.36}$$

where $\dot{\xi}^i$ is the generalized velocity of a particle in an arbitrary coordinate system, and p_i is the generalized Newtonian momentum. If one knows the representation of the metric tensor for a given set of coordinates, one knows how to express the kinetic energy in that coordinate representation. The contravariant velocities and the covariant momenta are related by the invertible matrix transformation

$$p_i(\xi, \dot{\xi}) = mg_{ij}(\xi)\dot{\xi}^j,$$
$$\dot{\xi}^i(\xi, p) = \frac{1}{m}g^{ij}(\xi)p_j. \tag{2.37}$$

The product of a generalized velocity and a conjugate momentum has the units of energy.

In the language of differential geometry, contravariant vectors are defined on the tangent space T of a differential manifold, while covariant vectors are defined on the *cotangent space* T^* of the manifold. After developing Lagrangian mechanics, it will become apparent that generalized forces and momenta are principally defined on the cotangent space of gradient functions, while generalized accelerations and velocities are principally defined on the tangent space of infinitesimal displacements.

2.2.5 Vector Transformations

Vectors are rank-1 tensors. The inner product of a covariant and contravariant vector is a scalar, which is a tensor of rank 0. The ordered, *outer product* of vectors can be used to generate tensors of increasing rank. Define the ordered outer product $\mathbf{T} = \mathbf{a} \otimes \mathbf{b} = \mathbf{a}\,\mathbf{b}$ of two vectors by

$$\mathbf{T} = \mathbf{a} \otimes \mathbf{b} = \mathbf{a}\,\mathbf{b} = (a^i b^j)\boldsymbol{\lambda}_i \boldsymbol{\lambda}_j = (a^i b_j)\boldsymbol{\lambda}_i \boldsymbol{\lambda}^j$$
$$= (a_i b^j)\boldsymbol{\lambda}^i \boldsymbol{\lambda}_j = (a_i b_j)\boldsymbol{\lambda}^i \boldsymbol{\lambda}^j. \tag{2.38}$$

This ordered product of vectors is historically referred to as a *dyad* (pair) of vectors. No contraction is implied in the outer product. A hidden (invisible) multiplication ($\otimes$) symbol is used to denote the outer product of two vectors, while a visible center dot ($\cdot$) symbol is used to denote the inner product of vectors and dyads. By analogy to standard geometric algebra notation, the antisymmetrized outer product will be denoted by the wedge symbol ($\wedge$). It results in a skew symmetric tensor of rank-2, called a *bivector*

$$\mathbf{T}^{(2)} = \mathbf{a} \wedge \mathbf{b} = \frac{1}{2}(\mathbf{a}\,\mathbf{b} - \mathbf{b}\,\mathbf{a}) = (a^i b^j)\boldsymbol{\sigma}_{ij},$$
$$\boldsymbol{\sigma}_{ij} = \frac{1}{2}(\boldsymbol{\lambda}_i \boldsymbol{\lambda}_j - \boldsymbol{\lambda}_j \boldsymbol{\lambda}_i). \tag{2.39}$$

The antisymmetric tensor basis of *bivectors* $\boldsymbol{\sigma}_{ij} = \boldsymbol{\lambda}_i \wedge \boldsymbol{\lambda}_j$ spans the space of surfaces. They play a special role in transformation theory. In Geometric Algebras,[6] defined in Chapter 7, where rotational motion is studied in detail, they form the generators for the special orthogonal group of similarity transformations in three space dimensions, referred to as the *spin group*, *Spin*(3). This group is isomorphic to the *special unitary group* in two complex dimensions referred to as $SU(2)$.

By projection, the components of the outer product $T^{ij} = \boldsymbol{\lambda}^i \cdot \mathbf{T} \cdot \boldsymbol{\lambda}^j$ represent a contravariant tensor of rank-2, $T_{ij} = \boldsymbol{\lambda}_i \cdot \mathbf{T} \cdot \boldsymbol{\lambda}_j$ represent a covariant tensor of rank-2, and $T_i{}^j = \boldsymbol{\lambda}_i \cdot \mathbf{T} \cdot \boldsymbol{\lambda}^j$, $T^i{}_j = \boldsymbol{\lambda}^i \cdot \mathbf{T} \cdot \boldsymbol{\lambda}_j$ are distinct rank-2 tensors composed of mixed covariant and contravariant components. The components of the tensor, found by projection, can be written as the square

---

[6]A Geometric Algebra is a Clifford Algebra defined on the field of real numbers. Geometric Algebras provide a complete description of the geometric structure of a metric space.

matrix with raised, lowered, or mixed indices; for example,

$$\{T^i{}_j\} = \{\boldsymbol{\lambda}^i \cdot \mathbf{T} \cdot \boldsymbol{\lambda}_j\} = \begin{bmatrix} T^1{}_1 & \cdots & T^1{}_n \\ \vdots & \ddots & \vdots \\ T^n{}_1 & \cdots & T^n{}_n \end{bmatrix}. \tag{2.40}$$

Sums of dyads can be used to construct a general rank-2 tensor called a *dyadic*. A dyadic is a somewhat old-fashioned, but concise, term used to represent a special class of linear tensor transformations that can be used to map a vector field into itself. Dyadics form an associative multiplicative group under the dyadic dot product, which is constructed by invoking the contraction of adjacent vector elements. The special orthogonal group of Dyadic transformations with determinant $+1$ in three-space dimensions is referred to $SO(3)$. $Spin(3)$ is a double-valued cover of $SO(3)$.

In the following, the dot product is assumed to have greater precedence than the dyadic product. The scalar product of two vectors is a scalar that commutes with the vector basis. Since dyadic multiplication is associative, given two dyads $\mathbf{a}\,\mathbf{b}$ and $\mathbf{c}\,\mathbf{d}$, the resulting product is a dyad

$$\mathbf{a}\,\mathbf{b} \cdot \mathbf{c}\,\mathbf{d} = \mathbf{a}(\mathbf{b} \cdot \mathbf{c})\mathbf{d} = (\mathbf{b} \cdot \mathbf{c})\mathbf{a}\,\mathbf{d}. \tag{2.41}$$

The *completeness relation* for a vector basis gives rise to the *identity tensor*

$$\mathbf{1} = \mathbf{e}^i\mathbf{e}_i = \boldsymbol{\lambda}^i\boldsymbol{\lambda}_i \tag{2.42}$$

The dot product of two dyadic tensors results in another dyadic tensor. Given two dyadic tensors $\mathbf{U}$, and $\mathbf{V}$, with a product $\mathbf{W}$, such that

$$\mathbf{W} = \mathbf{U} \cdot \mathbf{V}, \tag{2.43}$$

the components of the product are extracted by contraction with an appropriate vector basis

$$W^i{}_j = \mathbf{e}^i \cdot \mathbf{W} \cdot \mathbf{e}_j = \mathbf{e}^i \cdot \mathbf{U} \cdot \mathbf{1} \cdot \mathbf{V} \cdot \mathbf{e}_j = \mathbf{e}^i \cdot \mathbf{U} \cdot \mathbf{e}_k\mathbf{e}^k \cdot \mathbf{V} \cdot \mathbf{e}_j = U^i{}_k V^k{}_j. \tag{2.44}$$

The rule for the multiplication of dyadic tensors is equivalent to matrix multiplication of their tensor components, provided that one always contracts adjacent pairs of raised and lower tensor indices. A dyadic acting on a vector transforms it into another vector. Repeated multiplication of dyadics results in a chain of vector transformations.

A covariant tensor T_{ij} is said to be *symmetric*, if it is even under an exchange of indices (transposition), and *antisymmetric*, if it is odd under an exchange of indices. This transposition symmetry is preserved by similarity transformations.

2.2.6 Orthogonal Transformations

Reversion of dyadics is equivalent to the matrix transposition of its vector basis components

$$\mathbf{T} = T_{ij}\boldsymbol{\lambda}^i\boldsymbol{\lambda}^j \quad \rightarrow \quad \bar{\mathbf{T}} = T_{ij}\boldsymbol{\lambda}^j\boldsymbol{\lambda}^i = T_{ji}\boldsymbol{\lambda}^i\boldsymbol{\lambda}^j = (T^{\mathrm{T}})_{ij}\boldsymbol{\lambda}^i\boldsymbol{\lambda}^j. \tag{2.45}$$

Let the rank-2 tensor

$$\mathbf{A} = \mathbf{e}'_i\mathbf{e}^i = (A)^i{}_j\mathbf{e}_i\mathbf{e}^j = (\bar{A})_i{}^j\mathbf{e}_j\mathbf{e}^i. \tag{2.46}$$

denote a transformation of from one orthogonal Cartesian basis to another one, $\mathbf{e}_j \rightarrow \mathbf{e}'_j$, such that

$$\mathbf{A}\cdot\mathbf{e}_j = \mathbf{e}'_i\mathbf{e}^i\cdot\mathbf{e}_j = \mathbf{e}'_i\delta^i{}_j = \mathbf{e}'_j. \tag{2.47}$$

The linear transformation matrix $\mathbf{A}$ is clearly orthogonal, since

$$\bar{\mathbf{A}}\cdot\mathbf{A} = \mathbf{e}_i\mathbf{e}'^i\cdot\mathbf{e}'_j\mathbf{e}^j = \mathbf{e}_i\delta^i_j\mathbf{e}^j = \mathbf{e}_i\mathbf{e}^i = \mathbf{1}. \tag{2.48}$$

Under reversion, vectors and their adjoints transform in a complementary manner

$$\mathbf{v}' = \mathbf{A}\cdot\mathbf{v}, \quad \bar{\mathbf{v}}' = \bar{\mathbf{v}}\cdot\bar{\mathbf{A}}. \tag{2.49}$$

The previous transformation leaves the inner scalar product of vectors unchanged. Explicitly, one finds

$$\mathbf{a}'\cdot\mathbf{b}' = \bar{\mathbf{a}}'\cdot\mathbf{b}' = \bar{\mathbf{a}}\cdot\bar{\mathbf{A}}\cdot\mathbf{A}\cdot\mathbf{b} = \mathbf{a}\cdot\mathbf{b}. \tag{2.50}$$

An *orthogonal transformation* in n-space dimensions forms the group $A \in O(n)$ given by the requirement that $\mathbf{A}\cdot\bar{\mathbf{A}} = 1$. The requirement that $\det(A) = +1$ results in the special orthogonal group $SO(n)$. Orthogonal matrices satisfy the orthogonality relationship

$$A^i{}_k\bar{A}^k{}_j = A^i{}_k A_j{}^k = \delta^i{}_j. \tag{2.51}$$

Dyadic tensors transform by *similarity* under an orthogonal similarity transformation

$$\mathbf{T}' = \mathbf{A}\cdot\mathbf{T}\cdot\bar{\mathbf{A}}. \tag{2.52}$$

Therefore, a contraction of a tensor observable by any two vectors remains invariant under the group of orthogonal transformations

$$\mathbf{a}'\cdot\mathbf{T}'\cdot\mathbf{b}' = \mathbf{a}\cdot\bar{\mathbf{A}}\cdot\mathbf{A}\cdot\mathbf{T}\cdot\bar{\mathbf{A}}\cdot\mathbf{A}\cdot\mathbf{b} = \mathbf{a}\cdot\mathbf{T}\cdot\mathbf{b}. \tag{2.53}$$

It follows that the metric tensor is unaffected by an arbitrary orthogonal transformation. Explicitly, one finds

$$g'_{ij} = \boldsymbol{\lambda}'_i \cdot \boldsymbol{\lambda}'_j = \boldsymbol{\lambda}_i \cdot \mathbf{\bar{A}} \cdot \mathbf{A} \cdot \boldsymbol{\lambda}_j = \boldsymbol{\lambda}_i \cdot \boldsymbol{\lambda}_j = g_{ij}. \tag{2.54}$$

Projection of the components of the transformed vector gives

$$\boldsymbol{\lambda}_i \cdot \mathbf{A} \cdot \mathbf{a} = A_{ij}a^j = A_i^j a_j, \quad \boldsymbol{\lambda}^i \cdot \mathbf{A} \cdot \mathbf{a} = A^i{}_j a^j = A^{ij}a_j. \tag{2.55}$$

The group of special orthogonal transformations that can be generated continuously from the identity element is called the *special orthogonal group*. In three-space dimensions, the special orthogonal group $SO(3)$ is equivalent to the rotation group. This is Euler's Theorem, which will be demonstrated in Chapter 7. The invariance of the metric tensor under rotations results in the rotational symmetry of Euclidean space. As a consequence, the global rotation of all vectors in an inertial frame generates another inertial frame of reference. The rotational properties of Euclidean space will be examined further in Chapter 7.

2.2.7 Dirac Notation

A complementary notational scheme used for inner product spaces is the *bra-ket notation* due to Paul A.M. Dirac (1902–1984). He originally applied his notation to complex *Hilbert Spaces* in Quantum Mechanics [Dirac 1981], but his formalism can be as easily applied to any normed vector space. In his notation, a vector is designated as a *ket-vector*

$$\mathbf{V} = |V\rangle \tag{2.56}$$

and its adjoint under vector reversion is designated as *bra-vector*

$$\mathbf{\bar{V}} = \overline{|V\rangle} = \langle V| \tag{2.57}$$

For complex spaces, field conjugation is implied. Note that tensors in Dirac notation are not normally written in boldface notation. The bracket notation is sufficient to make it transparent that one is dealing with vectors and tensors. Standard math (italics) script is used to denote vectors in the Dirac notation. The contraction of a vector with its dual is written as a *bra-ket*. For complex fields, the inner product is

$$\langle a|b\rangle = \mathbf{\bar{a}} \cdot \mathbf{b} = \bar{a}_i b^i = a_i^* b^i, \tag{2.58}$$

where $\bar{a}_i = a_i$ for real-valued fields. Projection of covariant and contravariant components of vectors is given by contraction. In bra-ket notation one finds

$$\langle \lambda_i|a\rangle = a_i, \quad \langle \lambda^i|a\rangle = a^i. \tag{2.59}$$

The ordered outer product of two vectors is a dyad, denoted by reversing the bra and ket order

$$\mathbf{b} \otimes \bar{\mathbf{a}} = \mathbf{b}\,\bar{\mathbf{a}} \rightarrow |b\rangle\langle a| = b^i \bar{a}_j |\lambda_i\rangle\langle\lambda^j|. \tag{2.60}$$

The completeness relation in bra-ket notation is given by

$$|\lambda_i\rangle\langle\lambda^i| = 1. \tag{2.61}$$

The matrix components of a rank-2 tensor can be projected out by contraction using the basis vectors just as in the vector-tensor notation

$$\boldsymbol{\lambda}^i \cdot \mathbf{M} \cdot \boldsymbol{\lambda}_j = \langle\lambda^i|M|\lambda_j\rangle = M^m{}_n \langle\lambda^i|\lambda_m\rangle\langle\lambda^n|\lambda_j\rangle = M^m{}_n \delta^i{}_m \delta^n{}_j = M^i{}_j. \tag{2.62}$$

The inner product of tensors in the bra-ket notation does not need a dot symbol since matrix multiplication (contraction of adjacent indices) is always implied

$$\boldsymbol{\lambda}^i \cdot \mathbf{M}_a \cdot \mathbf{M}_b \cdot \boldsymbol{\lambda}_j = \langle\lambda^i|M_a M_b|\lambda_j\rangle = \langle\lambda^i|M_a|\lambda_k\rangle\langle\lambda^k|M_b|\lambda_j\rangle = M_a{}^i{}_k\, M_b{}^k{}_j \tag{2.63}$$

In this notation, an orthogonal transformation is given by

$$\langle b'| = \langle b|\bar{A}, \quad M' = AM\bar{A}, \quad |a'\rangle = A|a\rangle, \tag{2.64}$$

where the criteria $\bar{A}A = 1$ defines an orthogonal transformation. All inner products are preserved by orthogonal transformations

$$\langle b'|M'|a'\rangle = \langle b|\bar{A}AM\bar{A}A|a\rangle = \langle b|M|a\rangle. \tag{2.65}$$

Note the following identity

$$\langle e_i|e^j\rangle = \mathbf{e}_i \cdot \mathbf{e}^j = \delta_i^j. \tag{2.66}$$

By summation, one gets

$$\langle e_i|e^i\rangle = \mathbf{e}_i \cdot \mathbf{e}^i = \sum_i \delta_i^i = N_{\text{dim}} \tag{2.67}$$

where N_{dim} is the dimensionality of the vector space.

The Dirac bra-ket and dyadic-vector notations are entirely equivalent. It is a matter of taste and familiarity as to which should be preferred. Ideally, one should be equally conversant with both notations. In this text, the Gibbs vector notation is primarily used, in deference to the dominant notation found in elementary textbooks.

2.3 VECTOR OPERATORS

One of the legacy issues involved in the use of the Gibbs Vector Calculus is the deliberate obscuring of the geometric differences between vectors and surface elements. Vectors and axial-vectors are treated equivalently. This is possible only because of a peculiarity of three-dimensional space. The number of linearly independent surfaces elements in an n-dimensional space is given by the combinatorial $\binom{n}{2} = n(n-1)/2$. In three dimensions, and only in three dimensions, the number of surface elements equals the number of vector dimensions. This results in a special "dual" relationship between vectors and surface elements, allowing surface elements to be treated as axial-vectors. While axial-vectors indeed behave as vectors under rotations, they have different behavior under mirror-reflection or space-inversion. These issues can be avoided if the cross product is replaced by an equivalent dyadic tensor transformation.

In Chapter 7, an elementary introduction to *Geometric Algebra* will be presented to clarify the differences, and demonstrate how to systematically generate geometric tensors of higher tensor rank. For now, however, the standard, if limiting, definition of the cross product of vectors will be used.

In spaces of higher dimensions, the cross product $\mathbf{a} \times \mathbf{b}$ is replaced by the *exterior (wedge) product*, written as $a \wedge b$, to provide a dimension-independent way of denoting elements of two-surface. The rules for constructing the wedge product will be introduced formally in Chapter 7. Briefly, $a \wedge b$ denotes the antisymmetric element of area swept out by two vectors in a space of arbitrary dimensions, and $a \wedge b \wedge c$ denotes the totally antisymmetric element of three-volume swept out by three vectors. Unlike the cross product, the exterior product is associative, and it can be extended to spaces of arbitrary dimension. Successive use of the exterior product can be used to generate the entire geometric structure of the space.

Using Euclidean basis vectors, the cross product can be defined as the determinant

$$\mathbf{c} = \mathbf{a} \times \mathbf{b} = \mathbf{e}^i(\varepsilon_{ijk}a^j b^k) = \begin{vmatrix} \mathbf{e}^i & \mathbf{e}^2 & \mathbf{e}^3 \\ a^1 & a^2 & a^3 \\ b^1 & b^2 & b^3 \end{vmatrix}. \tag{2.68}$$

The cross product has the skew matrix representation

$$\begin{bmatrix} c_1 \\ c_2 \\ c_3 \end{bmatrix} = \begin{bmatrix} 0 & -a_3 & a_2 \\ a_3 & 0 & -a_1 \\ -a_2 & a_1 & 0 \end{bmatrix} \begin{bmatrix} b_1 \\ b_2 \\ b_3 \end{bmatrix}. \tag{2.69}$$

The cross product can therefore be replaced by the dot product of a rank-2, skew-symmetric, dyadic tensor transformation acting on a vector. The latter representation can be applied to spaces of arbitrary dimension.

2.3.1 Vector Products

The cross product of two vectors **a** and **b** can be interpreted geometrically as the surface of the parallelogram given by the exterior product $a \wedge b$ having vectors **a** and **b** as its sides and θ as the angle between them (See Figure 2.1). Its orientation unit vector **n** is orthogonal to the area of the surface generated using the right-hand-rule to sweep **a** into **b**. The cross product of two vectors is treated as a vector, but it actually signifies an antisymmetric rank-2 tensor $(\mathbf{a}\,\mathbf{b} - \mathbf{b}\,\mathbf{a})$. The magnitude of the cross product can be derived from Lagrange's identity

$$|\mathbf{a} \times \mathbf{b}|^2 = a^2 b^2 - |\mathbf{a} \cdot \mathbf{b}|^2; \tag{2.70}$$

therefore,

$$\mathbf{a} \times \mathbf{b} = |\mathbf{a}||\mathbf{b}| \sin \theta \, \mathbf{n}. \tag{2.71}$$

The *triple scalar product* $\mathbf{r} \cdot (\mathbf{g} \times \mathbf{b})$ actually signifies a *pseudoscalar* which denotes a rank-3 totally antisymmetric tensor. Geometrically, the triple scalar product behaves like an element of three-volume given by the exterior

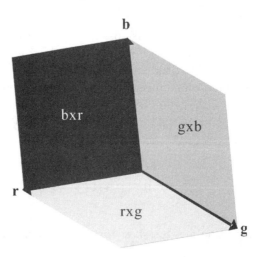

FIGURE 2.1   The areas of the faces of a parallelepiped formed by three vectors $(\mathbf{r}, \mathbf{g}, \mathbf{b})$ are given by the three cross products $(\mathbf{g} \times \mathbf{b}, \mathbf{b} \times \mathbf{r}, \mathbf{r} \times \mathbf{g})$. The volume of the parallelepiped is given by the triple scalar product $\mathbf{r} \cdot \mathbf{g} \times \mathbf{b}$.

product $r \wedge g \wedge b$. Its magnitude is the volume element of the parallelepiped with sides $(\mathbf{r}, \mathbf{g}, \mathbf{b})$ shown in Figure 2.1. The triple scalar product can be defined, using Cartesian coordinates, as the determinant

$$\mathbf{c} \cdot \mathbf{a} \times \mathbf{b} = \varepsilon_{ijk} c^i a^j b^k = \begin{vmatrix} c^1 & c^2 & c^3 \\ a^1 & a^2 & a^3 \\ b^1 & b^2 & b^3 \end{vmatrix}. \tag{2.72}$$

The *triple vector product* of three vectors in E^3 is, perhaps surprisingly, actually a vector. It is defined as

$$\mathbf{a} \times (\mathbf{b} \times \mathbf{c}) = (\mathbf{a} \cdot \mathbf{b})\mathbf{c} - \mathbf{b}(\mathbf{a} \cdot \mathbf{c}).$$

Reordering terms gives the equivalent dyadic representation

$$\mathbf{a} \times (\mathbf{b} \times \mathbf{c}) = (\mathbf{a} \cdot \mathbf{b} - \mathbf{b}\,\mathbf{a}) \cdot \mathbf{c}. \tag{2.73}$$

2.3.2 Gradient Operator

The gradient operator, defined as

$$\nabla = \mathbf{e}^i \frac{\partial}{\partial x^i} = \boldsymbol{\lambda}^j \frac{\partial}{\partial \xi^j}, \tag{2.74}$$

serves as the prototype for all covariant vectors. Let the potential $V(x, y, z)$ denote a scalar function of position. Then, by the chain rule

$$dV = d\xi^i \frac{\partial V}{\partial \xi^i} = d\mathbf{r} \cdot \nabla V, \tag{2.75}$$

where dV is a scalar formed by the contraction of the contravariant displacement elements $d\xi^i$ and the covariant directional derivatives $\partial V / \partial \xi^i$. The *gradient* of a function V is defined as the vector function

$$\nabla V = \boldsymbol{\lambda}^i \frac{\partial V}{\partial \xi^i}. \tag{2.76}$$

For example, the work done by a conservative force $\mathbf{F} = -\nabla V$ when a particle moves from one point to another is given by the change in potential between two points

$$dW = \mathbf{F} \cdot d\mathbf{r} = F_i d\xi^i = (-\nabla V) \cdot d\mathbf{r} = \left(-\frac{\partial V(\xi)}{\partial \xi^i} \right) d\xi^i = -dV, \tag{2.77}$$

$$\int_1^2 dW = -\int_1^2 dV = -V|_1^2 = -\Delta V. \tag{2.78}$$

The gradient of a scalar function $\nabla f = \mathbf{grad}(f)$ is a vector field directed along the path of greatest change in the function f.

2.3.3 Divergence Operator

The *divergence* of a vector field is the scalar density field generated by contracting the gradient operator on a vector field. The divergence of the field measures the magnitude of a vector field's source or sink at a given point. In electrostatics, this leads to *Poisson's Equation*

$$\text{div}(\mathbf{E}) = \nabla \cdot \mathbf{E} = \varepsilon_0 \rho, \tag{2.79}$$

where $\mathbf{E}$ is the electrostatic field, ε_0 is the electric permeability of free space, and ρ is the electric charge density.

2.3.4 Divergence Theorem

The *divergence theorem* of vector calculus

$$\oint_\sigma \mathbf{E} \cdot d\boldsymbol{\sigma} = \int_\tau \nabla \cdot \mathbf{E} \, d\tau \tag{2.80}$$

can be used to calculate the *flux* exiting a volume of space τ through a closed boundary surface $\sigma = \partial \tau$.

$$\Phi_{\mathrm{E}} = \oint \mathbf{E} \cdot d\boldsymbol{\sigma} = \int \nabla \cdot \mathbf{E} \, d\tau = \varepsilon_0 \int \rho d\tau = \varepsilon_0 q_e, \tag{2.81}$$

where Φ_{E} is the electric flux crossing a surface, and q_e is the net charge enclosed in the volume τ. This is known as *Guass's Law for Electrostatics*, which states that the net electric flux exiting or entering a closed volume of space is proportional to the net enclosed charge.

In converting from Cartesian coordinates to general curvilinear coordinate systems, the *volume element* transforms by including the *Jacobian of the transformation*

$$d\tau = dx^1 dx^2 dx^3 = J d\xi^1 d\xi^2 d\xi^3. \tag{2.82}$$

In matrix form, the Jacobian reduces to evaluating the determinant of the transformation matrix

$$J = \left| \frac{\partial x^i}{\partial \xi^j} \right| = \begin{vmatrix} \dfrac{\partial x^1}{\partial \xi^1} & \dfrac{\partial x^1}{\partial \xi^2} & \dfrac{\partial x^1}{\partial \xi^3} \\[2mm] \dfrac{\partial x^2}{\partial \xi^1} & \dfrac{\partial x^2}{\partial \xi^2} & \dfrac{\partial x^2}{\partial \xi^3} \\[2mm] \dfrac{\partial x^3}{\partial \xi^1} & \dfrac{\partial x^3}{\partial \xi^2} & \dfrac{\partial x^3}{\partial \xi^3} \end{vmatrix}. \tag{2.83}$$

2.3.5 Curl Operator

The skew-symmetric derivative of a vector potential $\mathbf{A}$ is given by the totally anti-symmetric rank-2 tensor $\mathbf{B}^{(2)} = \nabla \wedge \mathbf{A}$, defined as

$$B^{(2)}{}_{ij} = \frac{\partial A_j}{\partial \xi^i} - \frac{\partial A_i}{\partial \xi^j}, \tag{2.84}$$

For example, in Electromagnetic theory, the *magnetic induction field* $B^{(2)} = \frac{1}{2}B_{ij}(\mathbf{e}^i\mathbf{e}^j - \mathbf{e}^j\mathbf{e}^i)$ denotes a totally antisymmetric rank-2 tensor or a surface-tensor.[7] Alternatively, in three-dimensions only, one can define an *axial-vector* $\mathbf{B} = B^k\mathbf{e}_k$, such that

$$\mathbf{B} = \nabla \times \mathbf{A} = \mathbf{curl}(\mathbf{A}). \tag{2.85}$$

where $\nabla \times \mathbf{A}$ is referred to as the *curl* of $\mathbf{A}$. The curl of a vector flow is a vector that points along the axis of the rotational flow and whose length corresponds to the rotation speed of the flow. The rank-2 tensor and axial-vector representations are related by

$$B^{(2)}_{ij} = \varepsilon_{ijk}B^k, \quad B^k = \frac{1}{2}\varepsilon^{ijk}B^{(2)}_{ij}. \tag{2.86}$$

where ε^{ijk} denote the totally antisymmetric *Levi-Civita permutation symbols*

$$\varepsilon^{ijk} = \mathbf{e}^k \cdot (\mathbf{e}^i \times \mathbf{e}^j), \tag{2.87}$$

By comparing notations, it can be seen that the magnetic force on a particle of charge q and velocity $\mathbf{v}$ can be written in two equivalent ways

$$m\,\mathbf{a} = q\,\mathbf{v} \times \mathbf{B} = q\,\mathbf{B}^{(2)} \cdot \mathbf{v}. \tag{2.88}$$

If one defines an instantaneous angular velocity vector as $\boldsymbol{\omega}_B = -q\,\mathbf{B}/m$, where $\boldsymbol{\omega}_B$ is the *cyclotron frequency*, the equations of motion can be rewritten as

$$\dot{\mathbf{v}} = \boldsymbol{\omega}_B \times \mathbf{v} = \boldsymbol{\omega}_B^{(2)} \cdot \mathbf{v}. \tag{2.89}$$

The effect of a magnetic field on a charge particle is to apply a rotation to the instantaneous direction of the velocity vector. No work is done by the magnetic force, since, by contraction, the power transfer identically vanishes

$$\mathbf{v} \cdot \mathbf{F} = q\,\mathbf{v} \cdot (\mathbf{v} \times \mathbf{B}) = q(\mathbf{v} \cdot \mathbf{B}^{(2)} \cdot \mathbf{v}) = 0. \tag{2.90}$$

This result follows from the skew-symmetric property of the rank-2 magnetic field tensor.

---

[7]In general, a totally antisymmetric tensor of rank n will be denoted as $T^{\langle n \rangle}$.

2.3.6 Stokes's Theorem

Stokes's theorem can be used to relate the surface integral of the curl of a vector field with the closed line integral of the vector integrated along its bounding curve $C = \partial\sigma$

$$\int_\sigma \nabla \times \mathbf{F} \cdot d\boldsymbol{\sigma} = \oint_C \mathbf{F} \cdot d\mathbf{l} \tag{2.91}$$

If $\mathbf{F}$ is a conservative force, its curl vanishes. The proof is straightforward. To use a known example from electrodynamics, the electric force on a charge q is given by $q\,\mathbf{E}$. Its effect is to accelerate a positively charged particle in the direction of the field. For electrostatics, $\nabla \times \mathbf{E} = 0$, so that the electrostatic field can be derived from a time-independent *electric potential* Φ

$$\mathbf{E} = -\nabla\Phi. \tag{2.92}$$

Integrating an electrostatic field over a closed path results in no net change in potential

$$\oint_C \mathbf{E} \cdot d\mathbf{l} = -\oint_C \nabla\Phi \cdot d\mathbf{l} = -\oint_C d\Phi \equiv 0. \tag{2.93}$$

Since this result is independent of the contour, it follows that the curl identically vanishes.

2.3.7 Laplacian Operator

The scalar *Laplacian operator* is defined as $\nabla^2 = \nabla \cdot \nabla$. When acting on a vector field $\mathbf{f}$ the result is given by *Lagrange's formula*

$$\nabla^2 \mathbf{f} = \nabla(\nabla \cdot \mathbf{f}) - \nabla \times (\nabla \times \mathbf{f}). \tag{2.94}$$

A number of additional useful vector identities are listed in Table 2.2.

2.4 CURVILINEAR COORDINATE SYSTEMS

A coordinate system with generalized coordinates (ξ^1, ξ^2, ξ^3) is composed of intersecting surfaces, created by holding one coordinate at a time constant. If the intersections are all at right angles, then the curvilinear coordinates are said to form an orthogonal coordinate system. If not, they form a skew coordinate system, as shown in Figure 2.2.

TABLE 2.2   A summary of useful vector identities: Here, $\mathbf{a}, \mathbf{b}, \mathbf{c}$ denotes vector fields and ψ denotes a scalar field [Jackson 1999]

$\mathbf{a} \cdot (\mathbf{b} \times \mathbf{c}) = \mathbf{b} \cdot (\mathbf{c} \times \mathbf{a}) = \mathbf{c} \cdot (\mathbf{a} \times \mathbf{b})$ (cyclic)	(a)
$\mathbf{a} \times (\mathbf{b} \times \mathbf{c}) = (\mathbf{a} \cdot \mathbf{c})\mathbf{b} - (\mathbf{a} \cdot \mathbf{b})\mathbf{c}$	(b)
$(\mathbf{a} \times \mathbf{b}) \cdot (\mathbf{c} \times \mathbf{d}) = (\mathbf{a} \cdot \mathbf{c})(\mathbf{b} \cdot \mathbf{d}) - (\mathbf{a} \cdot \mathbf{d})(\mathbf{b} \cdot \mathbf{c})$	(c)
$\nabla \times \nabla \psi = 0$	(d)
$\nabla \cdot (\nabla \times \mathbf{a}) = 0$	(e)
$\nabla \times (\nabla \times \mathbf{a}) = \nabla(\nabla \cdot \mathbf{a}) - \nabla^2 \mathbf{a}$	(f)
$\nabla \cdot (\psi \mathbf{a}) = \mathbf{a} \cdot \nabla \psi + \psi \nabla \cdot \mathbf{a}$	(g)
$\nabla \times (\psi \mathbf{a}) = \nabla \psi \times \mathbf{a} + \psi \nabla \times \mathbf{a}$	(h)
$\nabla(\mathbf{a} \cdot \mathbf{b}) = (\mathbf{a} \cdot \nabla)\mathbf{b} + (\mathbf{b} \cdot \nabla)\mathbf{a} + \mathbf{a} \times (\nabla \times \mathbf{b}) + \mathbf{b} \times (\nabla \times \mathbf{a})$	(i)
$\nabla \cdot (\mathbf{a} \times \mathbf{b}) = \mathbf{b} \cdot (\nabla \times \mathbf{a}) - \mathbf{a} \cdot (\nabla \times \mathbf{b})$	(j)
$\nabla \times (\mathbf{a} \times \mathbf{b}) = \mathbf{a}(\nabla \cdot \mathbf{b}) - \mathbf{b}(\nabla \cdot \mathbf{a}) + (\mathbf{b} \cdot \nabla)\mathbf{a} - (\mathbf{a} \cdot \nabla)\mathbf{b}$	(k)

2.4.1 Orthogonal Coordinate Systems

A point $\mathbf{r}$ in three-dimensional Euclidean space can be parameterized by three parameters $\mathbf{r}(\xi^1, \xi^2, \xi^3)$. The transformation $\xi^i(x, y, z)$ defines a coordinate map. In curvilinear coordinates, the parameters often form an orthogonal coordinate system. An orthonormal set of unit vectors $\hat{\boldsymbol{\xi}}^i$, satisfying $\hat{\boldsymbol{\xi}}_i \cdot \hat{\boldsymbol{\xi}}_j = \delta_{ij}$, can then be constructed from the differential displacements by defining appropriate scale factors h^i to the tensor basis vectors

$$d\mathbf{r}(\xi^1, \xi^2, \xi^3) = \frac{\partial \mathbf{r}}{\partial \xi^i} d\xi^i = h^i \hat{\boldsymbol{\xi}}_i d\xi^i. \tag{2.95}$$

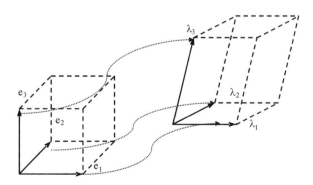

FIGURE 2.2   **A mapping of a Euclidean basis to a skew-coordinate basis.**

By expanding, $d\mathbf{r} = d\xi^i \boldsymbol{\lambda}_i = h_i d\xi^i \hat{\boldsymbol{\xi}}_i$, it becomes clear that the *scale factor* represents the magnitude of the basis vector. Notice the common convention of using the name of the coordinate variable in labeling its unit vector. The hat symbol is used to distinguish the unit vector from the coordinate. The scale factors are related to the metric tensor by

$$d\mathbf{r} \cdot d\mathbf{r} = g_{ij} d\xi^i d\xi^j = \sum_i (h^i d\xi^i)^2. \tag{2.96}$$

Since the basis is orthonormal, there is no difference in using raised or lowered indices for vector quantities

$$\mathbf{a} = a_i \hat{\boldsymbol{\xi}}^i = a^i \hat{\boldsymbol{\xi}}^i = a^i \hat{\boldsymbol{\xi}}_i. \tag{2.97}$$

Therefore, lower indices will be used for vector components throughout the following sections. The gradient operator can be written using orthogonal curvilinear coordinates as

$$\nabla = \hat{\boldsymbol{\xi}}^i h_i^{-1} \frac{\partial}{\partial \xi^i}. \tag{2.98}$$

The divergence in orthogonal curvilinear coordinates is

$$\nabla \cdot \mathbf{A} = \frac{1}{h_1 h_2 h_3} \left(\frac{\partial}{\partial \xi^1} A_1 h_2 h_3 + \frac{\partial}{\partial \xi^2} A_2 h_3 h_1 + \frac{\partial}{\partial \xi^3} A_3 h_1 h_2 \right). \tag{2.99}$$

The curl in orthogonal curvilinear coordinates is

$$\nabla \times \mathbf{A} = \frac{1}{h_1 h_2 h_3} \begin{vmatrix} h_1 \hat{\boldsymbol{\xi}}^i & h_2 \hat{\boldsymbol{\xi}}^i & h_3 \hat{\boldsymbol{\xi}}^i \\ \partial/\partial \xi^1 & \partial/\partial \xi^2 & \partial/\partial \xi^3 \\ h_1 A_1 & h_2 A_2 & h_3 A_3 \end{vmatrix}. \tag{2.100}$$

Likewise, the scalar Laplacian operator reduces to

$$\nabla^2 = \frac{1}{h_1 h_2 h_3} \left[\frac{\partial}{\partial \xi_1} \left(\frac{h_2 h_3}{h_1} \frac{\partial}{\partial \xi_1} \right) + \frac{\partial}{\partial \xi_2} \left(\frac{h_1 h_3}{h_2} \frac{\partial}{\partial \xi_2} \right) + \frac{\partial}{\partial \xi_3} \left(\frac{h_1 h_2}{h_3} \frac{\partial}{\partial \xi_3} \right) \right]. \tag{2.101}$$

2.4.2 Polar Coordinate Systems

Motion in a two dimensional plane, shown in Figure 2.3, can be thought of as motion in three dimensions with the third coordinate z constrained to a constant value, usually set to zero. *Plane polar coordinates* alternately can be defined as the limit $\sin \theta = 1$ of a spherical polar frame. That is, the

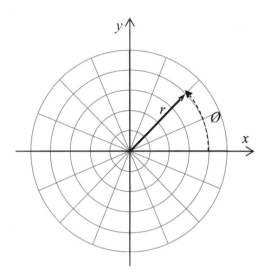

FIGURE 2.3   **Plane polar coordinates.**

plane represents the equatorial plane of a spherical coordinate system. The radial coordinate r and the azimuthal coordinate ϕ, are defined in terms of Cartesian coordinates by

$$x = r \cos \phi, \quad y = r \sin \phi, \quad z = 0, \qquad (2.102)$$

$$r = \sqrt{x^2 + y^2} \geq 0, \quad \phi = \tan^{-1}(y/x), \qquad (2.103)$$

where r is the radial distance from the origin, and φ is the counterclockwise rotation angle from the $\hat{\mathbf{x}}$-axis. As in other polar coordinate systems, such as the spherical polar and cylindrical polar maps, this map is singular along the axis of rotation, where the azimuthal angle is undefined. Functions of the azimuthal angle are periodic on interval 2π.

The position vector and line elements are

$$\mathbf{r} = x\hat{\mathbf{x}} + y\hat{\mathbf{y}} = r \cos \phi \, \hat{\mathbf{x}} + r \sin \phi \, \hat{\mathbf{y}} = r \hat{\mathbf{r}}, \qquad (2.104)$$

$$d\mathbf{r} = dx \, \hat{\mathbf{x}} + dy \, \hat{\mathbf{y}} = dr \, \hat{\mathbf{r}} + r d\phi \, \hat{\boldsymbol{\varphi}}. \qquad (2.105)$$

giving derivatives

$$\dot{\mathbf{r}} = \dot{r}\hat{\mathbf{r}} + r\dot{\phi} \, \hat{\boldsymbol{\varphi}}, \quad \ddot{\mathbf{r}} = (\ddot{r} - r\dot{\phi}^2)\hat{\mathbf{r}} + (2\dot{r}\dot{\phi} + r\ddot{\phi})\hat{\boldsymbol{\varphi}}. \qquad (2.106)$$

The metric of the two-dimensional manifold is given by

$$\{g_{ij}\} = \begin{bmatrix} 1 & 0 \\ 0 & r^2 \end{bmatrix}. \qquad (2.107)$$

The element of length is

$$ds^2 = dr^2 + r^2 d\phi^2, \tag{2.108}$$

and the scale factors are

$$g_{rr} = h_r^2 = 1, \quad g_{\phi\phi} = h_\phi^2 = r^2. \tag{2.109}$$

The unit vectors are

$$\hat{\mathbf{r}} = \frac{1}{h_r}\frac{\partial \mathbf{r}}{\partial r} = \cos\phi\,\hat{\mathbf{x}} + \sin\phi\,\hat{\mathbf{y}}, \quad \hat{\boldsymbol{\phi}} = \frac{1}{h_\phi}\frac{\partial \mathbf{r}}{\partial \phi} = -\sin\phi\,\hat{\mathbf{x}} + \cos\phi\,\hat{\mathbf{y}}, \tag{2.110}$$

with partial derivatives

$$\frac{\partial \hat{\mathbf{r}}}{\partial r} = 0, \quad \frac{\partial \hat{\boldsymbol{\phi}}}{\partial r} = 0,$$
$$\frac{\partial \hat{\mathbf{r}}}{\partial \phi} = \hat{\boldsymbol{\phi}}, \quad \frac{\partial \hat{\boldsymbol{\phi}}}{\partial \phi} = -\hat{\mathbf{r}}. \tag{2.111}$$

It is straightforward to calculate the velocity and acceleration vectors giving

$$\dot{\hat{\mathbf{r}}} = \dot{\phi}\,\hat{\boldsymbol{\phi}}, \quad \dot{\hat{\boldsymbol{\phi}}} = -\dot{\phi}\,\hat{\mathbf{r}}, \tag{2.112}$$

Cylindrical coordinates (ρ, ϕ, z), shown in Figure 2.4, are an extension of two-dimensional plane polar coordinates to three dimensions by letting $\rho = r\sin\theta$ so that $r = \rho$, when $\theta = \pi/2$, and then extruding a third axis with a height dimension $z = r\cos\theta$. These coordinates are most useful if the

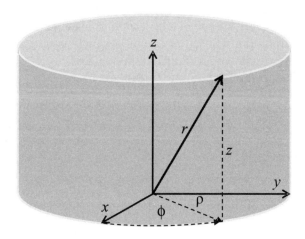

FIGURE 2.4 **Cylindrical polar coordinates.**

geometry has an axis of symmetry, such as motion on the surface of a cone or on a cylinder. Unfortunately, there are a number of different notations commonly used for cylindrical coordinates. Either r or ρ is commonly used to refer to the radial coordinate in the plane, and either θ or ϕ is used for the azimuth coordinate. Arfken [Arfken 2000], for example, uses (ρ, ϕ, z), while the *CRC Standard Mathematical Tables* [Beyer 1987] uses (r, θ, z).

In terms of the Cartesian coordinates, cylindrical coordinates are given by

$$x = \rho \cos \phi, \quad y = \rho \sin \phi, \quad z = z. \tag{2.113}$$

A position vector is given in cylindrical coordinates by

$$\mathbf{r} = \rho \cos \phi \, \hat{\mathbf{x}} + \rho \sin \phi \, \hat{\mathbf{y}} + z \, \hat{\mathbf{z}} = \rho \hat{\boldsymbol{\rho}} + z \, \hat{\mathbf{z}}. \tag{2.114}$$

The infinitesimal line element is

$$d\mathbf{r} = d\rho \, \hat{\boldsymbol{\rho}} + \rho d\phi \, \hat{\boldsymbol{\varphi}} + dz \, \hat{\mathbf{z}}, \tag{2.115}$$

with measure of length given by

$$ds^2 = d\rho^2 + \rho^2 d\phi^2 + dz^2. \tag{2.116}$$

The non-vanishing metric elements and corresponding scale factors are given by

$$g_{\rho\rho} = h_\rho^2 = 1, \quad g_{\phi\phi} = h_\phi^2 = \rho^2, \quad g_{zz} = h_z^2 = 1. \tag{2.117}$$

The unit vectors are

$$\hat{\boldsymbol{\rho}} = \frac{1}{h_\rho} \frac{\partial \mathbf{r}}{\partial \rho} = \cos \phi \, \hat{\mathbf{x}} + \sin \phi \, \hat{\mathbf{y}}, \quad \hat{\boldsymbol{\varphi}} = \frac{1}{h_\phi} \frac{\partial \mathbf{r}}{\partial \phi} = -\sin \phi \, \hat{\mathbf{x}} + \cos \phi \, \hat{\mathbf{y}}, \quad \hat{\mathbf{z}} = \hat{\mathbf{z}}, \tag{2.118}$$

with time derivatives of the unit vectors are given by

$$\dot{\hat{\boldsymbol{\rho}}} = \dot{\phi} \, \hat{\boldsymbol{\varphi}}, \quad \dot{\hat{\boldsymbol{\varphi}}} = -\dot{\phi} \, \hat{\boldsymbol{\rho}}, \quad \dot{\hat{\mathbf{z}}} = 0, \tag{2.119}$$

Spherical polar coordinates, shown in Figure 2.5, are a system of curvilinear coordinates that are natural for describing positions on a spherical surface or for problems involving spherical symmetry. Define $0 \le \phi < 2\pi$ to be the azimuthal angle in the $(x \wedge y)$ plane measured from the $\hat{\mathbf{x}}$-axis; define $0 \le \theta \le \pi$ to be the polar angle measured from the $\hat{\mathbf{z}}$-axis (the colatitude); and define r to be the distance (radius) from the origin to a point.[8]

---

[8]Mathematics textbooks and programming software often use ϕ for the polar angle and θ for the azimuthal angle. This more or less standard math notation is reversed from the usual notation in physics.

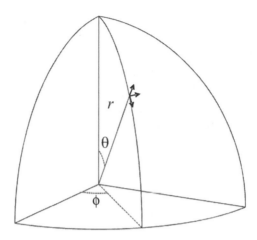

FIGURE 2.5   Spherical polar coordinates.

The spherical coordinates (r, θ, ϕ) are related to the Cartesian coordinates (x, y, z) by the reciprocal coordinate maps

$$x = r \cos \phi \sin \theta, \quad y = r \sin \phi \sin \theta, \quad z = r \cos \theta, \tag{2.120}$$

$$r = \sqrt{x^2 + y^2 + z^2}, \quad \cos \theta = z / \sqrt{x^2 + y^2 + z^2}, \quad \cos \phi = x / \sqrt{x^2 + y^2}. \tag{2.121}$$

The position vector and differential element of displacement are

$$\mathbf{r} = r\,\hat{\mathbf{r}} = r \sin \theta \cos \phi\,\hat{\mathbf{x}} + r \sin \theta \sin \phi\,\hat{\mathbf{y}} + r \cos \theta\,\hat{\mathbf{z}}. \tag{2.122}$$

$$d\mathbf{r} = dr\,\hat{\mathbf{r}} + rd\theta\,\hat{\boldsymbol{\theta}} + r \sin \theta d\phi\,\hat{\boldsymbol{\varphi}}. \tag{2.123}$$

The measure of length is $ds^2 = dr^2 + r^2 d\theta^2 + r^2 \sin^2 \theta d\phi^2$, with nonvanishing diagonal metric tensor components

$$g_{rr} = h_r^2 = 1, \quad g_{\theta\theta} = h_\theta^2 = r^2, \quad g_{\phi\phi} = h_\phi^2 = r^2 \sin^2 \theta. \tag{2.124}$$

The unit vectors for a spherical coordinate system are.

$$\hat{\mathbf{r}} = \frac{1}{h_r} \frac{\partial \mathbf{r}}{\partial r} = \cos \phi \sin \theta\,\hat{\mathbf{x}} + \sin \phi \sin \theta\,\hat{\mathbf{y}} + \cos \theta\,\hat{\mathbf{z}},$$

$$\hat{\boldsymbol{\theta}} = \frac{1}{h_\theta} \frac{\partial \mathbf{r}}{\partial \theta} = \cos \phi \cos \theta\,\hat{\mathbf{x}} + \sin \phi \cos \theta\,\hat{\mathbf{y}} - \sin \theta\,\hat{\mathbf{z}}, \tag{2.125}$$

$$\hat{\boldsymbol{\varphi}} = \frac{1}{h_\phi} \frac{\partial \mathbf{r}}{\partial \phi} = -\sin \phi\,\hat{\mathbf{x}} + \cos \phi\,\hat{\mathbf{y}}.$$

TABLE 2.3   Basic formulas for the position of a point and its derivatives in common orthogonal coordinate systems

	Cartesian	Cylindrical	Spherical
$\mathbf{r}$	$x\,\hat{\mathbf{x}} + y\,\hat{\mathbf{y}} + z\,\hat{\mathbf{z}}$	$\rho\,\hat{\boldsymbol{\rho}} + z\,\hat{\mathbf{z}}$	$r\,\hat{\mathbf{r}}$
$\mathbf{v} = \dot{\mathbf{r}}$	$\dot{x}\,\hat{\mathbf{x}} + \dot{y}\,\hat{\mathbf{y}} + \dot{z}\,\hat{\mathbf{z}}$	$\dot{\rho}\,\hat{\boldsymbol{\rho}} + \rho\dot{\phi}\,\hat{\boldsymbol{\varphi}} + \dot{z}\,\hat{\mathbf{z}}$	$\dot{r}\,\hat{\mathbf{r}} + r\dot{\theta}\,\hat{\boldsymbol{\theta}} + r\sin\theta\dot{\phi}\,\hat{\boldsymbol{\varphi}}$
$d\mathbf{r}\cdot d\mathbf{r}$	$dx^2 + dy^2 + dz^2$	$d\rho^2 + \rho^2 d\phi^2 + dz^2$	$dr^2 + r^2 d\theta^2 + r^2\sin^2\theta d\phi^2$
$\mathbf{v}^2$	$\dot{x}^2 + \dot{y}^2 + \dot{z}^2$	$\dot{\rho}^2 + \rho^2\dot{\phi}^2 + \dot{z}^2$	$\dot{r}^2 + r^2\dot{\theta}^2 + r^2\sin^2\theta\dot{\phi}^2$
$\mathbf{a} = \ddot{\mathbf{r}}$	$\ddot{x}\,\hat{\mathbf{x}} + \ddot{y}\,\hat{\mathbf{y}} + \ddot{z}\,\hat{\mathbf{z}}$	$(\ddot{\rho} - \rho\dot{\phi}^2)\hat{\boldsymbol{\rho}}$	$(\ddot{r} - r\dot{\theta}^2 - r\sin^2\theta\dot{\phi}^2)\hat{\mathbf{r}}$
		$+(2\dot{\rho}\dot{\phi} + \rho\ddot{\phi})\hat{\boldsymbol{\varphi}} + \ddot{z}\,\hat{\mathbf{z}}$	$+(2\dot{r}\dot{\theta} + r\ddot{\theta} - r\sin\theta\cos\theta\dot{\phi}^2)\hat{\boldsymbol{\theta}}$
			$+(2\sin\theta\dot{r}\dot{\phi} + 2r\cos\theta\dot{\phi}\dot{\theta}$
			$+r\sin\theta\ddot{\phi})\hat{\boldsymbol{\varphi}}$

Their partial derivatives are given by

$$\frac{\partial\hat{\mathbf{r}}}{\partial r} = 0 \quad \frac{\partial\hat{\boldsymbol{\theta}}}{\partial r} = 0 \quad \frac{\partial\hat{\boldsymbol{\varphi}}}{\partial r} = 0$$

$$\frac{\partial\hat{\mathbf{r}}}{\partial\theta} = \hat{\boldsymbol{\theta}}, \quad \frac{\partial\hat{\boldsymbol{\theta}}}{\partial\theta} = -\hat{\mathbf{r}}, \quad \frac{\partial\hat{\boldsymbol{\varphi}}}{\partial\theta} = 0, \tag{2.126}$$

$$\frac{\partial\hat{\mathbf{r}}}{\partial\phi} = \sin\theta\,\hat{\boldsymbol{\varphi}}, \quad \frac{\partial\hat{\boldsymbol{\theta}}}{\partial\phi} = \cos\theta\,\hat{\boldsymbol{\varphi}}, \quad \frac{\partial\hat{\boldsymbol{\varphi}}}{\partial\phi} = -\cos\theta\,\hat{\boldsymbol{\theta}} - \sin\theta\,\hat{\mathbf{r}}.$$

Table 2.3 summarizes the representation of a point and it derivatives in Cartesian, cylindrical and spherical coordinate systems.

Likewise, Table 2.4 summarizes the representation of basic vector operators with respect the most common orthonormal coordinate systems.

2.4.3   *Non Orthogonal Coordinate Systems

As an example of a skewed coordinate system, consider a charged particle *Vertical Drift Chamber* (VDC) detector shown in Figure 2.6. Physical considerations often lead to a non-orthogonal choice of crossing angles for two planes of wires referred to as u and v planes. The individual wires define contours of constant u or v respectively. When a particle crosses the planes of the chamber, ionized knockoff electrons drift to the nearest set of u and v wires, triggering an avalanche event. The readout of the (u, v) wire planes are converted into orthogonal (x, y) coordinates by the track analyzer. A point in the plane is described in terms of (u, v) coordinates as

$$\mathbf{r}(u, v) = x\,\mathbf{e}_1 + y\,\mathbf{e}_2 = \cos\alpha(u - v)\mathbf{e}_1 + \sin\alpha(u + v)\mathbf{e}_2. \tag{2.127}$$

TABLE 2.4   Vector operators for common coordinate systems

Gradient Operator

Cartesian:
$$\nabla\psi = \hat{\mathbf{x}}\frac{\partial}{\partial x} + \hat{\mathbf{y}}\frac{\partial}{\partial y} + \hat{\mathbf{z}}\frac{\partial}{\partial x}$$

Cylindrical:
$$\nabla\psi = \hat{\boldsymbol{\rho}}\frac{\partial\psi}{\partial\rho} + \hat{\boldsymbol{\varphi}}\frac{1}{\rho}\frac{\partial\psi}{\partial\phi} + \hat{\mathbf{z}}\frac{\partial\psi}{\partial z}$$

Spherical:
$$\nabla\psi = \hat{\mathbf{r}}\frac{\partial\psi}{\partial r} + \hat{\boldsymbol{\theta}}\frac{1}{r}\frac{\partial\psi}{\partial\theta} + \hat{\boldsymbol{\varphi}}\frac{1}{r\sin\theta}\frac{\partial\psi}{\partial\phi}$$

Divergence Operator

Cartesian:
$$\nabla\cdot\mathbf{a} = \frac{\partial a_x}{\partial x} + \frac{\partial a_y}{\partial y} + \frac{\partial a_z}{\partial z}$$

Cylindrical:
$$\nabla\cdot\mathbf{a} = \frac{1}{\rho}\frac{\partial(\rho a_\rho)}{\partial\rho} + \frac{1}{\rho}\frac{\partial a_\phi}{\partial\phi} + \frac{\partial a_z}{\partial z}$$

Spherical:
$$\nabla\cdot\mathbf{a} = \frac{1}{r^2}\frac{\partial}{\partial r}(r^2 a_r) + \frac{1}{r\sin\theta}\frac{\partial}{\partial\theta}(\sin\theta a_\theta) + \frac{1}{r\sin\theta}\frac{\partial a_\phi}{\partial\phi}$$

Curl Operator

Cartesian:
$$\nabla\times\mathbf{a} = \hat{x}\left(\frac{\partial a_z}{\partial y} - \frac{\partial a_y}{\partial z}\right) + \hat{y}\left(\frac{\partial a_x}{\partial z} - \frac{\partial a_z}{\partial x}\right) + \hat{z}\left(\frac{\partial a_y}{\partial x} - \frac{\partial a_x}{\partial y}\right)$$

Cylindrical:
$$\nabla\times\mathbf{a} = \hat{\boldsymbol{\rho}}\left(\frac{1}{\rho}\frac{\partial a_z}{\partial\phi} - \frac{\partial a_\phi}{\partial z}\right) + \hat{\boldsymbol{\varphi}}\left(\frac{\partial a_\rho}{\partial z} - \frac{\partial a_z}{\partial\rho}\right) + \hat{\mathbf{z}}\frac{1}{\rho}\left(\frac{\partial(\rho a_\phi)}{\partial\rho} - \frac{\partial a_\rho}{\partial\phi}\right)$$

Spherical:
$$\hat{\boldsymbol{\varphi}}\times\mathbf{a} = \hat{\mathbf{r}}\frac{1}{r\sin\theta}\left(\frac{\partial}{\partial\theta}(\sin\theta a_\phi) - \frac{\partial a_\theta}{\partial\phi}\right) + \hat{\boldsymbol{\theta}}\left(\frac{1}{r\sin\theta}\frac{\partial a_r}{\partial\phi} - \frac{1}{r}\frac{\partial}{\partial r}(r a_\phi)\right)$$
$$+\hat{\boldsymbol{\varphi}}\frac{1}{r}\left(\frac{\partial}{\partial r}(r a_\theta) - \frac{\partial a_r}{\partial\theta}\right)$$

Laplacian Operator

Cartesian
$$\nabla^2\psi = \frac{\partial^2\psi}{\partial x^2} + \frac{\partial^2\psi}{\partial y^2} + \frac{\partial^2\psi}{\partial z^2}$$

Cylindrical
$$\nabla^2\psi = \frac{1}{\rho}\frac{\partial}{\partial\rho}\left(\rho\frac{\partial\psi}{\partial\rho}\right) + \frac{1}{\rho^2}\frac{\partial^2\psi}{\partial\phi^2} + \frac{\partial^2\psi}{\partial z^2}$$

Spherical
$$\nabla^2\psi = \frac{1}{r^2}\frac{\partial}{\partial r}\left(r^2\frac{\partial\psi}{\partial r}\right) + \frac{1}{r^2}\left(\frac{1}{\sin\theta}\frac{\partial}{\partial\theta}\left(\sin\theta\frac{\partial\psi}{\partial\theta}\right) + \frac{1}{\sin^2\theta}\frac{\partial^2\psi}{\partial\phi^2}\right)$$

2.5   MANIFOLDS

A manifold is an abstract space in which, locally, every point has a neighborhood resembling a flat n-dimensional Euclidean space. Globally, the structure can be more complicated. A manifold, for example, might have

u-v vertical drift chamber

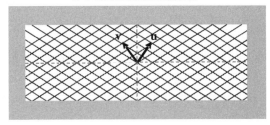

FIGURE 2.6   A crossed-wire vertical drift chamber, with two (u, v) wire planes that are not orthogonal to each other.

multiple connections or even have disconnected regions. A manifold need not be closed or finite. An open line, for example, minus its discontinuous endpoints, is a manifold. The concept of dimension is critical to the description of manifolds. Curves in $\mathbb{R}^3$ form a one-dimensional manifold, and surfaces form two-dimensional manifolds. In a one-dimensional manifold, every point has a neighborhood that looks like a straight line segment. Examples of one-dimensional manifolds include a curved line, a circle, and a linked pair of circles. In a two-dimensional manifold, every point has a neighborhood that looks like a flat plane. Nontrivial examples include the surface of a sphere and the surface of a torus. A conic section, defined by the fixed polar angle $\theta = \theta_0$, has two disconnected regions for $z > 0$ and $z < 0$. The vertex of the cone is not a point on the manifold since the metric tensor is singular there. A differentiable manifold allows one to construct a local coordinate system. Calculus on differentiable manifolds is known as differential geometry.

The coordinates specifying a manifold are defined over some region called a *map* or a *coordinate chart*. The collection of overlapping maps needed to cover a manifold is called an *atlas*. For example, there is no way to parameterize the entire surface of a sphere with a single flat coordinate chart (ask any map maker). To continue a manifold from one overlapping chart to another, a *coordinate patch* is used to map the coordinates of the initial map to the coordinates of an overlapping map. Cylindrical and spherical coordinates are coordinate charts for flat Euclidean space that are valid nearly everywhere, except along the $\hat{\mathbf{z}}$-axis where the azimuthal angle is undefined.

Consider a one dimensional manifold embedded in the Euclidean Space E^3 given by the one-parameter family of curves

$$x = x(s), \quad y = y(s), \quad z = z(s). \tag{2.128}$$

The basis vector is defined to be tangent to the direction of motion

$$\boldsymbol{\lambda}_s = d\mathbf{r}/ds. \tag{2.129}$$

Likewise, a two-parameter family of points is given by the parametric equations

$$x = x(u, v), \quad y = y(u, v), \quad z = z(u, v). \tag{2.130}$$

This defines a two-dimensional surface (assuming linear independence) with two basis vectors

$$\boldsymbol{\lambda}_u = \partial \mathbf{r}(u, v)/\partial u, \quad \boldsymbol{\lambda}_v = \partial \mathbf{r}(u, v)/\partial v. \tag{2.131}$$

In general, any geometrical space can be considered a differential *manifold* with a *metric measure* of length. A manifold M is essentially a curved m-dimensional surface embedded in some higher-order, n-dimensional, real, flat Euclidean (or pseudo-Euclidean) space of definite (or indefinite) signature. Over a small interval about a point, the manifold is smooth and looks like a region of pseudo-Euclidean space, but, globally, its structure may be more complex.

The concept of a manifold is most easily introduced by analogy to the well-known properties of three-dimensional flat Euclidean Space E^3. Ordinary, smooth surfaces are two-dimensional manifolds embedded in E^3, and smooth curves are one-dimensional manifolds in E^3. An example of a two-dimensional manifold is illustrated in Figure 2.7, where a point is constrained to move on the two-dimensional surface of a *torus* defined by the parametric equations

$$\begin{aligned}
x^1(\theta, \phi) &= x(\theta, \phi) = (a + b\cos\phi)\cos\theta, \\
x^2(\theta, \phi) &= y(\theta, \phi) = (a + b\cos\phi)\sin\theta, \\
x^3(\theta, \phi) &= z(\theta, \phi) = b\sin\phi.
\end{aligned} \tag{2.132}$$

Here, (x, y, z) are Cartesian coordinates. The coordinates of points in the manifold are given by

$$\mathbf{P}(\theta, \phi) = (a + b\cos\phi)\cos\theta\, \mathbf{e}_1 + (a + b\cos\phi)\sin\theta\, \mathbf{e}_2 + b\sin\phi\, \mathbf{e}_3. \tag{2.133}$$

Note that the origin $\mathbf{O} = (0, 0, 0)$ of E^3 is not a point on the manifold of the torus. In principle, any manifold can be embedded in some (pseudo) Euclidean space of higher dimension.

A point in the Euclidean cover for the manifold can be written as

$$\mathbf{P}(\theta, \phi) = x^i(\theta, \phi)\mathbf{e}_i. \tag{2.134}$$

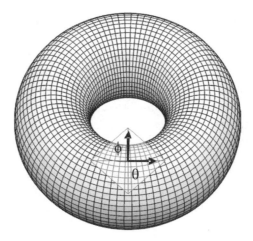

FIGURE 2.7 **A torus is a two-dimensional manifold embedded in Euclidean space. The instantaneous direction of a particle moving on the surface of a torus is constrained to lie in the tangent plane touching the surface at any point on the surface.**

Points on a manifold are not vectors, although differences of points in flat Euclidean spaces are vectors. Vectors are defined, locally, by the allowed directions of infinitesimal displacement of points within the manifold. The infinitesimal displacements

$$d\mathbf{P} = d\xi^i(\partial\mathbf{P}/\partial\xi^i) = d\xi^i\,\boldsymbol{\lambda}_i \qquad (2.135)$$

define the prototypes for all vectors. $d\xi^i$ is an element of generalized displacement, and $\boldsymbol{\lambda}_i = \partial\mathbf{P}/\partial\xi^i$ defines a vector basis. Infinitesimal displacements are elements of a tangent vector space attached to the manifold at every point of the manifold.

On a manifold, the configuration space of coordinate points and the tangent space of differential displacements do not necessarily coincide. This can be clearly seen by considering the torus shown in Figure 2.7. The coordinate space of the torus is the locus of points $\mathbf{P} \in Q$, where Q is the configuration space defined by the parameterized surface of the manifold. The space of *infinitesimal displacements* $d\mathbf{r} \in T$, where T is the tangent space defined by the set of directional derivatives tangent to the surface at every point. A vector basis in the tangent space can be created by use of the chain rule $d\mathbf{P} = d\xi^i(\partial\mathbf{P}/\partial\xi^i) = d\xi^i\boldsymbol{\lambda}_i$. For the torus, this gives

$$\begin{aligned}
\boldsymbol{\lambda}_\theta &= \partial\mathbf{P}/\partial\theta = (a - b\cos\phi)(-\sin\theta\,\hat{\mathbf{x}} + \cos\theta\,\hat{\mathbf{y}}), \\
\boldsymbol{\lambda}_\phi &= \partial\mathbf{P}/\partial\phi = b\sin\phi(\cos\theta\,\hat{\mathbf{x}} + \sin\theta\,\hat{\mathbf{y}}) + b\cos\phi\,\hat{\mathbf{z}}.
\end{aligned} \qquad (2.136)$$

As the particle moves, the attached tangent space T rotates with the coordinate point. Note that the tangent space has the same dimensionality (degrees of freedom) as the configuration space. The *velocity phase space* of coordinates and their velocities $(\mathbf{r}, \dot{\mathbf{r}})$ is given by the product space TQ, and is referred to as the *tangent bundle*. The *momentum phase space* of coordinates and their momentum $(\mathbf{r}, \mathbf{p})$ is given by the product space T^*Q and is referred to as the *cotangent bundle*.

The straight-line displacement in $\mathbb{R}^3$ between two separated points lying on the surface of the torus does not generally lie in the tangent plane. It is neither a vector (an allowed direction of motion), nor does it describe an allowed path. *Geodesics* on manifolds are paths that minimize the distance between points while remaining on the manifold. They define the notion of parallel transport on the manifold.

2.5.1 Motion on a Cycloid

For example, a particle constrained to move on a wire bent into the shape of cycloid in the (x, y) plane moves on the path

$$x(\phi) = a(\phi - \sin\phi), \quad y(\phi) = a(1 - \cos\phi). \tag{2.137}$$

This is the path taken by a point on a wheel of radius a as it rolls along the $\hat{\mathbf{x}}$-axis through an angle ϕ. The tensor basis vector with respect to the parameter ϕ is given by

$$\boldsymbol{\lambda}_\phi = \frac{\partial \mathbf{r}}{\partial \phi} = a(1 - \cos\phi)\hat{\mathbf{x}} + a\sin\phi\,\hat{\mathbf{y}}, \tag{2.138}$$

which is everywhere tangent to the curve. The kinetic energy of a particle restricted to following this curve is given by

$$T(\phi, \dot{\phi}) = \frac{1}{2}m(\dot{x}^2 + \dot{y}^2) = \frac{ma^2}{2}\dot{\phi}^2[2 - 2\cos\phi]. \tag{2.139}$$

The distance traveled by a particle moving on this curve between two points on the cycloid is given by the integral over elements of length

$$\begin{aligned}
\Delta s &= \int_1^2 ds = \int_1^2 \sqrt{dx^2 + dy^2} \\
&= \int_1^2 \sqrt{\left(\frac{dx}{d\phi}\right)^2 + \left(\frac{dy}{d\phi}\right)^2}\, d\phi \\
&= \sqrt{2}a \int_1^2 \sqrt{1 - \cos\phi}\, d\phi.
\end{aligned} \tag{2.140}$$

2.5.2 Motion on the Surface of the Earth

The surface of the Earth is a two-dimensional manifold as can be seen in Figure 2.8. By zooming into a local neighborhood (here selected to be Paris, France) the regional map seen to be approximately Euclidean. The surface of the Earth is approximately a sphere, referred to as the *two-sphere* S^2 in the language of topology, since it has two parameters. On a two-sphere, the sum of the angles of a triangle is not equal to 180°. The surface geometry of the two-sphere is definitely not Euclidean, but for local neighborhoods, the laws of the Euclidean geometry are good approximations. For a small enough triangle on the face of the Earth, the sum of the angles is very nearly 180°. A sphere can be represented by a collection of two-dimensional maps; therefore, a two-sphere is a two-dimensional manifold embedded in a flat three-dimensional Euclidean space.

The *degrees of freedom*, or dimensionality of a manifold, are the number of parameters that can be independently varied. For example, a point constrained to move on a two-sphere of radius r_0 has two coordinate degrees of freedom; $\mathbf{r} = \mathbf{r}(\theta, \phi)$ in spherical coordinates. The constraint condition on the radial direction can be written as

$$c_r = r - r_0 = 0. \tag{2.141}$$

This defines the two-dimensional manifold

$$\mathbf{r}(\theta, \phi) = r_0\hat{\mathbf{r}} = r_0(\cos\phi\sin\theta\,\hat{\mathbf{x}} + \sin\phi\sin\theta\,\hat{\mathbf{y}} + \cos\theta\,\hat{\mathbf{z}}). \tag{2.142}$$

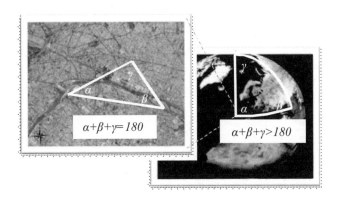

FIGURE 2.8 The surface of a sphere is a two-dimensional manifold embedded in a three-dimensional Euclidean space. The regional map shows Paris, France, on Planet Earth. The Maps of planet Earth and of Paris were generated using NASA open source *World Wind* software and the French *Geoportail* plug-in [WWW: NASA 2007].

or

$$x(\theta, \phi) = r_0 \cos\phi \sin\theta, \quad y(\theta, \phi) = r_0 \sin\phi \sin\theta, \quad z(\theta, \phi) = r_0 \cos\theta.$$
$$(2.143)$$

This spherical polar coordinate map defines the surface of a sphere everywhere except along the polar axis where the azimuthal angle is undefined. A minimum of two coordinates maps are needed to cover the sphere completely. Motion is constrained to lie in the $\hat{\theta} \wedge \hat{\phi}$ surface, which is everywhere perpendicular to $\hat{\mathbf{r}}$.

$$d\mathbf{r} = r_0(d\theta \, \hat{\boldsymbol{\theta}} + \sin\theta d\phi \, \hat{\boldsymbol{\varphi}}), \quad d\mathbf{r} \cdot \hat{\mathbf{r}} = 0. \qquad (2.144)$$

The metric tensor for this coordinate map is

$$\{g_{ij}\} = r_0^2 \begin{bmatrix} 1 & 0 \\ 0 & \sin^2\theta \end{bmatrix}, \quad \{g^{ij}\} = \frac{1}{r_0^2} \begin{bmatrix} 1 & 0 \\ 0 & 1/\sin^2\theta \end{bmatrix}. \qquad (2.145)$$

Note that the metric is invertible everywhere except at the points along the $\hat{\mathbf{z}}$-axis ($\sin\theta = 0$), which must be excluded. A second map about another axis can be used as a patch to reach these points in a continuous manner.

The time derivatives of the position vector can be obtained by using spherical polar coordinates and applying the constraint condition $\dot{r} = \ddot{r} = 0$.

$$\mathbf{v} = \dot{\mathbf{r}} = r_0(\dot{\theta} \, \hat{\boldsymbol{\theta}} + \sin\theta\dot{\phi} \, \hat{\boldsymbol{\varphi}}), \qquad (2.146)$$

$$\mathbf{a} = \ddot{\mathbf{r}} = -r_0(\dot{\theta}^2 + \sin^2\theta\dot{\phi}^2)\hat{\mathbf{r}} + r_0(\ddot{\theta} - \sin\theta\cos\theta\dot{\phi}^2) \, \hat{\boldsymbol{\theta}}$$
$$+ r_0(2\cos\theta\dot{\phi}\dot{\theta} + \sin\theta\ddot{\phi}) \, \hat{\boldsymbol{\varphi}}. \qquad (2.147)$$

The gradient and Laplacian operators for the two-sphere are given by

$$\nabla = \frac{1}{r_0} \left(\hat{\boldsymbol{\theta}} \frac{\partial}{\partial\theta} + \hat{\boldsymbol{\varphi}} \frac{1}{\sin\theta} \frac{\partial}{\partial\phi} \right), \qquad (2.148)$$

$$\nabla^2 = \frac{1}{r_o^2} \left(\frac{1}{\sin\theta} \frac{\partial}{\partial\theta} \sin\theta \frac{\partial}{\partial\theta} + \frac{1}{\sin^2\theta} \frac{\partial^2}{\partial\phi^2} \right). \qquad (2.149)$$

2.6 *COVARIANT DIFFERENTIATION

Covariant differentiation of vectors and tensors gets messy because the variation of the vector basis must be taken into account, in addition to the explicit variation of the vector components. This requires the introduction of *Christoffel symbols*. Fortunately, nearly all of the dynamical systems in this textbook can be derived from a scalar *Lagrangian* prescription. There will normally be no need to calculate the Christoffel symbols explicitly. The resulting *Euler-Lagrange* or *Hamiltonian* equations of motion will generate the necessary affine connections, required for *covariant differentiation*, automatically.

Since much of what follows is independent of the dimension of the space, the indices will be denoted by the letters $abc\ldots$ or $\alpha\beta\gamma\ldots$. In general, Latin letters will denote a basis offset starting with 1 for the first element, while Greek letters will be used to denote a basis offset starting with 0 for the first element. The choice of offset is solely a matter of convention.

Covariant differentiation is the same as ordinary differentiation when applied to invariant objects. For example, let $\mathbf{V}$ denote a vector invariant that can be expanded in a tensor basis $\mathbf{V} = V^a\boldsymbol{\lambda}_a$. Let D_b denote covariant differentiation with respect to a coordinate and $\partial_b = \partial/\partial\xi^b$ denote ordinary partial derivatives; then,

$$D_b\mathbf{V} = \partial_b\mathbf{V} = \frac{\partial\mathbf{V}}{\partial\xi^b} \tag{2.150}$$

transforms in a covariant manner since $\mathbf{V}$ is an invariant object. Now expand the partial derivatives using the chain rule

$$D_b\mathbf{V} = \partial_b\mathbf{V} = \partial_b(V^a\boldsymbol{\lambda}_a) = \partial_b(V^a)\boldsymbol{\lambda}_a + V^a\partial_b(\boldsymbol{\lambda}_a) = (D_bV^a)\boldsymbol{\lambda}_a. \tag{2.151}$$

Let the local connection of a basis under a small change in position be denoted by the three index symbols Γ^c_{ab} (for a tensor basis these reduce to Christoffel symbols, defined in the next section)

$$\frac{\partial}{\partial\xi^b}(\boldsymbol{\lambda}_a) = \partial_b(\boldsymbol{\lambda}_a) = \Gamma^c_{ab}\boldsymbol{\lambda}_c, \tag{2.152}$$

then

$$D_b\mathbf{V} = (D_bV^a)\boldsymbol{\lambda}_a = (\partial_bV^a + \Gamma^a_{cb}V^c)\boldsymbol{\lambda}_a, \tag{2.153}$$

which defines the covariant derivative of a vector as

$$D_bV^a = \partial_bV^a + \Gamma^a_{cb}V^c. \tag{2.154}$$

Note that

$$D_b\boldsymbol{\lambda}_a \equiv 0. \tag{2.155}$$

Therefore, covariant differentiation satisfies the product rule (Leibniz's law)

$$D_\mu(AB) = D_\mu(A)B + AD_\mu(B). \tag{2.156}$$

Covariant differentiation takes into account the implicit change in direction and magnitude of a basis as well as the explicit change in the coordinates.

The concept of covariant differentiation can be extended to tensor invariants of any tensor rank. The basic rule is that contracted indices are scalars and require only ordinary differentiation.

A basis is said to be *Riemannian* if it can be derived in terms of an allowed coordinate mapping from another Riemannian basis. In this case, the connection is called the affine connection and takes on a special form given by the Christoffel symbols. Accelerated bases used in rotational motion are not Riemannian bases and require a more complicated connection.

If the equations to be solved can be expressed in terms of a scalar Lagrangian, only ordinary partial derivatives are needed to get the equations of motion, since the affine connection will be computed automatically by the formalism.

2.6.1 *Christoffel Symbols

The partial derivatives of a Riemann tensor basis are expressed in terms of the three-index *Christoffel symbols*[9] of the second kind Γ_{ab}^{c}. *Christoffel symbols of the second kind* arise naturally in the calculation of the geodesics of manifolds. These are defined by [Arfken 2000] as

$$\Gamma_{ab}^{c} = \left\{ \begin{matrix} c \\ ab \end{matrix} \right\} = \lambda^{c} \cdot \frac{\partial \lambda_{a}}{\partial \xi^{b}} = \frac{1}{2} g^{cc'} \left(\frac{\partial g_{c'a}}{\partial \xi^{b}} + \frac{\partial g_{c'b}}{\partial \xi^{a}} - \frac{\partial g_{ab}}{\partial \xi^{c'}} \right). \tag{2.157}$$

Contraction of the symbols gives [Misner 1970, p. 222]

$$\Gamma_{ab}^{b} = \frac{\partial \ln \sqrt{\|g\|}}{\partial \xi^{a}}, \tag{2.158}$$

where $\|g\|$ denotes the absolute value of the determinant of the metric tensor g_{ab}. The Christoffel symbols are also known as the *affine connection* [Weinberg 1972]. Christoffel symbols are symmetric under the interchange of the final two indices

$$\Gamma_{ab}^{c} = \Gamma_{ba}^{c}. \tag{2.159}$$

The affine connection keeps vectors from leaving the space of the manifold when they are varied. It defines the *parallel transport* of a vector field on a manifold. The Christoffel symbols vanish identically for a constant Euclidean basis.

---

[9]Named for Elwin Bruno Christoffel (1829–1900).

Closely related are the *Christoffel symbols of the first kind*, $\Gamma_{kij} = [ij, k]$, which are linear differential functions of the metric tensor:

$$\Gamma_{cab} = [ab, c] = g_{cd}\Gamma^d_{ab} = \lambda_c \cdot \frac{\partial \lambda_a}{\partial \xi^b} = \frac{1}{2}\left(\frac{\partial g_{ca}}{\partial \xi^b} + \frac{\partial g_{cb}}{\partial \xi^a} - \frac{\partial g_{ab}}{\partial \xi^c}\right). \quad (2.160)$$

The Christoffel symbols for the reciprocal basis can be deducted from the Kronecker delta normalization condition:

$$\frac{\partial \delta^c_a}{\partial \xi^b} = 0 = \frac{\partial(\lambda_a \cdot \lambda^c)}{\partial \xi^b} = \frac{\partial \lambda_a}{\partial \xi^b}\cdot\lambda^c + \lambda_a \cdot \frac{\partial \lambda^c}{\partial \xi^b} = \Gamma^c_{ab} + \tilde{\Gamma}^c_{ab}, \quad (2.161)$$

which requires that

$$\tilde{\Gamma}^c_{ab} = \lambda_a \cdot \frac{\partial \lambda^c}{\partial \xi^b} = -\Gamma^c_{ab}. \quad (2.162)$$

This relationship guarantees that covariant differentiation applied to contracted indices gives the same result as ordinary differentiation. Although the Christoffel symbols are written using in the same index notation as tensors, they are not tensors, since they do not transform as tensors in general.

Summarizing, the *covariant derivative* of vectors involves the use of Christoffel symbols. An invariant vector can be written in terms of its covariant components as $\mathbf{v} = v_a\lambda^a$. Using the chain rule, the covariant derivative of a vector can be written variously as

$$D_b\mathbf{v} = \partial_b\mathbf{v} = (\partial_b v_a)\lambda^a + v_a(\partial_b\lambda^a)$$
$$= (\partial_b v_a - \Gamma^c_{ab}v_c)\lambda^a = (D_b v_a)\lambda^a, \quad (2.163)$$

Applying this to a contraction between two vectors gives the expected result

$$D_b(\mathbf{u}\cdot\mathbf{v}) = D_b(u_a v^a) = (D_b u_a)v^a + u_a(D_b v^a)$$
$$= (\partial_b u_a)v^a - \Gamma^c_{ab}u_c v^a + u_a(\partial_b v^a) + \Gamma^a_{cb}u_a v^c$$
$$= (\partial_b u_a)v^a + u_a(\partial_b v^a) = \partial_b(u_a v^a) = \partial_b(\mathbf{u}\cdot\mathbf{v}). \quad (2.164)$$

The symmetry of Christoffel symbols can be used to show that the antisymmetrized covariant derivative of a vector field is the same as the ordinary derivative. For example, let $\mathbf{A}$ denote a vector field, then

$$D_a A_b - D_a A_b = \partial_a A_b - \partial_a A_b - \Gamma^c_{ab}A_c + \Gamma^c_{ab}A_c$$
$$= \partial_a A_b - \partial_b A_a. \quad (2.165)$$

The covariant derivative of a vector transforms like a rank-2 covariant tensor. Likewise, the covariant derivative of a rank-2 tensor is a tensor of rank-3 given by

$$D_c T_{ab} = \partial_c T_{ab} - \Gamma_{ac}^d T_{db} - \Gamma_{bc}^d T_{ad}, \qquad (2.166)$$

where the affine connection is applied to all non-contracted indices. This can be easily seen by applying the derivative to the invariant form of the tensor and using the chain rule

$$\begin{aligned}
\partial_c \mathbf{T} &= \partial_c (T_{ab} \boldsymbol{\lambda}^a \boldsymbol{\lambda}^b) \\
&= (\partial_c T_{ab}) \boldsymbol{\lambda}^a \boldsymbol{\lambda}^b + T_{ab} (\partial_c \boldsymbol{\lambda}^a) \boldsymbol{\lambda}^b + T_{ab} \boldsymbol{\lambda}^a (\partial_c \boldsymbol{\lambda}^b) \\
&= (\partial_c T_{ab} - \Gamma_{ac}^d T_{db} - \Gamma_{bc}^d T_{ad}) \boldsymbol{\lambda}^a \boldsymbol{\lambda}^b = D_c T_{ab} \boldsymbol{\lambda}^a \boldsymbol{\lambda}^b, \qquad (2.167)
\end{aligned}$$

As an example, calculate the Christoffel symbols of the second kind for plane polar coordinates, letting $(\xi^1 = r, \xi^2 = \phi)$. The metric tensor in plane polar coordinates is

$$\{g_{ij}\} = \begin{bmatrix} 1 & 0 \\ 0 & r^2 \end{bmatrix} = \begin{bmatrix} g_{rr} & 0 \\ 0 & g_{\phi\phi} \end{bmatrix}, \qquad (2.168)$$

where the indices are denoted by the labels $(1, 2) \leftrightarrow (r, \phi)$ for convenience. Only the metric component $g_{\phi\phi}$ has a non zero derivative $\partial g_{\phi\phi}/\partial r = 2r$. First calculate the Christoffel symbols of the first kind, which can be expressed as linear functions of the metric tensor (Equation (2.159)). The Christoffel symbols are symmetric on the last two indices $\Gamma_{cab} = \Gamma_{cba}$. They can be expressed in matrix form as

$$\Gamma_r = \begin{bmatrix} 0 & 0 \\ 0 & -r \end{bmatrix}, \quad \Gamma_\phi = \begin{bmatrix} 0 & r \\ r & 0 \end{bmatrix}. \qquad (2.169)$$

Christoffel symbols of the second kind are given by using the inverse metric tensor as a raising operator

$$\Gamma_{ab}^c = g^{cc'} \Gamma_{cab} \rightarrow \Gamma^c = g^{cc'} \Gamma_c, \qquad (2.170)$$

giving

$$\Gamma^r = \begin{bmatrix} 0 & 0 \\ 0 & -r \end{bmatrix}, \quad \Gamma^\phi = \begin{bmatrix} 0 & 1/r \\ 1/r & 0 \end{bmatrix}. \qquad (2.171)$$

The nonvanishing components are $\Gamma_{\phi\phi}^r = -r$ and $\Gamma_{r\phi}^\phi = \Gamma_{\phi r}^\phi = 1/r$. A summary of Christoffel Symbols of the Second Kind for common curvilinear coordinate systems is presented in Table 2.5.

TABLE 2.5   Christoffel Symbols of the Second Kind for common curvilinear coordinate systems

Plane Polar	$\Gamma^r = \begin{bmatrix} 0 & 0 \\ 0 & -r \end{bmatrix}, \Gamma^\phi = \begin{bmatrix} 0 & 1/r \\ 1/r & 0 \end{bmatrix}$
Cylindrical	$\Gamma^r = \begin{bmatrix} 0 & 0 & 0 \\ 0 & -r & 0 \\ 0 & 0 & 0 \end{bmatrix}, \Gamma^\phi = \begin{bmatrix} 0 & 1/r & 0 \\ 1/r & 0 & 0 \\ 0 & 0 & 0 \end{bmatrix}, \Gamma^z = \begin{bmatrix} 0 & 0 & 0 \\ 0 & 0 & 0 \\ 0 & 0 & 0 \end{bmatrix},$
Spherical Polar	$\Gamma^r = \begin{bmatrix} 0 & 0 & 0 \\ 0 & -r\sin^2\theta & 0 \\ 0 & 0 & -r \end{bmatrix}, \Gamma^\theta = \begin{bmatrix} 0 & 0 & 1/r \\ 0 & -\sin\theta\cos\theta & 0 \\ 1/r & 0 & 0 \end{bmatrix},$ $\Gamma^\phi = \begin{bmatrix} 0 & 1/r & 0 \\ 1/r & 0 & \cot\theta \\ 0 & \cot\theta & 0 \end{bmatrix}.$

2.6.2 *Divergence Theorem

Equation (2.157) leads to the following remarkable and useful result

$$\partial_a(\sqrt{\|g\|}\,J^a) = \sqrt{\|g\|}D_aJ^a. \tag{2.172}$$

Therefore, the divergence of a vector can be written as

$$D_aJ^a = \frac{1}{\sqrt{\|g\|}}\partial_a(\sqrt{\|g\|}\,J^a). \tag{2.173}$$

Likewise, it can be shown that the divergence of an antisymmetric tensor can be written as [Misner 1970, p. 222]

$$D_bF^{ab} = \frac{1}{\sqrt{\|g\|}}\partial_b(\sqrt{\|g\|}F^{ab}). \tag{2.174}$$

Consider the divergence of a vector J in n-dimensions

$$\nabla \cdot J = D_aJ^a. \tag{2.175}$$

The n-dimensional divergence theorem is

$$\int_{\mathcal{R}} d^n\xi\sqrt{\|g\|}D_\mu J^\mu = \int_{\mathcal{R}} d^n\xi\,\partial_\mu(\sqrt{\|g\|}J^\mu) = \oint_{\partial\mathcal{R}} \sqrt{\|g\|}(d^{n-1}\xi)_\mu J^\mu. \tag{2.176}$$

where $\mathcal{R}$ is a region of n-space and $\partial\mathcal{R}$ is its $n-1$ dimensional boundary, resulting in

$$\int_{\mathcal{R}} dV\, \nabla \cdot \mathbf{J} = \oint_{\partial\mathcal{R}} d\boldsymbol{\sigma} \cdot \mathbf{J}. \qquad (2.177)$$

where $dV = \sqrt{\|g\|}d^n\xi$ is an n-dimensional volume element and $d\sigma_\mu = \sqrt{\|g\|}(d^{n-1}\xi)_\mu$ is an $(n-1)$-dimensional boundary element.

2.6.3 *Riemann Curvature Tensor

Manifolds are curved spaces in general. The *intrinsic curvature* of a space, independent of any *extrinsic space* in which it is embedded, can be determined by evaluating the *Riemann tensor*,[10] which depends only on the metric tensor through the affine connection

$$R^a_{bcd} = \partial_d\Gamma^a_{bc} - \partial_c\Gamma^a_{bc} + \Gamma^e_{bc}\Gamma^a_{ed} - \Gamma^e_{bd}\Gamma^a_{ec}. \qquad (2.178)$$

The Riemann tensor R^a_{bcd} is also known as the *Riemann-Christoffel curvature tensor* [Misner 1973, p. 273]. It is important for the study of *General Relativity*. For flat spaces such as Euclidean Space, the Riemann tensor identically vanishes. A surface embedded in E^3 may have a non-vanishing intrinsic curvature, however.

The Riemann tensor has a number of symmetries. It is antisymmetric under the exchange of the last two indices, for example. Its list of symmetries include

$$R_{abcd} = R_{cdab}, \quad R_{abcd} = -R_{abdc} = -R_{bacd}, \quad R_{abcd} + R_{acdb} + R_{adbc} = 0. \qquad (2.179)$$

The number of independent components of the Riemann tensor is restricted by the above symmetries. In n-dimensional space, there are $C_n = n^2(n^2-1)/12$ independent entries in the Riemann tensor. In one dimension, the sole component R^1_{111} identically vanishes. That is, a curve, if even if closed, can always be cut, and the resulting line is topologically equivalent to a straight line on a flat metric space. Curved lines therefore have no intrinsic curvature. In two-dimensional spaces, there is only one-independent non-vanishing component, given by $R_{1212} = R_{2121} = -R_{1221} = -R_{2112}$, describing the intrinsic curvature of a point on the surface. The surface of a cylinder is topologically flat. However, there is no way to take a sphere, with

---

[10]Georg Friedrich Bernhard Riemann (1826–1866) made important contributions to analysis and differential geometry. He extended the concept of surface to higher dimensions, developing non-Euclidean geometry.

its non-zero intrinsic curvature, and map it point to point to a flat plane having zero curvature. Map makers are well aware of this. At least two maps are required to cover the globe of the Earth. Things get complicated quickly for higher dimensions. There are 6 independent components in three-dimensional space, and 20 in four-dimensional space. In three-dimensional Euclidean space in which the majority of this text resides, the curvature tensor identically vanishes, and therefore study of the Riemann tensor is a largely intellectual exercise until the general theory of relativity is introduced.

Closely associated with the Riemann tensor are the symmetric *Ricci*[11] *tensor* $R_{\alpha\beta}$ and the *Ricci scalar R*, which are defined by repeated contraction of the Riemann tensor with the metric tensor

$$R_{ab} = g^{cd} R_{acbd}, \quad R_{ab} = R_{ba}, \tag{2.180}$$

$$R = g^{ab} R_{ab}. \tag{2.181}$$

A useful formula for calculating the Ricci tensor is [Misner 1970, p. 222]

$$R_{ab} = \frac{1}{\sqrt{\|g\|}} \partial_c (\sqrt{\|g\|} \Gamma^c_{ab}) - \partial_b \partial_a \ln (\sqrt{\|g\|}) - \Gamma^c_{da} \Gamma^d_{bc}. \tag{2.182}$$

2.6.4 *Curvature of the Two-Sphere

For the two-dimensional manifold of the two-sphere, for example, the surface of the Earth, the calculation of the Riemann tensor is relatively simple. The metric of the manifold in spherical coordinates is

$$g = r_0^2 \begin{bmatrix} 1 & 0 \\ 0 & \sin^2 \theta \end{bmatrix}, \tag{2.183}$$

where r_o is the radius of the sphere. The non-vanishing Christoffel symbols are given by

$$\Gamma^2_{12} = \Gamma^2_{21} = \frac{\cos \theta}{\sin \theta}, \quad \Gamma^1_{22} = -\sin \theta \cos \theta, \tag{2.184}$$

yielding the following non-zero Riemann tensor elements

$$R_{\theta\phi\theta\phi} = R_{\phi\theta\phi\theta} = -R_{\theta\phi\phi\theta} = -R_{\phi\theta\theta\phi} = r_0^2 \sin^2 \theta. \tag{2.185}$$

The Ricci tensor is

$$R_{ab} = \frac{g_{ab}}{r_0^2}, \tag{2.186}$$

---

[11]Gregorio Ricci-Curbastro (1853–1925) invented the tensor calculus.

and the Ricci scalar is

$$R = g^{\theta\theta}R_{\theta\theta} + g^{\phi\phi}R_{\phi\phi} = \frac{2}{r_0^2}. \tag{2.187}$$

Note that the Ricci scalar decreases with increasing radius, becoming nearly flat for large radii. Euclidean spaces are flat and, therefore, have vanishing Riemann curvature everywhere.

2.7 SUMMARY

Curves and surfaces of arbitrary dimensions can be considered as manifolds embedded in some larger flat Euclidean space. A point on a manifold is defined by one or more maps in terms of a set n linearly-independent, continuous and differential parameters that define a coordinate system

$$\mathbf{P}(\xi^i). \tag{2.188}$$

The number of independent parameters n defines the dimensionality of the manifold. An infinitesimal displacement on the manifold is a tangent vector defined as

$$d\mathbf{r} = d\xi^i \frac{\partial \mathbf{P}}{\partial \xi^i} = d\xi^i \boldsymbol{\lambda}_i, \tag{2.189}$$

where $\boldsymbol{\lambda}_i$ defines a covariant basis with respect to the contravariant infinitesimal displacements. The inner product of the covariant basis vectors with each other defines the metric tensor of the space

$$\boldsymbol{\lambda}_i \cdot \boldsymbol{\lambda}_j = g_{ij}. \tag{2.190}$$

Geometric manifolds are metric spaces with the invariant element of length ds defined as

$$ds^2 = d\mathbf{r} \cdot d\mathbf{r} = g_{ij}d\xi^i d\xi^j.$$

Associated with a covariant basis is the inverse, contravariant basis defined as

$$\boldsymbol{\lambda}^i = g^{ij}\boldsymbol{\lambda}_j, \quad \boldsymbol{\lambda}^i \cdot \boldsymbol{\lambda}_j = \delta^i_j. \tag{2.191}$$

The gradient operator is defined in terms of this basis and is given by

$$\nabla = \boldsymbol{\lambda}^i \frac{\partial}{\partial \xi^i}, \tag{2.192}$$

where the operators $\partial/\partial\xi^i$ define the components of a covariant vector operator.

The infinitesimal displacement vector $d\mathbf{r}$ defines the prototype for all contravariant vectors and the gradient operator ∇ defines the prototype for all covariant vectors. The metric tensor can be used to convert between covariant and contravariant representations of vectors.

$$V_i = g_{ij}V^j, \quad V^j = g^{ji}V_i. \tag{2.193}$$

The inner dot product between "raised" contravariant vector components and "lowered" covariant components is a scalar invariant

$$\text{scalar}\,(a,b) = \mathbf{a}\cdot\mathbf{b} = a^i b_i. \tag{2.194}$$

In bra-ket notation this is written as

$$\text{scalar}\,(a,b) = \langle a|b\rangle = a^i b_j \langle e_i|e^j\rangle = a^i b_i. \tag{2.195}$$

The outer (ordered) product of two vectors is a second rank tensor called a dyad

$$\mathbf{dyad}\,(a,b) = \mathbf{a}\,\mathbf{b} = a^i b_j\,\mathbf{e}_i\,\mathbf{e}^j. \tag{2.196}$$

In bra-ket notation this is written as

$$\text{dyad}(a,b) = |a\rangle\langle b| = a^i b_j |e_i\rangle\langle e^j|. \tag{2.197}$$

The sum of dyads is called a dyadic. Dyadic multiplication involves contraction of adjacent elements and is equivalent to matrix multiplication of its components

$$\mathbf{W} = \mathbf{U}\cdot\mathbf{V}, \quad W^i{}_j = \mathbf{e}^i\cdot\mathbf{W}\cdot\mathbf{e}_j = U^i{}_k V^k{}_j. \tag{2.198}$$

Orthogonal coordinate systems have diagonal metric tensors where

$$g_{ij} = 0, \quad \text{if } i \neq j. \tag{2.199}$$

In orthogonal coordinate systems, one can define a scale factor

$$h_i = \sqrt{g_{ii}}, \quad h^i = \sqrt{g^{ii}} = h_i^{-1}. \tag{2.200}$$

In these coordinates, one can express the elementary tensor operators in terms of dimensionless unit vectors

$$d\mathbf{r} = h_i\hat{\boldsymbol{\xi}}_i d\xi^i, \quad \nabla = \frac{1}{h_i}\hat{\boldsymbol{\xi}}^i\frac{\partial}{\partial\xi^i}, \quad \hat{\boldsymbol{\xi}}^i = \hat{\boldsymbol{\xi}}_i. \tag{2.201}$$

Common examples of orthogonal coordinate systems are the Cartesian, cylindrical-polar, and spherical-polar coordinate systems.

The covariant derivative of a tensor invariant is the same as the ordinary partial derivative. The covariant derivative of a tensor basis vanishes. Covariant derivatives satisfy Leibniz's law

$$D_\mu(AB) = D_\mu(A)B + AD_\mu(B). \tag{2.202}$$

Applying the covariant derivative to a vector invariant gives

$$D_b\mathbf{V} = (D_b V^a)\lambda_a = (\partial_b V^a + \Gamma^a_{cb} V^c)\lambda_a,$$
$$\partial_b\mathbf{V} = (\partial_b V^a)\lambda_a + V^a(\partial_b \lambda_a), \tag{2.203}$$

where the Christofell Symbols of the Second Kind are defined as

$$\Gamma^c_{ab} = \left\{ \begin{matrix} c \\ ab \end{matrix} \right\} = \lambda^c \cdot \frac{\partial \lambda_a}{\partial \xi^b} = \frac{1}{2} g^{cc'} \left(\frac{\partial g_{c'a}}{\partial \xi^b} + \frac{\partial g_{c'b}}{\partial \xi^a} - \frac{\partial g_{ab}}{\partial \xi^{c'}} \right). \tag{2.204}$$

2.8 EXERCISES

1. Derive the law of cosines $c^2 = a^2 + b^2 + 2|\mathbf{a}||\mathbf{b}|\cos\theta$ for a triangle $\mathbf{c} = \mathbf{a} + \mathbf{b}$ in $\mathbb{R}^2$ starting from the definition of the cosine in terms of the dot product $\mathbf{a} \cdot \mathbf{b} = |\mathbf{a}||\mathbf{b}|\cos\theta$. (See Figure 2.9.)
2. Derive the law of sines $\sin\alpha/|a| = \sin\beta/|b| = \sin\gamma/|c|$ for the triangle $\mathbf{c} = \mathbf{a} + \mathbf{b}$, illustrated in Figure 2.9.
3. Find the surfaces of constraint $c(x, y, z) = 0$, defined by the following constraint conditions with position $\mathbf{r}$ and where $\mathbf{r}_0$ is a fixed point:
   a. $(\mathbf{r} - \mathbf{r}_0) \cdot \mathbf{r} = 0$.
   b. $(\mathbf{r} - \mathbf{r}_0) \cdot \mathbf{r}_0 = 0$.

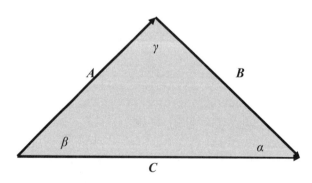

FIGURE 2.9   Figure of a triangle.

Sketch the curves satisfying the above constraint conditions in the (x, y) plane if $\mathbf{r}_0 = (0, 0, 1)$.

4. Three points on the surface of a sphere of radius a are located at $\{(a, 0, 0), (0, a, 0), (0, 0, a)\}$. Find the interior angles of the spherical triangle formed by connecting these points with great circles on the sphere. Show that the sum of the interior angles is greater than $180°$.

5. By explicitly calculating the Jacobian of the transformation, show that a volume element in spherical polar coordinates can be written as $dx\, dy\, dz = r^2 \sin\theta\, dr\, d\theta\, d\phi$.

6. Show that the electrostatic field distribution $\mathbf{E} = E_0(\hat{\mathbf{r}} \cos\theta - \hat{\boldsymbol{\theta}} \sin\theta)$ has vanishing curl, and find the electric potential Φ, where $\mathbf{E} = -\nabla\Phi$.

7. The vector potential for a magnetic dipole $\mathbf{m}_\mathrm{D}$ is given by

$$\mathbf{A} = \frac{\mu_0}{2\pi} \frac{\mathbf{m}_\mathrm{D} \times \mathbf{r}}{r^3}.$$

Find the magnetic induction field $\mathbf{B} = \nabla \times \mathbf{A}$, for $r > 0$.

8. A particle of charge q, moving in the (x, y) plane with a velocity of magnitude v, sees a constant magnetic field $\mathbf{B}$ in the z-direction. Solve the equations of motion and show that the motion is a circle of radius $r = mv/qB$ and that its angular frequency of rotation is given by the cyclotron frequency $\omega = qB/m$. The inverse of the radius of curvature is called the magnetic rigidity of the particle.

9. Given the equation of motion $\dot{\mathbf{v}} = \boldsymbol{\omega} \times \mathbf{v}$ for a charged particle rotating in a magnetic field with cyclotron frequency $|\boldsymbol{\omega}|$, show that the equation can be rewritten in any of the following equivalent forms

$$\dot{v}^i = \Omega^i_j v^j, \quad \dot{\mathbf{v}} = \Omega \cdot \mathbf{v}, \quad |\dot{v}\rangle = \Omega|v\rangle,$$

where $\Omega_{ij} = \varepsilon_{ijk}\omega^k$ is an antisymmetric rank-2 tensor.

10. A particle of mass m confined to the surface of the two-sphere has the two-dimensional coordinate map, valid in the northern hemisphere, given by

$$x = x, \quad y = y, \quad z = +\sqrt{R^2 - x^2 - y^2}$$

where R is the radius of the sphere. Find the kinetic energy of the particle in terms of this coordinate mapping.

11. Calculate the two-dimensional metric tensor for the torus given by

$$x(\theta, \phi) = (a + b\cos\phi)\cos\theta, \quad y(\theta, \phi) = (a + b\cos\phi)\sin\theta,$$
$$z(\theta, \phi) = b\sin\phi.$$

12. Find the area and circumference of an ellipse with semimajor axis a and semiminor axis b. The result for the circumference can be expressed as a complete elliptic integral of the second kind.

13. A point particle of mass m is constrained to move on a wire that is in the shape of an ellipse

$$x = a \cos \phi, \quad y = b \sin \phi, \quad z = 0.$$

a. Find an expression for its kinetic energy in terms of the parameter ϕ.

b. Find an integral expression for its path length for one complete cycle around an ellipse. (Hint: the arc-length of an ellipse can be written in terms of a complete elliptic integral of the second kind.)

14. Consider a point in a two-dimensional plane $\mathbf{P}(x, y) = x\,\mathbf{e}_1 + y\,\mathbf{e}_2$ that is mapped to a skewed coordinate system given by

$$u = (x/2 \cos \alpha + y/2 \sin \alpha), \quad v = (-x/2 \cos \alpha + y/2 \sin \alpha),$$

where $\alpha = 30°$ is a constant parameter.

a. Find the covariant basis vectors $\boldsymbol{\lambda}_i$ for the new coordinate system.

b. Find the metric tensor with respect to the new basis.

c. Find the contravariant basis vectors $\boldsymbol{\lambda}^i$.

15. Parity is a symmetry property of physical observables under space inversion, which corresponds to their even or odd symmetry under the reflection of all three Cartesian coordinates in $\mathbb{R}^3$. Determine the parity of each of the three unit vectors $(\hat{\mathbf{r}}, \hat{\boldsymbol{\theta}}, \hat{\boldsymbol{\phi}})$ in spherical coordinates.

16. Use the fact that the metric tensor is symmetric to show that a rank-2 covariant tensor has the same symmetry under index exchange as a rank-2 contravariant tensor.

17. Let S^{ab} be a symmetric rank-2 contravariant tensor $S^{ab} = S^{ba}$, and let A^{ab} be an antisymmetric rank-2 contravariant tensor $A^{ab} = -A^{ba}$. Show that $S^{ab} A_{ab} \equiv 0$.

18. Calculate the Christoffel symbols of the second kind for cylindrical coordinates, letting $(\xi^1 = r, \xi^2 = \phi, \xi^3 = z)$.

19. Calculate the Christoffel symbols of the second kind for spherical polar coordinates letting $(\xi^1 = r, \xi^2 = \theta, \xi^3 = \phi)$.

20. A parabolic coordinate system in the (x, y) plane is given by the coordinate mapping

$$x = uv, \quad y = \frac{1}{2}(u^2 - v^2).$$

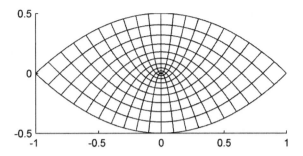

FIGURE 2.10 **A parabolic coordinate system.**

    a. Find the metric tensor in (u, v) coordinates and write down the kinetic energy for a particle of mass m in these coordinates.

    b. Show that the coordinate system is orthogonal with scale factors $h_u = h_v = \sqrt{(u^2 + v^2)}$. [NOTE: The curves shown in Figure 2.10 are parabolas $2y = x^2/v^2 - v^2$, which open upward, and $2y = -x^2/u^2 + u^2$, which open downward. The foci of all the parabolas are at the origin $(x = y = 0)$.]

21. Elliptic cylindrical coordinates (u, v, z) are orthogonal curvilinear coordinates. The v coordinates are the asymptotic angle of confocal hyperbolic cylinders symmetrical about the $\hat{x}$-axis. The u coordinates are confocal elliptic cylinders centered on the origin. The mapping is given by the transformations

$$x = a \cosh u \cos v, \quad y = a \sinh u \sin v, \quad z = z.$$

    a. Find the covariant basis vectors for the new coordinate system.

    b. Show that the scale factors for the metric tensor are given by

$$g_{uu} = (h_u)^2 = a^2(\sin^2 v + \sinh^2 u),$$
$$g_{vv} = (h_v)^2 = a^2(\sin^2 v + \sinh^2 u), \quad g_{zz} = 1.$$

    c. Find the kinetic energy function of a particle of mass m in these coordinates.

    d. Draw some representative contours of constant u and v in the plane $z = 0$.

22. Verify the following formula for the Christoffel symbols of the first kind, assuming that they are symmetric in the last two indices:

$$\Gamma_{kij} = \lambda_k \cdot \frac{\partial \lambda_i}{\partial \xi^j} = \frac{1}{2}\left(\frac{\partial g_{ik}}{\partial \xi^j} + \frac{\partial g_{jk}}{\partial \xi^i} - \frac{\partial g_{ij}}{\partial \xi^k} \right).$$

23. Calculate the Christoffel symbols of the first and second kind for the two-dimensional manifold of a right circular cylinder of constant radius a, Show by explicit calculation from the Christoffel symbols that the Riemann tensor vanishes, and that therefore the surface of a cylinder is topologically flat. It is sufficient to calculate a single element R_{1212}, since there is only one linearly independent element of the Riemann tensor for a two-dimensional manifold.

Chapter **3**

LAGRANGIAN AND HAMILTONIAN DYNAMICS

I n this chapter, the Lagrangian and Hamiltonian formulations of mechanics will be developed. The principle of virtual work will be used to eliminate point constraints, allowing a system to be expressed in terms of a reduced set of generalized coordinates. Every generalized coordinate has a canonical momentum associated with it. Physical observables will be expressed as functions of the generalized coordinates and momenta in a phase space representation of the dynamics of a mechanical system.

3.1 GENERALIZED COORDINATES AND CONSTRAINTS

Lagrangian dynamics is an independent formulation of classical dynamics developed by Joseph-Louis Lagrange (1736–1813). Lagrange developed the theory of differential equations and applied variational principles to mechanics, reformulating Newton's Laws of motion. His masterpiece, *Mécanique analytique* [Lagrange 1999], was the basis for future work in the field. Lagrange also played a critical role in the development of the Calculus of Variations, which was originated by Johann Bernoulli (1667–1748), although it was not called by that name that until Leonhard Euler[1] (1707–1783) referred to it as such in 1766. In this chapter, Lagrange's equations of motion will be derived using D'Alembert's Principle of Least Work.[2] In Chapter 4 (Hamilton's Principle) a more elegant formation using the *Calculus of Variations* will be developed. Lagrange's equations of motion provide an alternative set of equations of motion, entirely equivalent to Newton's but with several significant advantages:

- The equations of motion are derived from a scalar Lagrangian function, which can then be transformed into any frame of reference,

---

[1] Leonhard Euler made enormous contributions to a broad range of mathematics and physics and is responsible for establishing much of modern mathematical notation.
[2] The Principle of Virtual Work in its modern form was largely developed by Lagrange, but the earliest statements of the principle are credited to Jean Le Rond d'Alembert (1717–1783).

including accelerating frames, in a form-invariant manner, provided the Lagrangian is originally described in some inertial frame.

– The resulting mathematical formulation is both more general and easier to use than the tensor calculus (one does not have to keep track of Christoffel symbols).

– The formulation provides a mechanism for eliminating constraints between the coordinates. This reduces the number of degrees of freedom to only those that are free to vary.

In a Cartesian inertial frame, Newton's laws for a system of particles can be written as a set of coupled differential equations of the form

$$\mathbf{F}_j(x, \dot{x}, t) = m_j \ddot{\mathbf{x}}_j \tag{3.1}$$

where $\mathbf{F}_j(x, \dot{x}, t)$ is the net Newtonian force acting on a particle j with mass m_j. Here, let

$$x = \{x^i : i = 1, n\} = \{x_j^k : k = 1, 2, 3; j = 1, N\} \tag{3.2}$$

denote an N-dimensional configuration space, consisting of the $3N$ Cartesian coordinates of an N-particle system. If the force is a known function of coordinates, velocities, and the time, the result is a coupled set of second-order differential equations. Specification of all the coordinates and velocities at any given time determines the evolution of the system for all later times.

An important class of transformations for this purpose is the invertible set of time-dependent *point transformations* from one set of coordinates to another. Point transformations are time-dependent coordinate transformations in an n-dimensional configuration space Q. These point mappings can be defined to and from a set of Cartesian coordinates x^i to a set of generalized coordinates q^i in an redundant, configuration space. At first the mapping will be taken as one-to-one; later, the constrained coordinates will be eliminated. Configuration space can be represented as an n-tuple column vector

$$\begin{bmatrix} \langle 1|x \rangle \\ \vdots \\ \langle n|x \rangle \end{bmatrix} = \begin{bmatrix} x^1 = x^1(q, t) \\ \vdots \\ x^n = x^n(q, t) \end{bmatrix}. \tag{3.3}$$

Point transformations are defined as time-dependent, invertible coordinate maps of the form

$$x^i = x^i(q, t) \leftrightarrow q^i = q^i(x, t). \tag{3.4}$$

In the new coordinate system, one or more of the generalized coordinates may turn out to be *frozen* or inactive due to constraint conditions on the coordinates. Any explicit time dependence of the mapping affects the total derivative of a point function of the coordinates and time, which can be shown by use of the chain rule:

$$\frac{d}{dt}q^j(x,t) = \dot{x}^i\frac{\partial q^j(x,t)}{\partial x^i} + \frac{\partial q^j(x,t)}{\partial t} = \dot{q}^j(x,\dot{x},t). \tag{3.5}$$

In the preceding equation, the generalized velocities $\dot{q}^j(x,\dot{x},t)$ are a function of the coordinates, their velocities, and the time.

3.1.1 Functionals

In the Lagrangian formulation, physical observables are defined on a $2n$-dimensional product space TQ, which is called the *tangent bundle*. A physical observable or *functional* is a description of a classical physical observable as a function of its defined, explicit, dynamical dependence on its velocity phase space coordinates (positions and velocities) and the time.

The resulting equations of motion, derived from the Lagrangian functional, are called Lagrange's equations of motion. These are form invariant expressions; that is, the equations take on the same form in any coordinate system. They are based on taking partial derivatives of a scalar function called the Lagrangian, which is a functional of the generalized coordinates, their velocities and the time. After the equations of motion are evaluated, the velocities can be related to the coordinates by taking the time derivatives of the generalized coordinates. Then the flow in the product phase space TQ collapses to trajectories in configuration space Q.

For example, the kinetic energy of a single particle in Cartesian coordinates is given by the functional

$$T(\dot{x},\dot{y},\dot{z}) = \frac{1}{2}m\mathbf{v}^2 = \frac{m}{2}(\dot{x}^2 + \dot{y}^2 + \dot{z}^2). \tag{3.6}$$

Note that the kinetic energy does not depend explicitly on the Cartesian coordinates or the time, but only on the velocities. The kinetic energy is said to be cyclic in a coordinate q^i if $\partial T/\partial q^i \equiv 0$. The kinetic energy of a particle in Cartesian coordinates is cyclic in all three coordinates and the time. Now, consider a time-independent coordinate mapping of Euclidean space into spherical coordinates given by

$$x = r\sin\theta\cos\phi, \quad y = r\sin\theta\sin\phi, \quad z = r\cos\theta. \tag{3.7}$$

Since the kinetic energy is a scalar function, it transforms by invariance. The kinetic energy functional written in terms of the new coordinate system has the functional representation

$$T(r,\theta,\phi,\dot{r},\dot{\theta},\dot{\phi},t) = T(r,\theta,\dot{r},\dot{\theta},\dot{\phi}) = \frac{m}{2}(\dot{r}^2 + r^2\dot{\theta}^2 + r^2\sin^2(\theta)\dot{\phi}^2). \quad (3.8)$$

The particle's kinetic energy in spherical polar coordinates does not explicitly depend on the azimuthal angle or the time. The kinetic energy function is said to be cyclic in those variables.

The total derivative of an arbitrary functional $f(q,\dot{q},t)$ in an arbitrary set of generalized coordinates is given by

$$df(q,\dot{q},t) = \frac{\partial f}{\partial q^j}dq^j + \frac{\partial f}{\partial \dot{q}^j}d\dot{q}^j + \frac{\partial f}{\partial t}dt. \quad (3.9)$$

The value of a functional at a given instance on a given path is a real number given by substitution of the instantaneous values for the system's coordinates and velocities at the given time.

3.1.2 Independent Degrees of Freedom

Newton's equations of motion assume that particles can move freely anywhere in space. This is not generally true. Free space is an idealization. The atoms in a crystalline solid, for example, are constrained to remain at or near their lattice sites. Other examples of physical constraints on the motion include

- An amusement-park rollercoaster ride constrained (hopefully) to motion on a rail. This is an example of a holonomic constraint yielding a one-dimensional curve or manifold. Holonomic constraints are constraint equations on the coordinates and the time.
- The point of contact of a tire with the road, constrained to roll without slipping. This is an example of a velocity-dependent constraint, which may, or may not, be convertible into a holonomic constraint.
- Billiard balls moving on a pool table constrained to the boundaries of the table. This is an example of a nonholonomic constraint defined by an inequality relation.

How does one accommodate constraints in the equation of motion? The answer depends on the type of the constraint.

3.1.3 Holonomic Constraints

Holonomic constraints are functional restrictions on the coordinates of the system, which can be expressed as a generalized coordinate that is constant for all time

$$q^k = c^k(x,t) = 0, \qquad (3.10)$$

where c^k defines some generalized constant coordinate, which is convention-
ally defined to be zero along the allowed path.[3] The constraint conditions
may be time dependent in general. Holonomic constraints represent frozen,
inactive degrees of freedom.

Examples of holonomic constraints include

- A particle constrained to move on the radius of a sphere: $c_r = r - a = 0$.
- A particle moving along the radius of a wire rotating with constant
  angular velocity: $c_\phi = \phi - (\phi_0 + \omega t) = 0$.
- A particle constrained to move in the $\hat{x} \wedge \hat{y}$ plane: $c_z = z - z_0 = 0$.
- Rigid body motion with the lengths of all relative displacements fixed:
  $c_{ij} = (\mathbf{r}_i - \mathbf{r}_j)^2 - a_{ij}^2 = 0$.

Holonomic constraints can be applied to functionals prior to attempting to
derive the equations of motion. They effectively reduce the dimensionality
of the problem.

3.1.4 Nonholonomic Constraints

All other constraints are called nonholonomic constraints. Nonholonomic
constraints cannot be eliminated prior to solving for the equations of motion.
Examples of nonholonomic constraints include inequalities, such as $z \geq 0$,
which restricts the motion to the positive z-axis. Local differential constraints
are often nonintegrable and therefore nonholonomic. For example, the constraint that a ball rolls without slipping on a surface is a differential constraint
that requires that the velocity of the point of contact with the surface be
instantaneously at rest with respect to the surface. In general, this leads to a
constraint of the form

$$f_i(x,t)dx^i - g(x,t)dt = 0. \qquad (3.11)$$

If an integrating factor $\Lambda(x^j, t)$ can be found, the preceding equation can be
converted into a holonomic constraint $c(x^j, t) = 0$:

$$dc(x,t) = \frac{\partial c}{\partial x^i}dx^i + \frac{\partial c}{\partial t}dt = \Lambda(x,t)(f_i(x,t)dx^i - g(x,t)dt) = 0. \qquad (3.12)$$

---

[3] Setting $c^k(x,t) = 0$ is a convenience, not a necessary. Most textbook treatments do
not impose this convention on constrained coordinates. The rationale is that when one
sets $c^k = 0$, any constants remaining in the equations of motion represent constant
parameters (c-numbers) rather than a residual of a frozen coordinate.

For example, differential constraints, such as a wheel "rolling-without-slipping," may be either holonomic (integrable) or nonholonomic (nonintegrable) in character. If the constraints are holonomic, then the resulting frozen degrees of freedom can be eliminated. In the following chapter, Lagrange's method for dealing with nonholonomic differential constraints will be developed.

3.1.5 Independent Generalized Coordinates

Each holonomic constraint reduces the number of independent coordinates by one. For example, if the Cartesian coordinate z is constrained $(c_z = z - z_0 = 0)$, one is left with only x and y as independent degrees of freedom, yielding a two-dimensional Euclidean subspace $(x, y) \in \mathbb{R}^2$. For more complicated constraints, one may be able to solve the holonomic constraint conditions $c^k(x, t) = 0$ for one or more variables. One can then eliminate these variables.

To accomplish this goal, one may have to switch to a new coordinate system. For example, for a particle constrained to move on the surface of a sphere of radius $r = a$, a good choice for independent coordinates would be a spherical polar basis, allowing one to trivially eliminate the radial coordinate, by defining a constrained coordinate $c_r = r - a = 0$. Mapping to a new set of variables $(x, y, z) \to (\theta, \phi)$ gives a system with two independent generalized coordinates, yielding the manifold equations of the two-sphere

$$x(\theta, \phi) = a \sin \theta \cos \phi, \quad y(\theta, \phi) = a \sin \theta \sin \phi, \quad z(\theta, \phi) = a \cos \theta. \quad (3.13)$$

The term "generalized coordinates" is usually applied to the reduced set of independent degrees of freedom that are active after eliminating the constraint conditions. For example, a particle system with N particles has $3N$ Cartesian coordinates in an inertial frame denoted by the configuration space vector $x^i = \{x^1, \ldots, x^{3N}\}$. If one introduces k independent holonomic constraints, one can reduce the system to $3N - k$ degrees of freedom. One can then define $n = 3N - k$ linearly independent of active generalized coordinates

$$q^i(x, t) = \{q^1, \ldots, q^n\}, \quad (3.14)$$

and k null-valued constrained coordinates

$$c^j(x, t) = \{c^1, \ldots, c^k\} = 0. \quad (3.15)$$

The Cartesian coordinates can now be expressed in terms of a reduced number of degrees of freedom, ignoring the constrained coordinates by setting $c^j = 0$

$$x^i(q, c, t) = x^i(q, t). \tag{3.16}$$

3.1.6 Constraint Forces

The primary difficulty with constraint forces is that their functional dependence is not known a priori. They simply acquire the value needed to enforce the constraint. Thus, they represent a type of boundary condition. Applying the constraint conditions allows one to remove these coordinate degrees of freedom. The effect of the constraint forces on the equations of motion can be eliminated systematically by applying the *principle of virtual work*. The concept behind the principle of virtual work is simple enough: To enforce a constraint condition, at any given time a force of constraint must act along the direction of constraint, which is orthogonal to the allowed directions of motion of the particle. In such cases, the force of constraint does no *virtual work* on the system, provided one considers virtual variations to be variations consistent with the constraints that are defined for constant time. This is the idea behind the principle of virtual work that will be developed fully in the following section.

Forces of constraint are idealizations. A holonomic constraint implies an infinite response force, or an infinitely high potential wall. Reality is always more complicated. Real constraints smear out the motion in the constraint direction; e.g., the potential of a diatomic molecule might result in a strong, semiclassical, binding force constraining the motion in the radial direction. The two particles are otherwise free to orbit each other, as indicated by the potential energy diagram shown in left side of Figure 3.1. The radial binding

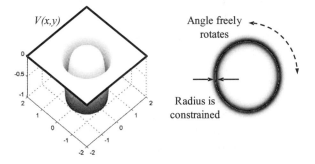

FIGURE 3.1 A two-particle system with a deep and narrow radial potential well can rotate freely in the relative angular direction but is constrained in its relative radial motion; if the radial amplitude is small enough, the motion can be treated as a holonomic radial constraint, requiring that the particles move in circular orbits about each other.

force in this case makes the radial direction special. If the potential well is deep, and the free energy of radial vibrations small, the radial coordinate can be treated as a positional constraint as shown in the right side of the figure.

Holonomic constraints are extreme simplifications, which, however, can often be applied fruitfully to greatly reduce the complexity of a system. For example, the number of molecular degrees of freedom in one gm-mol of water (about 1 cm^3 of liquid or solid water) is a very large number considering that there are Avogadro's number ($N_A \simeq 6.022 \times 10^{23}$) of molecules in a gm-mol of matter. If the water is in the form of frozen ice, however, from an empirical point of view, the system can be treated as a rigid body, reducing the active degrees of freedom to only six, consisting of the three translational and three rotational degrees of freedom available to a rigid body.

In the absence of interactions, the energy of a system is purely kinetic. The Cartesian system is often the simplest choice of coordinates to use in this case. Forces break the symmetry of empty space. Some coordinate systems work better than others in describing the resulting problem. Generalized coordinates offer a natural way of handling systems with constraint conditions or exhibiting a special symmetry.

3.2   PRINCIPLE OF VIRTUAL WORK

They are two aspects of holonomic constraints: a) Each constraint reduces the dimension of the problem by one degree of freedom (good, so far); b) but the constraint forces are not known a priori (not so good). What is needed is a method for eliminating both the constrained coordinates while simultaneously eliminating their forces. To accomplish this, one can partition the forces conceptually into applied forces $\mathbf{F}^a$ whose functional behaviors are known a priori, and forces of constraint $\mathbf{F}^c$ that are known only through their effects (constraint conditions)

$$\mathbf{F} = \mathbf{F}^a + \mathbf{F}^c. \tag{3.17}$$

A dynamical system can be treated as an instantaneously static system if the dynamic response is treated as an inertial force. This gives rise to the following force equations written in a Cartesian basis

$$\mathbf{F}^a_i + \mathbf{F}^c_i - \dot{\mathbf{p}}_i = 0, \tag{3.18}$$

where the dynamical term $-\dot{\mathbf{p}}_i$ is interpreted as a fictitious "inertial force." If one only considers virtual variations of the system at constant time, the dynamical system can then be treated instantaneously as a static system.

The approach to be employed is to eliminate the constraints in an n-dimensional configuration space, using a Cartesian basis $\{x^i : i = 1, n = 3N\}$, then show that the results can be written in a coordinate-independent manner. Since Newton's laws are valid in a fixed Euclidean configuration space, and the masses of the individual particles are constants, the momenta are defined as

$$p_i = g_{ij} m_j \dot{x}^j, \quad \dot{p}_i = \frac{d}{dt} g_{ij} m_j \dot{x}^j = g_{ij} m_j \ddot{x}^j, \quad (3.19)$$

where the metric tensor $g_{ij} = \delta_{ij}$ is constant for the special case of Cartesian coordinates in an inertial frame. (Here, the mass suffix m_j denotes a particle label, not a tensor index.) Covariant and contravariant vectors are identical in these coordinates. The kinetic energy of the particle system in this frame is given by

$$T = \frac{1}{2} m_j g_{ij} \dot{x}^i \dot{x}^j = \frac{1}{2} p_i \dot{x}^i. \quad (3.20)$$

Next, consider a variation of the path in configuration space subject to the constraints on the allowed motion of the system. Imagine moving all the particles slightly along the instantaneous allowed directions of motion, while simultaneously maintaining the constraint conditions $\delta c^i = \delta \dot{c}^i = 0$

$$\begin{aligned} x^i &\to x^i + \delta x^i, \quad \delta t = 0, \\ q^i &\to q^i + \delta q^i, \quad c^i \to c^i = 0. \end{aligned} \quad (3.21)$$

Note that the variations δx^i are not independent, since they must satisfy a number of constraint conditions of the form

$$\delta c^i(x, t) = \frac{\partial c^i}{\partial x^j} \delta x^j + \frac{\partial c^i}{\partial t} \delta t = \frac{\partial c^i}{\partial x^j} \delta x^j = 0. \quad (3.22)$$

Fortunately, the reduced set of independent generalized coordinates q^i can be independently varied.

This virtual variation is made while holding the time constant. Why? The forces of constraint F_i^c act to satisfy the constraint conditions c^i and are instantaneously orthogonal to the allowed direction of motion. Note that this virtual variation does not describe any real motion of the system since $\delta t = 0$, while time varies for real motions. Infinitesimal variations are implicitly smooth-varying functions; one can assume that all the necessary derivatives exist. A virtual displacement is perpendicular to the constraint forces; therefore, the constraint forces do no virtual work.

$$F_i^c \delta x^i = 0. \quad (3.23)$$

The *principle of virtual work* states that the constraint forces acting on a system do no work under a virtual displacement satisfying holonomic constraint conditions. Therefore, the total virtual work done on a system is given by the virtual work done by the applied forces alone

$$\delta W = F_i \delta x^i = F_i^{\mathrm{a}} \delta x^i = \dot{p}_i \delta x^i, \tag{3.24}$$

which implies

$$(F_i^{\mathrm{a}} - \dot{p}_i) \delta x^i = 0. \tag{3.25}$$

The result is a scalar equation in which all the constraint forces have vanished identically from the formalism. One cannot deduce that the coefficients of the variations individually vanish, $(F_i^{\mathrm{a}} - \dot{p}_i) \neq 0$, however, since the variations δx^i are not linearly independent; they are subject to constraint conditions. Now make the substitution

$$\delta x^i(q, c, t)|_{c=0} = \frac{\partial x^i}{\partial q^j} \delta q^j, \tag{3.26}$$

where the q^j are independent coordinates and use has been made of the fact that the constraint coordinates are not to be varied. Equation (3.25) can be rewritten as

$$\left((F_i^{\mathrm{a}} - \dot{p}_i) \frac{\partial x^i}{\partial q^j} \right) \delta q^j = 0. \tag{3.27}$$

Since the variations with respect to generalized coordinates are both independent and arbitrary, this results in $3N - k$ independent equations of motion

$$(F_i^{\mathrm{a}} - \dot{p}_i) \frac{\partial x^i}{\partial q^j} = 0. \tag{3.28}$$

This achieves the goal of eliminating the constraint conditions. Now, one wants to reformulate this result in a covariant manner. Define the covariant generalized forces as

$$Q_j \equiv F_i^{\mathrm{a}} \frac{\partial x^i}{\partial q^j}, \tag{3.29}$$

then, integration by parts results in

$$\begin{aligned}
Q_j &\equiv F_i^{\mathrm{a}} \frac{\partial x^i}{\partial q^j} = \frac{dp_i}{dt} \frac{\partial x^i}{\partial q^j} = \frac{d}{dt}\left(p_i \frac{\partial x^i}{\partial q^j} \right) - p_i \frac{d}{dt}\left(\frac{\partial x^i}{\partial q^j} \right) \\
&= \frac{d}{dt}\left(p_i \frac{\partial x^i}{\partial q^j} \right) - p_i \left(\frac{\partial \dot{x}^i}{\partial q^j} \right),
\end{aligned} \tag{3.30}$$

where the last step comes from interchanging the order of derivatives as shown next

$$\left(\frac{\partial \dot{x}^i}{\partial q^j}\right) = \frac{\partial}{\partial q^j}\left(\frac{\partial x^i(q,t)}{\partial q^k}\dot{q}^k + \frac{\partial x^i(q,t)}{\partial t}\right) = \frac{\partial^2 x^i(q,t)}{\partial q^j \partial q^k}\dot{q}^k + \frac{\partial^2 x^i(q,t)}{\partial q^j \partial t}, \quad (3.31)$$

$$\frac{d}{dt}\left(\frac{\partial x^i}{\partial q^j}\right) = \frac{\partial^2 x^i(q,t)}{\partial q^k \partial q^j}\dot{q}^k + \frac{\partial^2 x^i(q,t)}{\partial t \partial q^j}. \quad (3.32)$$

Therefore,

$$\left(\frac{\partial \dot{x}^i}{\partial q^j}\right) = \frac{d}{dt}\left(\frac{\partial x^i}{\partial q^j}\right). \quad (3.33)$$

The total time derivatives of the Cartesian coordinates can be replaced by

$$\dot{x}^i = \frac{dx^i(q,t)}{dt} = \frac{\partial x^i(q,t)}{\partial q^j}\dot{q}^j + \frac{\partial x^i(q,t)}{\partial t}, \quad (3.34)$$

justifying the critical substitution

$$\frac{\partial \dot{x}^i}{\partial \dot{q}^j} = \frac{\partial x^i}{\partial q^j}. \quad (3.35)$$

The equations of motion (3.30) can now be rewritten as

$$Q_j = \frac{d}{dt}\left((m\dot{x})_i \frac{\partial \dot{x}^i}{\partial \dot{q}^j}\right) - (m\dot{x})_i\left(\frac{\partial \dot{x}^i}{\partial q^j}\right). \quad (3.36)$$

But, the masses are constant; moreover, contravariant and covariant velocities are identical when expressed in Cartesian coordinates; therefore

$$Q_j = \frac{d}{dt}\frac{\partial}{\partial \dot{q}^j}(T) - \left(\frac{\partial T}{\partial q^j}\right), \quad (3.37)$$

where

$$T(q, \dot{q}, t) = \frac{1}{2}m_{(i)}\delta_{ij}\dot{x}^i\dot{x}^j. \quad (3.38)$$

But the kinetic energy is a scalar invariant. Equation (3.37) are the desired form-independent, covariant equations of motion expressed in independent generalized coordinates written in terms of a scalar kinetic energy functional $T(q, \dot{q}, t)$.

3.3   LAGRANGE'S EQUATIONS OF MOTION

Often, some, if not all, of the forces can be derived from a velocity dependent scalar potential functional U of the form

$$Q_j = \left[\frac{d}{dt} \frac{\partial U(q, \dot{q}, t)}{\partial \dot{q}^i} - \frac{\partial U(q, \dot{q}, t)}{\partial q^i} \right] + Q_j^*, \tag{3.39}$$

where Q_j^* denotes additional time dependent, frictional, or dissipative forces that cannot be written in terms of such a generalized potential. One can then define a scalar Lagrangian $L = T - U$ by combining the kinetic energy and potential terms, giving Lagrange's equations of motion for non-monogenic systems

$$Q_j^* = \frac{d}{dt} \frac{\partial L(q, \dot{q}, t)}{\partial \dot{q}^j} - \frac{\partial L(q, \dot{q}, t)}{\partial q^j}. \tag{3.40}$$

If the extra, irregular, force terms Q_j^* vanish, the equations of motion depend only on a single scalar Lagrangian function. In such cases, the equations of motion are said to be *monogenic* (i.e. they are derived from a single function), and Lagrange's equations takes on the elegant form

$$\frac{d}{dt} \frac{\partial L(q, \dot{q}, t)}{\partial \dot{q}^j} - \frac{\partial L(q, \dot{q}, t)}{\partial q^j} = 0. \tag{3.41}$$

When written in this homogeneous form, the equations are called the Euler-Lagrange equations of motion and are derivable from a minimization principle. These equations are *form invariant* in the sense that they have the same differential form in any coordinate representation. They presuppose, however, that the scalar Lagrangian be first written in some inertial frame and then transcribed by invariance to any other frame by making a point transformation.

When working with generalized coordinates, the canonical form[4]

$$p_j = \partial L(q, \dot{q}, t)/\partial \dot{q}^j \tag{3.42}$$

replaces the Newtonian momentum. It is referred to as the *canonical momentum* associated with a generalized coordinate q^i. Lagrange's equations of motion, when expressed in terms of the canonical momenta, become

$$\dot{p}_j = \frac{\partial L(q, \dot{q}, t)}{\partial q^j} + Q_j^*. \tag{3.43}$$

---

[4]The word canonical is used in the sense of meaning "regular" or "accepted."

3.3.1 Eliminating Holonomic Constraints

The two examples presented below illustrate how holonomic constraints are eliminated.

3.3.1.1 Contact Constraints

Figure 3.2 shows a block sliding down an inclined wedge that is free to move on the flat horizontal surface of a table. Assume that the motion is constrained to the $\hat{\mathbf{x}}$ (horizontal axis) and $\hat{\mathbf{y}}$ (vertical axis) plane. There are normal forces of constraint acting between the block and wedge and between the block and table. The Lagrangian is the difference of the kinetic and potential energy, defined, in an inertial frame, as

$$L = \frac{1}{2}m_W \mathbf{v}_W^2 + \frac{1}{2}m_B \mathbf{v}_B^2 - m_B g y_B. \tag{3.44}$$

The Lagrangian can now be transcribed by invariance into a non-inertial frame

$$\begin{aligned} L &= \frac{1}{2}m_W \mathbf{v}_W^2 + \frac{1}{2}m_B(\mathbf{v}_W + \mathbf{v}_B')^2 - m_B g y_B \\ &= \frac{1}{2}m_W \dot{x}_W^2 + \frac{1}{2}m_B((\dot{x}_W + \dot{x}_B')^2 + (\dot{y}_W + \dot{y}_B')^2) - m_B g y_B \end{aligned} \tag{3.45}$$

Note that the velocity of the block $\mathbf{v}_B$ is measured with respect to a fixed inertial frame, while the velocity $\mathbf{v}_B'$ is measured relative to the accelerating frame of the wedge. A common mistake is to use $\frac{1}{2}m_B \mathbf{v}_B'^2$, the kinetic energy in the accelerating frame, for the total kinetic energy of the block in the defining

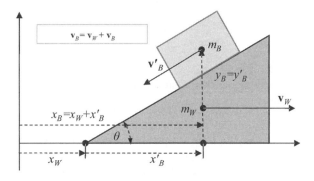

FIGURE 3.2 A block sliding down an inclined wedge that is free to slide in the horizontal plane.

Equation (3.44). The Lagrangian must be valid in a special inertial frame of reference. It can then be transcribed into any other frame of reference.

The constraint conditions for this example are

- the wedge is constrained to move in the horizontal direction $c_W = z_W = 0$, and
- the block slides along the surface of the plane $c_B = y'_B - x'_B \tan\theta = 0$.

Eliminating the constraint conditions gives a Lagrangian system with two independent degrees of freedom, which can be taken to be (x_B, x'_B). The Lagrangian, with respect to independent degrees of freedom, can then be written as

$$L(x_B, \dot{x}_B, \dot{x}_W) = \frac{1}{2}m_W \dot{x}_W^2 + \frac{1}{2}m_B((\dot{x}_W + \dot{x}'_B)^2 + \dot{x}_B'^2 \tan^2\theta) - m_B g x'_B \tan\theta.$$

(3.46)

The resulting Lagrangian equations of motion for this system are

$$\frac{dp_w}{dt} = (m_w + m_B)\ddot{x}_W + m_B \ddot{x}'_B = 0,$$

$$\frac{dp'_B}{dt} = m_B((\ddot{x}_W + \ddot{x}'_B[1 + \tan^2\theta])) = -m_B g \tan\theta,$$

(3.47)

The first equation is equivalent to the conservation of total momentum in the x-direction, since the external forces act only in the y-direction. The solution to this problem is straightforward, since all the accelerations are constant. Solving for the accelerations gives

$$\ddot{x}'_B = \frac{-g\tan\theta}{[1 + \tan^2\theta] - \frac{m_B}{m_B + m_W}},$$

(3.48)

$$\ddot{y}'_B = \ddot{x}'_B \tan\theta, \quad \ddot{x}_W = -\frac{m_B}{m_B + m_W}\ddot{x}'_B.$$

3.3.1.2 Time-Dependent Constraints

Now consider a particle moving, without friction, on a wire rotating in a horizontal plane with constant angular velocity $\dot{\phi} = \omega$, as illustrated in Figure 3.3. Integration yields a holonomic time-dependent constraint of the form

$$c_\theta = \phi - \phi_0 - \omega t = 0.$$

(3.49)

There is no potential term, so the Lagrangian is due solely to the kinetic energy contribution in the plane of motion. Using polar coordinates and

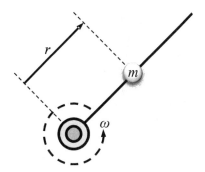

FIGURE 3.3   A bead sliding on a uniformly rotating wire in the horizontal plane.

applying the constraint gives

$$L(r, \dot{r}, t) = T = \frac{m}{2}(\dot{r}^2 + r^2\dot{\phi}^2) = \frac{m}{2}(\dot{r}^2 + r^2\omega^2). \tag{3.50}$$

There is only one independent degree of freedom resulting in the equation of motion:

$$\left(\frac{d}{dt}\frac{\partial T}{\partial \dot{r}} - \frac{\partial T}{\partial r}\right) = m\ddot{r} - m\omega^2 r = 0, \tag{3.51}$$

The constraint condition reduces the problem to one freedom by introducing an effective potential energy term

$$V_{\mathit{eff}}(r) = -mr^2\omega^2/2 \tag{3.52}$$

The general solution can be given in terms of exponential functions

$$\begin{aligned} r(t) &= Ae^{+\omega t} + Be^{-\omega t} \geq 0, \\ \dot{r}(t) &= A\omega e^{+\omega t} - B\omega e^{-\omega t}, \end{aligned} \tag{3.53}$$

where the coefficients (A, B) are determined by the initial conditions

$$r_0 = A + B \geq 0, \quad \dot{r}_0 = \omega(A - B). \tag{3.54}$$

If the particle is initially moving toward the pivot, the geometry requires a minimum value of the radial coordinate given by the turning point $\dot{r} = 0$ for some $r_{\min} > 0$. Otherwise, there is a collision with the pivot when $r = 0$. The velocity for an outward-moving particle increases exponentially until it flies off the end of the wire with some constant final momentum.

This problem has several interesting features. The Lagrangian problem is initially set up in terms of the kinetic energy of the particle in an inertial frame, then a time-dependent coordinate transformation is applied to reduce the problem to an effective one-dimensional problem in a rotating frame. The mechanical energy of the particle $E = T$ is purely kinetic in character and is not conserved. This is a case where the force of constraint does no virtual work but does actual work on the system. Energy is pumped into the bead by the driving mechanism of the rotating pivot. In the accelerating frame, there is first constant of integration leading a conversed "Hamiltonian energy"

$$H = T_r + V_{eff} = \frac{1}{2}m(\dot{r}^2 - r^2\omega^2), \tag{3.55}$$

but this energy cannot be identified with the mechanical energy of the system $H \neq E$.

3.3.2 Dissipative Forces

Dissipative forces are terms that do not easily fit into a Lagrangian framework. They originate from the stochastic effects of ignored internal degrees of freedom (heating due to friction, drag in a medium, etc.) and depend on the relative motion of the system. (Time-dependent driving forces may also be treated as dissipative terms, since they do not conserve mechanical energy, although they can often be as easily incorporated into a time-dependent potential.)

Dissipative forces can often be derived from a velocity-dependent scalar dissipative function D given by

$$Q_j^* = Q_j^D = -\frac{\partial D(q, \dot{q}, t)}{\partial \dot{q}^j}. \tag{3.56}$$

Such forces are empirical in nature: resulting from collective effects of internal degrees of freedom. The equations of motion are no longer monogenic, taking the covariant form

$$\frac{d}{dt}\frac{\partial L(q, \dot{q}, t)}{\partial \dot{q}^j} - \frac{\partial L(q, \dot{q}, t)}{\partial q^j} = -\frac{\partial D(q, \dot{q}, t)}{\partial \dot{q}^j}. \tag{3.57}$$

An example is the damped driven oscillator of Chapter 1, where the standard SHO Lagrangian is given by

$$L_{SHO}(x, \dot{x}) = \frac{1}{2}m\dot{x}^2 - \frac{1}{2}m\omega^2x^2, \tag{3.58}$$

and the remainder of the forces can be put into the dissipative function

$$D(\dot{x}, t) = \frac{1}{2}m\gamma\dot{x}^2 - F(t)\dot{x}. \tag{3.59}$$

The resulting equations of motion are

$$m(\ddot{x} + \gamma \dot{x} + \omega^2 x) = F(t). \tag{3.60}$$

It is easy to show that the mechanical energy $E = T + V$ of the damped, driven oscillator is not conserved. Its time derivative is equal to the power flow from the nonconservative forces

$$\frac{dE}{dt} = Q^D \dot{x} = -m\gamma \dot{x}^2 + F(t)\dot{x}. \tag{3.61}$$

3.3.3 Transformation of the Lagrangian

The choice of Lagrangian is not unique. Many possible Lagrangians result in the same equations of motion. Physical theories attempt to prescribe a "canonical" or standard form of the Lagrangian. Even so, it is straightforward to show that the addition of any total derivative of a function of the coordinates and the time, $F(q,t)$, to a Lagrangian has no effect on the equations of motion. That is

$$\frac{d}{dt}\left(\frac{\partial}{\partial \dot{q}^j}\left(\frac{dF(q,t)}{dt}\right)\right) \equiv \frac{\partial}{\partial q^j}\left(\frac{dF(q,t)}{dt}\right). \tag{3.62}$$

The proof starts with substituting for the total derivative by using the chain rule

$$\frac{dF(q,t)}{dt} = \frac{\partial F}{\partial q^j}\dot{q}^j + \frac{\partial F}{\partial t}. \tag{3.63}$$

Differentiating both sides of Equation (3.62) gives

$$\begin{aligned}
\frac{\partial}{\partial q^i}\frac{dF(q,t)}{dt} &= \dot{q}^j \frac{\partial^2 F}{\partial q^i \partial q^j} + \frac{\partial^2 F}{\partial q^i \partial t}, \\
\frac{d}{dt}\frac{\partial}{\partial \dot{q}^i}\frac{dF(q,t)}{dt} &= \frac{d}{dt}\frac{\partial F}{\partial q^i} = \dot{q}^j \frac{\partial^2 F}{\partial q^j \partial q^i} + \frac{\partial^2 F}{\partial t \partial q^i}.
\end{aligned} \tag{3.64}$$

These two terms are identical and cancel in the original equation since the order of partial derivatives can be exchanged. Therefore, the transformation to a new Lagrangian $L \to L'$ of the form

$$L' = L + \frac{dF(q,t)}{dt}, \tag{3.65}$$

generates the same equations of motion as the original Lagrangian.

3.3.4 Canonical Momentum

One can define a covariant canonical momentum p_j associated with a contravariant-generalized coordinate q^j by

$$p_j \equiv \frac{\partial L}{\partial \dot{q}^j}. \tag{3.66}$$

This results in the following fundamental relationship for monogenic systems

$$\frac{dp_j}{dt} = \frac{\partial L}{\partial q^j}. \tag{3.67}$$

The *conservation of canonical momentum* states that if a monogenic Lagrangian does not explicitly depend on a generalized coordinate (it is said to be cyclic in that coordinate) then the corresponding conjugate momentum is a constant of the motion.

A note on units: Generalized coordinates do not always have the units of displacement, and generalized momenta do not always have the units of momentum. However, integrals of the form $\int \dot{p}_i \delta q^i$ always have units of work. This implies the products of the form $p_i \dot{q}^i$ also have units of work (or angular momentum). For example, the kinetic energy of a free particle in an inertial frame can be written as $T = p_i \dot{q}^i / 2$.

3.3.5 Conservative Systems

For conservative systems, the applied forces can be derived from a scalar potential function of the coordinates only. Let

$$F_i^{(a)} = -\nabla_i V(x) = -\frac{\partial V}{\partial x^i}. \tag{3.68}$$

In this case, we have

$$Q_j = \sum_i Q_j = \sum_i -\nabla_i V \frac{\partial x^i}{\partial q^j} = -\frac{\partial V}{\partial q^j}. \tag{3.69}$$

This results in a particularly simple form of Lagrange's equations

$$\left\{ \frac{d}{dt} \frac{\partial}{\partial \dot{q}^j} - \frac{\partial}{\partial q^j} \right\} (T(q, \dot{q}, t) - V(q, t)) = 0. \tag{3.70}$$

Consider, for example, the spherical pendulum shown in Figure 3.4. The length $r = l$ of the pendulum is constant. In spherical coordinates, centered at its pivot, the polar angle θ is the only noncyclic coordinate.

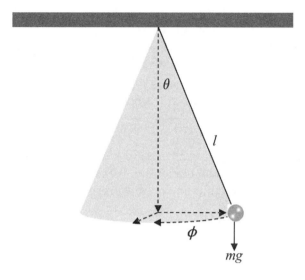

FIGURE 3.4 **The spherical pendulum.**

The kinetic energy is

$$T = \frac{1}{2}m\dot{\mathbf{r}}^2 = \frac{1}{2}m(\dot{r}^2 + r^2\dot{\theta}^2 + r^2\sin^2\theta\dot{\phi}^2) = \frac{1}{2}ml^2(\dot{\theta}^2 + \sin^2\theta\dot{\phi}^2). \quad (3.71)$$

The potential energy is

$$V = -mgz = mgl(1 - \cos\theta). \quad (3.72)$$

The Lagrangian for this system is

$$L(\theta, \dot{\theta}, \dot{\phi}) = \frac{1}{2}ml^2(\dot{\theta}^2 + \sin^2\theta\dot{\phi}^2) - mgl(1 - \cos\theta). \quad (3.73)$$

where $I = ml^2$ is the moment of inertia of the pendulum. One first integral of the equation of motion is equivalent to a statement that the mechanical energy is conserved. Another first integral involves the azimuthal component of the angular momentum. This gives two first integrals of the motion

$$E = \frac{1}{2}ml^2\left(\dot{\theta}^2 + \frac{p_\phi^2}{m^2l^4\sin^2\theta}\right) + mgl(1 - \cos\theta), \quad (3.74)$$

$$p_\phi = (ml^2\dot{\phi})\sin^2\theta. \quad (3.75)$$

The complete solution for the time-dependence of the motion can be found by reduction to quadrature.

3.3.6 Reduction to Quadrature

If the dissipative terms vanish, the equations of motion are said to be monogenic, i.e., they are derivable from a single scalar Lagrangian function $L(q, \dot{q}, t)$.

$$\frac{d}{dt}\frac{\partial L}{\partial \dot{q}^j} - \frac{\partial L}{\partial q^j} = 0 \to \dot{p}_j = \frac{\partial L}{\partial q^j}. \tag{3.76}$$

In this case, if the Lagrangian is cyclic in a particular coordinate, then the corresponding canonical momentum is conserved, and one finds an immediate first integral of the motion in the corresponding canonical momentum

$$\frac{\partial L}{\partial q^j} = 0 \to \dot{p}_j = 0. \tag{3.77}$$

The study of monogenic systems is particularly important for dynamics, since all the known fundamental forces of nature can be derived in terms of monogenic functions. Dissipative forces originate from approximating complicated processes, involving many internal degrees of freedom, by simple empirical expressions, such as engineering formulae for frictional and drag forces.

The analytic solution of the equations of motion is a process of reducing the equations of motion to quadrature, or formal integration. As an example of reducing Lagrange's equation to quadrature, consider a point particle moving in a plane under the influence of a central force potential. The equations of motion are monogenic and can be derived from a Lagrangian of the form

$$L = T - V = \frac{1}{2}mv^2 - V(r) = \frac{1}{2}m(\dot{r}^2 + r^2\dot{\phi}^2) - V(r). \tag{3.78}$$

Lagrange's equation of motion can be written as

$$\frac{dp_\phi}{dt} = \frac{d}{dt}(mr^2\dot{\phi}) = 0, \tag{3.79}$$

$$\frac{dp_r}{dt} = m\ddot{r} = mr\dot{\phi}^2 - \frac{\partial}{\partial r}V(r). \tag{3.80}$$

The first equation, states that the angular momentum of the particle is conserved, giving a first integral of the motion $p_\phi = \ell$. Substituting into the second equation gives

$$m\ddot{r} = \frac{\ell^2}{mr^3} - \frac{d}{dr}V(r). \tag{3.81}$$

This equation can be reduced to quadrature by multiplying both sides by $\dot{r} = dr/dt$:

$$m\dot{r} \cdot \frac{d\dot{r}}{dt} = \left(\frac{\ell^2}{mr^3} - \frac{d}{dr}V(r) \right) \frac{dr}{dt},$$

$$\int m\dot{r}\,d\dot{r} = \int \left(\frac{\ell^2}{mr^3} - \frac{d}{dr}V(r) \right) dr + C. \tag{3.82}$$

Carrying out the integral and rearranging terms gives

$$\frac{1}{2}m\dot{r}^2 + \frac{\ell^2}{2mr^2} + V(r) = T + V = E. \tag{3.83}$$

which can be recognized as conservation of mechanical energy for this system. Carrying out the process further, the resulting first integrals are themselves reducible to quadrature. This will be demonstrated in Chapter 5: "Central force motion."

A coupled set of n second order differential equations is said to be completely integrable if $2n$ constants of the motion can be found. Complete integrability of a dynamical system of equations cannot be taken for granted. The study of classical chaos began when Poincaré demonstrated that the three-body Earth-Moon-Sun system was not completely integrable, and that therefore the long-term stability of the orbits could not be guaranteed.

3.3.7 Hamiltonian Function

An important physical observable is given by the Hamiltonian function defined by

$$H(q,\dot{q},t) = \dot{q}^i \frac{\partial L(q,\dot{q},t)}{\partial \dot{q}^i} - L(q,\dot{q},t) = \dot{q}^i p_i(q,\dot{q},t) - L(q,\dot{q},t). \tag{3.84}$$

Calculating the time derivative of the Hamiltonian gives

$$\frac{dH}{dt} = \ddot{q}^i p_i + \dot{q}^i \dot{p}_i - \frac{\partial L}{\partial \dot{q}^i}\ddot{q}^i - \frac{\partial L}{\partial q^i}\dot{q}^i - \frac{\partial L}{\partial t},$$

$$= \ddot{q}^i \left(p_i - \frac{\partial L}{\partial \dot{q}^i} \right) + \dot{q}^i \left(\dot{p}_i - \frac{\partial L}{\partial q^i} \right) - \frac{\partial L}{\partial t}$$

$$= \dot{q}^i Q_j^{\mathrm{D}} - \frac{\partial L}{\partial t}. \tag{3.85}$$

If the system is monogenic $Q_j^{\mathrm{D}} \equiv 0$ and the Lagrangian is cyclic in the time (i.e. the Lagrangian does not explicitly depend on the time $\partial L/\partial t \equiv 0$), then the Hamiltonian of the system is a constant of the motion $dH/dt = 0$.

3.3.8 Energy Conservation

Often (but not always) the Hamiltonian can be identified with the mechanical energy of the system. This will be often be true, for example, if the Lagrangian of the system can be derived from a standard free particle Lagrangian that is quadratic in the velocities when written with respect to a Cartesian basis. Let

$$L(x, \dot{x}, t) = T - A_i \dot{x}^i - V$$

$$= \sum_{ij} \frac{1}{2} M_{ij} \dot{x}^i \cdot \dot{x}^j - \sum_i \dot{x}^i A_i(x, t) - V(x, t), \qquad (3.86)$$

where

$$M_{ij} = m_{(i)} g_{ij}$$

is the diagonal *mass matrix* of the Cartesian basis. The matrix M is non-singular, since the inertia masses are positive definite $m_i > 0$. $A_i(x, t)$ are components of some vector potential and V is some scalar potential.

Now consider some time-independent mapping to some set of generalized coordinates. The Cartesian velocities are linear homogeneous functions of the generalized velocities. The kinetic energy is a quadratic homogeneous function of the generalized velocities that transforms by invariance. The Lagrangian can be written as a quadratic function in the generalized velocities. Let $x^i = x^i(q)$, then the Cartesian velocities are linear homogenous functions of the generalized velocities

$$\dot{x}^i = \frac{\partial x^i}{\partial q^j} \dot{q}^j = \dot{x}^i(q, \dot{q}).$$

The Lagrangian transforms by invariance, becoming

$$L(q, \dot{q}) = T(q, \dot{q}) - \dot{q}^j A'_j(q) - V(q),$$

$$T(x, \dot{x}) = T(q, \dot{q}) = \frac{1}{2} M'_{ij}(q) \dot{q}^i \dot{q}^j, \qquad (3.87)$$

$$V(q) = V(x),$$

where the vector potential and tensor mass matrix transform covariantly

$$A'_j(q) = \frac{\partial x^i}{\partial q^j} A_i(x),$$

$$M'_{ij}(q) = \frac{\partial x^m(q)}{\partial q^i} \frac{\partial x^n(q)}{\partial q^j} M_{mn}, \qquad (3.88)$$

then, by explicit construction, the Hamiltonian can be identified as the mechanical energy in the new coordinate frame

$$H = \dot{q}^i \frac{\partial L}{\partial \dot{q}^i} - L = M'_{ij}\dot{q}^i\dot{q}^j - \dot{q}^j A'_j - \left(\frac{1}{2}M'_{ij}\dot{q}^i\dot{q}^j - \dot{q}^j A'_j - V \right),$$

$$H(q,\dot{q},t) = \frac{1}{2}M'_{ij}(q)\dot{q}^i\dot{q}^j + V(q,t) = T + V = E.$$

(3.89)

The vector potential term does not contribute to the energy. The only remaining explicit time dependence is in the potential energy. If that is time independent, the Hamiltonian function can be identified with the conserved total mechanical energy of the system.

Note that the identification of the Hamiltonian with the energy requires two assumptions: i) the original Lagrangian has a standard quadratic form, and ii) the coordinate transformation is not time dependent. If the transformation is time dependent, the identification of the Hamiltonian function with the energy cannot be made. If there are dissipative forces not included in the Lagrangian, then one needs to include the work done by these forces. If the dissipative forces are derivable from a scalar dissipation function D, the time rate of change of the mechanical energy is given by

$$\frac{dE}{dt} = -\frac{\partial L}{\partial t} - \frac{\partial D}{\partial \dot{q}^i}\dot{q}^i.$$

(3.90)

3.4 ELECTROMAGNETIC INTERACTIONS

The general form of the Lagrangian allows for velocity-dependent potentials

$$L(q,\dot{q},t) = T(q,\dot{q},t) - U(q,\dot{q},t).$$

(3.91)

For example, the potential for a particle of charge q_e moving in an electromagnetic field is given by

$$U(\mathbf{r},\mathbf{v},t) = q_e\Phi(\mathbf{r},t) - q_e\mathbf{v}\cdot\mathbf{A}(\mathbf{r},t),$$

(3.92)

$$L = \frac{1}{2}m\mathbf{v}^2 - U(\mathbf{r},\mathbf{v},t),$$

(3.93)

where q is the charge of the particle, Φ is the electromagnetic scalar potential, and $\mathbf{A}$ is the electromagnetic vector potential. If one restricts oneself to time-independent coordinate transformations, the canonical momentum can be written as a vector

$$\mathbf{p} = m\mathbf{v} + q_e\mathbf{A}.$$

(3.94)

This, of course, is not the same as Newton's mechanical momentum.

Solving Lagrange's equations gives rise to the *Lorentz force equation*

$$\frac{d(m\dot{\mathbf{r}})}{dt} = q_e(\mathbf{E} + \mathbf{v} \times \mathbf{B}). \tag{3.95}$$

The proof, using Cartesian coordinates, is straightforward, and it is left for an exercise.[5] Here, the electric field $\mathbf{E}$ and magnetic field $\mathbf{B}$ are defined as

$$\mathbf{E} = -\nabla\Phi - \frac{\partial \mathbf{A}}{\partial t}, \tag{3.96}$$

$$\mathbf{B} = \nabla \times \mathbf{A}, \tag{3.97}$$

The equations of motion are invariant under the gauge transformation

$$\mathbf{A} \rightarrow \mathbf{A} + \nabla\psi(\mathbf{r}, t), \quad \Phi \rightarrow \Phi - \partial\psi(\mathbf{r}, t)/\partial t, \tag{3.98}$$

$$U \rightarrow U - q_e\frac{d\psi(\mathbf{r}, t)}{dt}. \tag{3.99}$$

The mechanical energy of the system given by

$$E = H = T + V = T + q_e\Phi = \frac{1}{2m}(\mathbf{p} - q_e\mathbf{A})^2 + q_e\Phi. \tag{3.100}$$

Particle velocities and kinetic energy are gauge-independent concepts. Therefore, the following combinations of observables are independent of the choice of electromagnetic gauge

$$m\mathbf{v} = \mathbf{p} - q_e\mathbf{A}, \quad T = H - q_e\Phi. \tag{3.101}$$

3.5   HAMILTON'S EQUATIONS OF MOTION

If the defining equations for the canonical momenta can be inverted to solve for the generalized velocities in terms of the generalized momentum, then Lagrange's equations can be put a form suitable for numerical integration using Runge-Kutta methods. Formally, one looks for a coupled set of first-order differential equations in momentum phase space

$$\dot{p}_i = \frac{\partial L(q, \dot{q}, t)}{\partial q^j} + Q_j^* = f_i(q, p, t), \quad \dot{q}^i = v^i(q, p, t). \tag{3.102}$$

The systematic way to accomplish this is obtained by applying a Legendre transformation to convert from generalized velocities to generalized

---

[5]This equation remains valid at relativistic speeds if the rest mass m is replaced by the relativistic mass $m_r = (1 - v^2/c^2)^{-1/2}m$.

momenta as independent degrees of freedom. This results in Hamilton's equations of motion.[6]

The generalized momenta are defined as the partial derivative

$$p_i = \frac{\partial L}{\partial \dot{q}^i} = p_i(q, \dot{q}, t).$$ (3.103)

Assume that these equations are non-singular and invertible for the generalized velocities

$$\dot{q}^i = \dot{q}^i(q, p, t).$$ (3.104)

Taking the total differential of the Hamiltonian function, given by Equation (3.84), gives

$$
\begin{aligned}
dH &= p_i d\dot{q}^i + \dot{q}^i dp_i - \frac{\partial L}{\partial q^i} dq^i - \frac{\partial L}{\partial \dot{q}^i} d\dot{q}^i - \frac{\partial L}{\partial t} dt \\
&= \dot{q}^i dp_i - \frac{\partial L}{\partial q^i} dq^i - \frac{\partial L}{\partial t} dt, \\
&= \dot{q}^i dp_i - (\dot{p}_i - Q_i^*) dq^i - \frac{\partial L}{\partial t} dt,
\end{aligned}
$$ (3.105)

where the last step involves using Lagrange's equations of motion. But, alternatively, by applying the chain rule to the Hamiltonian as a function of coordinates and momenta gives

$$dH(q, p, t) = \frac{\partial H}{\partial q^i} dq^i + \frac{\partial H}{\partial p_i} dp_i + \frac{\partial H}{\partial t} dt$$ (3.106)

Comparing terms gives in Equations (3.106) and (3.105) gives,

$$\dot{p}_i = -\frac{\partial H(q, p, t)}{\partial q^i} + Q_i^*,$$ (3.107)

$$\dot{q}^i = \frac{\partial H(q, p, t)}{\partial p_i},$$ (3.108)

$$\frac{\partial H(q, p, t)}{\partial t} = -\frac{\partial L(q, \dot{q}, t)}{\partial t}.$$ (3.109)

The first two equations are Hamilton's equations of motion, including possible non-standard dissipative force terms. They are the desired first order differential equations of the motion in a form suitable for numerical

---

[6]William Rowan Hamilton (1805–1865) made numerous important contributions to the fields of optics, dynamics, and algebra. He is best known for his discovery of quaternion algebra.

integration. The last equation relates the explicit time dependences of the Hamiltonian function in momentum phase space coordinates to the Lagrangian function in velocity phase space coordinates. Note that the Hamiltonian is cyclic in the time if the Lagrangian is cyclic in the time.

For monogenic systems, the dissipative terms Q^* are not present, and one gets the more familiar form of Hamilton's canonical equations of motion:

$$\dot{p}_i = -\frac{\partial H(q,p,t)}{\partial q^i}, \quad \dot{q}^i = \frac{\partial H(q,p,t)}{\partial p_i}, \tag{3.110}$$

where the time dependence of the Hamiltonian is given by

$$\frac{dH}{dt} = \frac{\partial H}{\partial t} = -\frac{\partial L}{\partial t}. \tag{3.111}$$

It must be emphasized Hamilton's equations are entirely equivalent to Lagrange's equations of motion. The two approaches can be considered equally fundamental starting points for describing particle dynamics.

The Hamiltonian method is better suited for numerical calculations and for phase space visualizations. However, unless there are cyclic coordinates, the direct use of Hamilton's equations does not significantly reduce the complexity of the problem from an analytical point of view. The true power of the Hamiltonian method only becomes apparent when one begins to study Canonical Transformation Theory starting in Chapter 9.

3.5.1 Prescription for Generating Hamilton's Equations

Hamilton's equations of motion can be generated from a Lagrangian framework by applying the following prescription:

- Given a Lagrangian, calculate the canonical momenta using

$$p_i(q,\dot{q},t) = \partial L(q,\dot{q},t)/\partial \dot{q}^i. \tag{3.112}$$

- Invert $p_i = p_i(q,\dot{q},t)$ to obtain the generalized velocities as functions of the momenta and coordinates

$$\dot{q}^i = \dot{q}^i(q,p,t). \tag{3.113}$$

- Use $\dot{q}^i = \dot{q}^i(q,p,t)$ to eliminate the velocities from the Hamiltonian

One can now derive Hamilton's equations of motion.

$$\dot{p}_i = -\frac{\partial H(q,p,t)}{\partial q^i}, \quad \dot{q}^i = \frac{\partial H(q,p,t)}{\partial p_i}. \tag{3.114}$$

Not all Lagrangian systems have a Hamiltonian form. (The converse is true as well. Not all Hamiltonian systems have a Lagrangian formulation.) In particular, step two of the prescription fails unless the *Hessian*, defined as

$$M_{ij} = \partial^2 L / \partial \dot{q}^i \partial \dot{q}^j, \tag{3.115}$$

is nondegenerate $|M_{ij}| \neq 0$. The Hessian is sometimes referred to as the *mass matrix*, since for free particle systems, using Cartesian coordinates, it is diagonal in the particle masses

$$M_{ij} = \frac{\partial}{\partial \dot{x}^i} \frac{\partial}{\partial \dot{x}^j} T = \frac{\partial}{\partial \dot{x}^i} \frac{\partial}{\partial \dot{x}^j} \sum_i \frac{1}{2} m_{(i)} \delta_{ij} \dot{x}^i \dot{x}^j = m_{(i)} \delta_{ij}. \tag{3.116}$$

3.5.2 Electromagnetic Hamiltonian

For standard single particle interactions, Hamilton's equations can often be directly derived without the intermediate step of first generating a Lagrangian. For example, the single particle Hamiltonian for a particle in an electromagnetic field can be identified with its mechanical energy in an inertial frame

$$\begin{aligned} H = T + V &= \frac{1}{2m} (\mathbf{p} - q_e \mathbf{A})^2 + q_e \Phi \\ &= \frac{g^{ij}}{2m} (p_i - q_e A_i)(p_j - q_e A_j) + q_e \Phi, \end{aligned} \tag{3.117}$$

If one knows how to write the metric tensor for a coordinate system, then the Hamiltonian can be constructed in that coordinate system. For example in spherical coordinates, with generalized coordinates $(q^1, q^2, q^3) = (r, \theta, \phi)$, the metric tensor is

$$g_{ij} = \begin{pmatrix} 1 & 0 & 0 \\ 0 & r^2 & 0 \\ 0 & 0 & r^2 \sin^2 \theta \end{pmatrix}, \quad g^{ij} = \begin{pmatrix} 1 & 0 & 0 \\ 0 & 1/r^2 & 0 \\ 0 & 0 & 1/r^2 \sin^2 \theta \end{pmatrix},$$

resulting in the Hamiltonian

$$H = \frac{1}{2m} \left((p_r - q_e A_r)^2 + \frac{1}{r^2} (p_\theta - q_e A_\theta)^2 + \frac{1}{r^2 \sin^2 \theta} (p_\phi - q_e A_\phi)^2 \right) + q_e \Phi.$$

Hamilton's equations of motion can now be directly derived.

3.5.3 Momentum Phase Space

Each point of phase space corresponds to a definite mechanical state of the system. Since the motion in phase space is a single-valued function of the time, trajectories in phase space never cross. The motion or flow of the representative point in phase space describes a curve called the phase trajectory. For monogenic and autonomic systems one can define a time-independent Hamiltonian *flow*.

$$\dot{p}_i = -\nabla_{q^i} H(q,p) = -\frac{\partial H(q,p)}{\partial q^i},$$
$$\dot{q}^i = \nabla_p^i H(q,p) = \frac{\partial H(q,p)}{\partial p_i}.$$
(3.118)

The preceding equations treat coordinates and momentum in a nearly symmetric way. This flow function can be inserted into a Runge-Kutta integration routine to numerically integrate the problem. The skew-symmetry, due to the relative sign change in the previous equations, gives rise to a special symplectic[7] character to the coupling. This results in the conservation of the canonical phase space and other interesting features that will be studied further in Chapter 9. The momentum is a covariant variable defined on the cotangent space T^* of functional derivatives and the Hamiltonian flow is defined in the product space T^*Q.

In the study of the dynamical systems, the expansion of the motion of the system about *stationary points* is important, as is an analysis of the boundaries between different types of motion. The points for which $\dot{p}_i$ and $\dot{q}^i$ simultaneously vanish represent the *stationary points* of a dynamical system. A system placed at one of these points will remain there. If the system is restorative with respect to small variations of all coordinates and momenta, the stationary point is both stable and static. If there are cyclic coordinates, the system can still be considered as a stationary point with respect to its noncyclic coordinates. However, the system is free to move uniformly with respect to its cyclic coordinates.

For example, consider a particle constrained to move on a circle of radius $r = r_0$. The free particle Hamiltonian is $H = p_\phi^2/2I$ where $I = mr_0^2$. Angular momentum and energy are conserved. The phase space for this problem is the cylinder formed by attaching the momentum p_ϕ to the circle $r = r_0$. The motion in phase space are circles of constant angular momentum $p_\phi = \ell$. The constant energy contours are shown in Figure 3.5.

---

[7]Greek for skewed or twisted.

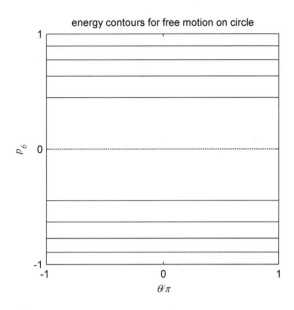

FIGURE 3.5 Constant energy contours for the free motion of a particle moving on a circle. The phase space is the surface of a cylinder with the angular momentum forming the $\hat{z}$-axis. The dashed line is the line of neutral equilibrium for zero angular momentum.

3.5.4 Attractors

If the system of equations is derivable from a monogenic Hamiltonian, phase space is conserved. However, when dissipative forces are included, the phase space decays to an invariant subspace called the *attractor*. For example, the linear damped oscillator has a dissipative force $Q_D = -m\Gamma\dot{x} = -\gamma p$ giving rise to the equations of motion

$$\dot{p} = -kx - \gamma p, \quad \dot{x} = p/m. \tag{3.119}$$

This system decays to a stable equilibrium point at $x = p = 0$ for $\gamma > 0$. Other types of attractors are possible. For example, the damped nonlinear oscillator equation, given by

$$m\ddot{x} + m\gamma\dot{x}\left(\left(\frac{x}{x_0}\right)^2 - 1\right) + w_0^2 x = 0, \tag{3.120}$$

decays to a one-dimensional subspace referred to as a *limit cycle*. The standard phase space form of the equation is

$$\dot{p} = -mw_0^2 x - \gamma p((x/x_0)^2 - 1), \quad \dot{x} = p/m. \tag{3.121}$$

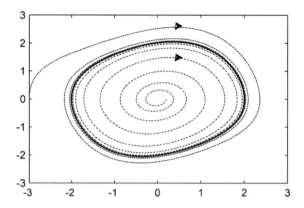

FIGURE 3.6   A phase portrait of a non-linear damped oscillator shows the system evolving towards a limit cycle.

For $\gamma > 0$, the stationary point at $x = p = 0$ is unstable. The nonlinearity in the dissipative term pumps energy into the system for $x^2 < x_0^2$ and takes energy out of the system for $x^2 > x_0^2$. Independent of the starting conditions, the system evolves toward a limiting curve shown in Figure 3.6.

3.5.5 Chaos in Classical Systems

Non-linear dynamical systems with at least three degrees of freedom can exhibit chaotic behavior. A chaotic system is one that exhibits extreme sensitive to initial conditions such that the long-term behavior becomes completely unpredictable. That is, very small changes in the system's initial conditions extrapolate to wildly divergent results when the equations of motion are numerically integrated over a long time scale. Associated with chaotic behavior are *strange attractors* of fractal dimension.

3.5.6 Lorenz Butterfly Equations of Motion

The Lorenz butterfly effect was discovered in 1961 by meteorologist Edward Lorenz [Lorenz 1963]. In its initial manifestation, it described a thermodynamical system, not a mechanical one. However, a mechanical analog can be constructed with the same mathematical behavior. This provides a simple example of a classical system that exhibits known chaotic behavior. Consider a one-particle system with a velocity dependent Lagrangian of the form

$$L = \frac{1}{2m}(m\mathbf{v} - \lambda\mathbf{A})^2 = \frac{1}{2}m v^2 - \lambda\mathbf{v} \cdot \mathbf{A} + \frac{\lambda^2}{2m}\mathbf{A}^2. \tag{3.122}$$

This Lagrangian has a vector potential and scalar potential given by $V = \lambda^2 \mathbf{A}^2/2m$. The equations of motion can be written as

$$\dot{p}_j = \lambda p_i \partial A^i/\partial x^j, \quad \mathbf{p} = (m\mathbf{v} - \lambda \mathbf{A}). \tag{3.123}$$

The generalized momentum vanishes if $m\mathbf{v} = \lambda \mathbf{A}(x, y, z)$. Substituting the following explicit form for the nonlinear potential

$$\frac{\lambda}{m}\mathbf{A}(x, y, z) = \frac{1}{\tau}\left[\begin{array}{c} \sigma(x - y) \\ x(\rho - z/a) - y \\ xy/a - \beta z \end{array}\right] \tag{3.124}$$

gives the Lorenz butterfly equations of motion for $\mathbf{p} = 0$, which leads to chaotic behavior for certain choices of the parameters.

The Lorenz butterfly equations are valid for $\beta > 0$; using dimensionless units ($m = \lambda = a = 1$), the parameters are typically chosen to be $\sigma = 10$, $\beta = 8/3$ while ρ is varied.[8] The system exhibits chaotic behavior for $\rho = 28$. The strange attractor for this case is a fractal, with a Hausdorff dimension of 2.06. That is, the space that it fills is somewhere between a 2-surface and a 3-volume. However, it displays knotted periodic orbits for other values of ρ. For example, for $\rho = 99.96$, it becomes a $T(3, 2)$ torus knot, a special kind of knot which lies on the surface of an unknotted torus given by $(\rho - 2)^2 + z^2 = 1$ in cylindrical coordinates. The $T(p, q)$−torus knot has the parameterization

$$x = \left(2 + \cos\left(\frac{q\phi}{p}\right)\right)\cos\phi,$$

$$y = \left(2 + \cos\left(\frac{q\phi}{p}\right)\right)\sin\phi, \tag{3.125}$$

$$z = \sin\left(\frac{q\phi}{p}\right).$$

Typical time evolution plot for $\rho = 28$ is illustrated in Figure 3.7. For the set of chosen parameters, the graph of a Lorenz attractor shows spiral-like patterns alternating between two stationary points, but the pattern never repeats itself. Two initial states, no matter how close initially will eventually diverge dramatically. The flow does not form limit cycles nor does it ever reach a steady state. The motion switches in an unpredictable manner between orbits that appear to be localized about one of the two attractors. That is, a particle may appear to be in a stable orbit with respect to one of the attractors

---

[8]In fluid mechanics, σ is called the *Prandtl number* and ρ is called the *Rayleigh number*.

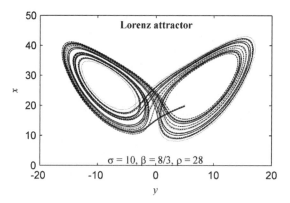

FIGURE 3.7   The Lorenz butterfly equation in the $x - y$ plane for $\sigma = 10$, $\beta = 8/3$, $\rho = 28$.

for a long time. Then, suddenly, it appears to orbit the other attractor. The flip-over-time is very sensitive to the initial conditions, a characteristic of deterministic chaos. Very small differences in the initial conditions lead to widely divergent paths in phase space. For example, Figure 3.8 shows two initial trajectories that differ initially by only one part in 10^{-5}. For some time, the trajectories track each other closely, but eventually they diverge with the range of divergence becoming as large as the available phase space permits.

This long-term unpredictability is one of the criteria for chaotic behavior and is called the *Butterfly Effect*, a term coined by Lorenz. Because of the

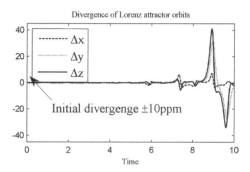

FIGURE 3.8   Lorenz attractor for $\sigma = 10$, $\beta = 8/3$, $\rho = 28$. The nominal starting value is located at [20, 5, -5]. The fractional difference between the starting positions of the two trajectories is only one part in 10^{-5}. After some time, the deviations grow to encompass the available phase space. The result demonstrates the extreme sensitivity to initial conditions seen under chaotic conditions.

chaotic behavior of weather patterns, a solitary butterfly flapping its wings in the jungles of West Africa could give rise to a perturbation that in time may grow to be a category five hurricane beating the shores of some Caribbean island paradise. The onset of chaos will be examined in more detail in the context of the examining the stability of planetary orbits in Chapter 5.

3.6   ROUTH'S PROCEDURE FOR ELIMINATING CYCLIC VARIABLES

One can systematically remove the cyclic coordinates from the Lagrangian by making use of a transformation due to Routh.[9] Routh's method is similar to the method for obtaining Hamilton's equations of motion, except that the Legendre transformation of the Lagrangian is applied only to the cyclic coordinates. First, separate the coordinates into non-cyclic coordinates

$$\{q^i\}; \quad i = 1, N - N_c, \tag{3.126}$$

and cyclic coordinates

$$\{q_c^k\}; \quad k = 1, N_c. \tag{3.127}$$

The Lagrangian of the system can be written as $L(q^i, \dot{q}^i, \dot{q}_c^k, t)$, where $p_k^c = \partial L / \partial \dot{q}_c^k$ are constants of the motion since $(\partial L / \partial q_c^k \equiv 0)$.

One can now define a reduced Lagrangian, or *Routhian*, that depends only on the non-cyclic coordinates and the conserved momenta by making a Legendre transform to eliminate the cyclic coordinates[10]

$$L_R(q^i, \dot{q}^i, p_k^{(c)}, t) = L(q^i, \dot{q}^i, \dot{q}_c^k, t) - p_k^c \dot{q}_c^k. \tag{3.128}$$

The transformation is obtained by taking the total derivative of both sides of the equation and comparing terms. The left-hand side gives

$$dL_R(q^i, \dot{q}^i, p_k^c, t) = \frac{\partial L_R}{\partial q^i} dq^i + \frac{\partial L_R}{\partial \dot{q}^i} d\dot{q}^i + \frac{\partial L_R}{\partial p_k^c} dp_k^c + \frac{\partial L_R}{\partial t} dt, \tag{3.129}$$

---

[9]Edward John Routh (1831–1907) was an exceptional teacher and author of mathematics and physics texts who contributed significantly to the analysis of the stability of stationary motion.

[10]Some textbooks, such as [Goldstein 2002] define the Routhian with an overall sign difference compared to the convention used here.

while the right hand side of the expression yields

$$dL(q^i, \dot{q}^i, \dot{q}_c^k, t) - p_k^c d\dot{q}_c^k - \dot{q}_c^k dp_k^c$$
$$= \frac{\partial L}{\partial q^i} dq^i + \frac{\partial L}{\partial \dot{q}^i} d\dot{q}^i + \frac{\partial L}{\partial \dot{q}_c^k} d\dot{q}_c^k + \frac{\partial L}{\partial t} dt - p_k^c d\dot{q}_c^k - \dot{q}_c^k dp_k^c, \quad (3.130)$$

where substituting Lagrange's equations of motion gives

$$\frac{\partial L_R}{\partial q^i} dq^i + \frac{\partial L_R}{\partial \dot{q}^i} d\dot{q}^i + \frac{\partial L_R}{\partial p_k^c} dp_k^c + \frac{\partial L_R}{\partial t} dt$$
$$= \dot{p}_i dq^i + p_i d\dot{q}^i + p_k^c d\dot{q}_c^k + \frac{\partial L}{\partial t} dt - p_k^c d\dot{q}_c^k - \dot{q}_c^k dp_k^c$$
$$= \dot{p}_i dq^i + p_i d\dot{q}^i + \frac{\partial L}{\partial t} dt - \dot{q}_c^k dp_k^c. \quad (3.131)$$

Matching terms, the equations of motion in the Routhian system are

$$\dot{q}_c^j = -\frac{\partial L_R}{\partial p_j^c}, \quad \dot{p}_j^c = \frac{\partial L_R}{\partial q_c^j} = 0,$$
$$\frac{d}{dt}\left(\frac{\partial L_R}{\partial \dot{q}^i}\right) = \frac{\partial L_R}{\partial q^i}, \quad \frac{\partial L_R}{\partial t} = \frac{\partial L}{\partial t}. \quad (3.132)$$

Note that the Routhian behaves like a Lagrangian for the noncyclic coordinates and like a Hamiltonian for the cyclic coordinates. The noncyclic variables in the Routhian can now be expanded about a point of stationary equilibrium, using the method of small oscillations. In effect, the dimensionality of the equations of motion has been reduced by elimination of the cyclic coordinates. Once one knows the time behavior of the noncyclic coordinates $q^i(t)$, the cyclic coordinates can be integrated to give

$$q_c^j(t) - q_c^j(t_0) = -\int_{t_0}^t \frac{\partial L_R(q^i, \dot{q}^i, p_k^c, t)}{\partial p_j^c} dt. \quad (3.133)$$

3.7   SUMMARY

D'Alembert's principle states that no virtual work is done by holonomic constraints. This allows a system to be expressed in terms of a reduced set of n linearly independent generalized coordinates $q^i(x, t)$ that are point functions of the Newtonian coordinates and time. The principle allows Newton's equations of motion to be expressed in the covariant form

$$\frac{d}{dt}\left(\frac{\partial T}{\partial \dot{q}^j}\right) - \left(\frac{\partial T}{\partial q^j}\right) = Q_j, \tag{3.134}$$

where T is the kinetic energy, and Q_j are generalized forces acting on the generalized coordinates, given by

$$Q_j = F_i \frac{\partial x^i(q,t)}{\partial q^j}. \tag{3.135}$$

If the forces can be derived from a generalized potential of the coordinates and the velocities of the form

$$Q_j = \frac{d}{dt}\frac{\partial U}{\partial \dot{q}^j} - \frac{\partial U}{\partial q^j}, \tag{3.136}$$

the result is the Euler-Lagrange equations of motion in terms of a monogenic Lagrangian function $L = T - U$

$$\frac{d}{dt}\left(\frac{\partial L}{\partial \dot{q}^j}\right) - \left(\frac{\partial L}{\partial q^j}\right) = 0. \tag{3.137}$$

A particularly important generalized potential is the electromagnetic potential acting on charged particles given by

$$U(r,v,t) = q_e \Phi(r,t) - q_e \mathbf{v} \cdot \mathbf{A}(r,t), \tag{3.138}$$

Every generalized coordinate has a canonical momentum associated with it, defined as

$$p_j = \left(\frac{\partial L}{\partial \dot{q}^j}\right). \tag{3.139}$$

The generalized momentum associated with a coordinate is a constant of the motion if the Lagrangian does not explicitly depend on the coordinate. Likewise, if the Lagrangian does not explicitly depend on time, the Hamiltonian function

$$H = \dot{q}^i \left(\frac{\partial L}{\partial \dot{q}^j}\right) - L \tag{3.140}$$

is a constant of the motion.

By making a Legendre transformation, the Euler-Lagrange equations of motion can be written as $2n$ first order differential equations of the motion known as Hamilton's Equations of Motion

$$\dot{p}_i = -\frac{\partial H(q,p,t)}{\partial q^i}, \tag{3.141}$$

$$\dot{q}^i = \frac{\partial H(q,p,t)}{\partial p_i}, \tag{3.142}$$

In this form the equations of motion are particularly well-suited for numerical integration. The following relationships apply to the total time derivative of the Hamiltonian function

$$\frac{dH}{dt} = \frac{\partial H(q,p,t)}{\partial t} = -\frac{\partial L(q,\dot{q},t)}{\partial t}. \tag{3.143}$$

Phase space portraits of the system give a unique insight into the dynamics. Lines of flow in phase space never cross. Of importance to the dynamical analysis of a system are evaluations of the stationary points of motion and of the invariant surfaces that separate distinct dynamical domains.

Cyclic coordinates can be eliminated by constructing the Routhian of the system, defined as

$$L_R(q^i, \dot{q}^i, p_k^c, t) = L(q^i, \dot{q}^i, \dot{q}_c^k, t) - p_k^c \dot{q}_c^k, \tag{3.144}$$

where q_c^k are cyclic coordinates. This results in Routh's equations of motion

$$\dot{p}_j^c = \frac{\partial L_R}{\partial q_c^j} = 0, \quad \dot{q}_c^j = -\frac{\partial L_R}{\partial p_j^c},$$

$$\frac{d}{dt}\left(\frac{\partial L_R}{\partial \dot{q}^i}\right) = \frac{\partial L_R}{\partial q^i}, \quad \frac{\partial L_R}{\partial t} = \frac{\partial L}{\partial t}. \tag{3.145}$$

3.8   EXERCISES

1. A particle is free to move on a surface of a torus given by

$$x(\theta, \phi) = (a + b\cos\phi)\cos\theta,$$
$$y(\theta, \phi) = (a + b\cos\phi)\sin\theta,$$
$$z(\theta, \phi) = b\sin\phi.$$

   a. Find a suitable Lagrangian for this problem.
   b. Find a suitable Hamiltonian for this problem.
   c. Find two first integrals of the motion.

2. Verify that a particle moving under the influence of the generalized electromagnetic potential given by $U(\mathbf{r}, \mathbf{v}, t) = q_e \Phi(\mathbf{r}, t) - q_e \mathbf{v} \cdot \mathbf{A}$ $(\mathbf{r}, t)$ obeys the Lorentz force equation $m\ddot{\mathbf{r}} = q_e(\mathbf{E} + \mathbf{v} \times \mathbf{B})$.

3. A particle of charge q_e in an electromagnetic field sees a generalized potential

$$U = q_e\Phi - q_e\mathbf{v} \cdot \mathbf{A}.$$

Show that under the gauge transformation

$$\mathbf{A}' = \mathbf{A} + \nabla\psi(\mathbf{r}, t), \quad \Phi' = \Phi - \frac{\partial\psi(\mathbf{r}, t)}{\partial t},$$

the Lagrangian changes by a total time derivative of the coordinates and time given by

$$L \to L' = L + q_e\frac{d}{dt}\psi(\mathbf{r}, t).$$

What effect does this transformation have on the Lorentz force equation?

4. A bead slides without friction on a rotating wire as shown in Figure 3.9. The straight wire is oriented at constant polar angle θ, rotating about a vertical axis with constant angular velocity $\Omega = \dot{\phi}$. The apparatus is in a uniform gravitational field aligned with the vertical axis (see Figure 3.9).

    a. Construct the Lagrangian using the radial distance r from the pivot $(z = r\cos\theta)$ as the independent coordinate.

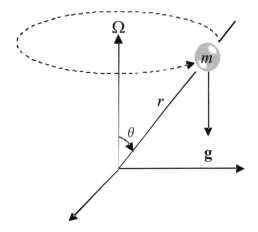

FIGURE 3.9 **A bead sliding on a rotating wire.**

    b. Show that the condition for an equilibrium circular orbit is given by

$$r_0 = g \cos \theta / (\Omega \sin \theta)^2.$$

    c. Discuss the stability of the orbit against small displacements along the wire. [Hint: expand $r(t) = r_0 + \eta(t)$, where η is a small quantity.]

5. Complete the exercise of the spherical pendulum shown in Figure 3.4 by reducing the solution to quadrature.

6. Find Hamilton's equations of motion for the spherical pendulum of Exercise 3.5.

7. Consider a particle of mass m falling vertically in the z-direction under a constant gravitational force $-mg\hat{z}$ that sees a drag force due to air resistance. Assume that this drag force can be derived from a velocity-dependent dissipation function given by

$$D = \gamma T = \frac{1}{2} m \gamma \dot{z}^2.$$

If the particle is initially at rest, solve for the velocity as a function of time and determine its terminal velocity.

8. A particle of mass m move s under the influence of a generalized potential that has a central force component and a spin-angular momentum coupling of the form

$$U(\mathbf{r}, \dot{\mathbf{r}}) = V(r) + \mathbf{s} \cdot \mathbf{L}, \quad \mathbf{L} = \mathbf{r} \times \mathbf{p},$$

where $\mathbf{s} = s_z \mathbf{e}_3$ is a constant vector that points in the z-direction. Obtain Lagrange's equations of motion in spherical polar coordinates. Find the generalized momentum in these coordinates. Evaluate the Hamiltonian function $H(\mathbf{r}, \dot{\mathbf{r}})$ in these coordinates, and convert the Hamiltonian to a function of the coordinates and generalized momenta. Hint: the second term can be converted to a vector potential by using

$$m(\mathbf{s} \cdot \mathbf{r} \times \dot{\mathbf{r}}) = m(\mathbf{s} \times \mathbf{r} \cdot \dot{\mathbf{r}}) = \dot{\mathbf{r}} \cdot \mathbf{A}(\mathbf{r}).$$

9. Two particles of masses m_1 and m_2 are connected by a string of length $l = r + z$ that passes through a hole on a horizontal flat table, as shown in Figure 3.10. Particle 1 with mass m_1 is constrained to move in the $\hat{x} \wedge \hat{y}$ plane of the table. Particle 2 hangs below the table and is constrained to move in the vertical $\hat{z}$ direction. Assume a constant acceleration of gravity $\mathbf{g} = g\hat{z}$ directed vertically downwards. Eliminate the holonomic constraint and find Lagrange's equations

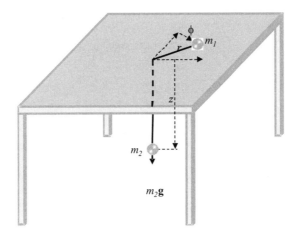

FIGURE 3.10 Two masses attached by a string passing through a hole in a table. The upper mass m_1 is free to move in the horizontal plane of the table. The lower mass is constrained to move in the vertical z-direction.

of motion. Find two first integrals of the motion and interpret their meaning in terms of common physical observables.

10. For the table-top arrangement of Exercise 3.9, let the two masses be equal, with $m_1 = m_2 = 1$ kg, and find the conditions that the energy and angular momentum must satisfy for uniform rotation circular motion at a radius of 1 m. For fixed angular momentum, generate contour plots of constant energy verses (r, p_r). Let $g = 9.8$ m/s^2. Contour plots provide a simple way of generating a phase portrait of the motion at fixed angular momentum.

11. Show that a generalized impulse is given by

$$\left(\frac{\partial L}{\partial \dot{q}^i}\right)_f - \left(\frac{\partial L}{\partial \dot{q}^i}\right)_i = \int_{\Delta t}\left(\frac{\partial L}{\partial q^i} + Q_i\right)dt \approx \int_{\Delta t} Q_i dt,$$

where Q_i is a large, generalized impulsive force that acts for a sufficiently short period of time.

12. Figure 3.11 shows a plane pendulum of length l and mass m_1 suspended from a mass point m_2 that is free to slide without friction in the horizontal $\hat{x}$-direction measured from a fixed origin. The gravitational acceleration $\mathbf{g} = -g\hat{y}$ is directed vertically downward and the pendulum as it swings makes an angle θ with respect to the vertical. Construct the Lagrangian for this system, and derive Lagrangian equations of motion in terms of x and θ degrees of freedom. Find two first integrals for the motion.

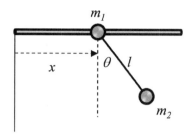

FIGURE 3.11   A plane pendulum with mass m_1 attached to a second mass m_2 free to slide in the $\hat{x}$-direction.

13. For the sliding pendulum of Exercise 3.12, apply the constraint that the conserved x-component of total linear momentum is zero. Then generate a phase portrait of the system by plotting contours of constant energy $E(\theta, p_\theta)$. Let $m_1 = m_2 = 1$ kg, $g = 9.8$ m/s^2 and $l = 1$ m.

14. A particle of mass m is constrained to move on the surface of a cone of revolution $z = r \cos \alpha$ with constant polar angle θ. It is being acted on by a constant gravitational field given by $\mathbf{g} = -g\hat{z}$ along the axis of the cone.
    a. Find the Lagrangian of the system and the Lagrangian equations of motion.
    b. From the symmetries of the Lagrangian, find two constants of the motion.
    c. Find the angular frequency of a circular orbit at $r = z_0 / \cos \alpha$.
    d. What is the frequency of small radial oscillations?
    e. What is the criteria on the angle α for the orbits to be closed?

15. Consider the system defined in Exercise 3.14, but this time find the Hamiltonian of the system and use it to develop Hamilton's equations of motion. From the symmetries of the Hamiltonian, find two constants of the motion.

16. Find the Lagrangian for the double-plane pendulum described in Figure 3.12 in terms of the angles θ_1 and θ_2, which are measured with respect to the vertical. The length of the upper pendulum is l_1 and its mass is m_1. The lower pendulum is hung from mass m_1 and has a length l_2 and attached mass m_2.

17. Two masses moving in the horizontal $\hat{x}$ direction are attached by springs to each other and a wall as indicated in Figure 3.13. Use a suitable Lagrangian to generate the equations of motion. Find the Hamiltonian for this system and express it in terms of the canonical momenta and coordinates.

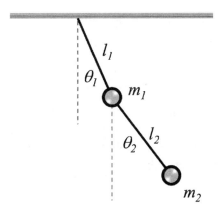

FIGURE 3.12 **A double pendulum system with the lower pendulum swinging from the upper mass point.**

18. Consider a particle of mass m constrained to move on a parabola of revolution $z = kr^2$. The symmetry axis is in the vertical direction with respect to a uniform gravitational field given by $\hat{\mathbf{g}} = -g\hat{\mathbf{z}}$. Use a Lagrangian to derive the equations of motion. Find two first integrals of the motion.

19. Derive Lagrange's equations of motion for a charged particle of charge q_e moving in the $\hat{x} \wedge \hat{y}$ plane subject to a uniform magnetic field $\mathbf{B} = B_0\hat{\mathbf{z}}$ that is perpendicular to the plane. Show that an appropriate vector potential is $\mathbf{A} = rB\hat{\boldsymbol{\varphi}}/2$. Show that the particle undergoes uniform circular motion about its center of gyration with a radius of gyration given by $r_g = |mv_\perp|/q_e B$ and that the motion has a cyclotron angular frequency of $\omega_c = -q_e B/m$. Evaluate the conserved canonical momentum associated with the ϕ degree of freedom and show that $p_\phi = mr^2\omega_c/2$.

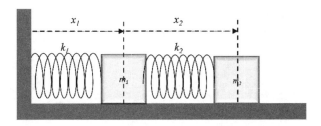

FIGURE 3.13 **Two blocks attached to a wall and to each other by springs.**

20. The Lagrangian for the reduced central force problem in plane polar coordinates is given by

$$L = \frac{1}{2}m(\dot{r}^2 + r^2\dot{\phi}^2) - V(r).$$

Construct the Routhian that eliminates the cyclic azimulthal coordinate and find Routh's equations of motion of this system.

21. The motion the symmetric top with one point fixed in a uniform gravitational field will be analyzed in detail in Chapter 8. Two of the three coordinates are cyclic when the Lagrangian is written in terms of the Euler angles

$$L(\dot{\theta}, \dot{\psi}, \dot{\phi}, \theta) = \frac{1}{2}I_1[\dot{\theta}^2 + \dot{\phi}^2\sin^2\theta] + \frac{1}{2}I_3(\dot{\phi}\cos\theta + \dot{\psi})^2 - mgl\cos\theta.$$

Find the Routhian $L_R(\dot{\theta}, \theta, p_\psi, p_\phi)$ for this system.

Chapter 4

HAMILTON'S PRINCIPLE

I n this chapter, the Euler-Lagrange equations of motion will be derived from a minimization principle. The calculus of variations will be extended to non-mechanical systems, and the method of undetermined multipliers will be used to solve for contact constraints. The role of symmetry in physics will then be explored. Noether's theorem will be used to demonstrate the close correspondence between the continuous symmetries of a system and the existence of conserved physical observables. The magnetic optics of charged particles will be discussed.

4.1 HAMILTON'S ACTION PRINCIPLE

Hamilton's Principle states that the physical path taken by a particle system moving between two fixed points in configuration space is one for which the action integral is stationary under a virtual variation of the path. The action (or action integral) is defined as

$$S = \int_1^2 L(q, \dot{q}, t) dt \tag{4.1}$$

Hamilton's Principle is sometimes also called the *Principle of Least Action*, although this term is sometimes also used for another variational principle as well. Hamilton's Principle was originally developed by Hamilton to describe optical systems [Hamilton 1834]. Lagrange would latter apply the method to mechanics. The action is a scalar invariant under coordinate transformations. The units of the action are the same as that of angular momentum:

$$[\text{action}] = [\text{energy} \cdot \text{time}] = [\text{angular-momentum}] \tag{4.2}$$

For monogenic systems, Hamilton's principle asserts that the action is an extremum given by

$$\delta S = \delta \int_1^2 L(q, \dot{q}, t) dt = 0. \tag{4.3}$$

By using the methods of the calculus of variations, it will be shown that this statement implies the validity of the *Euler-Lagrange equations of motion*

$$\frac{d}{dt}\left(\frac{\partial L}{\partial \dot{q}^j}\right) = \frac{\partial L}{\partial q^j}. \tag{4.4}$$

Note that Hamilton's principle does not explicitly refer to any specific set of coordinates. It is a coordinate-free statement about the behavior of a physical system. Everything, all dynamics, is contained in the scalar action. The action integral of a physical system is required to be stationary for the actual path taken by the particle.

The use of a single scalar function to define all the system's dynamics helps to insure the self-consistency of the derived equations of motion. Since all known fundamental physical forces can be derived from a Lagrangian function, Hamilton's principle is considered a fundamental approach. It is the de facto starting point for the development of most modern physical theories.

The action principle allows one to explore the role that symmetry plays in developing physical law. The fundamental relationship between symmetries of a system and its constants of the motion will be formalized by the derivation of Noether's theorem in the latter sections of this chapter.

4.2 ROLE OF THE ACTION IN QUANTUM FIELD THEORY

A generalization of the action principle to Quantum Mechanics due to Richard Feynman (1918–1988) states that a particle can take any path between two points with some probability determined by the action path integral

$$S = \int_1^2 L dt, \tag{4.5}$$

where the probability amplitude of some path history is given by

$$e^{iS/\hbar}. \tag{4.6}$$

Substituting for the Lagrangian, written in terms of the Hamiltonian, and treating time as a fourth coordinate gives

$$S = \int_1^2 (p_i \dot{q}^i - H)dt = \int_1^2 p_u dq^u, \tag{4.7}$$

where the fourth component of spacetime phase space is given by

$$p_0 = -H, \quad q^0 = t. \tag{4.8}$$

The action for a free particle, along the classical trajectory, has constant four-momentum, resulting in an oscillatory phase factor

$$e^{ip_u(x^u(2)-x^u(1))/\hbar}. \tag{4.9}$$

As the time interval between endpoints increases, this phase oscillates ever more rapidly for deviations from the stationary path. The phases of the action add coherently only for small deviations from the stationary paths, i.e., paths of constant phase, resulting in the classical trajectory $\delta S = 0$ being the most probable one. Feynman's construction effectively treats classical physics as the geometric ray limit of a probabilistic wave theory.

4.3 CALCULUS OF VARIATIONS

The virtual displacements referred to in the statement of Hamilton's principle are variations that satisfy the following three criteria:

1. The displacements are independent variations that satisfy the holonomic constraint conditions.
2. The displacements are made at constant time

$$\delta t = 0. \tag{4.10}$$

3. The displacements vanish at the end points of the varied path:

$$\delta q^i(t_1) = \delta q^i(t_2) = 0. \tag{4.11}$$

Consider two paths that are infinitesimally close to each other as shown in Figure 4.1. The variation can be expanded as a perturbation in terms of a small parameter α. The variation is stationary if the difference of the action integrals is zero to first-order in the variations of $q^i(t)$. The stationary path is often a local minimum. However, it could as well be a maximum or even a saddle-point in configuration space.

Variations of $\delta q^j(t)$ about the physical path $q_{(0)}^j(t)$ are almost arbitrary, subject only to smoothness criteria. The variations δq^j are required to be well-behaved functions of the time, with continuous first and second derivatives,

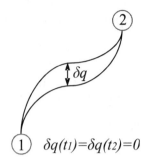

FIGURE 4.1   An infinitesimal variation in two paths in configuration space.

that shrink to zero as the limit of a small perturbation parameter α goes to zero:

$$q^j(t) = q^j_{(0)}(t) + \alpha \eta^j(t), \tag{4.12}$$

$$\dot{q}^j(t) = \dot{q}^j_{(0)}(t) + \alpha \dot{\eta}^j(t). \tag{4.13}$$

It follows that that the variations satisfy the following "pass-through" criteria for variations and differentiation

$$\delta \dot{q}^j = \delta \left(\frac{dq^j}{dt} \right) = \frac{d}{dt}(\delta q^j) = \delta \alpha \dot{\eta}^j(t). \tag{4.14}$$

That is, the order of the time and variational derivatives can be interchanged. The variations, moreover, are required to identically vanish at the end points; therefore, they satisfy the boundary conditions

$$\eta^j(t_1) = \eta^j(t_2) = 0. \tag{4.15}$$

Since the end points of integration are fixed, in Hamilton's principle the order of the operations of variation and integration can be interchanged

$$\delta S = \delta \int_1^2 L(q, \dot{q}, t) dt = \int_1^2 \delta L(q, \dot{q}, t) dt = 0, \tag{4.16}$$

where, by the chain rule

$$\delta L(q, \dot{q}, t) = \frac{\partial L}{\partial q^j} \delta q^j + \frac{\partial L}{\partial \dot{q}^j} \delta \dot{q}^j. \tag{4.17}$$

No variation of the time is permitted. Interchanging the order of variation, using Equation 4.14, gives

$$\delta L(q, \dot{q}, t) = \left(\frac{\partial L}{\partial q^j} \delta q^j + \frac{\partial L}{\partial \dot{q}^j} \frac{d}{dt} \delta q^j \right), \tag{4.18}$$

and integration by parts yields

$$\delta L(q,\dot{q},t) = \left(\frac{\partial L}{\partial q^j}\delta q^j - \frac{d}{dt}\left(\frac{\partial L}{\partial \dot{q}^j} \right)\delta q^j + \frac{d}{dt}\left(\frac{\partial L}{\partial \dot{q}^j}\delta q^j \right) \right). \qquad (4.19)$$

The last term is a total differential. But the variations vanish at the end points. Integrating over the fixed limits of integration eliminates the contribution from the total differential giving

$$\int_1^2 \left(\frac{\partial L}{\partial q^j} - \frac{d}{dt}\left(\frac{\partial L}{\partial \dot{q}^j} \right) \right) \delta q^j dt = 0. \qquad (4.20)$$

Moreover, the variations are linearly independent and arbitrary. Therefore, the terms in the integrand must individually vanish, yielding Lagrange's equations of motion

$$\frac{\partial L}{\partial q^j} - \frac{d}{dt}\left(\frac{\partial L}{\partial \dot{q}^j} \right) = 0. \qquad (4.21)$$

The choice of Lagrangian is not unique. Many Lagrangians yield the same equations of motion. As was pointed out in the previous chapter, one can modify the Lagrangian by adding a total time derivative of a function $F(q,t)$ to the Lagrangian

$$L'(q,\dot{q},t) = \left(L(q,\dot{q},t) + \frac{dF(q,t)}{dt} \right). \qquad (4.22)$$

Since the variation of $F(q,t)$ vanishes at the endpoint of the integral, the two actions lead to the same equations of motion:

$$\delta S' = \delta \int_1^2 L'dt = \delta \int_1^2 \left(L + \frac{dF}{dt} \right) dt$$

$$= \delta \int_1^2 Ldt + \delta(F_2 - F_1) = \delta \int_1^2 Ldt = \delta S. \qquad (4.23)$$

Therefore, it follows that

$$\left(\frac{d}{dt}\frac{\partial}{\partial \dot{q}^j} - \frac{\partial}{\partial q^j} \right) \frac{dF(q,t)}{dt} \equiv 0. \qquad (4.24)$$

This ability to add an arbitrary total derivate of the coordinates and time to the Lagrangian will be latter prove to be an essential starting point for developing Canonical Transformation Theory, which will be discussed in Chapter 9.

Hamilton's Principle is but a single example of the application of the *Calculus of Variations*, which was first developed by John Bernoulli (1654–1748) to solve the minimum time problem. The calculus of variations states that the solution to the variation of a general integral of the form

$$\delta \int_1^2 f(y^i, \dot{y}^i, x) dx = 0. \tag{4.25}$$

where x is some independent variable and $y^i, \dot{y}^i = dy^i/dx$ are dependent variables, is given by the family of curves that satisfy the equations

$$\frac{d}{dx} \frac{\partial f(y, \dot{y}, x)}{\partial \dot{y}^i} = \frac{\partial f(y, \dot{y}, x)}{\partial y^i}, \tag{4.26}$$

The calculus can be extended to functionals containing higher order differentials but this form is sufficient for the present purposes. These equations are identical to the Euler-Lagrange equations, where the Lagrangian $L(q, \dot{q}, t)$ is replaced by some generic functional $f(y, \dot{y}, x)$: the coordinates q by y, dq/dt by dy/dx, and the time t by some independent parameter x.

4.3.1 Fermat's Principle

In geometric optics, the velocity of light moving in a medium with index of refraction $n(x^1, x^2, x^3)$ is given by

$$v(x) = c/n(x). \tag{4.27}$$

Fermat's Principle states that a ray of light in a medium follows the path that minimizes the time traveled between two fixed points relative to small variations of the path. The transit time between the two points can be written the path integral

$$T = \int_1^2 dt = \int_1^2 \frac{ds}{v}. \tag{4.28}$$

Using $ds = \sqrt{dx^i dx_i}$ gives

$$f = \frac{1}{v} = \frac{n(x)}{c} \sqrt{g_{ij} \frac{dx^i}{ds} \frac{dx^i}{ds}}.$$

Therefore, the path can be found by solving the Euler-Lagrange equations for the functional f

$$\frac{d}{ds} \frac{\partial f}{\partial (dx^i/ds)} - \frac{\partial f}{\partial x^i} = 0. \tag{4.29}$$

Note that if f is cyclic in the independent variable, the quantity

$$C = f - \frac{\partial f}{\partial \dot{x}^i} \dot{x}^i, \quad \dot{x}^i = \frac{dx^i}{ds}, \tag{4.30}$$

is a first integral of the variational equations. It plays a role analogous to the Hamiltonian in particle dynamics. Likewise, if f is cyclic in some x^i, the quantity $\partial f / \partial \dot{x}^i$ is a first integral of the equations. The functional derivatives $\partial f / \partial \dot{x}^i$ play a role similar to the conjugate momenta of particle dynamics.

4.3.2 Brachistochrone Problem

The question Bernoulli set out to solve that led to the development of the calculus of variations was to find the path taken by an object, falling without friction in a gravitational field, which would minimize the transit time between two fixed points in configuration space. This is referred to as the *Brachistochrone* or *minimum time* problem in mechanics. In general, the transit time depends on the path taken by the object. For example a pendulum of fixed length has a period that depends on the amplitude. The path of a falling particle in the Brachistochrone problem can be thought of as the arc traced by a pendulum of variable length, or as the path of a bead sliding without friction on some curved wire in a uniform gravitational field, where the path is adjusted to make the period independent of the amplitude. Assuming the path is frictionless, energy is conserved along the path. Let the y-axis be vertically downward. This provides an energy constraint on the functional form of the velocity

$$E = -mgy_0 = \frac{1}{2}mv^2 - mgy, \tag{4.31}$$

$$v(y) = \sqrt{2g(y - y_0)}. \tag{4.32}$$

The variational problem can be stated as the variation of the constrained path that minimizes the time integral

$$\delta t = \delta \int_1^2 ds/v(y, \dot{y}, \dot{x}) = \delta \int_1^2 \sqrt{\frac{dy^2 + dx^2}{2g(y - y_0)}}$$

$$= \delta \int_1^2 \sqrt{\frac{(dy/dx)^2 + 1}{2g(y - y_0)}} dx = \delta \int_1^2 f\left(y, \frac{dy}{dx}\right) dx. \tag{4.33}$$

The solution of the variational problem is given by the Euler-Lagrange equations (with x replacing t as the independent variable)

$$f(y, \dot{y}) = \sqrt{\frac{\dot{y}^2 + 1}{2g(y - y_0)}}. \tag{4.34}$$

The first integral of this equation can be found by noting that the function is cyclic in the independent variable. This gives a constant of the motion similar to the Hamiltonian in particle dynamics. Letting $y_0 = 0, \dot{y} = dy/dx$ gives

$$C = f - \frac{\partial f}{\partial \dot{y}} \dot{y} = \frac{1}{\sqrt{2gy}} \frac{1}{\sqrt{1 + \dot{y}^2}} = \frac{1}{\sqrt{4ga}}. \tag{4.35}$$

where a is some constant of integration. The solution of this problem, using Lagrange's equations, is given by the parametric equations for a cycloid (see Figure 4.2).

$$x(\phi) = a(\phi - \sin \phi), \quad y(\phi) = a(1 - \cos \phi). \tag{4.36}$$

The proof is left as an exercise.

A cycloid can be visualized as the path taken by a point mass attached to a massless hoop of radius a constrained to roll without slipping along the $\hat{x}$ axis, as illustrated in Figure 4.2. Not only is the time spread minimized, but the time needed to reach the bottom of the cycloid of given radius a is independent of the height above the minimum of potential energy from which the object is released. That is, the Brachistochrone curve is the modification needed to the path of a pendulum to get it to keep constant time. The period of oscillations can be shown to be $T = 4\pi \sqrt{a/g}$, independent of amplitude. This is also left as an exercise.

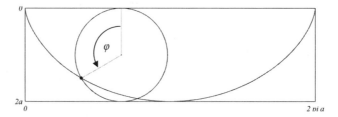

FIGURE 4.2   A cycloid can be parameterized the angle of rotation of a point mass fixed to a point on a hoop of radius a constrained to roll without slipping along the horizontal axis.

4.3.3 Geodesics

Geodesics are defined as the shortest allowed path between two points in a manifold. More precisely, the path is defined as an extremum relative to all nearby varied paths. In Euclidean space, parallel lines never cross, and there is only one straight line connecting any two points. On the surface of a sphere, the geodesics are great circles and there are two possible directions of motion, corresponding to clockwise or counter-clockwise rotations. In the absence of applied forces, the free motion of a particle on any surface of constraint can also be treated as geodesics on a curved space.

Define an element of arc length in flat Euclidean space using Cartesian coordinates as

$$ds = \sqrt{d\mathbf{r} \cdot d\mathbf{r}} = \sqrt{\delta_{ij} dx^i dx^j} = \sqrt{g_{ij} \dot{x}^i \dot{x}^j} ds, \tag{4.37}$$

where $\dot{x} = dx/ds$ is derivative with respect to the independent variable s. A geodesic is defined as the path that makes the path length an extremum

$$\delta \int_1^2 ds = \delta \int_1^2 \sqrt{\delta_{ij} \dot{x}^i \dot{x}^j} ds = 0. \tag{4.38}$$

This gives rise to the Euler-Lagrange equations of motion in terms of the independent parameter

$$\frac{d}{ds} \left(\frac{\partial \sqrt{\delta_{ij} \dot{x}^i \dot{x}^j}}{\partial \dot{x}^j} \right) = 0. \tag{4.39}$$

The equations are cyclic in all three coordinates giving the equation

$$\frac{d}{ds} \left(\frac{\dot{x}^i}{\sqrt{\mathbf{n}^2}} \right) = 0, \tag{4.40}$$

with solution

$$\dot{x}^i = \frac{dx^i}{ds} = n^i, \tag{4.41}$$

where the directional derivatives $n^i = dx^i/ds$ defines the components of unit vector $\mathbf{n} = \hat{\mathbf{v}}$.

The equation for a geodesic is a straight line in Euclidean space, in the direction of the directional derivative connecting the two points, giving the following parameterized equations of motion,

$$x^i(s) = x_0^i + n^i s. \tag{4.42}$$

In vector notation, the equation of a straight line between two points separated by a distance s is given by the displacement vector

$$\Delta \mathbf{x} = \mathbf{x}(s) - \mathbf{x}(0) = \mathbf{n}\,s, \tag{4.43}$$

where s is the magnitude of the displacement vector, and $\mathbf{n}$ is its unit orientation vector.

It is useful to compare this result to the trajectory of a free particle in free space. The free particle Lagrangian is $L = T = \frac{1}{2}mv^2$, or in Cartesian coordinates

$$L = \frac{m}{2}\delta_{ij}\frac{dx^i}{dt}\frac{dx^j}{dt}, \tag{4.44}$$

$$\therefore \delta \int_1^2 \left(\frac{m}{2}\delta_{ij}\dot{x}^i\dot{x}^j\right)dt = 0, \tag{4.45}$$

giving Lagrange equations

$$\frac{d}{dt}(m\dot{x}^i) = 0. \tag{4.46}$$

The particle moves with constant velocity $\mathbf{v} = v_0\,\mathbf{n} = (ds/dt)\mathbf{n}$ along the geodesic

$$\mathbf{x}(s) = \mathbf{x}_0 + \mathbf{n}\,s, \quad \mathbf{x}(t) = \mathbf{x}_0 + \mathbf{n}\,vt. \tag{4.47}$$

where $s = v_0 t$. Free particles move along geodesics with uniform velocity.

It can be shown, in general, that motion of the free particle in an arbitrary curved space is always a geodesic. In affine spaces, geodesics are defined to be curves whose tangent vectors remain parallel if they are transported along it. In Euclidian geometry, there is only one geodesic (straight line) connecting any two points in space. Its length is the shortest distance between the two points. In non-Euclidian geometries, multiple straight lines may connect two points. This can be seen by considering geodesics on the surface of a cylinder.

4.3.4 Geodesics on a Cylinder

Given a right circular cylinder of radius a, the element of displacement on its surface is given by

$$ds = \sqrt{a^2(d\phi)^2 + (dz)^2} = \sqrt{(du)^2 + (dz)^2}, \tag{4.48}$$

where $u = a\phi$. In terms of u and z coordinates, the cylinder has the same metric as a flat two-dimension plane. The surface of a cylinder is locally and topologically flat (one can cut it along the line $\phi = 0$ and lay it flat out on a table without distortion). The variational equation for a geodesic is the same

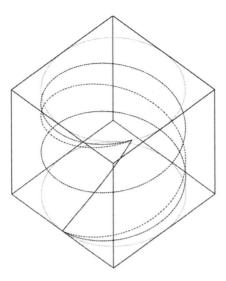

FIGURE 4.3 Geodesic curves on a cylinder: There are an infinite number of geodesics between two points on a cylinder, differing in the number of twists (winding number) or cycles of revolution.

as in flat two-dimensional Euclidean space, but now there are an infinite number of geodesics between any two points. This is because a cylindrical surface, although topological flat, has multiple leaves. Note that the Cartesian coordinates

$$(u, z) = (a(\phi \pm 2m\pi), z) \tag{4.49}$$

represents the same point on the cylinder where m is the number of twists (winding number) of a geodesic between two points on the cylinder as illustrated in Figure 4.3.

4.3.5 Geodesics in Curved Space

Applying a variation to minimize the path in an arbitrary metric space, gives

$$\delta \int_1^2 ds = \delta \int_1^2 \sqrt{g_{ij} \frac{dq^i}{ds} \frac{dq^j}{ds}} ds = 0 \tag{4.50}$$

where $n^i = dq^i/ds = \dot{q}^i$ are the directional derivatives and

$$n = \sqrt{n^j n_j} = \sqrt{g_{jk} \frac{dq^j}{ds} \frac{dq^k}{ds}} = 1 \tag{4.51}$$

is a constant of the motion. In general, the metric tensor is a function of the coordinates, giving rise to an affine connection. The Euler-Lagrange equations of motion are

$$\frac{1}{2}\frac{d}{ds}\left(\frac{g_{jk}}{n}\frac{\partial \dot{q}^j}{\partial \dot{q}^i}\frac{dq^k}{ds} + \frac{g_{jk}}{n}\frac{\partial \dot{q}^k}{\partial \dot{q}^i}\frac{dq^j}{ds}\right) = \frac{1}{2n}\frac{\partial g_{jk}}{\partial q^i}\frac{dq^j}{ds}\frac{dq^k}{ds} \qquad (4.52)$$

using $\partial \dot{q}^j/\partial \dot{q}^i = \delta_i^j$ gives

$$\frac{d}{ds}\left(\frac{g_{ik}}{n}\frac{dq^k}{ds} + \frac{g_{ji}}{n}\frac{dq^j}{ds}\right) = \frac{1}{n}\frac{\partial g_{jk}}{\partial q^i}\frac{dq^j}{ds}\frac{dq^k}{ds} \qquad (4.53)$$

Next, expand $dg_{ij}/ds = (\partial g_{ij}/\partial q^k)(dq^k/ds)$ to get

$$2g_{ij}\ddot{q}^j + \frac{\partial g_{ij}}{\partial q^k}\dot{q}^k\dot{q}^j + \frac{\partial g_{ik}}{\partial q^j}\dot{q}^j\dot{q}^k - \frac{\partial g_{jk}}{\partial q^i}\dot{q}^j\dot{q}^k = 0. \qquad (4.54)$$

The solution of the Euler-Lagrangian equations for the geodesic then becomes

$$\frac{d\dot{q}^i}{ds} + \Gamma^i_{jk}\dot{q}^j\dot{q}^k = 0, \qquad (4.55)$$

where the affine connections Γ^i_{jk} are the Christoffel symbols of tensor analysis given by

$$\Gamma^i_{jk} = \frac{1}{2}g^{im}\left(\frac{\partial g_{mj}}{\partial x^k} + \frac{\partial g_{mk}}{\partial x^j} - \frac{\partial g_{jk}}{\partial x^m}\right). \qquad (4.56)$$

Einstein, in developing his general theory of relativity, made note of the fact that the equations for the geodesics of a manifold also describe the path taken by a particle moving freely on the manifold.

4.4   LAGRANGE'S METHOD OF UNDETERMINED MULTIPLIERS

If, instead of eliminating constrained coordinates, one keeps them, one gets (after interchanging $\delta\dot{q}^j = d(\delta q^j)/dt$ and integrating by parts)

$$\int_1^2 \left(-\frac{d}{dt}\left(\frac{\partial L}{\partial \dot{q}^i}\right) + \frac{\partial L}{\partial q^i}\right)\delta q^i dt = 0 \quad \text{for } i = 1,\dots,N_0. \qquad (4.57)$$

However, this does not directly imply the Euler–Lagrange equations, since the variations are no longer independent, but subject to a set of holonomic constraints of the form

$$\tilde{q}^k(q,t) = c^k = 0; \quad \text{for } k = 1, \dots, N_k. \tag{4.58}$$

The holonomic constraint conditions can be separately varied to give the differential constraints

$$\frac{\partial \tilde{q}^k}{\partial q^i} dq^i + \frac{\partial \tilde{q}^k}{\partial t} dt = B_i^k dq^i - C^k dt = 0, \tag{4.59}$$

where

$$B_i^k = \frac{\partial \tilde{q}^k}{\partial q^i}, \quad C^k = \frac{\partial \tilde{q}^k}{\partial t}. \tag{4.60}$$

For virtual variations made at constant time, this is equivalent to

$$B_i^k \delta q^i = 0. \tag{4.61}$$

Multiplying each of these constraint equations by an unknown function $\Lambda_k(q,t)$, called an *undetermined multiplier* and summing over k gives the virtual work done by the forces of constraint

$$\Lambda_k B_i^k \delta q^i = Q_i^{(c)} \delta q^i = 0, \tag{4.62}$$

where

$$Q_i^{(c)} = \Lambda_k B_i^k. \tag{4.63}$$

Modifying the action variation by adding zero virtual work to it gives

$$\int_1^2 \left(-\frac{d}{dt}\left(\frac{\partial L}{\partial \dot{q}^i}\right) + \frac{\partial L}{\partial q^i} + Q_i^{(c)} \right) \delta q^j dt = 0 \quad \text{for } i = 1, \dots, n. \tag{4.64}$$

The undetermined multipliers can now be adjusted to make n_k constraint terms identically vanish. The remaining $q^i = \{q^1, \dots, q^{n-n_k}\}$ coordinates are then linearly independent. Therefore, term by term, the coefficients in Equation 4.64 can be made to vanish. This gives the following set of n simultaneous equations to solve

$$\frac{d}{dt}\left(\frac{\partial L}{\partial \dot{q}^i}\right) - \frac{\partial L}{\partial q^i} = Q_i^{(c)} = \Lambda_k B_i^k, \tag{4.65}$$

subject to the n_k differential constraint conditions

$$B_i^k \frac{dq^i}{dt} - C^k = 0, \tag{4.66}$$

where the $Q_i^{(c)}$ are interpreted as the generalized forces of constraint. There are n coordinates in total and n_k undetermined multipliers, for a total of $n + n_k$ equations in $n + n_k$ unknowns $\{q^i, \Lambda_k\}$ to solve. Note that the method of undetermined multipliers does not require the constraints to be holonomic, but only the lesser condition that they be first-order linear differential equations in the velocities of the general form

$$B_i^k \frac{dq^i}{dt} - C^k = 0. \qquad (4.67)$$

Such differential equations are not always exact integrals that can be converted to point constraints on the coordinates. Therefore, the method of undetermined multipliers can be extended to apply to nonholonomic local contact constraints on the velocities, such as when a body is constrained to roll without slipping on a surface with a single point of contact.

4.4.1 Atwood's Machine

Consider an Atwood's machine (Figure 4.4), consisting of a frictionless pulley with radius a, moment of inertia I, and two hanging masses m_1, m_2. In addition to the kinetic energy of translation of the two masses, there is additional rotational kinetic energy in the pulley. The total kinetic energy of the system is

$$T = \frac{1}{2}m_1\dot{x}_1^2 + \frac{1}{2}m_1\dot{x}_2^2 + \frac{1}{2}I\dot{\theta}^2. \qquad (4.68)$$

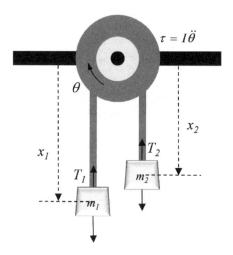

FIGURE 4.4 An Atwood's Machine where the pulley has radius a and momentum of inertia I.

where $x_{1,2}$ are the coordinates of the masses and θ is the rotation angle of the pulley wheel. Including the potential energy contributions of the two weights, the Lagrangian is given by

$$L = \frac{1}{2}m_1\dot{x}_1^2 + \frac{1}{2}m_2\dot{x}_2^2 + \frac{1}{2}I\dot{\theta}^2 - m_1gx_1 - m_2gx_2. \qquad (4.69)$$

The length l of the string joining the two masses is fixed, giving the holonomic constraint

$$x_1 + x_2 + \pi a - l = 0. \qquad (4.70)$$

If the pulley turns the string without slipping, there is a second differential constraint relating position to angle

$$\dot{x}_1 + a\dot{\theta} = 0, \qquad (4.71)$$

where θ is the rotation angle of the pulley. This constraint can be integrated to get the holonomic constraint

$$x_1 + a(\theta - \theta_o) = 0. \qquad (4.72)$$

The constraint conditions can be eliminated, giving a system with one coordinate degree of freedom. Clearly, this is the easiest way to solve this problem (see Exercise 4.1); but, to illustrate the method of undetermined multipliers, it will be considered instead as a system with two differential contact constraints

$$a\,d\theta + dx_1 = 0, \quad a\,d\theta - dx_2 = 0. \qquad (4.73)$$

The differential $d\theta$ is defined to be positive for a clockwise sense of rotation. The variational constraint conditions are

$$T_1(a\delta\theta + \delta x_1) = 0, \quad T_2(-a\delta\theta + \delta x_2) = 0, \qquad (4.74)$$

where $T_{1,2}$ are undetermined multipliers. The equations of motion are

$$m_1\ddot{x}_1 = -T_1 + m_1g, \quad m_1\ddot{x}_2 = -T_2 + m_2g, \quad I\ddot{\theta} = (T_2 - T_1)a. \qquad (4.75)$$

By inspection, the multipliers $T_{1,2}$ can be identified as the sting tensions acting vertically upward on the two masses. The term $(T_2 - T_1)a$ is the torque acting on the pulley. Eliminating the constraints gives the following equations of motion for the accelerations

$$\ddot{x}_2 = -\ddot{x}_1 = a\ddot{\theta} = \left(\frac{(m_2 - m_1)}{(m_1 + m_2) + I/a^2}\right)g. \qquad (4.76)$$

The purpose of the Atwood's machine is to reduce the effective strength of the acceleration of gravity. The direction of acceleration is given by the heavier of the two masses. Once the acceleration is known, the tensions in the strings can be solved, giving

$$T_1 = m_1(g + \ddot{x}_2), \quad T_2 = m_2(g - \ddot{x}_2). \tag{4.77}$$

Note that the tension on the string whose mass is accelerating upward is increased while the tension on the string that is accelerating downward is decreased, relative to the gravitational weights acting on the strings. The difference in the two tensions is given by

$$T_2 - T_1 = \left(\frac{I(m_2 - m_1)}{(m_1 + m_2)a^2 + I} \right) g. \tag{4.78}$$

The tensions in the two string segments are equal only if the two hanging masses are equal or if the pulley has no moment of inertia.

4.4.2 Wheel Rolling Down an Inclined Plane

As another example of a contact constraint, consider a wheel of radius a, mass m, and moment of inertia I_0 with respect to its axis of rotation rolling down an inclined plane under the action of gravity. See Figure 4.5 for a diagram of the system. Let α be the angle of the incline relative to the horizontal. The wheel is moving in a two-dimensional plane, so it has two translational degrees of freedom (x, y), and a single rotational degree of freedom labeled θ.

Let y be direction normal to the direction of motion, which is labeled as x. Clearly,

$$dy = 0 \tag{4.79}$$

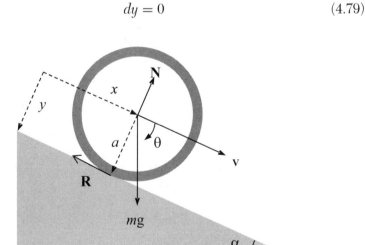

FIGURE 4.5   A wheel rolling without slipping down an incline.

is a constraint on the motion. In addition, the condition that the wheel rolls without slipping implies that the point of contact with the incline is always instantaneously at rest. This leads to a contact constraint given by

$$dx - a d\theta = 0. \tag{4.80}$$

Variation gives two undetermined multipliers that will be labeled as N and R, respectively

$$N(dy) = 0, \quad R(dx - a d\theta) = 0. \tag{4.81}$$

In these coordinates, the acceleration of gravity is given by

$$\mathbf{g} = g(\sin \alpha \, \mathbf{e}_1 + \cos \alpha \, \mathbf{e}_2). \tag{4.82}$$

The Lagrangian of the system is given by

$$L = \frac{1}{2}m(\dot{x}^2 + \dot{y}^2) + \frac{1}{2}I\dot{\theta}^2 + mg(x \sin \alpha + y \cos \alpha). \tag{4.83}$$

Lagrange's equations, using undermined multipliers to include the constraint conditions, are

$$m\ddot{x} = mg \sin \alpha - R, \quad m\ddot{y} = mg \cos \alpha - N, \quad I\ddot{\theta} = Ra. \tag{4.84}$$

One would obtain these same equations from an elementary force analysis. From the equations, it is clear how to interpret the multipliers. N is the normal force on the wheel needed to keep it in the plane $\dot{y} = \ddot{y} = 0$. R is a contact force parallel to the surface needed to keep the wheel from slipping. Ra is the corresponding constraining torque causing the wheel to undergo angular acceleration. The contact force R does no net work because the point of contact is instantaneously at rest

$$R dx - Ra d\theta = 0. \tag{4.85}$$

Eliminating the constraints gives the solution

$$\ddot{x} = \frac{ma^2 g \sin \alpha}{ma^2 + I} = a\ddot{\theta},$$

$$N = mg \cos \alpha, \tag{4.86}$$

$$R = \frac{I}{a^2}\ddot{x} = \frac{mI}{ma^2 + I}g \sin \alpha.$$

Undetermined multipliers are neither deep nor mysterious. They represent the contributions from standard Newtonian forces that can be eliminated by applying the constraint conditions. These constraint conditions, after all, are simply a statement of the observed consequences of these forces. The solution to a physical system can often be expedited by eliminating holonomic constraints a priori in the Lagrangian. However, it is sometimes useful to keep these coordinates active, when necessary, to assist with evaluating the constraint forces.

4.5   NOETHER'S THEOREM

Momenta and coordinates come in conjugate pairs. The following mappings are required to be invertible

$$p_i = p_i(q, \dot{q}, t) \Leftrightarrow \dot{q}^i = \dot{q}^i(q, p, t). \tag{4.87}$$

where the generalized or canonical momentum is defined by $p_j = \partial L / \partial \dot{q}^j$, which may (or may not) look or behave like a linear momentum. If the generalized potential $U(q, \dot{q}, t)$ depends on velocity, the canonical momentum p_i will differ dynamically from Newtonian momentum by additional vector potential contributions. For monogenic systems, the total time derivatives of the momenta are given

$$\dot{p}_j = \frac{\partial L(q, \dot{q}, t)}{\partial q^j} = -\frac{\partial H(q, p, t)}{\partial q^j}. \tag{4.88}$$

Note that if the Lagrangian, written as a function of coordinates, velocities, and time, is cyclic in a coordinate, the associated canonical momentum is conserved. This is also true if the Hamiltonian is cyclic in the coordinate when written as a function of coordinates, momenta, and time.

The intimate connection between symmetries and conservation laws is a general feature of Lagrangian systems, which is expressed by Noether's theorem [Noether 1918].[1] *Noether's Theorem* states that for any continuous symmetry of the motion, there exists a corresponding constant of the motion [Noether1918]. Note that Noether's theorem applies only to continuous (proper) symmetries. Discrete (improper) symmetries do not generate conservation laws. In general, there is a profound relationship between the continuous symmetries and conservation laws that is summarized by Noether's theorem. A dynamic observable or functional f is called a constant of motion if its total time derivative vanishes whenever the coordinates $q^i(t)$

---

[1]Amalie Emmy Noether (1882–1935) was a mathematician whose primary work in abstract algebra, particularly with ideals and ring theory.

satisfy Lagrange's equations, i.e., the observable is a constant along the actual physical path

$$\frac{df(q,\dot{q},t)}{dt} = \frac{\partial f}{\partial q^i}\dot{q}^i + \frac{\partial f}{\partial \dot{q}^i}\ddot{q}^i + \frac{\partial f}{\partial t} = 0. \tag{4.89}$$

This means that f is constant along the physical trajectories allowed by the dynamics. Consider, for example, a one-parameter family of maps

$$q^i(t) \rightarrow q^i(s,t) \quad \text{for } s \in \mathbb{R}, \tag{4.90}$$

such that $q^i(0,t) = q^i(t)$. This transformation is said to represent a continuous symmetry of the Lagrangian if the Lagrangian is cyclic in the parameter s, i.e., $\partial L/\partial s = 0$.

To simplify the proof of Noether's theorem, consider only the case where the parameter s is time-independent. Expand the partial derivative using the chain rule

$$\left.\frac{\partial L}{\partial s}\right|_{s=0} = \left.\left(\frac{\partial L}{\partial \dot{q}^i}\frac{\partial \dot{q}^i}{\partial s}\right)\right|_{s=0} + \left.\left(\frac{\partial L}{\partial q^i}\frac{\partial q^i}{\partial s}\right)\right|_{s=0} = 0. \tag{4.91}$$

Applying Lagrange's equations $\dot{p}_i = \partial L/\partial q^i$ and making use of the pass through property of variations $\partial \dot{q}^i/\partial s = d(\partial q^i/\partial s)/dt$, gives

$$\left.\frac{\partial L}{\partial s}\right|_{s=0} = \left.\left(\frac{\partial L}{\partial \dot{q}^i}\frac{d}{dt}\left(\frac{\partial q^i}{\partial s}\right)\right)\right|_{s=0} + \left.\left(\frac{d}{dt}\left(\frac{\partial L}{\partial \dot{q}^i}\right)\frac{\partial q^i}{\partial s}\right)\right|_{s=0}$$

$$= \left.\frac{d}{dt}\left(\frac{\partial L}{\partial \dot{q}^i}\frac{\partial q^i}{\partial s}\right)\right|_{s=0} = \frac{dC_s}{dt}. \tag{4.92}$$

Therefore, the physical observable

$$C_s = \left.\left(\frac{\partial L}{\partial \dot{q}^i}\frac{\partial q^i}{\partial s}\right)\right|_{s=0}. \tag{4.93}$$

evaluated at $s = 0$, is constant for all time if

$$\left.\frac{\partial L}{\partial s}\right|_{s=0} = 0. \tag{4.94}$$

The converse is also true, proving Noether's Theorem.

All conservation laws in nature can be related to symmetries of nature through Noether's theorem. This includes the conservation of electric charge, which is due to phase invariance. Furthermore, the fundamental force laws of physics can be associated with the gauge fields that enforce the known symmetries of physical systems. Conservation of total linear momentum, for example, is associated with the homogeneity of space, and conservation of total angular momentum is associated with the isotropy of space.

The conservation of the generalized momentum associated with cyclic coordinates is a direct manifestation of Noether's theorem. Let the Lagrangian be cyclic in the coordinate q^i. The infinitesimal generalized displacements associated with that degree of freedom result in a coordinate shift Δ^i

$$q^j(\Delta^i, t) = q^j(t) + \delta_i^j \Delta^i, \quad \dot{q}^j(\Delta^i, t) = \dot{q}^j(t). \tag{4.95}$$

By Noether's theorem, the conserved quantity associated with this invariance is the associated canonical momentum

$$C_{\Delta^i} = \frac{\partial L}{\partial(\dot{q}^j)} \frac{\partial(q^j + \delta_i^j \Delta^i)}{\partial \Delta^i}\bigg|_{\Delta^i=0} = \frac{\partial L}{\partial \dot{q}^j} \delta_i^j = p_i. \tag{4.96}$$

Consequently, the following statements are equivalent:

- A system has a symmetry or invariance with respect to a generalized coordinate.
- The coordinate is cyclic, i.e., it does not appear in the Lagrangian.
- The associated conjugate momentum is a constant of the motion.

Cyclic coordinates q_c are often referred to as *ignorable coordinates*. They can be ignored in the sense that their velocities can be removed from the equations of motion by replacing them with their dependence on their conserved momenta and the noncyclic degrees of freedom

$$\dot{q}_c^i = \dot{q}_c^i(q_{nc}, \dot{q}_{nc}, p_c, t). \tag{4.97}$$

The resulting reduced equations of motion for the noncyclic coordinates q_{nc} are now completely independent of any explicit reference to the cyclic variables or their velocities. Once one solves for the time behavior of the noncyclic coordinates along a given path, the cyclic coordinates can be integrated as a function of the time

$$q_c^j(t) - q_c^j(t_0) = \int_{t_0}^t \dot{q}_c^i(t)dt. \tag{4.98}$$

4.5.1 Homogeneity of Space

Consider a closed, isolated system of particles with a Lagrangian depending only on relative coordinates

$$L = \sum_i \frac{1}{2}m\dot{\mathbf{r}}_i \cdot \dot{\mathbf{r}}_i - V(|\mathbf{r}_i - \mathbf{r}_j|). \tag{4.99}$$

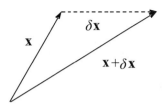

FIGURE 4.6   **The translation of a vector by a displacement.**

This Lagrangian has the symmetry of translation with respect to a constant global change of origin by some arbitrary amount s and direction $\hat{\mathbf{n}}$ (see Figure 4.6)

$$\mathbf{r}_i(s) = \mathbf{r}_i(0) + s\,\hat{\mathbf{n}}. \tag{4.100}$$

If the Lagrangian depends only on relative coordinates, one has

$$\begin{aligned} \mathbf{r}_i(s) - \mathbf{r}_j(s) &= \mathbf{r}_i(0) - \mathbf{r}_j(0), \\ \dot{\mathbf{r}}_i(s) &= \dot{\mathbf{r}}_i(0), \end{aligned} \tag{4.101}$$

since relative displacements are independent of the origin. Therefore, The Lagrangian is cyclic in the displacement $s\,\hat{\mathbf{n}}$.

$$L(\mathbf{r}(s,t), \dot{\mathbf{r}}(s,t), t) = L(\mathbf{r}(0,t), \dot{\mathbf{r}}(0,t), t), \tag{4.102}$$

and $\partial L/\partial s = 0$. The theorem can be extended to a generalized potential depending only on relative coordinates and velocities. In the language of Noether's Theorem, space is homogeneous if a uniform translation of the system by a constant vector $\delta\mathbf{r} = s\,\mathbf{n}$ does nothing to the equations of motion. (Global translations are elements of the Galilean group of inertial frame mappings.)

Since all system points are translated, the conserved quantities are found by summing over all particles and all possible orientations

$$C_s = \sum_i \sum_{j=1,2,3} \frac{\partial L}{\partial \dot{x}_i^j} n^j = \sum_{i,j} p_{ij} n^i = \mathbf{P}_{\text{total}} \cdot \hat{\mathbf{n}}. \tag{4.103}$$

But, since the direction $\hat{\mathbf{n}}$ is arbitrary, all components of the total momentum of the system are separately conserved, and the total momentum of the system as a whole is conserved.

$$\mathbf{P}_{\text{total}} = \sum_i \frac{\partial L}{\partial \dot{x}_i^j} \mathbf{e}^j = \sum_i \mathbf{p}_i. \tag{4.104}$$

The homogeneity of space, which is due to the translational invariance of the Lagrangian under a global translation of all particle points, results in the conservation of total canonical linear momentum of a closed isolated system.

4.5.2 Isotropy of Space

The isotropy of space is the statement that a closed system, described by a scalar Lagrangian

$$L = \sum_i \frac{1}{2} m \, \dot{\mathbf{r}}_i \cdot \dot{\mathbf{r}}_i - U(\mathbf{r}, \dot{\mathbf{r}}), \tag{4.105}$$

is invariant under uniform rotations about any fixed axis, where all the vectors are rotated by the same infinitesimal amount $\delta\phi$ about some arbitrary axis $\hat{\mathbf{n}}$

$$\mathbf{r}_i \to \mathbf{r}_i + \delta\mathbf{r}_i = \mathbf{r}_i + \delta\phi \, \hat{\mathbf{n}} \times \mathbf{r}_i. \tag{4.106}$$

Invariance follows from demanding that the Lagrangian be a scalar invariant under global rotations. In a global rotation, the scalar products of all vectors remain unchanged. This will be true if the Lagrangian is composed only of inner products of vectors, since scalars (and scalar dot products) are invariant under rotations. The conserved quantities are

$$\sum_i \frac{\partial L}{\partial \dot{x}_i^j} (\hat{\mathbf{n}} \times \mathbf{x}_i)^j = \sum_i \mathbf{p}_i \cdot (\hat{\mathbf{n}} \times \mathbf{r}_i) = \hat{\mathbf{n}} \cdot \sum_i (\mathbf{r}_i \times \mathbf{p}_i) = \hat{\mathbf{n}} \cdot \mathbf{L}_{\text{total}}. \tag{4.107}$$

This is the component of the total angular momentum in the direction of the axis of rotation. The orientation of this axis is arbitrary, leading to the conservation of total angular momentum about all three axes. The isotropy of space (expressed by the invariance of the Lagrangian under global rotations) results in the conservation of total angular momentum of a closed isolated system.

Rotational invariance about a fixed axis is shown in Figure 4.7. Note that the invariance of the Lagrangian with respect to an axis implies that the associated azimuthal coordinate is cyclic

$$L(\mathbf{r} + \mathbf{n} \times \mathbf{r} \, \delta\phi) = L(\mathbf{r}) \to \frac{\partial L}{\partial \phi} = 0. \tag{4.108}$$

This implies the conservation of the associated component of angular momentum

$$p_\phi = \frac{\partial L}{\partial \dot{\phi}}. \tag{4.109}$$

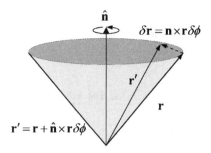

FIGURE 4.7   Rotation of a vector about an axis.

Since the rotation axis can be chosen arbitrarily, this implies conservation of total angular momentum of the system as a whole

$$\mathbf{L}_{\text{total}} = \sum_i (\mathbf{r}_i \times \mathbf{p}_i). \tag{4.110}$$

4.5.3   Homogeneity of Time

The restricted proof of Noether's theorem derived above assumed that the symmetries were time independent. Fortunately, the physical observable associated with time invariance is already known. It is the Hamiltonian of the system. In mathematical language, the homogeneity of time means the Lagrangian is invariant under a time translation $t \to t + \delta t$. This is equivalent to asserting that the Lagrangian is cyclic in time

$$L(t + \delta t) = L(t) \to \frac{\partial L}{\partial t} = 0. \tag{4.111}$$

The associated conserved quantity is the Hamiltonian given by

$$\frac{dH}{dt} = \frac{\partial H(q,p,t)}{\partial t} = -\frac{\partial L(q,\dot{q},t)}{\partial t} = 0. \tag{4.112}$$

The existence of a conserved quantity, called the energy, can be traced to the homogeneous passage of time. Note the analogy: time is to energy, as space is to momentum. That is, the Hamiltonian can be thought of as the representing the canonical momentum associated with the time coordinate.

In special relativity, energy and three-momentum combine to form a four-momentum vector. Since the four-momentum is conserved in the absence of four-force, it follows that its fourth component of momentum is also conserved. Energy conservation of an isolated system is a trivial consequence of a four-dimensional relativistic approach. However, the link between energy/momentum and time/space exists even in the non-relativistic framework of Newtonian physics.

4.6   *SYMMETRY AND SCALE IN EXPERIMENTAL DESIGN

The symmetries of a system have a close relationship with the resulting dynamics. Factoring in desired symmetries into the initial design of experimental equipment is one way of insuring that the apparatus will perform as expected. In a well-designed experiment, the known properties of the measurement device are used to study the unknown properties of some phenomena under investigation. It is also important to know how to change the scale of a working design while preserving the desired dynamical response. An important example is the design of systems for charged particle transport. A basic introduction to the optics of charged particles can be found in [Wollnik 1987]. Realistic magnetic spectrometer design demands a tradeoff between acceptance and resolution, tempered with the limitations of material properties, and the practicality of the design. Here, only the role played by imposing structural symmetries is explored.

4.6.1   *Particle Transport in a Static Magnetic Field

The important case of charged particle transport though a static magnetic field can be used to exhibit the use of scaling and symmetry in experimental design. For simplicity, consider a particle in a static magnetic field given by $\mathbf{B} = \nabla \times \mathbf{A}$. The Lagrangian includes a kinetic energy term and a vector potential contribution

$$L = \frac{1}{2}m\mathbf{v} \cdot \mathbf{v} + q_e\mathbf{v} \cdot \mathbf{A}. \tag{4.113}$$

There is no scalar potential. The equations of motion can be written as

$$\frac{d\mathbf{v}}{dt} = -\frac{q_e}{m}\mathbf{B} \times \mathbf{v} = \boldsymbol{\omega} \times \mathbf{v}. \tag{4.114}$$

It should be clear, from the right-hand side of the equation, that the effect of a magnetic field is to cause a rotation of the direction of motion of a charged particle, where the instantaneous axis of rotation given by the unit axial vector $\hat{\mathbf{n}} = \boldsymbol{\omega}/|\boldsymbol{\omega}|$. The magnitude of the axial vector $\omega_c = -q_eB/m$ is the local cyclotron frequency of the magnetic field. The magnitude of the velocity of a charged particle is a constant of the motion

$$\frac{1}{2}\frac{dv^2}{dt} = \mathbf{v} \cdot \frac{d\mathbf{v}}{dt} = \mathbf{v} \cdot \boldsymbol{\omega} \times \mathbf{v} = 0, \tag{4.115}$$

which can also be deduced from conservation of mechanical energy

$$E = \frac{1}{2}mv^2 = \frac{1}{2m}(p - q_e A)^2. \tag{4.116}$$

Since the magnitude of the velocity v is a constant, the equations of motion describe how the unit directional derivative rotates in time. Time can be uniformly parameterized by the distance traveled $s = vt$. Define the unit tangent vector in the direction of motion as $\hat{\mathbf{v}}$, then the directional derivative is

$$\hat{\mathbf{v}} = \frac{dx^i}{ds}\mathbf{e}_i = n^i \mathbf{e}_i, \tag{4.117}$$

where $ds = \sqrt{g_{ij}dx^i dx^i}$.

It is useful to convert to natural units, where B_0 is some natural unit for magnetic field, p_0 is some characteristic scale for momentum (often taken as the momentum of the central ray in magnetic transport analysis) and [2]

$$s_0[\text{meters}] = p_0/q_e B_0 \tag{4.118}$$

is some mean value of the *radius of curvature*, which sets the laboratory scale. Define

$$\mathbf{p} = (1 + \delta)\mathbf{p}_0, \quad \tilde{B} = B/B_0, \tag{4.119}$$

where $\tilde{B}$ is the magnetic field in dimensionless units and δ denotes the fractional change in momentum from some nominal design value. The equations of motion can then be written as the following coupled first-order differential equations in velocity phase space

$$\frac{d\hat{\mathbf{v}}}{ds} = \frac{-1}{(1 + \delta)s_0}\tilde{\mathbf{B}} \times \hat{\mathbf{v}}, \quad \frac{d\mathbf{r}}{ds} = \hat{\mathbf{v}}. \tag{4.120}$$

Note that time has been parameterized in terms of the distance $ds = vdt$. The equations become dimensionless if s is measured in units of s_0. For constant B_0, the laboratory dimensions scale linearly with the design's central momentum of interest.

---

[2]Relativistic corrections can be applied by letting $p = (1 + \delta)p_0 = mv/\sqrt{1 - v^2/c^2}$.

4.6.2 *Motion in the Midplane of a Symmetric Magnet

The motion simplifies further if one limits oneself to motion in the midplane of a symmetric magnet, where the magnetic field is orthogonal to the plane of motion. Any particle moving in this plane remains in the plane. Define the symmetry plane as the $\hat{x} \times \hat{y}$ plane for specificity; then, by symmetry, the magnetic field lies in the $\hat{z}$-direction. The directional derivatives in the plane of motion can be expressed in terms of the laboratory angle θ of the velocity vector with respect to the $\hat{x}$-axis.

$$\hat{\mathbf{v}} = \mathbf{e}_1 \cos \theta + \mathbf{e}_2 \sin \theta. \tag{4.121}$$

Making the indicated substitutions, the equations of motion become

$$\frac{dx}{ds} = \cos \theta, \quad \frac{d\theta}{ds} = -\frac{\tilde{B}_z(x,y)}{(1+\delta)s_0},$$
$$\frac{dy}{ds} = \sin \theta, \quad \frac{d\delta}{ds} = 0, \tag{4.122}$$

where the transit time is given by $\Delta t = \Delta s / v$. In this form, the particle trajectories can be calculated numerically as a function of (x, y, θ, δ) using Runge-Kutta techniques and a set of initial conditions. Note that numerical energy dissipation is not an issue, since the conservation of kinetic energy is an explicit constraint, $d\delta = 0$.

4.6.2.1 *Magnetic Chicane

As a simple example, consider a parallel transport chicane magnet, which has a magnetic field that is cyclic in the y coordinate

$$\mathbf{B} = B_z(x)\hat{\mathbf{z}}. \tag{4.123}$$

The magnetic field can be derived from a vector potential of the form $\mathbf{A} = A_y(x)\hat{\mathbf{y}}$. The Lagrangian is

$$L = \frac{1}{2}m(\dot{x}^2 + \dot{y}^2) + q_e \dot{y} A_y(x). \tag{4.124}$$

The Lagrangian is cyclic in y, giving a conserved canonical momentum

$$p_y = m\dot{y} + q_e A_y(x) = C_y. \tag{4.125}$$

The second constant of integration is the total energy in the particle

$$E = T = \frac{1}{2}m(\dot{x}^2 + \dot{y}^2) = \frac{1}{2m}(p_x^2 + (p_y - q_e A_y(x))^2). \tag{4.126}$$

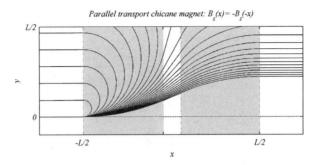

Parallel transport chicane magnet: $B_z(x) = -B_z(-x)$

FIGURE 4.8 A parallel transport magnetic chicane. The magnetic pole faces consist of two parallel sections with equal and opposite magnetic fields. The magnetic flux through the first section is returned by the second section. The incident beam line is located at $y = 0$. High momentum particles are transmitted by the magnet. Low momentum particles are reflected. The symmetry of the design dictates that all transmitted particles preserve their original direction of motion in the magnetic mid-plane. The transverse deflection of the particles relative to their initial direction can be related to their incident momentum.

If the magnet is designed such that all magnetic flux is required to be returned in the active area of the magnetic midplane, $\Delta A = 0$ for transmitted particles (see Exercise 4.19), it is straightforward to show that transmitted particles exit parallel to their original directions. Such a magnet is illustrated in Figure 4.8, where a number of parallel rays incident on the magnet are shown with a range of different incident beam energies. The most energetic particles undergo the least deflection in the magnet. The momentum of a charged particle can be determined by the transverse displacement produced by the magnet.

4.6.2.2 *Koerts Magnet

As more complicated example, consider next an axially symmetric *Koerts Magnet*. An example of such a magnet is illustrated in Figure 4.9. A Koerts magnet is a donut shaped magnet with a cylindrical symmetric magnetic field in the mid-plane

$$\mathbf{B} = B_{\underset{\sim}{z}}(\rho)\hat{\mathbf{z}}. \tag{4.127}$$

The vector potential can be written in the form

$$\mathbf{A} = A_\phi(\rho)\lambda_\phi. \tag{4.128}$$

In this case, the Lagrangian is cyclic in the azimuthal angle ϕ, and the associated canonical momentum is conserved. All magnetic flux is returned within the active cross-sectional area of the donut so that $\Delta A_\phi = 0$ for particles that

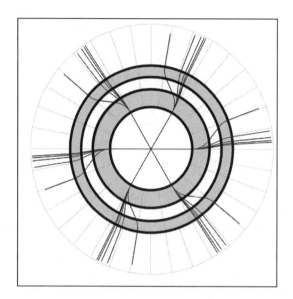

FIGURE 4.9 In a donut-shaped Koerts magnet, angular momentum is conserved in the midplane. The inner magnetic pole faces consist of an azimuthal symmetric ring whose magnetic flux is returned through the outer ring. Selected particle trajectories are shown for several scattering angles. At each scattering angle, different momenta are split by the magnetic field. All transmitted particles scattered from a target at the center of the magnet point directly back to the target, since they have zero angular momentum. Low-momentum particles are trapped in the magnet and loop back to the target. The momentum of the transmitted particles can be determined by measurement of the net rotation angle generated by the magnetic field.

completely transverse the magnet. This implies that the Newtonian angular momentum is also conserved at the entrance and exit of the magnet (see Exercise 4.18). The second constant of integration is the particle's kinetic energy

$$E = T = \frac{1}{2}m(\dot{\rho}^2 + \rho^2\dot{\phi}^2) = \frac{1}{2m}\left(p_\rho^2 + \frac{1}{\rho^2}(p_\phi - q_eA_\phi(\rho))^2\right). \quad (4.129)$$

High-momentum particles are transmitted. Low-momentum particles are reflected back to the target. The deflection angle for particles that are transmitted is a function of their momenta. The momentum of a charged particle scattered from the origin can be determined experimentally by its net rotation angle in the magnet. This result is independent of the detailed field distributions, provided that cylindrical and midplane symmetries are maintained in the manufacturing process.

4.7 SUMMARY

Hamilton's principle states that of all possible paths that a system could choose, the dynamical path actually selected is the path that makes the action an extremum

$$\delta S = \delta \int_1^2 L(q, \dot{q}, t) dt = 0. \tag{4.130}$$

The trajectory of a particle satisfying this criterion is given by the solution of the Euler-Lagrange equations of motion

$$\frac{d}{dt} \left(\frac{\partial L}{\partial \dot{q}^j} \right) = \frac{\partial L}{\partial q^j}. \tag{4.131}$$

The solution of the Euler-Lagrangian equations for the free motion of a particle on a manifold is a geodesic

$$\frac{d\dot{q}^i}{ds} + \Gamma^i_{jk} \dot{q}^j \dot{q}^k = 0. \tag{4.132}$$

The preceding analysis assumes that the coordinate degrees of freedom are all independent. If, instead, there are a number of differential constraints of the form

$$B^k_i dq^i - C^k dt = 0 \tag{4.133}$$

acting on the system, the equations of motion using Lagrange's method of undermined multipliers become

$$\frac{d}{dt} \left(\frac{\partial L}{\partial \dot{q}^i} \right) - \frac{\partial L}{\partial q^i} = Q_i^{(c)} = \Lambda_k B^k_i, \tag{4.134}$$

where $Q_i^{(c)}$ are the unknown constraint forces and Λ_k are undermined multipliers.

Noether's theorem states that for any continuous symmetry of the motion, there exists a corresponding conserved quantity. Let $q^i(s, t)$ be some continuous variation of the coordinates with respect to a time-independent parameter s. If a Lagrangian is cyclic in s

$$\frac{\partial L}{\partial s} = 0, \tag{4.135}$$

then the following physical observable is a constant of the motion

$$C_s = \left(\frac{\partial L}{\partial \dot{q}^i} \frac{\partial q^i}{\partial s} \right)\Bigg|_{s=0}. \qquad (4.136)$$

Symmetry plays a major role in the design and analysis of experiments.

4.8   EXERCISES

1. Solve the Atwood's machine problem described in Section 4.4.1 by eliminating the constraints, giving a Lagrangian with one coordinate degree of freedom. Compare the solution with the solution given in the text using the method of undermined multipliers.

2. Solve the wheel rolling down an incline problem described in Section 4.4.2 by eliminating the constraints, giving a Lagrangian with one coordinate degree of freedom. Compare the solution with the solution given in the text using the method of undermined multipliers.

3. Prove that the cycloid curve defined by

$$x(\phi) = a(\phi - \sin \phi), \quad y(\phi) = a(1 - \cos \phi)$$

is a solution to the Minimum Time (Brachistochrone) problem.

4. Find the period of a particle in a uniform gravitational field constrained to move on a single arc of a cycloid, and show that the period is independent of the point from which the particle is released.

5. By using the method of undetermined multipliers and the constraint condition $r^2 = a^2$, show that the geodesics of the sphere satisfy the equations

$$\frac{d}{ds}\left(\frac{d\mathbf{r}}{ds} \right) = -2\Lambda\mathbf{r}, \quad \frac{d\mathbf{r}}{ds} = \mathbf{e}_i \frac{dx^i}{ds},$$

where ds is an element of length defined along the path, dx^i/ds are the directional derivatives, and $\hat{\mathbf{v}} = d\mathbf{r}/ds$ is a unit vector tangent to the direction of motion. Solve explicitly for the undetermined multiplier $\Lambda(r)$. Show thereby that the geodesics of a sphere are great circles. Replace $s = vt$ to show that a particle moving with constant velocity $\mathbf{v} = v\,\hat{\mathbf{v}}$ experiences a centripetal acceleration $\ddot{\mathbf{r}} = -v^2\hat{\mathbf{r}}/r$, which is exactly the amount required to keep the particle moving on a circular path on the sphere.

6. Apply Fermat's principle to light moving in the atmosphere of the Earth. Assume that the index of refraction of air is a function only of its radial distance from the center of the Earth.

   a. Derive the following expression for a ray of light in the atmosphere:

   $$\frac{dr}{d\phi} = r(k^2 n^2 r^2 - 1)^{1/2},$$

   where k is a constant, and (r, ϕ) are the two-dimensional polar coordinates of a point on the ray with the center of the Earth as an origin.

   b. If $n(r)$ is proportional to r^m, find the value of m such that a ray lying initially along a tangent direction (parallel to $\hat{\phi}$) remains at a constant distance from the center of the Earth for any value of r.

7. A point mass m is constrained to move without friction on a two-dimensional surface in the absence of external forces. The surface is defined by a set of generalized coordinates (q^1, q^2) such that the square of the distance ds^2 between two infinitesimally close points is given by

   $$ds^2 = \sum_{i=1}^{2} \sum_{j=1}^{2} g_{ij}(q^1, q^2) dq^i dq^j,$$

   where the metric tensor is symmetric and depends on position coordinates.

   a. Construct the Lagrangian for a free particle moving on this surface.

   b. Show that the equations of motion of the particle are given by

   $$\sum_{j=1}^{2} g_{ij} \frac{d^2 q^j}{dt^2} + \frac{1}{2} \sum_{j,k} \left(\frac{\partial g_{ij}}{\partial q^k} + \frac{\partial g_{ik}}{\partial q^j} - \frac{\partial g_{jk}}{\partial q^i} \right) \frac{dq^j}{dt} \frac{dq^k}{dt} = 0.$$

   $$j, k = 1, 2.$$

   c. Introduce the inverse g^{ij} of the metric tensor, such that $g^{ik} g_{kj} = \delta^i_j$, and derive the equivalent equations of motion

   $$\frac{d^2 q^i}{dt^2} + \sum_{j,k} \Gamma^i_{jk} \frac{dq^j}{dt} \frac{dq^k}{dt} = 0, \quad j, k = 1, 2;$$

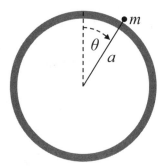

FIGURE 4.10   Particle sliding on a hoop.

where the affine connection is defined by

$$\Gamma^{i}_{jk} = \frac{1}{2}\sum_{m} g^{im}\left(\frac{\partial g_{mj}}{\partial q^{k}} + \frac{\partial g_{mk}}{\partial q^{j}} - \frac{\partial g_{jk}}{\partial q^{m}}\right).$$

   d. Prove that the kinetic energy is conserved and that the particle has a constant speed. Therefore, show that the motion of a free particle on a manifold is a geodesic.

8. A particle of mass m in a constant vertically oriented gravitational field **g** slides without friction on the surface of a fixed, vertically oriented cylindrical hoop with radius a, as shown in Figure 4.10. Using the method of undetermined multipliers, find the polar angle at which the particle falls off, assuming that it starts at rest from the top of the hoop. What is its speed at that point?

9. Solve the block sliding down a movable wedge problem described in Section 3.3.1.1 and Figure 3.2 using the method of undetermined multipliers to extract the normal force of constraint between the block and wedge.

10. The Lagrangian for a heavy symmetrical top spinning with one point fixed in a uniform gravitational field is given by

$$L = T - V = \frac{1}{2}I_{1}[\dot{\theta}^{2} + \dot{\phi}^{2}\sin^{2}\theta] + \frac{1}{2}I_{3}(\dot{\phi}\cos\theta + \dot{\psi})^{2} - mg\ell\cos\theta,$$

where (ϕ, θ, ψ) are the Euler angles and $I_{3}, I_{1} = I_{2}$ are constant momentums of inertia. By the symmetries of the problem, there are three constants of the motion. Identify them and represent them as functionals of the angular coordinates and velocities.

11. A two dimensional system is a solution to the Lagrangian

$$L = \frac{1}{2}\dot{q}_1\dot{q}_2 - \frac{\omega_0^2}{2}q_1q_2.$$

    a. Find a solution to the equations of motion and give a physical interpretation of the system.

    b. This Lagrangian is invariant under the continuous scale transformation

$$q_1 \rightarrow e^{\lambda}q_1, \quad q_2 \rightarrow e^{-\lambda}q_2.$$

Use Noether's theorem to find the conserved quantity associated with this invariance. Interpret its meaning.

12. Neither energy nor momentum are conserved for the Lagrangian

$$L = e^{\gamma t}\left(\frac{m\dot{q}^2}{2} - \frac{kq^2}{2}\right),$$

since the Lagrangian is neither cyclic in time nor in the generalized coordinate. Find the equations of motion and describe the system. Make the transformation $s = e^{\gamma t/2}q$. Show that in the new system of coordinates there is a conserved "energy." Solve the resulting equations of motion and interpret the results.

13. A uniform ladder of mass m and length L leaning against a vertical wall is free to slide in the absence of friction (see Figure 4.11). If the top of the ladder is at a height H_0 with respect to the level floor when it begins to slide, show that the ladder loses contact when it height has dropped to $\frac{2}{3}H_o$.

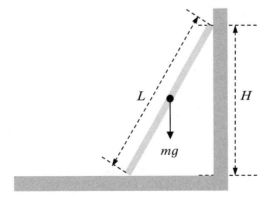

FIGURE 4.11 **Ladder sliding without friction along wall and floor.**

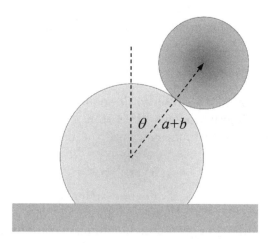

FIGURE 4.12   **A solid ball is rolling without slipping along the surface of a fixed ball.**

14. A solid rubber ball of mass m and radius a is placed on top of a larger
fixed sphere with radius b, as shown in Figure 4.12. It sees a uniform
gravitational field g. It begins to roll without slipping along the surface
of the larger sphere. Assume its path is a great circle. Find the polar
angle on the fixed sphere at which the ball loses contact with the
surface. This angle is independent of the ratio of the two radii. (The
momentum of inertia of a solid sphere is $\frac{2}{5}mr^2$ where r is its radius.)

15. A particle is free to move on the surface of a right circular cone with
half-vertex angle θ_0 (see Figure 4.13). The position of the particle is
given in spherical polar coordinates by the radial distance from the
vertex r and the azimuthal angle ϕ.

a. Show that the geodesics for this surface satisfy the equation

$$r\frac{d^2r}{d\phi^2} - 2\left(\frac{dr}{d\phi}\right)^2 = r^2 \sin^2 \theta_0.$$

b. Show that the solution to this equation is given by $r = r_0 \sec(\phi - \phi_0)\sin^2 \theta_0$.

16. The Lagrangian for a free particle in generalized curvilinear coordinates is $L = mg_{ij}\dot{q}^i\dot{q}^j/2$.

Find the canonical momenta $p_i = \partial L/\partial \dot{q}^i$ and show that the Hamiltonian can be written as $H = g^{ij}p_ip_j/2m$.

17. Use the result of exercise (4–16) to find the Hamiltonian of a free
particle in spherical coordinates as a function of the generalized

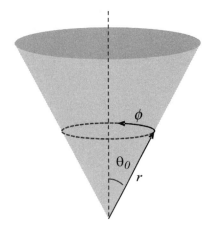

FIGURE 4.13   **A particle is free to move without friction on the surface of a cone of revolution.**

coordinates and momenta. Find Hamilton's equations of motion and identify two constants of the motion for this coordinate system.

18. A charged particle of mass m and charge e is moving in the (x, y) mid-plane of a cylindrically symmetric magnet $\mathbf{B} = B_z(\rho)\hat{\mathbf{z}}$. The vector potential can be written in the form $\mathbf{A} = A_\phi(\rho)\boldsymbol{\lambda}^\phi = \rho^{-2}A_\phi(\rho)\boldsymbol{\lambda}_\phi$. Derive the Lagrangian for this system in (ρ, ϕ) coordinates. Find two constants of the motion. Show that if $\int_1^2 dA_\phi = \int_1^2 \rho B_z(\rho)d\rho = 0$, that the Newtonian angular momentum $m\rho^2\dot{\phi}$ of a particle is conserved between the two points.

19. A charged particle of mass m and charge e is moving in the (x, y) mid-plane of a symmetric magnet $\mathbf{B} = B_z(x)\hat{\mathbf{z}}$ where the y coordinate is cyclic. The vector potential can be written in the form $\mathbf{A} = A_y(x)\hat{\mathbf{y}}$. Find the Lagrangian for this system in (x, y) coordinates. Find two constants of the motion. Show that if $\int_1^2 dA_y = \int_1^2 B_z(x)dx = 0$, that the Newtonian momentum $m\dot{y}$ is conserved between the two points.

CENTRAL FORCE MOTION

Chapter **5**

In this chapter, the two-body central force problem will be reduced to an effective one-body problem. Dynamical features common to all central force problems will be examined before turning to the important case of the inverse-square force law that accounts for the Kepler orbits of the planets. The restricted three-body problem will then be examined as a vehicle for exploring stability and chaos in deterministic Hamiltonian systems. Basic properties of the gravitational field will be discussed.

5.1 CENTRAL FORCE INTERACTIONS

Neither the geocentric nor the heliocentric view of the solar system has any special claim to represent unvarnished truth. If Galileo is correct, only relative motion is physically relevant, and celestial objects, in effect, orbit each other. The two paradigms represent different points of view, not underlying reality. From the viewpoint of Newtonian dynamics, it is preferential to view the sun-planet system from a special inertial frame of reference called the *center of mass frame*. In the center of mass frame of an isolated two-body system, the motion of bound celestial objects are co-rotating ellipses moving about the stationary center of mass.

The center of mass transformation reduces the two-body problem to an effective one-body problem. If the force law is central, then angular momentum is conserved, and, as a consequence, the relative motion is in a plane perpendicular to the orientation of the angular momentum vector. Using integration constants, the dynamical problem can be further reduced to an effective one-dimensional problem, which is directly solvable by reduction to quadrature.

After Newton solved the two-body Kepler problem, the next logical step was to examine interactions involving three or more planetary bodies. This proved to be great deal more challenging, as Newton had predicted; so challenging, in fact, that the three-body problem has never been fully solved. Henri Poincaré (1854–1912) demonstrated that there were simply not enough constants of the motion to fully reduce the three-body system to quadrature [Barrow-Green 1997]. This left unanswered important questions as to the stability of the Sun-Moon-Earth system and the Sun-Earth-Jupiter systems.

In his studies, Poincaré discovered that, under certain initial conditions, the long-term behavior of these systems defy prediction. If the results become so sensitive to initial conditions that prediction becomes effectively impossible, the motion then said to be *chaotic*. The significance of Poincaré's work was not fully appreciated at the time, and his comments regarding chaos were largely ignored. Chaos is not an exotic anomaly, however, but rather a common feature of non-linear dynamical system with three or more degrees of freedom. The recent development of modern computational algorithms and sophisticated mathematical toolkits makes realistic simulations of these effects possible. The *Circular Restricted Three-Body Problem* (CRTBP) is a simple case study that can easily be understood, yet complex enough to illustrate some of the important underlying dynamical issues. The CRTBP will be examined in some detail, after first exploring the underlying two-body dynamics governing the motion of Kepler bodies.

5.2   LAGRANGIAN FORMULATION OF THE TWO BODY PROBLEM

The Lagrangian for a two-body system, assuming a general central force potential, is a straightforward sum of the kinetic and potential energy contributions

$$L = T - V = \frac{1}{2}m_1\dot{\mathbf{r}}_1 \cdot \dot{\mathbf{r}}_1 + \frac{1}{2}m_2\dot{\mathbf{r}}_2 \cdot \dot{\mathbf{r}}_2 - V(|\mathbf{r}_1 - \mathbf{r}_2|) \qquad (5.1)$$

The coordinates of the two bodies of masses m_1, m_2 in an arbitrary inertial frame are defined in Figure 5.1, where $\mathbf{r}_{\text{cm}}$ is the center of mass position from the origin and $\mathbf{r} = \mathbf{r}_1 - \mathbf{r}_2$ is the relative displacement vector of the two bodies.

5.2.1   Separation of the Motion

The two-body Lagrangian can be separated into two one-body Lagrangians by making a transformation into center of mass coordinates

$$\mathbf{r}_1 = \mathbf{r}_{\text{cm}} + \mathbf{r}_1', \quad \mathbf{r}_2 = \mathbf{r}_{\text{cm}} + \mathbf{r}_2', \qquad (5.2)$$

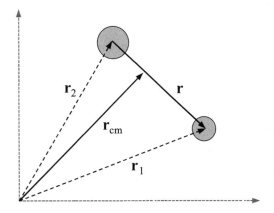

FIGURE 5.1   Center of mass coordinate system for the two-body problem.

where $\mathbf{r}_{\mathrm{cm}}$ is the center of mass point defined as

$$m_\Sigma \mathbf{r}_{\mathrm{cm}} = m_1 \mathbf{r}_1 + m_2 \mathbf{r}_2, \tag{5.3}$$

and where the total mass or mass sum is denoted by $m_\Sigma = m_1 + m_2$.

The vectors of the system relative to the center of mass point are given by

$$\mathbf{r}'_1 = (m_2/m_\Sigma)\mathbf{r}, \quad \mathbf{r}'_2 = (-m_1/m_\Sigma)\mathbf{r}, \tag{5.4}$$
$$\mathbf{r} = \mathbf{r}_1 - \mathbf{r}_2 = \mathbf{r}'_1 - \mathbf{r}'_2. \tag{5.5}$$

Here, $\mathbf{r}$ is the relative vector displacement of the two mass points. This transformation results in the separation of the Lagrangian

$$L = L_{\mathrm{cm}}(\mathbf{r}_{\mathrm{cm}}, \dot{\mathbf{r}}_{\mathrm{cm}}) + L'(\mathbf{r}, \dot{\mathbf{r}}), \tag{5.6}$$

where

$$L_{\mathrm{cm}}(\mathbf{r}_{\mathrm{cm}}, \dot{\mathbf{r}}_{\mathrm{cm}}) = \frac{1}{2} m_\Sigma \dot{\mathbf{r}}_{\mathrm{cm}} \cdot \dot{\mathbf{r}}_{\mathrm{cm}}, \tag{5.7}$$

and

$$L'(\mathbf{r}, \dot{\mathbf{r}}) = \frac{1}{2} m \dot{\mathbf{r}} \cdot \dot{\mathbf{r}} - V(r), \tag{5.8}$$

L_{cm} represents the one-body Lagrangian for a freely moving point particle of total mass m_Σ located at the center of mass, and $L'(\mathbf{r}, \dot{\mathbf{r}})$ is a effective one-body Lagrangian for a particle moving in a central force field with effective reduced mass m given by

$$m = m_1 m_2 / (m_1 + m_2). \tag{5.9}$$

If $m_1 \ll m_2$ the heavier body is nearly stationary and the reduced mass of the orbiting system can be approximated by mass of lighter particle $m \sim m_1$. Therefore, the lighter planets appear to orbit a stationary sun as preferred by the heliocentric theory of Kepler. This analysis illustrates an important result: If the Lagrangian separates into two additive pieces with respect to two sets of coordinates, the equations of motion also separate with respect to these sets of coordinates.

If one uses the Cartesian basis $\mathbf{r}_{\text{cm}} = x^i_{\text{cm}} \mathbf{e}_i$, the center of mass motion is cyclic in all three of the center of mass coordinates. The center of mass Lagrangian

$$L_{\text{cm}}(\dot{x}) = \frac{1}{2} m_\Sigma \delta_{ij} \dot{x}^i_{\text{cm}} \dot{x}^j_{\text{cm}}, \tag{5.10}$$

has three conserved components of linear momenta

$$p_i = \frac{\partial L}{\partial \dot{x}^i_{\text{cm}}} = \frac{\partial L_{\text{cm}}}{\partial \dot{x}^i_{\text{cm}}} = m_\Sigma \delta_{ij} v^i_{\text{cm}}, \tag{5.11}$$

where $\mathbf{p}$ is the total momentum of the system. The equations can be integrated immediately to give uniform linear motion of the center of mass

$$\dot{x}^i_{\text{cm}}(t) = \dot{x}^i_{\text{cm}}(0) + v^i_{\text{cm}} t, \tag{5.12}$$

where $\mathbf{v}_{\text{cm}} = \mathbf{p}/m_\Sigma$ is the velocity of the center of mass point.

Since the motion of the center of mass point is trivial and decouples from the internal motion, for the remainder of the chapter, one can safely assume an inertial frame in which the center of mass is at rest. One can then concentrate on solving the problem of the reduced one-body problem, which is a solution of the reduced Lagrangian

$$L'(\mathbf{r}, \dot{\mathbf{r}}) = \frac{1}{2} m \dot{\mathbf{r}} \cdot \dot{\mathbf{r}} - V(r). \tag{5.13}$$

The relative motion of the two bodies about their center of mass is given by an effective one-body interaction force law

$$m \ddot{\mathbf{r}} = \mathbf{F}(r) = -\nabla V(r), \tag{5.14}$$

where $\mathbf{F} = \mathbf{F}_1 = -\mathbf{F}_2$ by the law of action-reaction.

To complete the solution, we need to choose a coordinate system for the relative motion. The central-force interaction is spherically symmetric. The angular momentum in the center of mass frame $\mathbf{L} = m \, \mathbf{r} \times \mathbf{v}$ is conserved,

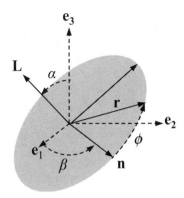

FIGURE 5.2   **Plane polar coordinate system for the reduced one-body problem.**

and the motion is restricted to the $\mathbf{r} \wedge \mathbf{v}$ plane, since $\mathbf{r} \wedge \mathbf{F} = 0$. In spherical coordinates $\mathbf{F} = f(r)\hat{\mathbf{r}}$, where $f(r) = -\partial V(r)/\partial r$.

The Lagrangian simplifies further if one uses plane polar coordinates, where the axis of the plane is in the direction of angular momentum $\mathbf{L}$, as shown in Figure 5.2. Without loss of generality, one can pick a coordinate system where $z' = 0$ defines the plane of motion and the origin coincides with the center of force. In these coordinates, the angular momentum of the reduced one-body problem is

$$\mathbf{L} = \ell \, \mathbf{e}_3'. \tag{5.15}$$

In the primed coordinate system, $\mathbf{e}_1'$ is the *line of nodes* where the plane of motion intersects the laboratory $\mathbf{e}_1 \wedge \mathbf{e}_2$ plane. The rotation angle ϕ of the planar trajectory can be measured relative to the line of nodes. The reference angles are defined by

$$\mathbf{e}_3 \cdot \mathbf{e}_3' = \cos \alpha, \quad \mathbf{e}_1 \cdot \mathbf{e}_1' = \cos \beta, \quad \mathbf{e}_1' \cdot \hat{\mathbf{r}} = \cos \phi. \tag{5.16}$$

In plane polar coordinates, the reduced Lagrangian is

$$L'(r, \dot{r}, \dot{\phi}) = \frac{m}{2}(\dot{r}^2 + r^2 \dot{\phi}^2) - V(r). \tag{5.17}$$

5.2.2 First Integrals of the Reduced One-Body Problem

The reduced Lagrangian is cyclic in both the azimuthal angle and time, so two immediate first-integrals of the motion are the total angular momentum and total mechanical energy of the reduced system. Lagrange's equations of motion are now

$$\dot{p}_\phi = \frac{d}{dt}\left(\frac{\partial L'}{\partial \dot{\phi}}\right) = \frac{d}{dt}(mr^2\dot{\phi}) = 0, \qquad (5.18)$$

$$\dot{p}_r = \frac{d}{dt}\left(\frac{\partial L'}{\partial \dot{r}}\right) = m\ddot{r} = -\frac{\partial V(r)}{\partial r} + mr\dot{\phi}^2. \qquad (5.19)$$

The first equation can be immediately integrated to get the constant of integration ℓ

$$p_\phi = mr^2\dot{\phi} = \ell. \qquad (5.20)$$

This equation then can be inverted to eliminate the angular velocity

$$\dot{\phi} = \frac{\ell}{mr^2}. \qquad (5.21)$$

The result, when applied to the radial equation of motion, is an effective one-dimensional potential $V_{\text{eff}}(\ell, r)$ when viewed in the corotating frame

$$m\ddot{r} = -\frac{\partial V_{\text{eff}}(\ell, r)}{\partial r}, \qquad (5.22)$$

where

$$V_{\text{eff}}(\ell, r) = V(r) + \frac{\ell^2}{2mr^2}. \qquad (5.23)$$

The second term in the effective potential is the contribution due to the fictional centrifugal force, which is always repulsive.

The second of Lagrange's equations (5.19) can now be integrated to get a second constant of integration. The result is equivalent to invoking energy conservation in the reduced system of coordinates

$$E = \frac{1}{2}m\dot{r}^2 + V_{\text{eff}}(r) = \frac{1}{2}m\dot{r}^2 + \frac{\ell^2}{2mr^2} + V(r). \qquad (5.24)$$

5.2.3 Energetic and Stability Considerations

Figure 5.3 shows the effective potential energy diagram for a particle trapped in a central force potential with a local minimum at r_1 and with the nearest maximum located at r_2. For bound orbits, there are two turning points (the apsides or apsidal turning points): one for the distance of closest approach r_{min} (which may be zero) and a second for the maximum radius r_{max}. There is a stationary solution, yielding a circular orbit for $r_{\text{min}} = r_{\text{max}} = r_1$, where

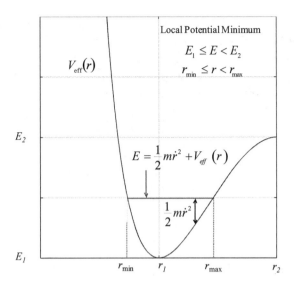

Local Potential Minimum

$E_1 \leq E < E_2$

$r_{min} \leq r < r_{max}$

$V_{eff}(r)$

$E = \dfrac{1}{2}m\dot{r}^2 + V_{eff}(r)$

$\dfrac{1}{2}m\dot{r}^2$

E_2

E_1

r_{min} r_1 r_{max} r_2

FIGURE 5.3 The effective potential plot for a bound particle bound in a region with a local minimum r_1. The closest maximum is located at r_2. If the energy of the particle is in the range $E_1 \leq E < E_2$, the motion has two apsidal turning points located at the limits of the interval $[r_{min}, r_{max}]$. When $E = E_1$, the particle motion is a circle with radius r_1.

the particle has a local energy minimum E_1. There are no physical solutions in regions for which $(E - V_{eff}(r)) < 0$.

Solving for the radial velocity gives two branches for the solution depending on whether the radial velocity is increasing or decreasing. (For the initial value problem, time is monotonically increasing.) The turning points occur when $E - V_{eff}(r)$ equals zero

$$\dot{r}^2 = \frac{2}{m}(E - V_{eff}(r)) \geq 0, \tag{5.25}$$

leading to the branch cuts

$$\frac{dr}{dt} = \pm\sqrt{2(E - V_{eff}(r))/m}. \tag{5.26}$$

Borrowing from the language of planetary motion, the two turning points correspond to the perihelion (minimum) and aphelion (maximum) distances from the center of force. These can be parameterized by

$$r_{min} = a(1 - e), \quad r_{min} + r_{max} = 2a, \quad r_{max} = a(1 + e), \tag{5.27}$$

where a is some average measure of the radius of the orbit, and e is the *eccentricity* of the orbit. The geometric mean is given by $r_{min}r_{max} = a^2(1 - e^2)$.

Note that this definition of eccentricity is a generalization of the usual definition defined for elliptical planetary orbits, and is valid for $0 \le e \le 1$.

Stationary points of the motion, occurring when $\dot{r} = \ddot{r} = 0$, are attractors occurring at critical points of the potential. They are found by evaluating the points at which the effective force vanishes

$$f_{\text{eff}}(r_1) = -\left.\frac{\partial V_{\text{eff}}}{\partial r}\right|_{r=r_1} = \frac{\ell^2}{mr_1^3} - \left.\frac{\partial V}{\partial r}\right|_{r=r_1} = 0. \tag{5.28}$$

The angular frequency for a circular orbit at the stationary point r_1 is given by solving the force equation assuming uniform circular motion

$$ma = -m\omega_\phi^2 r_1 = f(r_1) = -\left.\frac{\partial V}{\partial r}\right|_{r=r_1} = -\frac{\ell^2}{mr_1^3}, \tag{5.29}$$

yielding an angular period for rotational motion

$$T_\phi = \frac{2\pi}{\omega_\phi}. \tag{5.30}$$

The angular frequency can be found from Equation 5.29

$$\omega_\phi^2 = \frac{\ell^2}{m^2 r_1^4} = -\frac{f(r_1)}{mr_1}. \tag{5.31}$$

To compare the angular frequency of rotation with the angular frequency of radial oscillations, expand the effective potential in a power series $\eta = r - r_1$ about the energy minimum, giving

$$V_{\text{eff}}(r) = V_{\text{eff}}(r_1) + \left.\frac{\partial V_{\text{eff}}(r)}{\partial r}\right|_{r=r_1} \eta + \frac{1}{2}\left.\frac{\partial^2 V_{\text{eff}}(r)}{\partial r^2}\right|_{r=r_1} \eta^2 + \cdot \tag{5.32}$$

The first term is an ignorable constant, while the second term vanishes at a stationary point. In the small oscillation limit, the motion about a minimum is a simple harmonic in character

$$m\ddot{\eta} \cong -\left.\frac{\partial^2 V_{\text{eff}}}{\partial r^2}\right|_{r=r_1} \eta. \tag{5.33}$$

The criteria for stability of the orbits under small perturbations η is given by requiring that the effective force be restorative

$$\left.\frac{\partial^2 V_{\text{eff}}}{\partial r^2}\right|_{r=r_1} = m\omega_r^2 > 0. \tag{5.34}$$

The period for small radial oscillations is

$$T_r = \frac{2\pi}{\omega_r}.$$ (5.35)

Except for the special case of the Kepler force, the radial and angular periods are unlikely to be the same. The ratio of the periods determines whether the orbits are closed or open. If the ratio of periods are commensurate, satisfying

$$\left(\frac{T_r}{T_\phi}\right)^{-1} = \frac{\omega_r}{\omega_\phi} = \sqrt{\left.\frac{mr^4 \partial^2 V_{\text{eff}}/\partial r^2}{\ell^2}\right|_{r=r_0}} = \frac{n_r}{n_\phi},$$ (5.36)

where (n_r, n_ϕ) are integers, the orbits are closed. Otherwise, they are open and fill the space between the limits $[r_{\min}, r_{\max}]$. The second, space-filling, case is infinitely more likely, since the irrational numbers are dense compared to the rational numbers.

   This puts the exceptional character of the inverse-square law into prospective. In fact, there are only two central force laws that have closed orbits for all bound states: a) the Kepler potential $V = -kr^{-1}$, where the law of periods is given by Kepler's third law with $T_r = T_\phi$; and b) the central harmonic oscillator potential $V = -kr^2/2$, where all the periods are identical and $T_r = T_\phi/2$. In both cases, the orbits are ellipses as shown in Figure 5.4. And, in both cases, the universal closures of the bound orbits are indications of a deeper symmetry principle in action.

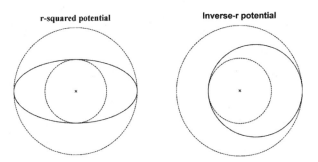

FIGURE 5.4   Comparison of bound orbits for the harmonic r-squared (left-hand side) and Kepler inverse-r (right-hand side) potentials: Note that the shapes of bound orbits are ellipses in either case, but that the position of the center of force is at the center of the ellipse for the harmonic case, and at one of the foci, for the Kepler case. The circles indicate the perihelion and aphelion distances for the cases shown.

5.2.4 Power Law Potentials

Power law potentials have the general form

$$V = kr^n/n, \quad f(r) = -kr^{n-1}. \tag{5.37}$$

The force is attractive if $k > 0$. Such potentials have a single minimum or maximum as illustrated by the three sample potentials shown in Figure 5.5. Note that if $n > 0$ only bound orbits are allowed, whereas for $n < 0$ both bound and unbound orbits are possible. Potentials with $n > -2$ have a single minimum, while for $n < -2$ there is a single maximum.

Analytic solutions to the orbit equation $r(\phi)$ can be found in terms of elementary (trigonometric and exponential) functions for $n = \{-2, -1, 2\}$, and in terms of elliptic functions for $n = \{-6, -4, -3, +1, +4, +6\}$. A number of fractional powers can be solved in terms of elliptic integrals, and a greater number of exponents can by solved using hypergeometric functions [Goldstein 2002, 88–89]. The maximum or minimum of a power law interaction are located at $r = r_0$ where

$$-f(r_o) = kr_0^{n-1} = \frac{\ell^2}{mr_0^3}. \tag{5.38}$$

The angular frequency of rotation for stationary circular orbits is given by

$$\omega_\phi^2 = \frac{k}{m}r_0^{n-2} = \frac{-f(r_0)}{mr_0}. \tag{5.39}$$

Note that for the harmonic potential $n = 2$, all circular orbits have the same angular frequency $\omega_\phi = \sqrt{k/m}$. In general, the law of angular periods for

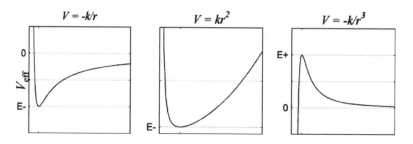

FIGURE 5.5   Three attractive power law potentials of the form $V = kr^n/n$ with $n = \{2, -1, -3\}$. All such potentials (except $n = -2$) have a single extremum of the effective potential $V_{\text{eff}} = (\ell^2/2mr^2 + V)$. If $n > -2$ the result is a local minimum, whereas, if $n < -2$ the result is a local maximum.

circular orbits is given by

$$T_\phi = \frac{2\pi}{\sqrt{k/m}} r_0^{-(n-2)/2}. \tag{5.40}$$

This yields the $r^{3/2}$ behavior expected for the Kepler potential. Likewise, the angular frequencies for small radial oscillations are given by

$$m\omega_r^2 = \left.\frac{\partial^2 V_{\text{eff}}}{\partial r^2}\right|_{r=r_1} = \frac{3\ell^2}{mr_0^4} + (n-1)kr_0^{n-2} = (2+n)\frac{\ell^2}{mr_0^4}$$

$$= -(2+n)\frac{f(r_0)}{r_0} > 0, \tag{5.41}$$

which requires that $n > -2$ for stable attractive power law potentials. Comparing the radial and angular periods for small oscillations gives

$$\lim_{\eta \to 0} \frac{T_\phi}{T_r} = \sqrt{2+n}, \tag{5.42}$$

giving $T_\phi/T_r = 1$ for the inverse r-squared law, and $T_\phi/T_r = 2$ for the spherically symmetric harmonic oscillator. If, instead of expanding the time-dependence, one expands the angular dependence, the small oscillation result is

$$\frac{\Delta\phi_r}{2\pi} = \frac{1}{\sqrt{2+n}}, \tag{5.43}$$

where $\Delta\phi_r$ is the net angular rotation needed for one radial period. *Bertrand's theorem* states that, for all orbits to close, this ratio must be a constant,[1] restricting the central force laws that have all bound orbits closed to the Kepler ($n = -1$) and Harmonic ($n = 2$) potential cases. A proof of the theorem can be found in [José 1998, 88–92].

5.2.5 Solution to the Orbit Equations

The radial equation can be integrated by solving the energy equation for the radial velocity

$$\frac{dr}{dt} = \pm\sqrt{2(E - V_{\text{eff}})/m}. \tag{5.44}$$

---

[1] If $\Delta\phi_r(E, \ell)$ were to be allowed to vary continuously, it must pass through irrational multiples of 2π unless $n = \{-1, 2\}$.

By choosing the positive branch of the integral for increasing $r(t) > r_{min}$, one gets

$$t - t_{min} = \int_{r_{min}}^{r} \frac{dr}{\sqrt{2(E - V_{eff})/m}} = \int_{1/r}^{1/r_{min}} \frac{du/u^2}{\sqrt{\frac{2}{m}(E - V(\frac{1}{u})) - \frac{\ell^2 u^2}{2m}}}, \quad (5.45)$$

where $u = 1/r$ is a commonly made, sometimes useful, substitution. The negative branch can be found by reflection at the turning points. Time-reversal invariance or reflection symmetry about the turning points can be used to solve for the second branch. Once $r(t)$ is known by inversion of the resulting integral for $t(r)$, one can then solve for the cyclic variable by direct integration

$$\phi - \phi_0 = \int_{t_0}^{t} \frac{\ell}{mr^2(t)} dt. \quad (5.46)$$

Therefore, the central force problem can be explicitly reduced to quadrature.

It is often more useful to eliminate the time and solve for the orbit $r(\phi)$ directly. This leads to first-order equations of the form

$$\frac{dr}{dt} = \frac{dr}{d\phi} \frac{\ell}{mr^2} = -\frac{du}{d\phi} \frac{\ell}{m} = \pm\sqrt{\frac{2}{m}(E - V_{eff}(r))}, \quad (5.47)$$

which lead to the following quadratures

$$\int d\phi = \pm \int \frac{dr}{\sqrt{\frac{2mr^4}{\ell^2}(E - V_{eff}(r))}}$$

$$= \mp \int \frac{du}{\sqrt{\frac{2m}{\ell^2}(E - V_{eff}(\frac{1}{u}))}}. \quad (5.48)$$

Alternatively, by directly substituting

$$\frac{d^2r}{dt^2} = \frac{-\ell}{m} \frac{d}{dt} \frac{du}{d\phi} = \frac{-\ell^2 u^2}{m^2} \frac{d^2u}{d\phi^2} \quad (5.49)$$

into Lagrange's second-order radial differential equation (5.22) one gets

$$\frac{d^2u}{d\phi^2} = -u - \frac{m}{\ell^2} \frac{\partial V(1/u)}{\partial u}. \quad (5.50)$$

The orbital trajectories for bound orbits of the central force motion may be closed or space filling within the allowed phase space, depending on whether

precession of orbits

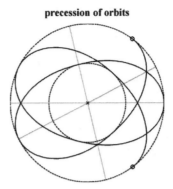

FIGURE 5.6 Orbital trajectories for central force motion are time-reversal invariant and have reflection symmetry about the turning points $dr/d\phi = 0$. The orbits may be closed or space filling within the allowed phase space, depending on whether or not the ratios of the radial and angular periods are commensurate rational fractions or incommensurate irrational fractions.

or not the ratios of the radial and angular periods are commensurate rational fractions or incommensurate irrational fractions of each other. The equations are time reversal invariant. As shown in Figure 5.6, this implies that the solutions are mirror symmetric about the turning points.

5.2.5.1 Phase Space formulation of the central force problem

The Hamiltonian for the effective one-body problem is given by

$$H = T + V = \frac{g^{ij}p_i p_j}{2m} + V(r) = \frac{p_r^2}{2m} + \frac{p_\phi^2}{2mr^2} + V(r), \qquad (5.51)$$

where

$$g^{ij} = \begin{pmatrix} 1 & 0 \\ 0 & 1/r^2 \end{pmatrix} \qquad (5.52)$$

in plane polar coordinates. Hamilton's method gives first-order differential equations, suitable for numerical integration. In plane polar coordinates, this results in four equations of motion given by

$$\dot{r} = p_r/m, \qquad \dot{p}_r = p_\phi^2/mr^3 - \partial V/\partial r,$$
$$\dot{\phi} = p_\phi/mr^2, \qquad \dot{p}_\phi = 0. \qquad (5.53)$$

Substituting $p_r = m\dot{r}$, $p_\phi = \ell$ reproduces the radial equation of motion previously obtained using Lagrange's formulation

$$m\ddot{r} = -\frac{\partial V_{\mathrm{eff}}(r, \ell)}{\partial r} = \frac{\ell^2}{mr^3} - \frac{\partial V}{\partial r}. \qquad (5.54)$$

5.2.5.2 Inverse Fourth Power Potential

The attractive r^{-5} force law, or inverse r-fourth potential, shown in Figure 5.7 has the form

$$V_{\mathrm{eff}}(r) = -k/4r^4 + \ell^2/2mr^2. \qquad (5.55)$$

V_{eff} has a single maximum. Particles with energy $0 < E < V_{\mathrm{eff(max)}}$ are either bounded or unbounded depending on the initial value of r. Bounded orbits spiral into the origin. The unbound orbits have a distance of closest approach. The analytic solution for the bound orbits of inverse r-fourth potential can be expressed in terms of elliptic functions. Let $x = r/r_1$, where r_1 is the maximum radius for bound orbits, then the motion is bounded by $0 \le x \le 1$. The equation to be solved is

$$dx/d\phi = \pm\sqrt{\frac{1}{1+\lambda^2}}\sqrt{(1-x^2)(1-\lambda^2 x^2)}, \quad \lambda^2 = (r_1/r_2)^2, \qquad (5.56)$$

which is an elliptic integral with solution

$$r(\phi) = r_1 sn\left(\sqrt{1+\lambda^2}\phi; \lambda^2\right). \qquad (5.57)$$

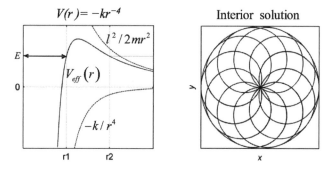

FIGURE 5.7   Motion in an inverse r-fourth potential with $E > 0$: The inner branch of the orbit $0 < r < r_1$, shown on the right, passes though the origin. The unbound exterior branch for $r > r_2$ is not shown.

This solution for the bound orbits simply continues the particle through the center of force. In a more realistic algorithm, a signal would be raised whenever there is a particle collision, which would then call a function to determine how to handle the event.

Special care has to be used in numerical simulations, since the particle velocity goes to infinity at the origin (but the dwell time goes to zero). Polar coordinates should be avoided, since they are undefined at the origin. Making the substitution $dt = ds/v$ leads to a well-behaved algorithm using a Cartesian coordinate system. One can make use of energy conservation,

$$v(E,r) = \sqrt{2(E - V(r))/m} > 0 \quad \text{for } \ell^2 > 0, \tag{5.58}$$

where the radial coordinate r is greater than or equal to zero. Let $V' = dV/dr$, then Newton's force law can be rewritten as

$$m\ddot{x} = m\frac{vd}{ds}\left(\frac{vdx}{ds}\right) = -V'(r)x/r,$$

$$m\ddot{y} = m\frac{vd}{ds}\left(\frac{vdy}{ds}\right) = -V'(r)y/r. \tag{5.59}$$

Next, define the directional derivatives as

$$\frac{dx}{ds} = \cos\theta, \quad \frac{dy}{ds} = \sin\theta, \tag{5.60}$$

where θ is the angle of the velocity vector with respect to the x-axis. This gives the following equations to be solved

$$\left(\cos\theta\frac{dv}{ds} - v\sin\theta\frac{d\theta}{ds}\right) = -\frac{V'(r)}{v}\frac{x}{r},$$

$$\left(\sin\theta\frac{dv}{ds} + v\cos\theta\frac{d\theta}{ds}\right) = -\frac{V'(r)}{v}\frac{y}{r}, \tag{5.61}$$

where

$$r(x,y) = \sqrt{x^2 + y^2}, \quad V'(r) = \frac{\partial V(r)}{\partial r}, \quad \sin\theta = \frac{v_x}{v},$$

$$v^2(E,r) = \sqrt{\frac{2}{m}(E - V(r))}, \quad \frac{dE}{ds} = 0, \quad \sin\phi = x/r. \tag{5.62}$$

$dE/ds = 0$ is just a restatement of energy conservation. Finally, by eliminating dv/ds in Equation 5.61, one gets the following numerical recipe for the first-order differential equations of motion:

$$\frac{dx}{ds} = \cos\theta, \quad \frac{dy}{ds} = \sin\theta,$$

$$\frac{d\theta}{ds} = \frac{V'(r)}{v^2(E,r)}\left(\frac{x}{r}\cos\theta - \frac{y}{r}\sin\theta\right), \quad \frac{dt}{ds} = \frac{1}{v(E,r)}. \tag{5.63}$$

If one is interested only in the orbit and not in the time-dependence, only the first three equations need to be integrated. More importantly, there are no branch cuts to apply. The special case of $\ell = 0$ can be handled separately, as the motion is then one-dimensional in the radial direction.

5.2.5.3 Spherical Harmonic Oscillator Potential

The spherical harmonic oscillator potential can be written as

$$V(r) = kr^2/2 \rightarrow \mathbf{F}(\mathbf{r}) = -k\mathbf{r}. \tag{5.64}$$

The three-dimensional oscillator separates in Cartesian coordinates, giving coupled simple harmonic equations of motion in the plane of motion

$$m\ddot{x} = -kx, \quad m\ddot{y} = -ky, \tag{5.65}$$

with solutions

$$x(t) = r\cos\phi = a\cos\omega(t - t_0),$$
$$y(t) = r\sin\phi = b\sin\omega(t - t_0). \tag{5.66}$$

The orbit equation is that of an ellipse centered at the origin

$$\left(\frac{x}{a}\right)^2 + \left(\frac{y}{b}\right)^2 = 1. \tag{5.67}$$

In plane polar coordinates this becomes

$$r^2(\phi) = \frac{b^2}{1 - e^2\cos^2\phi}. \tag{5.68}$$

All orbits are closed and have identical frequencies that are independent of the amplitude

$$\omega_x = \omega_y = \sqrt{k/m}. \tag{5.69}$$

The orbits form the simplest of Lissajous figures, a closed ellipse centered on the center of force as illustrated in the left-hand image in Figure 5.4.

Transforming the time-dependence into plane polar coordinates gives

$$r(t) = \sqrt{x^2 + y^2} = a\sqrt{1 - e^2 \sin^2 \omega(t - t_0)},$$
$$\tan \phi = \frac{y}{x} = (b/a) \tan (\omega(t - t_0)).$$

(5.70)

Note that the radial period is half that of the angular period, since the angular rotation from one radial maximum to the next radial maximum is π radians.

5.2.6 Viral Theorem

The bounded motion of a system of particles can be used to prove the *viral theorem*.[2] Consider the time-dependent observable

$$G(t) = \sum_i \mathbf{p}_i \cdot \mathbf{r}_i,$$

(5.71)

where the sum runs over all particles in the system. The time derivative of G is

$$\frac{dG(t)}{dt} = \sum_i \dot{\mathbf{p}}_i \cdot \mathbf{r}_i + \sum_i \mathbf{p}_i \cdot \mathbf{v}_i = \sum_i \mathbf{F}_i \cdot \mathbf{r}_i + 2T,$$

(5.72)

where T is the total kinetic energy of the system. Integrating over a time τ that is long compared to the characteristic times of oscillation, and taking the time average gives

$$\left\langle \frac{dG(t)}{dt} \right\rangle_{\text{ave}} = \frac{\int_0^\tau dG}{\tau} = \frac{G(\tau) - G(0)}{\tau} = \left\langle \sum_i \mathbf{F}_i \cdot \mathbf{r}_i \right\rangle_{\text{ave}} + \langle 2T \rangle_{\text{ave}},$$

(5.73)

$$\frac{G(\tau) - G(0)}{\tau} = \left\langle \sum_i \mathbf{F}_i \cdot \mathbf{r}_i \right\rangle_{\text{ave}} + \langle 2T \rangle_{\text{ave}}.$$

(5.74)

If the system is periodic on interval τ, the left-hand side of the preceding equation vanishes. But even if this were not the case, but the motion is bound to some finite region in phase space, the difference is less than some upper bound

$$|G(\tau) - G(0)| < B.$$

(5.75)

For long enough integration times, the left-hand side of Equation 5.74 can be made as small as one wants. Therefore, over a long time scale, one has the

---

[2]Developed by Rudolf Julius Emanuel Clausius (1822–1888), who was one of the pioneers of thermodynamics.

viral theorem

$$\langle T \rangle_{ave} = -\frac{1}{2} \left\langle \sum_i \mathbf{F}_i \cdot \mathbf{r}_i \right\rangle_{ave}. \tag{5.76}$$

The viral theorem is useful for relating the average kinetic and potential energies of power-law potentials. Let $V(r) = kr^n/n$ and $\mathbf{F}(r) = -kr^{n-1}\hat{\mathbf{r}}$, then for an effective single particle system in the center of mass frame

$$\langle T \rangle_{ave} = \frac{1}{2} \langle kr^n \rangle_{ave} = \frac{n}{2} \langle V \rangle_{ave}. \tag{5.77}$$

For the spherical harmonic oscillator potential ($n = 2$), kinetic and potential energy are shared equally $\langle T \rangle_{ave} = \langle V \rangle_{ave}$, while for an inverse-square force ($n = -1$), one gets $\langle T \rangle_{ave} = -\langle V \rangle_{ave}/2$. For a particle in a circular orbit, where the kinetic energy is purely rotational, the viral theorem reduces to

$$\frac{\ell^2}{2mr_0^2} = \frac{1}{2}kr_0^n = -\frac{1}{2}f(r_0)r_0, \quad \frac{\ell^2}{mr_0^3} + f(r_0) = 0, \tag{5.78}$$

which is just a restatement of the condition for a circular orbit expressed in Equation 5.38. The viral theorem plays an important role in the kinetic theory of gases where it can be used to derive the ideal gas law [Goldstein 2002, 85–86].

5.3   INVERSE-SQUARE FORCE LAW

In 1684, astronomer Edmond Halley (1656–1742) asked Newton to investigate the orbit that a body would follow under an inverse-square force law. Newton, generally secretive by nature, immediately replied that it would be an ellipse, a result that he obtained some five years earlier and had promptly buried in his notebooks. Halley quickly realized that Newton had solved one of the great problems of physics. He had successfully demonstrated the inverse-square law of attraction of gravitational bodies.[3] Due to Halley's urging and tact, Newton returned to his studies of gravitation, and the result was his *Principia*. Because the Royal Society had already spent its budget, Halley paid for the first printing of the *Principia* and proofread the copy himself. Halley, a great physicist in his own right, understood and facilitated the greater genius of Newton.

---

[3] In 1784, Charles Coulomb (1736–1806) would demonstrate the existence of a similar inverse-square law of interaction (attraction and repulsion) between electric charges using a torsion balance.

5.3.1 Laws of Planetary Motion

Newton's law of universal gravitation and his theoretical principles of particle dynamics had lead to an elegant result that had reproduced Kepler's empirical formulas and set the stage for a new age of rational analysis of physical phenomena. It is useful to restate Kepler's laws, starting with the second law, which is the easiest to demonstrate.

5.3.1.1 Kepler's Second Law

Kepler's *second law of planetary motion* states that the orbits of the planets sweep out equal areas in equal times. Kepler's second law is a direct consequence of angular momentum conservation for central force interactions. Kepler's second law is best demonstrated by geometric construction as illustrated in Figure 5.8. One can approximate the area swept out by an orbit in a small time interval by a triangle

$$\Delta A = \frac{1}{2}|\mathbf{r} \wedge \mathbf{v}|\Delta t = \frac{1}{2}r^2\dot{\phi}\Delta t. \tag{5.79}$$

Taking the limit as time goes to zero gives a constant areal velocity

$$\frac{dA}{dt} = \frac{1}{2}r^2\dot{\phi} = \frac{\ell}{2m}, \tag{5.80}$$

thereby validating Kepler's second law.

5.3.1.2 Kepler's First Law

Kepler's *first law of planetary motion* states that the orbits of the planets are ellipses with the sun at one focus. In retrospect, Newton's response to

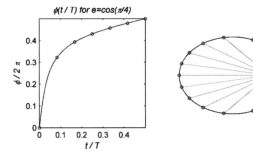

FIGURE 5.8 (Left-hand side) Calculation of angle verses time for an elliptical orbit with $e = 1/\sqrt{2}$ (see Section 5.3.1.4 for methods used). (Right-hand side) As a particle moves in this orbit, equal areas are swept out in equal times. Here, an ellipse is cut into twelve equal slices, in time and in area.

Halley was, in effect, that an inverse-square force law satisfies this empirical observation exactly. For a $1/r$ potential, the second-order orbit equation reduces to a simple harmonic oscillator when the radial equation is written as a function of $u = 1/r$, (see Equation 5.50)

$$\frac{d^2u}{d\phi^2} = -(u - u_0), \quad u_0 = \left(\frac{mk}{\ell^2}\right). \tag{5.81}$$

The solution for bound orbits is

$$u - \frac{mk}{\ell^2} = e \cos \phi, \tag{5.82}$$

where $e \geq 0$. Solving for the radial distance gives

$$r = \frac{r_p(1 + e)}{(1 + e \cos \phi)}. \tag{5.83}$$

where the phase has been chosen so that the perihelion distance r_p occurs when $\phi = 0$. When $e < 1$, the orbits are bound. The result is indeed an equation for an ellipse with the center of force at one focus. A diagram of an ellipse is shown in Figure 5.9. The sum of the distances from a point on the ellipse to each focus is a constant.

The semi-major axis is given by

$$a = (r_{\min} + r_{\max})/2, \tag{5.84}$$

while the semi-minor axis b is

$$b = a\sqrt{1 - e^2}. \tag{5.85}$$

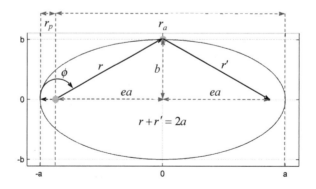

FIGURE 5.9 The geometry of an ellipse with semi-major axis *a* and semi-minor axis *b*. The sum of the distances from a point on the ellipse to each focus is a constant.

The eccentricity is given by

$$e = \sqrt{1 - (b/a)^2} = \sqrt{1 + 2E\ell^2/mk^2} = \sqrt{1 - \ell^2/mka}, \qquad (5.86)$$

and the area of the ellipse is

$$A = \pi ab. \qquad (5.87)$$

With some algebra, the following relations can be found relating the physical constants E and ℓ with the geometric constants a and e

$$E = -k/2a, \quad \ell = \sqrt{mkb^2/a} = \sqrt{mka(1 - e^2)}. \qquad (5.88)$$

Newton showed that all the solutions (bound and unbound) for an inverse-square force are *conic sections* with the sun at one focus. Conic sections are defined as the curves generated by the intersection of a cone with a plane as illustrated in Figure 5.10. The general equation for a conic section is a quadratic of the form

$$Ax^2 + Bxy + Cy^2 + Dx + Ey + F = 0. \qquad (5.89)$$

Conic sections can be classified by the value of the eccentricity e as shown in Table 5.1.

Figure 5.11 shows the effective potential energy diagram for the inverse-radius potential. Bound orbits are ellipses (or circles) with the center of force at one focus. The orbits are bounded for

$$V_{\text{eff}}|_{\min} \leq E < 0. \qquad (5.90)$$

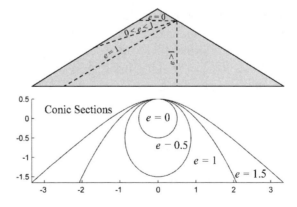

FIGURE 5.10   Top: Conic sections are planar cross sections cut through a cone. Bottom: Relative cross-sectional areas are shown for conic sections with the same perihelion distance but different eccentricities.

TABLE 5.1   Classification of conic sections

Shape	Eccentricity
Circle	$e = 0$
Ellipse	$0 < e < 1$
Parabola	$e = 1$
Hyperbola	$e > 1$

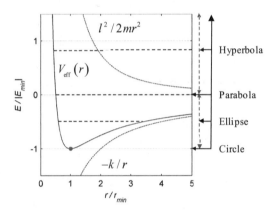

FIGURE 5.11   **Effective potential energy diagram for a $1/r$ potential: Negative energy states $E_{min} \leq E < 0$ are bound. Positive energy states $E \geq 0$ are unbound.**

For $E \geq 0$, the trajectories are open, and particles can escape the gravitational well, as indicated in Figure 5.12, which describes the geometry for hyperbolic orbits. The open trajectories are parabolas for $E = 0$, and hyperbolas for $E > 0$. The minimal escape velocity is given by setting $E = 0 = \frac{1}{2}mv^2 - k/r$, giving

$$v_{esc}(r) = \sqrt{2k/mr} = \sqrt{2(m_1 + m_2)G/r}. \tag{5.91}$$

The formula for the conic section remains essentially unchanged for unbounded motion, but now the eccentricity is greater than one. There is a minimum distance of approach, but no maximum. The total rotation angle from the perihelion is limited by

$$r(\phi) = \frac{r_{min}(1 + e)}{1 + e \cos\phi} \leq \infty. \tag{5.92}$$

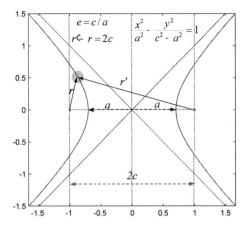

FIGURE 5.12 The unbound trajectories in a $1/r$ potential for $E > 0$ are hyperbolas $e > 1$. Note that in the figure r is the radial distance for attractive inverse r potentials, while r' is the radial distance for repulsive inverse r potentials. They are related by $r' - r = 2a$.

Therefore, the maximum rotation angle from the perihelion is given by

$$\cos \phi_{\max} = -1/e. \tag{5.93}$$

A view of the radial phase space for the sun's gravitational well in the vicinity of the Earth is shown in Figure 5.13. The lines shown are equally spaced constant energy contours for fixed orbital angular momentum. The separatrix between open and closed orbits is the zero energy contour, highlighted with superimposed dots. The stationary point denoted by the 'x' is the location of the nearly circular Earth orbit.

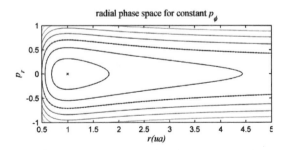

FIGURE 5.13 Radial phase space for the solar potential in the vicinity of the Earth's orbit. The stationary circular orbit of the Earth is indicated by the x. All orbits shown have the same angular momentum. The zero energy parabolic orbit line is highlighted with dots. It separates bound from unbound trajectories.

The cross section of the orbits of constant energy and angular momentum surrounding a stationary point are closed tori that envelop the stationary path. These are KAM tori, which will be discussed later in this chapter in the section on chaos.

5.3.1.3 Kepler's Third Law

Kepler's Third Law of Planetary Motion states that periods of the planetary orbits are proportional to the three-half power of the semi-major axis. Figure 5.14 shows a plot of planetary distance from the sun verses the orbital period, illustrating Kelper's Third Law.[4] See Appendix B for a tabulation of selected planetary data upon which this figure is based.

The area of an ellipse is πab. Therefore, the integral over the areal velocity gives

$$\oint dA = \pi ab = (\ell/2m)T. \tag{5.94}$$

Substituting $k = m_1 m_2 G$ and $m = m_1 m_2/(m_1 + m_2)$ and solving for the period T gives

$$T = 2\pi ab(m/\ell) = \frac{2\pi a^{3/2}}{\sqrt{(m_1 + m_2)G}}. \tag{5.95}$$

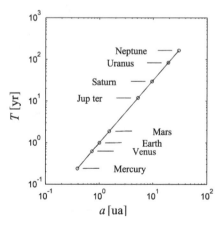

FIGURE 5.14   **A plot of period in years verses semimajor axis in astronomical units (1Astronomical Unit = 1 ua) for the major planets. The straight line through the points is the power law prediction of Kepler's Third Law.**

[4]Note: 1Astronomical Unit = 1 A.U. = 1 ua.

Since the sun is much more massive than any planet, the deviations due to planetary masses may be ignored giving Kepler's power law ratio for the periods of the planets

$$T \approx \frac{2\pi a^{3/2}}{\sqrt{m_{\text{sun}}G}}. \tag{5.96}$$

5.3.1.4 Time Dependence of the Motion

Given angular momentum conservation, and normalizing the time in units of the period given in Equation 5.96, one can analytically solve for the time-dependence of the orbits

$$\frac{1}{T}\int_{t_0}^{t} dt = \int_{\phi_0}^{\phi} \frac{mr^2(\phi)}{\ell T}d\phi = \frac{1}{2\pi}\int_{\phi_0}^{\phi} \frac{(1-e^2)^{3/2}d\phi}{(1+e\cos(\phi))^2}. \tag{5.97}$$

The solution can be expressed in terms of elementary trigonometry functions using a symbolic integrator

$$\frac{t(\phi)}{T} = \frac{1}{\pi}\tan^{-1}\left(\frac{1-e}{\sqrt{1-e^2}}\tan\frac{\phi}{2}\right) - \frac{e\sqrt{1-e^2}}{2\pi}\frac{\sin\phi}{1+e\cos\phi}, \tag{5.98}$$

$$t(T/2 + \phi) = T - t(\phi), \quad \text{for } [0 \le \phi \le \pi], \; t(\pi) = T/2.$$

Only $0 < t < T/2$ needs to be evaluated since the result is periodic in the period T and symmetric over half that interval.

Inversion of transcendental equations is not trivial. To find $\phi(t)$ directly, it is better to apply a numerical Runge Kutta integrator directly to the difference equations

$$d\phi = \frac{2\pi}{T}\frac{(1+e\cos\phi)^2}{\sqrt[3]{1-e^2}}dt. \tag{5.99}$$

Only the half period from minimum to maximum need be solved, the solution can be extended to other intervals by using

$$\phi(nT + t) = 2n\pi + \phi(t),$$
$$\phi(T/2 + t) = 2\pi - \phi(t), \quad \text{for } [0 \le t \le T/2]. \tag{5.100}$$

This was the method used to generate the plot of angle verses time shown in Figure 5.8. Knowing the time-dependence, one can correlate seasonal time in units of Earth-years with diurnal time. Time flows uniformly as measured by the apparent rotation of the stars in the celestial sphere. The angular motion of the Earth orbit about the sun speeds up as the Earth approaches its perihelion and slows down as it approaches its aphelion.

5.3.1.5 Orbital Dynamics

One straightforward way to move a satellite from an orbit about one planet to an orbit about a second is to use a *Hohmann transfer* orbit, illustrated in Figure 5.15. This is an elliptical trajectory where, for example, one has the Earth (1 ua) at perigee and Jupiter (5.2 ua) at apogee. By firing the rocket in low Earth orbit at the point of perigee, an elliptical crossing orbit with Jupiter is established. The rocket is fired once again on arrival at Jupiter, to drop the spacecraft down into a circular orbit about the planet or one of its moons. If the launch window it timed properly, the spacecraft will arrive at the aphelion right as Jupiter is passing by. Hohmann transfers are commonly used as a starting point for plotting Earth-moon and Earth-Mars missions.

Now consider using the mass of a nearby planet as a gravitational slingshot. A spacecraft, on a trajectory that will take it close to a planet, will accelerate as it approaches the planetary potential well. After passing the planet, gravity will continue acting on the spacecraft, slowing it down. The net effect on the orbit speed with respect to the planet is zero, but, importantly, (i) the direction may have changed in the process, and (ii) the speed is being measured with respect to a moving frame of reference.

The slingshot effect comes from the fact that the planets are moving in their orbits around the Sun. While the speed of the spacecraft has remained the same as measured with reference to Jupiter, the initial and final speeds may be quite different as measured in the sun's frame of reference. Depending on the direction of the outbound leg of the trajectory, the spacecraft can

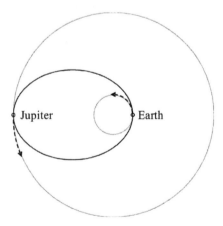

FIGURE 5.15   Illustration of Hohmann transfer orbit, with perihelion distance 1 ua and aphelion distance 5.2 ua, moving a spacecraft from an orbit about the Earth to one about Jupiter.

gain a significant fraction of the orbital speed of the planet. In the case of Jupiter, this is over 13 km/s. This gravitational slingshot effect was used by the Voyager deep space probes to make a grand tour of the planets of the solar system. For realistic mission planning, of course, three-body corrections need to be applied to the calculation of orbital trajectories.

5.3.2 Runge-Lenz Vector

A remarkable thing about the Kepler problem is that the radial and angular periods are degenerate. Consequently, the perihelion does not precess. Since this is such a familiar example, it would be easy to assume that closure of orbits is a common phenomenon. It is not; rather, it is exceptional. This degeneracy of frequencies implies the existence of an additional conserved vector quantity, unique to this interaction law, which is called the *Runge-Lenz Vector*. (The three-dimensional harmonic oscillator is the other outstanding exception, due to a similarly conserved rank-2 tensor.) The Runge-Lenz vector points in the direction of the perihelion and is defined as

$$\mathbf{A} = \mathbf{p} \times \mathbf{L} - mk\frac{\mathbf{r}}{r}. \tag{5.101}$$

Its conservation can be verified by taking its total time-derivative

$$\frac{d\mathbf{A}}{dt} = \mathbf{F} \times \mathbf{L} + \mathbf{p} \times \mathbf{N} - k\,\mathbf{p}/r + k(\hat{\mathbf{r}} \cdot \mathbf{p})\mathbf{r}/r^2$$
$$= -k\hat{\mathbf{r}} \times (\mathbf{r} \times \mathbf{p})/r^2 - k\,\mathbf{p}/r + k(\hat{\mathbf{r}} \cdot \mathbf{p})\mathbf{r}/r^2 = 0. \tag{5.102}$$

The magnitude of the Runge-Lenz vector can be found by evaluating some scalar products:

$$\mathbf{L} \cdot \mathbf{A} = 0, \tag{5.103}$$

$$\mathbf{p} \cdot \mathbf{A} = -mk\frac{\mathbf{p} \cdot \mathbf{r}}{r}, \tag{5.104}$$

$$\mathbf{r} \cdot \mathbf{A} = \mathbf{r} \cdot \left(\mathbf{p} \times \mathbf{L} - mk\frac{\mathbf{r}}{r}\right) = \mathbf{L} \cdot \mathbf{r} \times \mathbf{p} - mkr = \ell^2 - mkr. \tag{5.105}$$

The first equation states that the vector lies in the plane of motion. The second and third vector products are satisfied if the vector lies in the direction of the perihelion $\hat{\mathbf{r}}_P$ and its amplitude is given by

$$\mathbf{A} = -(mke)\hat{\mathbf{r}}_P. \tag{5.106}$$

Together, the Runge-Lenz vector $\mathbf{A}$ and the angular momentum vector $\mathbf{L}$ give six conserved constants of integration, sufficient to totally describe the state of the system without resorting to any need for direct integration. By substitution, one can directly recover the orbit equations

$$\mathbf{r} \cdot \mathbf{A} = rmke \cos \phi = \ell^2 - mkr \rightarrow$$
$$r = \ell^2/mk(1 + e \cos \phi)^{-1} = a(1 - e^2)(1 + e \cos \phi)^{-1}. \tag{5.107}$$

The energy can be expressed in terms of the magnitude of these vectors

$$E(A, \ell) = \frac{A^2 - (mk)^2}{2mL^2} = -(1 - e^2)\frac{mk^2}{2\ell^2}. \tag{5.108}$$

The conservation of both the angular momentum vector and the Runge-Lenz vector results from a special symmetry of the system that is unique to inverse-square central forces law. Other central forces conserve angular momentum, but not the Runge-Lenz vector. In Quantum Mechanics, this extra symmetry of the motion results in an additional degeneracy of the energy spectrum. (Degenerate frequencies result in degenerate energies using the quantum condition $E = \hbar\omega$.) In fact, the Runge-Lenz vector can be used to derive the energy spectrum of the hydrogen atom.

5.3.3 Spherical Coordinates

When calculating the orbits of planets and moons in a solar system, one does not have the luxury of assuming the orbits are all coplanar, lying in some common equatorial plane of the celestial sphere. One needs to account for the inclination of the planets and their satellites. It is therefore important to know how to solve the effective one-body problem in spherical coordinates. The plane of the motion, in general, has some arbitrary angle of inclination with respect to the polar basis, and there is an arbitrary angle of rotation from the line of nodes to the Runge-Lenz vector defining the direction of the perihelion.

The square of the angular momentum $\mathbf{L}^2$ and its component along the polar axis $\mathbf{L} \cdot \hat{\mathbf{z}}$ are individually conserved, along with the Hamiltonian E. This is sufficient to reduce the problem to quadrature. In a spherical basis, the Hamiltonian is given by

$$H = \frac{1}{2m}g^{ij}p_ip_j - \frac{k}{r} = \frac{1}{2m}\left(p_r^2 + \frac{p_\theta^2}{r^2} + \frac{p_\phi^2}{r^2 \sin^2 \theta}\right) - \frac{k}{r}. \tag{5.109}$$

The resulting equations of motion are

$$\begin{array}{ll} \dot{r} = p_r/m, & \dot{p}_r = -k/r^2 + (p_\theta^2 + p_\phi^2/\sin^2 \theta)/2mr^3, \\ \dot{\theta} = p_\theta/mr^2, & \dot{p}_\theta = p_\phi^2 \cos\theta/mr^2 \sin^3 \theta, \\ \dot{\phi} = p_\phi/mr^2 \sin^2 \theta, & \dot{p}_\phi = 0. \end{array} \tag{5.110}$$

The Hamiltonian is cyclic in both the azimuthal angle and the time, so the energy E and the z-component of the angular momentum ℓ_z are both conserved

$$H = E, \quad p_\phi = \ell_z. \tag{5.111}$$

Total angular momentum, given by

$$\ell^2 = |\mathbf{r} \times \mathbf{p}|^2 = p_\theta^2 + \frac{p_\phi^2}{\sin^2 \theta}, \tag{5.112}$$

is also conserved, although is it not as obvious in this coordinate system. It is easiest to demonstrate this by showing that its total time derivative vanishes. The proof will be left as an exercise. This results in the standard one-dimensional reduction for the energy equation for radial motion

$$E = \frac{1}{2}m\dot{r}^2 + \frac{\ell^2}{2mr^2} - \frac{k}{r}. \tag{5.113}$$

By rotating the z-axis into the direction of the total angular momentum, one gets the planar constraints

$$\ell = \ell_z, \quad \theta = \pi/2, \quad p_\theta = \dot{p}_\theta = 0. \tag{5.114}$$

The result is a reduction of the spherical basis to the plane polar coordinates originally used to describe the motion.

5.4 RESTRICTED THREE-BODY PROBLEM

After Newton's success, the next step was to solve for the detailed motion of the planets, taking into account their perturbative effects on each other's motion. For example, the Lagrangian of the three-body Earth-moon-sun system can be written as

$$L = \frac{1}{2}m_e\dot{\mathbf{r}}_e^2 + \frac{1}{2}m_s\dot{\mathbf{r}}_s^2 + \frac{1}{2}m_m\dot{\mathbf{r}}_m^2 - \frac{m_e m_s G}{r_{es}} - \frac{m_e m_m G}{r_{em}} - \frac{m_m m_s G}{r_{ms}}. \tag{5.115}$$

No exact solution for this problem has ever been found. Poincaré demonstrated finally that there are simply not enough constants of the motion to reduce the 3-body problem to quadrature. This leaves the question of the long-term stability of the system unanswered. More seriously, Poincaré showed that, for certain initial conditions, the long-term behavior of classical dynamical systems could be chaotic, that is, inherently unpredictable.

5.4.1 Circular, Restricted, Three-Body Problem

Treatment of the general three-body problem is beyond the scope of this text. Interesting new results are continually being discovered. For example, Chenciner and Montgomery [Chenciner 2000] recently discovered a new figure-eight orbit for three identical masses where every mass traces out the same trajectory, head chasing tail, so to speak. Its existence was surmised by using symmetry arguments. Hestenes [1986, 400–406] presents an introductory overview categorizing the types of motion to expect.

The three-body problem [Szebehely 1967; Moulton 1970, Chapter 8] is usually tackled by making some simplifying starting approximations. The restricted three-body problem is defined as the approximation that one of the three bodies has a negligible mass in comparison to the other two, which move in an elliptical two-body orbit in a fixed plane. The circular restricted three-body problem (Figure 5.16) adds the constraint that the two massive bodies are in uniform circular motion about each other. The problem can be analyzed using an effective potential in the corotating frame.

Consider the case of two massive bodies, (e.g., the sun and Jupiter) moving in circular orbits about their center of mass. Introduce a third body of negligible mass (e.g., the Earth), which has no significant effect on the orbits of the two massive bodies. For convenience, it is best to work in a noninertial frame of reference, which rotates about the center of mass at a rate equal to the orbital frequency of the two massive bodies. Natural units are commonly used. The motion of the massive bodies is circular with separation distance of 1 in dimensionless units. The rotational frequency Ω of the two heavy objects is also set to 1. Since the mass of the light body is taken to be negligible, it factors out in this approximation, and can therefore be set arbitrarily to 1 with no effect on the equations of motion.

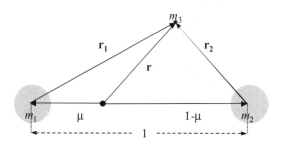

FIGURE 5.16 The restricted, circular, three-body system consisting of two heavy bodies and one light satellite. The distance to the light mass is measured from the stationary center of mass of the heavy bodies.

In the CRTBP, the two heavy masses are moving in circular orbits with angular frequency Ω given by Kepler's third law

$$\Omega^2 R^3 = (m_1 + m_2)G. \tag{5.116}$$

The light mass m_3 is light enough that its influence on the heavy masses is negligible. It sees a force given by

$$\mathbf{F}_3 = -\frac{m_3 m_1 G}{|\mathbf{r}_3 - \mathbf{r}_1|^3}(\mathbf{r}_3 - \mathbf{r}_1) - \frac{m_3 m_2 G}{|\mathbf{r}_3 - \mathbf{r}_2|^3}(\mathbf{r}_3 - \mathbf{r}_2). \tag{5.117}$$

Because the motion of the heavy masses is time-dependent, energy is not conserved, but one can define an effective energy in the rotating frame that is conserved.

The Lagrangian for the effective one-body problem in a rotating frame is given by

$$L = \frac{1}{2}m_3[(\dot{x} - \Omega y)^2 + (\dot{y} + \Omega x)^2] + m_3 m_1 G/|\mathbf{r}_3 - \mathbf{r}_1| + m_3 m_2 G/|\mathbf{r}_3 - \mathbf{r}_2|,$$

$$|\mathbf{r}_3 - \mathbf{r}_{1,2}|^2 = (x \pm mr/m_{1,2})^2 + y^2. \tag{5.118}$$

Here x is in the direction of the line joining the two heavy masses and y is in the other orthogonal direction for the plane of motion. Ω is the angular frequency of the corotating frame. In this frame, the time-dependence vanishes, leading to a conserved Hamiltonian. The Lagrangian equations of motion are

$$m_3\ddot{x} = 2m_3\Omega^2\dot{y} + m_3\Omega^2 x - \partial V/\partial x,$$
$$m_3\ddot{y} = -2m_3\Omega^2\dot{x} + m_3\Omega^2 y - \partial V/\partial y, \tag{5.119}$$
$$m_3\ddot{z} = -\partial V/\partial z.$$

In the rotating frame there is a conserved "energy" given by

$$E = \frac{1}{2}m_3\dot{\mathbf{r}}^2 + \frac{m_3\Omega^2 r^2}{2} - \frac{m_3 m_1 G}{|\mathbf{r}_3 - \mathbf{r}_1|} - \frac{m_3 m_2 G}{|\mathbf{r}_3 - \mathbf{r}_2|}. \tag{5.120}$$

Typically, the equations are expressed in velocity phase space using the reduced units $|\mathbf{r}_1 - \mathbf{r}_2| = 1, \Omega = 1, m_3 = 1$, giving the dimensionless equations

$$\dot{x} = v_x, \quad \dot{v}_x = 2v_y + x - (1 - \mu)(x + \mu)r_1^{-3} - \mu(x - 1 + \mu)r_2^{-3},$$
$$\dot{y} = v_y, \quad \dot{v}_y = -2v_x + y - (1 - \mu)yr_1^{-3} - \mu yr_2^{-3}, \tag{5.121}$$
$$\dot{z} = v_z, \quad \dot{v}_z = -(1 - \mu)zr_1^{-3} - \mu zr_2^{-3},$$

where

$$r_1 = \sqrt{(x + \mu)^2 + y^2 + z^2}, \quad r_2 = \sqrt{(x - 1 + \mu)^2 + y^2 + z^2}, \qquad (5.122)$$

and $\mu = m_2/(m_1 + m_2)$.

The only known first integral of the motion is called the *Jacobi integral*, which is defined as

$$C = (x^2 + y^2) + \frac{2(1 - \mu)}{r_1} + \frac{2\mu}{r_2} - (v_x^2 + v_y^2 + v_z^2) = -2E, \qquad (5.123)$$

where E is the energy associated with the conserved Hamiltonian in the corotating frame.

The CRTBP equations of motion are invariant under time-reversal. Given a plane of incidence, for every trajectory that moves forward in time, there is a symmetric trajectory on the other side that flows backwards in time. If one starts on this plane with a velocity vector that is normal to the plane, then the two trajectories are connected. This can be used to find periodic orbits. If the trajectory reenters the plane at some later time normal to the plane, the trajectory forms a closed loop and the orbit is therefore periodic.

5.4.2 Lagrange Points

Lagrange was the first to demonstrate that there are five points in the circular two body system where a test particle, placed at rest, would feel no net force in the co-rotating non-inertial frame. The *Lagrange points* of the CRTBP are the 5 Libration points of the problem for which all six derivatives of the coordinates and momenta of the test particle vanish simultaneously. They are stationary points of the system in the co-rotating frame, similar to a geosynchronous satellite in the Earth's rotating frame. The collinear stationary solutions with $y = z = 0$ for which $\dot{x} = \dot{y} = \dot{z} = 0$ are designated as the L1, L2, and L3 Lagrange points. The locations of these stationary points are illustrated graphically in Figure 5.17, which also shows the zero velocity contours of the Jacobi Integral. The precise locations of these three Lagrange points are given by the solutions to the nonlinear equations

$$L1: \quad x - \frac{1 - \mu}{(x + \mu)^2} + \frac{\mu}{(x + \mu - 1)^2} = 0,$$

$$L2: \quad x - \frac{1 - \mu}{(x + \mu)^2} - \frac{\mu}{(x + \mu - 1)^2} = 0, \qquad (5.124)$$

$$L3: \quad x + \frac{1 - \mu}{(x + \mu)^2} + \frac{\mu}{(x + \mu - 1)^2} = 0.$$

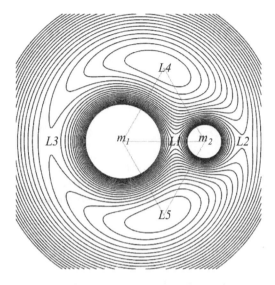

FIGURE 5.17 Lagrange points of the restricted three-body problem are shown on contour plot of the effective potential in the co-rotating frame for the two heavy masses in the restricted three-body problem. Mass scale ratio used for this figure is $(m_1 : m_2 :: 4 : 1)$.

There are two additional stationary solutions for $y \neq 0$ called the L4 and L5 Lagrange points. These are located at the vertices of an equilateral triangle in the co-rotating frame with respect to the first two masses.[5]

The L4 and L5 Lagrange points correspond to hilltops and L1, L2 and L3 correspond to saddles (i.e. points where the potential is curving up in one direction and down in the other). Surprisingly, the L4 and L5 Lagrange points, although potential maxima, turn out to be stable, due to the Coriolis effect. When a satellite parked at the L4 or L5 points starts to roll off the potential maximum it picks up speed. At this point, the Coriolis force kicks into action, sending the satellite into a stable orbit around the Lagrange point. There are no satellites at L4 and L5 points of the Earth-moon system, but several hundred asteroids have been found at the L4 and L5 points of Jupiter. These are called the *Trojan satellites* of Jupiter. An example of a Trojan orbit is shown in the left-side of Figure 5.18. On the right-side is phase space plot of this orbit captured at the zero-crossings of the sun-L5 axis. This is an example of a Poincaré section, which will be discussed further in Section 5.4.4.

---

[5]A nice derivation by of the locations and stability of the Lagrange points can be downloaded from the NASA website [Cornish undated].

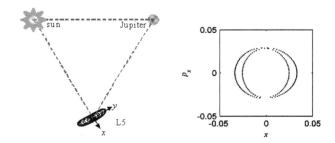

FIGURE 5.18   (Left) The orbital motion in the co-rotating frame of a Trojan object at the Lagrangian L5 point of Jupiter tracked for 20 Jovian years. (Right) A Poincaré section of the radial phase space of this orbit relative to the sun tracked for 200 Jovian years.

The L1 point of the Earth-Sun system affords an uninterrupted view of the sun and is currently occupied by the Solar and Heliospheric Observatory Satellite. The L2 point of the Earth-Sun system is home to the Wilkinson Microwave Anisotropy Probe and, by the year 2011, to the James Webb Space Telescope as well. The L1 and L2 points are unstable on a time scale of approximately 23 days, which requires satellites parked at these positions to undergo regular course corrections.

The two-body system is nearly completely surrounded by the two flanking hilltops given by the L4 and L5 Lagrange points. Figure 5.19 shows a zero velocity contour that nearly surrounds the two-body system. Particles with lower energy than this contour line can only enter or exit the system through the bottleneck given by the L2 Lagrange saddle point. Particle inside the well may be captured or may have a long circuitous route in finding their way out

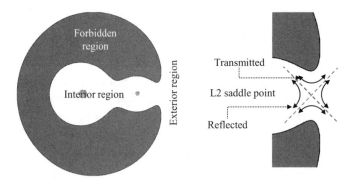

FIGURE 5.19   A zero velocity contour in the co-rotating frame taken in the hill region showing a restricted opening at the L2 saddle point.

of the system. Orbital dynamics studies for the space program [Gomez 2004] have given a lot of insights into the dynamical process. Particles approaching the saddle-point are transmitted (or reflected) depending on whether they are inside (or outside) of an invariant manifold called a *transfer tube*. In addition, particles can be trapped in periodic or quasiperiodic orbits within the saddle region. The asymptotic orbits separating transmission and reflection of particles connect these bound orbits to the outside world. By analyzing the manifold structure of the gravitational surface [Gomez 2004], one can inject or extract a spacecraft into a planetary or satellite orbit in a fuel efficient way.

5.4.3 Earth-Sun-Jupiter System

The Earth's orbit about the sun is nearly circular. If the hypothetical mass of Jupiter is increased, Jupiter's perturbation of the Earth's orbit turns it into a doughnut shape, as shown in Figure 5.20. As the perturbation increases, so do the cross sections of the KAM tori that are created by the perturbation. Eventually, the system becomes unstable and the tori are destroyed. Long before the instability point is reached, however, the induced temperature swings alone would make Earth uninhabitable.[6]

A torus represents the surface of an inner tube in phase space. When one plots the Earth's orbit in configuration space, a projection of this torus is created in the orbital plane, giving a slight thickness to the circle (or elliptical) orbit of the Earth, imperceptible at the scale shown in the diagram on the left side of Figure 5.20, where Jupiter at 0.001 solar masses is about its actual size. As Jupiter's mass increases, the torus in phase space deforms and expands.

FIGURE 5.20   Perturbed orbit of the Earth at 1ua with Jupiter at 5.203 ua, with mass set near its nominal value of ∼0.001 solar masses (left), then increased to 0.01 solar masses (centered), and 0.035 solar masses (right), showing the torus-like KAM surfaces. At larger mass values, the tori break up, and the Earth is likely ejected from the solar system, if it does not suffer an unfortunate collision with one of its neighbors first.

---

[6]The Computer simulation *Jupiter: The Three-Body Problem* allows one to interact with the Earth-sun-Jupiter system, illustrating a number of the effects that are discussed in the following sections, see [Sethna1996] for information on how to obtain a copy.

Their projection onto the Earth's (x, y) position coordinates fleshes out into a ring of visible thickness (center diagram), then into a fatter inner tube shape (shown on the right).

For every orbit that Jupiter makes, Earth spirals almost 12 orbits of the sun, but the two periods are not commensurate. Jupiter's year is an irrational number of Earth years. Thus, the orbit in phase space never closes on itself. After a sufficient number of cycles, the available phase space of the donut appears to be filled. This filling of phase space defines an invariant surface, which is crucial for the proof of the KAM theorem. The nature of these tori will become clearer after action-angle variables are discussed in Chapter 10.

5.4.4 Poincaré Sections

Visualizing periodic trajectories in configuration space is easy, but when the orbits are space filling or chaotic, it is more informative to view a Poincaré section of the dynamics. To create a Poincaré section, one defines a plane in the phase space, and draws a dot whenever the trajectory passes through the plane. This defines a mapping from the plane to itself called the *Poincaré first-return map* [Barrow-Green 1997]. For the restricted three-body problem, this cross-section is often given by the plane defined by the angular momentum and Runz-Lenz vector $\mathbf{L} \wedge \mathbf{A}$ of the two massive bodies. An example is the Poincaré section for the idealized Earth trajectory shown to the left of Figure 5.20. The associated Poincaré section of nearby orbits is shown in Figure 5.21. The enveloping perturbed orbits show a distinct KAM torus structure. Poincaré sections for this numerical simulation using Matlab were generated by setting an event signal to detect zero-crossings in either direction of the symmetry plane (near-side or far-side of the sun relative to Jupiter).

To study orbits crossing the L1, L2, L3 points of the CRTBP, the Poincaré plane is chosen to intersect the two-body axis. To study the L4 and L5 points, one rotates the Poincaré plane by $\pm 60°$. For example, the left-hand side of Figure 5.18 shows the motion of a randomly chosen low-velocity Trojan orbit in configuration space. The loop structure illustrates how the Coriolis Effect binds an asteroid to the L5 region. This simulation was tracked for 30 Jovian years. For longer periods, the region does not grow, but becomes black since the trajectory is space filling. The right-hand-side of the figure shows a Poincaré section of the radial phase space for the same orbit tracked for 200 Jovian years, revealing a well-defined KAM torus structure.

5.4.5 KAM Theorem

An important topological theorem due to Kolmogorov, Arnold, and Moser states that for small perturbations these tori are phase-space-invariant

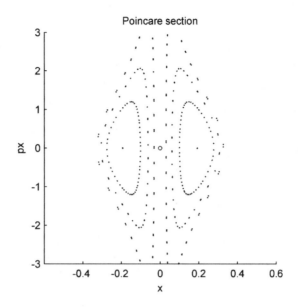

FIGURE 5.21 **Poincare section showing the KAM tori structure of the orbits centered on an idealized periodic orbit for the Earth at 1 ua. Jupiter has 0.001 solar masses in this simulation, close to its actual value. There are two $y = 0$ crossings for each orbit of the Earth about the sun as it intercepts the Sun-Jupiter plane: one on the near side of Jupiter and a return crossing on the far side of the sun from Jupiter.**

surfaces. The *KAM Theorem*[7] states that for sufficiently small perturbations of a periodic or quasiperiodic orbit, almost all tori (excluding those with rational frequency resonances) are preserved. The KAM theorem puts restrictions on the effects of chaos. The proof of the theorem depends on the irrationality (space filling character) of the Jovan perturbations. Using the KAM Theorem, the Earth's orbit under the small perturbation of Jupiter remains *quasiperiodic*,[8] provided that the Jovian year is "sufficiently irrational." The conditions for validity of the KAM Theorem require small, smooth perturbations and a sufficiently irrational winding number [Tabor 1989, p. 105]; see also [WWW: Wolfram Mathworld, listed under Kolmogorov-Arnold-Moser Theorem].

---

[7]The KAM Theorem was outlined by Kolmogorov in 1954 and later proven by Arnold in1963 for Hamiltonian systems and Moser in 1962 for Twist maps [Tabor 1989, p. 105].

[8]In periodic systems, periodic orbits involve integer multiples of some base frequency. Quasiperiodic orbits have multiple non-commensurate base frequencies.

5.4.6 Visualizing Chaos

In 1963, meteorologist Edward Lorenz [1963] rediscovered the chaotic deterministic behavior of Poincaré while studying a simple model of the atmosphere. This is the origin of the Lorenz Butterfly effect discussed in Chapter 3. A century earlier, Poincaré had already suggested that the difficulties in weather prediction were due to the intrinsic chaotic behavior of the atmosphere. Poincaré pointed out that apparently insignificant differences in the initial conditions may produce intractably large ones in the final outcomes. Over long time scales, prediction may prove to be impossible [Poincaré 2001].

An interesting aspect of Poincaré's original study into chaos was the complicated structure of the distributions in phase space about stable and unstable points, which defied simple categorization. It is now known that such distributions are fractal in character. However, the scientific investigation of fractals did not begin until Benoit Mandelbrot's work in 1975 [Mandelbrot 1977], nearly a century after Poincaré's first insights into the problem.

To illustrate Poincaré's point, consider a modified Earth formation scenario, illustrated in Figure 5.22, which results in a closed period-three mode-lock orbit due to perturbations by a more massive Jupiter. The orbit of interest is that shown to the left of the figure. It can be found by studying KAM trajectories that cross the plane normal to the surface, then adjusting the parameters to minimize the width of the torus, getting an orbit that closes on itself. The numerical conditions used to generate this orbit increased the Earth's perihelion radius to 1.24 ua and set the mass of Jupiter to 0.067 solar masses. Here, Jupiter's mass is still low enough so that the "Earth" remains bound to the solar system. The other two figures are snapshots of two other close-by orbits in the same region of phase space.

The three orbits have the same Jacobi energy in the corotating frame, each snapshot taken near islands of relative stability in the Poincaré map. Figure 5.23 shows the Poincaré first return map for this region of phase

FIGURE 5.22 A period-three mode-lock orbit (left) found by increasing the Earth's perihelion radius to 1.24 ua and setting the mass of Jupiter in CRTBP to 0.067 solar masses. The other two figures have the same Jacobi energy but represent different initial conditions. The simulation was run for 10 Jovian years for each case.

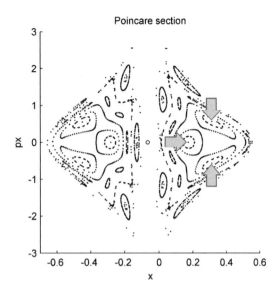

FIGURE 5.23 Poincare section showing the KAM tori structure surrounding the mode-lock period-three orbits (indicated by the 3 pointers) obtained by setting the perihelion of the Earth's orbit to 1.24 ua and increasing the mass of Jupiter to 0.067 solar masses.

space. The geometry, although complicated, shows a great deal of structure. If the Earth comes back to the same relative position every third time that it makes around Jupiter, this leads to a closed periodic orbit called a period-three mode-locked state. The arrows on the figure indict the locations of the three near-side zero-crossings of this orbit. These apparently isolated dots actually indicate multiple crossings taken over a 10-Jovian-year period.

Notice that there are a number of deformed loops scattered over the figure, some enveloping the mode-lock state. These represent KAM tori, consisting of nonchaotic *quasiperiodic* orbits. The tori, when they intersect the Poincaré section, cut it in a closed loop. Each periodic orbit is surrounded by enveloping KAM tori.

For larger radii, the tori become unstable and break up. At further distances, new tori may form, enveloping a larger region. One is clearly shown enveloping all three return points of the mode lock state in Figure 5.23. Trajectories in chaotic regions are sensitively dependent on their initial conditions. But the chaotic region is surprising rich in its structural detail. Islands of stability form, enclosed in their own tori. At every level of resolution, more islands of structure may become apparent, surrounded by a chaotic sea of instability.

5.5   GRAVITATION FIELD SOURCES

Newton's Law of Universal Gravitation states that all matter attracts all other matter with a force proportional to the product of their masses and inversely proportional to the square of the distance between them. In other words, one can define a vector force acting on a first object pointing to the center of mass of the second attracting object given by[9]

$$\mathbf{F}_1 = m_1 \mathbf{g}_2(r) = -\frac{m_1 m_2 G}{r^2}\hat{\mathbf{r}}. \tag{5.125}$$

Note that Newton equated gravitational mass to inertial mass. In modern language, the source of the Newtonian gravitational force is the *gravitational field* $\mathbf{g}$, which derivable from a scalar *gravitational potential* Φ

$$\mathbf{g}_2(r) = -\nabla \Phi_2(r). \tag{5.126}$$

The inverse-square character of the force law can be derived by assuming that the gravitational field satisfies *Gauss's Law* of gravito-statics, similar to Poisson's law for electrostatics, given by

$$\nabla \cdot \mathbf{g}_2(r) = -\nabla^2 \Phi_2(r) = -4\pi G \rho_2(r), \tag{5.127}$$

where Φ_2 is the gravitational potential due to particle 2 acting on particle 1. The volume integral of the mass density is the enclosed mass

$$\int \rho_2(r) dV = m_2. \tag{5.128}$$

For a point particle this results in the gravitational potential

$$\Phi_2(r) = -\frac{m_2 G}{r}. \tag{5.129}$$

Note that the interaction term of a particle with itself is ignored.

Newton's law of gravity can be derived from Guass's Law by constructing a closed Gaussian surface (see Figure 5.24) completely surrounding a spherically symmetric gravitational mass distribution. Using the integral form of Gauss's Law allows one to show that the gravitational field generated in the exterior region of any spherically symmetric object is identical to that of a

---

[9]The 2006 CODATA recommended value for G is $6.67428(67) \times 10^{-11}$ m^3·kg^{-1}·s^{-2} [Mohr 2007].

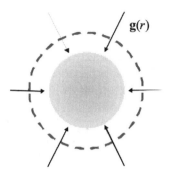

FIGURE 5.24 A Gaussian surface of constant radius surrounding an object with spherical symmetry. The net gravitational flux is radially inward into the surface and proportional to the enclosed mass.

point particle with all its mass concentrated at its center of mass point

$$\int \nabla \cdot \mathbf{g}_2 dV = \oint \mathbf{g}_2(r) \cdot d\mathbf{S} = \oint r^2 g_2(r) d\Omega = -4\pi G \int \rho_2 dV,$$

$$4\pi r^2 g_2(r) = -4\pi G m_2, \tag{5.130}$$

$$\mathbf{g}_2 = -m_2 G \hat{\mathbf{r}}/r^2.$$

The potential of a point mass also defines the free space *Green's Function* for an extended mass distribution of density $\rho(r')$

$$\Phi(r) = -\int dv' \rho(r') G(\mathbf{r} - \mathbf{r}') = -\int dv' \frac{\rho(r') G}{|\mathbf{r} - \mathbf{r}'|}. \tag{5.131}$$

Like the case for electromagnetism, the interaction of particle and field results a set of coupled equations. Matter, through its mass, is the source of a field in which other matter accelerates.

For moving distributions, the gravitational field acquires time dependence. Maxwell's equations specify the time dependence for the case of the electromagnetic force. For gravity, because of the weakness of experimental data, it is not entirely obvious at first how to derive the correct form of the force law. Gravity, for example, is always attractive, unlike electromagnetism, which has two signs of the charge. This rules out a vector coupling scheme, similar to Maxwell's equations, for gravity.

The gravitational field is the only field that couples universally to matter. Therefore, it is a candidate for being the gauge field associated with Poincaré invariance of physical law. The existence of such a field would allow one to move beyond the need for special inertial frames of reference. However, to do this it would also have to couple to photons, which are massless, but which carry energy and momentum. Therefore, by deduction, the relativistic generalization of Newton's theory needs to couple somehow to the rank-2 stress-energy tensor of the matter field, and not to the scalar mass density as Newton assumed. For non-relativistic speeds, the correction is expected to be small, since the total energy of a particle is dominated by the rest mass contribution under those conditions. The gravitational field will discussed further in Chapter 11 (Special Relativity) and Chapter 12 (Theory of Fields).

5.6 PRECESSION OF THE PLANETARY ORBITS

In the late 1800's, astronomers noted that there were slight differences between Mercury's observed orbit and that predicted by Newton's theory of gravitation. Mercury's elliptical path around the Sun precesses with each orbit, such that its perihelion advances by five arc minutes per century. Newton's theory could explain most, but not all, of the advance.

In principle, this small, but sensitive, effect provides a practical experimental way to look for deviations from Newton's law of universal gravitation. By expanding a gravitational potential in an inverse power series, to lowest order in perturbation theory, the potential can be written as

$$V(r) = -\frac{k}{a}\left(\frac{a}{r} - \eta\left(\frac{a}{r}\right)^2\right), \tag{5.132}$$

where η, which can be positive or negative, is some dimensionless scale parameter.

The orbit equation (5.50) becomes

$$\frac{d^2u}{d\phi^2} = -\left(1 - \frac{2mka}{\ell^2}\eta\right)u - \frac{mk}{\ell^2}. \tag{5.133}$$

The motion is still basically that of an ellipse, but now the ellipse precesses by a small amount with an analytic solution

$$r = \frac{r_p(1+e)}{1 + e\cos\left((1 - 2mka\eta/\ell^2)^{1/2}\phi\right)}, \tag{5.134}$$

where the angle needed to rotate from perihelion to perihelion is given by

$$\Delta\phi_r = \frac{2\pi}{(1 - 2mk\eta a/\ell^2)^{\frac{1}{2}}} \approx 2\pi\left(1 + \frac{1}{1-e^2}\eta\right). \tag{5.135}$$

This yields a precession rate for perihelion of

$$\frac{d\Omega}{dt} \approx \frac{2\pi}{T}\frac{1}{1-e^2}\eta,$$ (5.136)

where T is the orbital period and η is a dimensionless scale factor. In the case of the orbit of Mercury, after all known perturbations are included, the residual effect is only 42 arc-seconds per century. Nearly a third of this effect can be accounted for by relativistic corrections due to special relativity. (See Chapter 11 for an estimate.) The effect is completely accounted for by modification of the force law predicted by the *general theory of relativity* [Einstein 1920].

Another possible deviation from Newton's law of universal gravitation has been observed in the orbital rates of stars moving about the galactic plane in spiral galaxies, where the orbital rates are inconsistent with the visible mass distribution of stars about the galactic centers. This effect, however, is not currently attributed to a deviation of the force law from that predicted by general relativity, but rather to the existence of mysterious *dark matter*, needed to account for the *mass deficiency* of the observable universe. This dark matter, whatever its origin, appears to extend further in the galactic plane than the visible matter from stellar distributions.

Refinements to the Cavendish experiment, which was used originally to determine the strength of the gravitational coupling constant [Cavendish 1798; Clotfelter 1987], have looked, at the laboratory scale, for other deviations at small distances from an inverse-square law. Thus far, such experiments, sometimes labeled as searches for a *fifth force*, have returned null results [Adelberger 2003].

5.7 SUMMARY

The two-body central problem is completely reducible to quadrature. By transforming into center of mass coordinates, the Lagrangian

$$L = \frac{1}{2}m_\Sigma \dot{\mathbf{r}}_{\mathrm{cm}} \cdot \dot{\mathbf{r}}_{\mathrm{cm}} + \frac{1}{2}m\dot{\mathbf{r}} \cdot \dot{\mathbf{r}} - V(r),$$ (5.137)

separates into the trivial motion of the center of mass point and an effective one-body problem that contains the interesting dynamics, where $m_\Sigma = m_1 + m_2$ is the total mass of the two-body system, $m = m_1 m_2/m_\Sigma$ is the reduced mass of the effective one-body interaction, $\mathbf{r}_{\mathrm{cm}} = (m_1\mathbf{r}_1 + m_2\mathbf{r}_2)/m_\Sigma$

is the displacement of the center of mass, and $\mathbf{r} = \mathbf{r}_1 - \mathbf{r}_2$ is the relative displacement vector.

Angular momentum is conserved in central force motion. The motion lies in a plane, allowing the reduced one-body Lagrangian to be further separated in plane polar coordinates

$$L'(\mathbf{r}, \dot{\mathbf{r}}) = \frac{1}{2}m(\dot{r}^2 + r^2\dot{\phi}^2) - V(r), \tag{5.138}$$

where ϕ is a cyclic coordinate leading to a conserved angular momentum $p_\phi = \ell$. The first integrals of the motion are the angular momentum and the energy

$$\ell = mr^2\dot{\phi}, \quad E(\ell, r) = \frac{m\dot{r}^2}{2} + \frac{\ell^2}{2mr^2} + V(r). \tag{5.139}$$

The problem can now be reduced to quadrature to find $r(t)$ and $\phi(t)$. Of particular interest is the orbit equation $r(\phi)$, which can be found by inversion of the integral solution

$$\int d\phi = \pm \int \frac{dr}{\sqrt{\frac{2mr^4}{\ell^2}(E - V_{\text{eff}}(r))}} = \mp \int \frac{du}{\sqrt{\frac{2m}{\ell^2}\left(E - V_{\text{eff}}\left(\frac{1}{u}\right)\right)}}, \tag{5.140}$$

where $u = 1/r$ and $V_{\text{eff}} = \ell^2/2mr^2 + V$. Alternatively, it is sometimes easier to solve the second-order differential equation

$$\frac{d^2u}{d\phi^2} = -u - \frac{m}{\ell^2}\frac{\partial V(1/u)}{\partial u}. \tag{5.141}$$

The solutions for the orbit equation for the inverse-square force law are conic sections given by

$$r = \frac{r_p(1 + e)}{(1 + e\cos\phi)}, \tag{5.142}$$

where r_p is the perihelion distance, e is the eccentricity of the orbit, and the phase is chosen such that $\phi = 0$ is in the direction of the perihelion. For $0 \le e < 1$ the orbits are bound, and for $e \ge 1$, the orbits are open. The closure of the orbits is indicative of an additional symmetry, resulting in the conservation of the Runge-Lenz vector

$$\mathbf{A} = \mathbf{p} \times \mathbf{L} - mk\frac{\mathbf{r}}{r}. \tag{5.143}$$

The circular restricted three-body problem (CRTBP) consists of two heavy bodies moving in a circular orbit and a third body of negligible mass orbiting the first two. This system has five Lagrange points that are stationary in the corotating plane of the heavy masses. In the corotating frame there is a conserved energy given by the Jacobi integral

$$E = \frac{1}{2}m_3\dot{\mathbf{r}}^2 + \frac{m_3\Omega^2 r^2}{2} - \frac{m_3 m_1 G}{|\mathbf{r}_3 - \mathbf{r}_1|} - \frac{m_3 m_2 G}{|\mathbf{r}_3 - \mathbf{r}_2|}, \qquad (5.144)$$

where Ω is the angular velocity of the rotating frame. The CRTBP has a wide variety of solutions from periodic orbits to chaotic ones. The three-body problem cannot be reduced to quadrature and numerical ordinary differential equations solvers are required. Poincaré sections are often used to study the dynamics.

Newton's law of universal gravitation can be derived from Gauss's law of gravito-statics, where the divergence of the gravitational field couples to the mass density via the relation

$$\nabla \cdot \mathbf{g}(r) = -\nabla^2 \Phi(r) = -4\pi G\rho(r). \qquad (5.145)$$

5.8 EXERCISES

1. Sketch the effective potential for a particle in a $1/r^2$ potential and classify the orbits. Show that there are bound orbits that spiral into the origin for some maximum distance and that there are unbound orbits which spiral outward from some minimum distance. Note that the separatrix between these two types of behavior must be treated separately. Find analytical solutions for the orbit equations for each of these three cases making sure that the boundary conditions are satisfied.

2. A particle, acting under a central force, has a circular orbit with the center of force lying on the circumference of the circle. Show that the force varies as $1/r^5$ and find the period of the orbit.

3. The text shows that the solution for the orbit equation of the $1/r^4$ potential generally involves elliptic integrals. Examine the zero energy case and show that that the solution can be obtained in terms of elementary functions. Describe the resulting orbit.

4. For particles in an attractive $1/r$ potential show that
   a. If two orbits have the same angular momentum, the perihelion of the parabolic orbit is one-half the radius of the circular orbit.
   b. The speed of a particle at any point in a parabolic orbit is $\sqrt{2}$ faster than the speed of a particle in a circular orbit passing through the same point.

5. A particle of mass m moves in a singular central potential $V(r) = -kr^{-n}$ where $n > 2$. Reduce the equations of motion to an effective one-dimensional problem. Discuss the qualitative nature of the orbits for different values of the energy. For the bound orbits, show that the particle spirals into the origin in a finite time and requiring a finite number of revolutions.

6. At perihelion, a particle in an elliptical orbit in a gravitational $1/r$ potential receives a velocity boost Δv in the instantaneous direction of motion. Find the semi-major axis, eccentricity, and orientation angle of the new elliptical orbit in terms of the old orbit's parameters and the velocity boost. [This is basically how Hohmann transfer orbits are calculated, however getting the timing right is the hard part.]

7. At perihelion, a particle of mass m in an elliptical orbit in a gravitational $1/r$ potential receives an impulse $m\Delta v$ in the radial direction. Find the semi-major axis and eccentricity of the new elliptical orbit in terms of the old ellipse's parameters. By how much has the angle of the perihelion rotated?

8. The advance of its perihelion of mercury is 42 arc-seconds per century. This can be accounted for classically if one adds a small perturbation to the $1/r$ potential of the sun given by $V(r) = -k/r(1 - \eta a/r)$, where a is the semimajor axis of the orbit . How big would η have to be to account the observed discrepancy? The semimajor axis of Mercury's orbit is 0.387 ua, the orbital period 0.241 yr and its eccentricity is 0.206.

9. The Coulomb interaction between particles of charges q_1 and q_2 obeys an inverse square force law with a potential energy $V(r) = q_1 q_2 / 4\pi \varepsilon_0 r$. Oppositely signed charges attract, and similarly signed charges repel. The equations of motion are similar to the Kepler Problem, resulting in the replacement $k = -q_1 q_2 / 4\pi \varepsilon_0$ in the orbit equation. Find the solution to the orbit equation for the repulsive case, and show that the solutions are hyperbolas, with the center of force located at the further focus.

10. Find the approximate ratio of the mass of the Earth to that of the sun using only the periods of the Earth's orbit about the sun (365.25 days), the moon's orbit about the Earth (lunar month = 27.3 days), and the mean distances of the orbits of the Earth about the sun (1.49×10^8 km) and the moon about the Earth (3.8×10^5 km).

11. Derive Lagrange's equations of motion for a particle of mass m moving in an attractive linear central potential $V = kr$. Reduce the equation for the period of the orbit to quadrature, and find an analytic solution for the zero angular momentum case. Find the angular frequency

for a circular orbit. Compare this with the frequency for small radial oscillations about the circular orbit.

12. Formally solve the two-particle central force problem in the Hamiltonian framework by reduction to quadrature.

13. The effective one-body Hamiltonian for a single particle in a central well can be written in spherical coordinates as

$$H(p,q) = \frac{1}{2m}\left(p_r^2 + \frac{1}{r^2}\left(p_\theta^2 + \frac{1}{\sin^2\theta}p_\phi^2\right)\right) + V(r).$$

Derive Hamilton's equations of motion and prove that $(p_\theta^2 + p_\phi^2/\sin^2\theta) = \ell^2$ is a constant of the motion.

14. Use a computer to generate a radial phase space (r, p_r) map by finding lines of constant energy for the linear potential $V = kr$. Use dimensionless units, $m = k = 1$, and choose an angular momentum such that the minimum of the effective potential occurs at $r = 1$. Characterize the motion.

15. Draw an effective potential energy diagram for the central potential $V = -k/\cosh r/r_0$. Use dimensionless units, $m = k = r_0 = 1$ and pick a value of ℓ such that the effective potential energy is zero at $r = 1$. Characterize the motion in terms of the effective energy diagram. From the plot, determine the approximate radius for circular orbits and estimate its period.

16. Use a computer to generate a radial phase space (r, p_r) diagram for the central force potential $V = -k/\cosh r/r_0$ under the conditions described in Exercise 5.15.

17. Find an analytic solution for the orbit $r(\phi)$ in terms of elementary functions for a particle with zero energy in an attractive $1/r^4$ potential. Find the total rotation angle for a particle of this energy spiraling down to the center of force from its aphelion.

18. The spherical harmonic oscillator and the bound states of the Kepler potential both have ellipses for their orbital solutions. Find a transformation that maps the two sets of solutions into each other.

19. Starting from the Lagrangian for CRTBP find the Hamiltonian in Cartesian coordinates. Using reduced units $m_3 = \Omega = r_0 = 1$, find Hamilton's equations of motion in terms of canonical coordinates and momenta for this system in the (x, y) plane of the two-bodies.

20. Taking the ratio of the mass of the moon to that of Earth to be $\mu = 1/82.45$, numerically integrate the CRTBP Equations in the Earth-moon plane for ten Lunar cycles using the following data as initial input

$$[x, y, \dot{x}, \dot{y}] = [1.2, \quad 0, \quad 0, \quad -1.04935750983031990726].$$

The result should be a closed orbit in the corotating frame. (Sample orbital conditions taken from *Orbital Mechanics with MATLAB: The Circular-Restricted Three-Body Problem*, available at *http://www.cdeagle.com/ommatlab/crtbp.pdf.*)

21. Three stars with masses (m_1, m_2, m_3) are in a gravitationally bound triple-star system, where each of the stars is situated in the corners of an equilateral triangle with sides of length a. Determine the direction and magnitude of the rotational velocity which will leave the relative positions of the three stars unchanged.

22. Calculate the gravitational potential field of a thin ring of mass M and radius a.

Chapter 6

SMALL OSCILLATIONS

I n this chapter, critical points of Lagrangian systems will be analyzed for their stability under small perturbations from stationary equilibrium. The motion of stable systems is expanded in terms of their normal (collective) modes of oscillation about the equilibrium point. One dimensional lattice structures are then examined, which, in the continuum limit, give rise to the wave equation. The effects of damping and driving forces on the motion of the system are considered.

6.1 SMALL OSCILLATIONS ABOUT AN EQUILIBRIUM CONFIGURATION

Pioneering work on the treatment of small oscillations from equilibrium was carried out by Johann Bernoulli (1667–1748) and his son Daniel Bernoulli (1700–1782). The *Superposition Principle* of linear modes of oscillation is usually attributed to Daniel Bernoulli, who demonstrated in 1753 that the vibrations of a string can be constructed from a superposition of its normal modes of vibration. The general treatment of the small oscillations of a particle system was fully developed by Lagrange.

Consider a Lagrangian system in static equilibrium. If the system is deformable, then the *normal modes* of small oscillation of the system correspond to the resonant frequencies at which all components of the system move collectively, with fixed relative phase, such that the elements all pass through the equilibrium point at the same time. A *standing wave* is the continuum analog of a normal mode of oscillation of a discrete system. The actual motion of the system for small deviations from equilibrium can be expressed as a superposition of its normal modes of motion. A normal mode analysis of a critical point of the potential energy function will reveal whether the system is stable, unstable, or in neutral equilibrium with respect to small variations of a normal mode coordinate.

In structural design, damping is often applied to normal modes to limit their amplitude of oscillation under external driving forces. On the other hand, in resonant circuit design, a filter is often used to enhance a particular mode of oscillation, by damping competing nearby modes of oscillation.

If there are no cyclic coordinates, the stationary points are points of *static equilibrium*. If there are cyclic coordinates, the system is in neutral equilibrium with respect to variations of these coordinates. Cyclic coordinates are ignorable coordinates, however. They can be eliminated from the dynamical problem by replacing their velocities by their constant canonical momenta. This results in a dynamical system with fewer degrees of freedom. The reduced system may have points of *stationary motion*, where the system is in a minimum of energy with respect to the non-cyclic coordinates.

6.2   REVIEW OF THE ONE-DIMENSIONAL PROBLEM

In Chapter 1, one-dimensional periodic motions restricted to small variations about a minimum of potential were shown to be harmonic in character. In Chapter 5, small radial oscillations about stationary circular orbits were examined. In this second case, the problem reduces to an effective one-dimensional problem in the radial direction, and one expands the solution about the minimum of an effective one-dimensional potential. In this chapter, the more general case of small oscillations about stationary points of motion for multidimensional systems will be analyzed to determine their stability. If the perturbations remain small, the small oscillation solution may remain a good approximation to the general motion for a long time.

Reconsider the one-dimensional Lagrangian for a point particle with a local potential energy minimum: The Lagrangian of the system is

$$L = \frac{1}{2}m\dot{x}^2 - V(x), \tag{6.1}$$

and the resulting equation of motion is given by

$$m\ddot{x} = F(x) = -\frac{\partial V(x)}{\partial x}. \tag{6.2}$$

The Lagrangian is time independent; therefore, mechanical energy is conserved

$$E = \frac{1}{2}m\dot{x}^2 + V(x). \tag{6.3}$$

A point x_0 corresponding to a potential extremum is a *critical point* of the motion if its velocity and acceleration simultaneously vanish

$$\dot{x} = \sqrt{\frac{2}{m}[E - V(x_0)]} = 0,$$

$$\ddot{x} = \dot{p}/m = -\frac{1}{m}\frac{\partial V(x)}{\partial x} = 0. \tag{6.4}$$

The above conditions describe a *stationary point* that might be stable or unstable. To examine the behavior of the systems for small deviations about the equilibrium point, one undertakes a Taylor Expansion of the potential

$$V = V(x_0) + \frac{\partial V}{\partial x}\bigg|_{x=x_0} (x - x_0) + \frac{1}{2}\frac{\partial^2 V}{\partial x^2}\bigg|_{x=x_0} (x - x_0)^2 + \cdots \qquad (6.5)$$

The first term is an ignorable constant, while the second term vanishes at the stationary point, leaving the third term as the leading order, nonvanishing term with physical significance.

Typically, one makes a change in coordinates with respect to small perturbations relative to the equilibrium point, $\eta = x - x_0$, giving the potential expansion

$$V(\eta) = V(x_0) + \frac{1}{2}\frac{\partial^2 V}{\partial x^2}\bigg|_{x=x_0} \eta^2 + O(\eta^3). \qquad (6.6)$$

Dropping the constant term in the potential, the Lagrangian now can be rewritten to leading order as

$$L = \frac{1}{2}m\dot{\eta}^2 - \frac{1}{2}k\eta^2. \qquad (6.7)$$

In this approximation, the result reduces to the Lagrangian for a simple harmonic oscillator. Higher order terms in the Lagrangian lead to higher order corrections in the equations of motion. This justifies truncating the Lagrangian at second order in the perturbation provided that the small amplitude approximation remains valid for all times. If the effective spring constant of the system is positive

$$k = \frac{\partial^2 V}{\partial x^2}\bigg|_{x=x_0} > 0, \qquad (6.8)$$

the system is stable under small perturbations. In this case, the small amplitude behavior is restorative and oscillatory in character, leading to the harmonic solution

$$\eta(t) = \eta_0 \cos(\omega t - \phi_0), \qquad (6.9)$$

where the frequency of small oscillations is given by

$$\omega = \sqrt{k/m}. \qquad (6.10)$$

On the other hand, the equilibrium point is unstable when the second derivative is negative

$$k = \frac{\partial^2 V}{\partial x^2}\bigg|_{x=x_0} < 0. \qquad (6.11)$$

In this case, exponential growth or damping of the perturbation cannot be ruled out. Finally, if the second derivative of the expansion vanishes, giving

$$k = \left.\frac{\partial^2 V}{\partial x^2}\right|_{x=x_0} = 0, \tag{6.12}$$

then the earlier test for stability fails. However, if all derivatives with respect to x vanish, one is dealing with a cyclic coordinate, which can be explicitly eliminated or simply ignored.

6.3   NORMAL MODE ANALYSIS OF COUPLED OSCILLATORS

Now, consider a multiparticle system with a time-independent scalar potential. Using a standard particle Lagrangian, the kinetic energy is a quadratic function of the velocities, provided that the generalized coordinates are time-independent functions of the Cartesian coordinates. The kinetic energy can be written in matrix form as

$$T = \frac{1}{2}T_{ij}(q)\dot{q}^i\dot{q}^j = \frac{1}{2}\bar{\dot{\mathbf{q}}} \cdot \mathbf{T} \cdot \dot{\mathbf{q}}, \tag{6.13}$$

where the velocities can be treated as components of a vector in an N-dimensional configuration space

$$\dot{\mathbf{q}} = \dot{q}^i\mathbf{n}_i. \tag{6.14}$$

The orthonormal configuration space basis $\mathbf{n}_i$ has the real-valued, self-adjoint column vector representation

$$\mathbf{n}_1 = \begin{bmatrix} 1 \\ 0 \\ \vdots \\ 0 \end{bmatrix}, \quad \mathbf{n}_2 = \begin{bmatrix} 0 \\ 1 \\ \vdots \\ 0 \end{bmatrix}, \dots, \quad \mathbf{n}_n = \begin{bmatrix} 0 \\ 0 \\ \vdots \\ 1 \end{bmatrix}, \tag{6.15}$$

which satisfies the completeness relation

$$\mathbf{n}_i\mathbf{n}_i = 1, \tag{6.16}$$

and the normalization condition

$$\mathbf{n}_i \cdot \mathbf{n}_j = \delta_{ij}. \tag{6.17}$$

In this space, the inertia tensor $\mathbf{T}$ has components

$$T_{ij}(q) = \frac{\partial^2 T}{\partial \dot{q}^i \partial \dot{q}^j}. \tag{6.18}$$

The standard particle Lagrangian is given by

$$L(q, \dot{q}) = \frac{1}{2}T_{ij}(q)\dot{q}^i\dot{q}^j - V(q) = \frac{1}{2}\bar{\dot{\mathbf{q}}} \cdot \mathbf{T} \cdot \dot{\mathbf{q}} - V, \qquad (6.19)$$

and Lagrange's equations of motion are

$$\dot{\mathbf{p}} = \frac{d(\mathbf{T}(q) \cdot \dot{\mathbf{q}})}{dt} = -\nabla_q V(q). \qquad (6.20)$$

Now assume that the potential energy $V(\mathbf{q})$ has a local minimum at $\mathbf{q} = \mathbf{q}_0$

$$\mathbf{F}(\mathbf{q}_0) = -\nabla_q V|_{\mathbf{q}=\mathbf{q}_0} = 0. \qquad (6.21)$$

On can expand the Lagrangian in terms of small displacements about this supposed minimum by making the substitution

$$\begin{aligned} \boldsymbol{\eta} &= \Delta\mathbf{q} = \mathbf{q} - \mathbf{q}_0, \\ \dot{\boldsymbol{\eta}} &= \dot{\mathbf{q}}. \end{aligned} \qquad (6.22)$$

Since the kinetic energy is already a second-order function of the generalized velocities, to leading order, the inertia tensor can be replaced by its value at the minimum

$$M_{ij} = T_{ij}(q_0) = \frac{\partial^2 T}{\partial \dot{q}^i \partial \dot{q}^j}\bigg|_{\eta=0} \qquad (6.23)$$

Likewise, a Taylor Expansion of the potential energy results in the series

$$V(q) = V(q_0) + \frac{\partial V}{\partial q^i}\bigg|_{\eta=0} \eta^i + \frac{1}{2}\frac{\partial^2 V}{\partial q^i \partial q^j}\bigg|_{\eta=0} \eta^i\eta^j + O(\eta^3) \qquad (6.24)$$

which can be written in dyadic form as

$$V(\mathbf{q}) = V(\mathbf{q}_0) - \mathbf{F}(\mathbf{q}_0) \cdot \boldsymbol{\eta} + \frac{1}{2}\bar{\boldsymbol{\eta}} \cdot \mathbf{K} \cdot \boldsymbol{\eta} + O(\eta^3) \approx V(q_0) + \frac{1}{2}\bar{\boldsymbol{\eta}} \cdot \mathbf{K} \cdot \boldsymbol{\eta} \qquad (6.25)$$

Ignoring the constant term in the potential, the effective Lagrangian, to second order in the perturbation expansion, is given by the contracted matrix equation

$$L = \frac{1}{2}\dot{\bar{\boldsymbol{\eta}}} \cdot \mathbf{M} \cdot \dot{\boldsymbol{\eta}} - \frac{1}{2}\boldsymbol{\eta} \cdot \mathbf{K} \cdot \boldsymbol{\eta}, \qquad (6.26)$$

where

$$\mathbf{K}_{ij} = \left. \frac{\partial^2 V}{\partial q^i \partial q^j} \right|_{\eta=0}, \tag{6.27}$$

$$\mathbf{M}_{ij} = \left. \frac{\partial^2 T}{\partial \dot{q}^i \partial \dot{q}^j} \right|_{\eta=0}. \tag{6.28}$$

Both $\mathbf{K}$ and $\mathbf{M}$ are symmetric rank-2 tensors. Therefore, in the small amplitude approximation, the equations of motion are linear second-order ordinary differential equations with constant coefficients. The equations of motion can be expressed in the linear, homogenous matrix form

$$\mathbf{M} \cdot \ddot{\boldsymbol{\eta}} = -\mathbf{K} \cdot \boldsymbol{\eta}. \tag{6.29}$$

Problems arise if the critical point of the analysis is not a local minimum as assumed. If $\mathbf{q}_0$ is a saddle point or maximum, one cannot guarantee the deviations would not grow in an uncontrollable fashion. One can get either exponential damping or exponential growth depending on the initial conditions. The critical point is unstable in such cases. Therefore, one must test the solutions for stability. Furthermore, cyclic coordinates do not contribute to the potential energy and give rise to zero-frequency, neutral-equilibrium modes of motion in this analysis. One must distinguish true cyclic coordinates from cases where the second derivative of the potential with respect to a coordinate is zero, but higher order derivatives are present. The latter case must be expanded to include the first nonvanishing higher order term in the expansion. Such higher order expansions cannot be handled within the present linear framework.

6.3.1 Solutions to the Eigenvalue Problem

The linearized small oscillation problem can be solved by using two well-known theorems from linear matrix analysis:

- The symmetry under exchange of indices of a rank-2 matrix is unaffected by a similarity transformation.
- Any real-valued symmetric matrix can be diagonalized by an orthogonal transformation. The resulting elements of the diagonalized matrix are all real.

The proof of the first theorem follows from the fact that a real-valued symmetric matrix $\mathbf{T}_s$ is self-adjoint (even) under the operation of tensor reversion. Likewise, a real-valued antisymmetric matrix $\mathbf{T}_a$ is antiadjoint (odd) under reversion, where reversion is equivalent to transposition for real-valued rank-2 tensors

$$\mathbf{T}_s = \bar{\mathbf{T}}_s \rightarrow T_{sij} = T_{sji},$$
$$\mathbf{T}_a = -\bar{\mathbf{T}}_a \rightarrow T_{aij} = -T_{aji}. \tag{6.30}$$

The first theorem states that, under a similarity transformation, this property is preserved. Let $\mathbf{T}' = \mathbf{U} \cdot \mathbf{T} \cdot \bar{\mathbf{U}}$ denote a "unitary" similarity transform with $\bar{\mathbf{U}} = \mathbf{U}^{-1}$,[1] then

$$\mathbf{T}'_s - \bar{\mathbf{T}}'_s = \mathbf{U} \cdot (\mathbf{T}_s - \bar{\mathbf{T}}_s) \cdot \bar{\mathbf{U}} = 0,$$
$$\mathbf{T}'_a + \bar{\mathbf{T}}'_a = \mathbf{U} \cdot (\mathbf{T}_a + \bar{\mathbf{T}}_a) \cdot \bar{\mathbf{U}} = 0, \tag{6.31}$$

The proof of the second theorem follows from the solution of the eigenvalue problem. It is similar to the proof that Hermitian matrices in Quantum Mechanics are diagonalizable and have real eigenvalues.

Consider the eigenvalue problem for a real symmetric matrix and let its eigenvalues be possibly complex, where the adjoint operator under reversion is defined to have the property $\bar{\lambda}_i = \lambda_i^*$ for complex eigenvalues. The eigenvalue problem for a matrix $\mathbf{M}$ is to find the solutions to the equation

$$\mathbf{M} \cdot \mathbf{a}_i = \lambda_i \mathbf{a}_i, \tag{6.32}$$

where $\mathbf{a}_i$ is a (possibly complex) eigenvector and λ_i is a (possibly complex) eigenvalue. Complex numbers form an inner product space with a norm $\bar{c}c = c^*c \geq 0$. Complex conjugation defines a dual relationship for eigenvectors and their eigenvalues

$$\bar{\mathbf{a}}_i = \mathbf{a}_i^*, \tag{6.33}$$
$$\bar{\lambda}_i = \lambda_i^* \tag{6.34}$$

Note that it is necessary to consider the possibility of complex solutions because, by the *Fundamental Theorem of Algebra*, polynomial equations can be completely factored, in general, only by extending the real number field to the complex number field. Because the equations to be solved have real coefficients, however, the real and imaginary parts of the solutions will separately satisfy the equations of motion. One can always project out the real part of a complex solution to obtain a physically real solution. Let $c = a + bi$ denote

---

[1] The term "unitary" is often reserved for unitary transformations using the Hermitian adjoint $\mathbf{T}' = \mathbf{U} \cdot \mathbf{T} \cdot \mathbf{U}^\dagger$, where $\mathbf{U}^\dagger = \mathbf{U}^{-1}$. Here the concept is extended to the operation of reversion.

a complex number. Scalar projection is defined to result in the extraction of a real value

$$\text{Re}(c) = \langle c \rangle = a, \tag{6.35}$$

$$\text{Im}(c) = \langle \bar{i}c \rangle = \langle i^*c \rangle = b. \tag{6.36}$$

For a real-valued symmetric matrix $\mathbf{M}$, the tensor adjoint operator gives

$$\bar{\mathbf{a}}_i \cdot \bar{\mathbf{M}} = \bar{\mathbf{a}}_i \cdot \mathbf{M} = \bar{\mathbf{a}}_i \lambda_i^* \tag{6.37}$$

Since $\mathbf{M}$ is symmetric, one has

$$\bar{\mathbf{a}}_i \cdot \mathbf{M} \cdot \mathbf{a}_j - \bar{\mathbf{a}}_i \cdot \mathbf{M} \cdot \mathbf{a}_j \equiv 0 = (\lambda_i^* - \lambda_j)(\bar{\mathbf{a}}_i \cdot \mathbf{a}_j). \tag{6.38}$$

But for $i = j$, $\bar{\mathbf{a}}_i \cdot \mathbf{a}_i = |\mathbf{a}|^2 > 0$, therefore

$$\bar{\lambda}_i = \lambda_i^* = \lambda_i. \tag{6.39}$$

It follows that the eigenvalues of a real symmetric matrix are all real. This implies that the eigenvectors can also be chosen to be real-valued vectors, since their real and imaginary parts must separately satisfy the eigenvalue problem.

For $i \neq j$, the result of Equation (6.38) is

$$(\lambda_i - \lambda_j)(\bar{\mathbf{a}}_i \cdot \mathbf{a}_j) = 0 \tag{6.40}$$

Therefore, if the eigenvalues of a real symmetric tensor are distinct, the eigenvectors must be orthogonal to each other, since $\lambda_i \neq \lambda_j$ implies

$$\bar{\mathbf{a}}_i \cdot \mathbf{a}_j = 0. \tag{6.41}$$

However, even if the eigenvalues are not distinct, given a complete linearly-independent set of degenerate solutions, the *Gramm-Schmidt* orthonormalization procedure [Arfken 2001: pp 506–599] can be used to generate a complete orthonormal basis with the following properties

$$\hat{\bar{\mathbf{a}}}_i \cdot \hat{\mathbf{a}}_j = \delta_{ij}. \tag{6.42}$$

The proof follows from the fact that given m linearly independent vectors $\{\mathbf{a}_k : k = 1, m\}$ having the same eigenvalue, any linear combination of these vectors have the same eigenvalue. A set of orthonormal eigenvectors can be therefore be created by the following procedure:

1. Begin by normalizing the first vector in the degenerate set: $\mathbf{a}_1 \rightarrow \mathbf{a}_1/|\mathbf{a}_1|$.
2. Next, find the projection of the second vector that is orthogonal to the first vector and normalize it: $\tilde{\mathbf{a}}_2 = (\mathbf{a}_2 - \hat{\mathbf{a}}_1(\hat{\bar{\mathbf{a}}}_1 \cdot \mathbf{a}_2))$, $\mathbf{a}_2 \rightarrow \tilde{\mathbf{a}}_2/|\tilde{\mathbf{a}}_2|$.

3.   Repeat the second step as necessary, finding the component of the kth vector that is orthogonal to the first $k - 1$ vectors and normalizing it.

The set of vectors $\{\hat{\mathbf{a}}_k : k = 1, m\}$ now forms an orthonormal set.

The transformation that diagonalizes a symmetric matrix can be constructed from a complete orthonormal set of eigenvectors. Let $\mathbf{O}$ be the transformation matrix, then

$$\hat{\mathbf{a}}_i = \mathbf{O} \cdot \mathbf{n}'_i, \tag{6.43}$$

where $\mathbf{O}$ is orthogonal $\bar{\mathbf{O}} \cdot \mathbf{O} = \mathbf{1}$, and diagonalizes $\mathbf{M}$

$$\mathbf{M}' = \bar{\mathbf{O}} \cdot \mathbf{M} \cdot \mathbf{O} = M_i \mathbf{n}'_i \mathbf{n}'^i = \begin{bmatrix} M_1 & 0 & 0 & 0 \\ 0 & M_2 & 0 & 0 \\ 0 & 0 & \ddots & 0 \\ 0 & 0 & 0 & M_n \end{bmatrix} \tag{6.44}$$

and $\mathbf{n}'$ are basis vectors associated with the normal mode coordinates

$$\boldsymbol{\eta}(t) = \mathbf{n}_i \eta^i(t) = \mathbf{n}'_j \varsigma^j(t). \tag{6.45}$$

The transformation matrix $\mathbf{O}$ is given by

$$\mathbf{O} = \hat{\mathbf{a}}_i \mathbf{n}'^i = \begin{bmatrix} \hat{a}^1_1 & \hat{a}^1_2 & \cdots & \hat{a}^1_n \\ \hat{a}^2_1 & \hat{a}^2_2 & & \hat{a}^2_n \\ \vdots & \vdots & & \vdots \\ \hat{a}^n_1 & \hat{a}^n_2 & \cdots & \hat{a}^n_n \end{bmatrix}. \tag{6.46}$$

Thus, any real-valued symmetric matrix can be diagonalized by an orthogonal transformation, and the resulting eigenvalues of the diagonalized matrix are all real.

6.3.2 Simultaneous Diagonalization of Two Matrices

The small oscillation problem to be solved differs from the simple eigenvalue problem in that there are two matrices that must be simultaneously diagonalized. The procedure for solving the problem is similar, however.

First, one creates an eigenvalue equation by postulating the existence of normal mode solutions to the equation of motion (6.29)

$$\boldsymbol{\eta}(t) = \begin{pmatrix} \eta^1(t) \\ \vdots \\ \eta^n(t) \end{pmatrix} = \mathbf{A} \cdot \boldsymbol{\varsigma} = \mathbf{a}_j \varsigma^j(t) = \mathbf{A} \begin{pmatrix} c^1 \cos(\omega_1 t + \phi_1) \\ \vdots \\ c^n \cos(\omega_n t + \phi_n) \end{pmatrix}. \tag{6.47}$$

By linear independence, each normal mode separately satisfies the equation of motion with solutions given by the eigenvalue equations

$$(-\omega_i^2 \mathbf{M} + \mathbf{K}) \cdot \mathbf{a}_i = 0. \tag{6.48}$$

If a solution exists, the eigenvalues are given by

$$\omega_i^2 = \frac{\bar{\mathbf{a}}_i \cdot \mathbf{K} \cdot \mathbf{a}_i}{\bar{\mathbf{a}}_i \cdot \mathbf{M} \cdot \mathbf{a}_i}. \tag{6.49}$$

To find the solution, one looks for a transformation which will simultaneously diagonalize the matrices $\mathbf{M}$ and $\mathbf{K}$. Such a transformation can be constructed by applying two orthogonal transformations and one scale transformation, given by the ordered product

$$\mathbf{A} = \mathbf{O}_M \cdot \mathbf{S}_M \cdot \mathbf{O}_K. \tag{6.50}$$

This results in the simultaneously diagonalized forms

$$\tilde{\mathbf{M}} = 1 = \bar{\mathbf{A}} \cdot \mathbf{M} \cdot \mathbf{A} = \mathbf{n}_i \mathbf{n}_i = |i\rangle\langle i|, \tag{6.51}$$

$$\tilde{\mathbf{K}} = \omega_i^2 \mathbf{n}_i' \mathbf{n}^{\prime i} = \omega_i^2 |i\rangle\langle i|. \tag{6.52}$$

The prescription for generating this transformation is as follows:

1. First, diagonalize the inertia matrix, by applying an orthogonal transformation to it

    $$\mathbf{M}' = \bar{\mathbf{O}}_M \cdot \mathbf{M} \cdot \mathbf{O}_M = M_i |i\rangle\langle i|. \tag{6.53}$$

    Note that the eigenvalues are positive definite because the kinetic energy is positive definite, unless all the velocities identically vanish. The mass matrix has a non-zero determinant and is invertible. Under the transformation, the potential matrix remains symmetric, but is not diagonal in general.

2. Next, apply a scale transformation, converting the inertial matrix to the identity matrix

    $$\bar{\mathbf{S}} \cdot \mathbf{M}' \cdot \mathbf{S} = 1. \tag{6.54}$$

    By construction, this scale transformation has the diagonal representation

$$\mathbf{S} = \frac{1}{(M_i)^{1/2}} |i\rangle\langle i| = \begin{bmatrix} (M_1)^{-1/2} & 0 & 0 & 0 \\ 0 & (M_2)^{-1/2} & 0 & 0 \\ 0 & 0 & \ddots & 0 \\ 0 & 0 & 0 & (M_n)^{-1/2} \end{bmatrix}. \tag{6.55}$$

The transformed spring matrix remains symmetric, but is not diagonal in general.

3. Finally, make a second orthogonal transformation that diagonalizes the spring matrix

$$\tilde{\mathbf{K}} = \bar{\mathbf{O}}_K \mathbf{K}'' \mathbf{O}_K = \omega_i^2 |i\rangle \langle i| = \begin{bmatrix} \omega_1^2 & 0 & 0 & 0 \\ 0 & \omega_2^2 & 0 & 0 \\ 0 & 0 & \ddots & 0 \\ 0 & 0 & 0 & \omega_n^2 \end{bmatrix}. \qquad (6.56)$$

This last orthogonal transformation has no effect on the identity matrix, so the goal of simultaneous diagonalization of two symmetric matrices has been achieved. In the new coordinate system, the Lagrangian separates completely

$$L = \frac{1}{2}\dot{\varsigma} \cdot \dot{\varsigma} - \frac{1}{2}\varsigma \cdot \tilde{\mathbf{K}} \cdot \varsigma = \sum_i \left(\frac{1}{2}\dot{\varsigma}_i \dot{\varsigma}^i - \frac{\omega_i^2}{2}\varsigma_i \varsigma^i \right), \qquad (6.57)$$

where, for a orthonormal basis, $\varsigma^i \equiv \varsigma_i$. As a consequence, the mechanical energy of each normal mode is separately conserved

$$E = \sum_i E_i = \sum_i \left(\frac{1}{2}\dot{\varsigma}_i \dot{\varsigma}^i + \frac{\omega_i^2}{2}\varsigma_i \varsigma^i \right). \qquad (6.58)$$

The equations of motion completely decouple into independent simple harmonic oscillator equations of motion, provided that eigenvalues of $\tilde{\mathbf{K}}$ are all positive definite

$$k_i = \omega_i^2 > 0. \qquad (6.59)$$

This is the strong criteria that must be met if the stationary point is to represent a true local minimum of potential. If any of the eigenvalues are negative, the motion is unstable, and the small amplitude assumption must be rejected. If any of the frequencies vanish, then there are two possibilities: a) either the mode corresponds to a cyclic mode or b) the procedure fails. Note that cyclic modes can be handled by this method even though the coordinate has a linear time-dependence that grows with time. If necessary, a time-dependent transformation can be used to explicitly remove this growth.

In practice, one does not know the desired transformation until the eigenvalue problem has been solved, and one must determine the eigenvalues before solving for the eigenvectors.

The eigenvalue problem (6.48) has solutions if the secular equation for the determinant vanishes

$$\det(-\omega^2 \mathbf{M} + \mathbf{K}) = 0. \tag{6.60}$$

This leads to an nth order polynomial equation for the eigenvalues

$$\prod_i (\omega^2 - \omega_i^2) = 0. \tag{6.61}$$

Note that both signs of the angular frequencies $\pm\omega_i$ are allowed solutions. Provided that the roots $\omega_i^2 \geq 0$, where the zero frequency modes correspond to cyclic modes, the equations of motion yield a stationary solution.

For every eigenvalue ω_i^2, one can now solve the eigenvalue equation

$$(-\omega_i^2 \mathbf{M} + \mathbf{K}) \cdot \mathbf{a}_i = 0. \tag{6.62}$$

where the eigenvalues resulting are orthogonalized, if necessary, using a variant of the Gram-Schmidt procedure, and normalized with respect to weight $\mathbf{M}$

$$\bar{\mathbf{a}}_i \cdot \mathbf{M} \cdot \mathbf{a}_j = \delta_{ij}. \tag{6.63}$$

When there is degeneracy, the modified Gramm-Schmidt procedure is:

1. First, normalize the first degenerate vector in the set: $\mathbf{a}_1 \rightarrow \mathbf{a}_1 / \sqrt{\bar{\mathbf{a}}_1 \cdot \mathbf{M} \cdot \mathbf{a}_1}$.
2. Next, find the projection of the second vector that is orthogonal to the first vector with respect to weight $\mathbf{M}$ and normalize it: $\tilde{\mathbf{a}}_2 = (\mathbf{a}_2 - \mathbf{a}_1(\bar{\mathbf{a}}_1 \cdot \mathbf{M} \cdot \mathbf{a}_2))$, $\mathbf{a}_2 \rightarrow \tilde{\mathbf{a}}_2 / \sqrt{\tilde{\mathbf{a}}_2 \cdot \mathbf{M} \cdot \tilde{\mathbf{a}}_2}$.
3. Repeat the second step as necessary, finding the component of the kth vector that is orthogonal to the first $k-1$ vectors and normalizing it.

The transformation matrix that diagonalizes the matrices is called the *modal matrix*. It is given by the dyadic construction

$$\mathbf{A} = \mathbf{a}_i \hat{\mathbf{n}}^i = \begin{bmatrix} a_1^1 & a_2^1 & \cdots & a_n^1 \\ a_1^2 & a_2^2 & & a_n^2 \\ \vdots & \vdots & & \vdots \\ a_1^n & a_2^n & \cdots & a_n^n \end{bmatrix}. \tag{6.64}$$

In terms of the modal matrix, the general solution now can be written as

$$\eta(t) = \mathbf{A} \cdot \varsigma(t). \tag{6.65}$$

In matrix index notation this becomes

$$\eta^i(t) = A^i{}_j \varsigma^j(t), \tag{6.66}$$

or, more explicitly, in column-vector notation as

$$\boldsymbol{\eta}(t) = \begin{pmatrix} \eta^1(t) \\ \vdots \\ \eta^n(t) \end{pmatrix} = \mathbf{A} \begin{pmatrix} c^1 \cos(\omega_1 t + \phi_1) \\ \vdots \\ c^n \cos(\omega_n t + \phi_n) \end{pmatrix}, \tag{6.67}$$

where the constants $\{c^j, \phi^j\}$ are determined from the initial conditions.

Since the modal matrix is invertible, one can express the normal modes in terms of the physical degrees of freedom

$$\boldsymbol{\varsigma} = \mathbf{A}^{-1} \cdot \boldsymbol{\eta} = \mathbf{A}^{-1} \begin{pmatrix} \eta^1(t) \\ \vdots \\ \eta^n(t) \end{pmatrix} = \begin{pmatrix} c^1 \cos(\omega_1 t + \phi_1) \\ \vdots \\ c^n \cos(\omega_n t + \phi_n) \end{pmatrix}, \tag{6.68}$$

where, for cyclic variables, the oscillatory solution is replaced by the zero-frequency solutions

$$\varsigma^j_{\text{cyclic}}(t) = c^j + v^j t. \tag{6.69}$$

The constants $\{c^j, v^j\}$ can be determined from the initial conditions.

6.3.3 Vibrations of a Triatomic Molecule

As a first example of this procedure, consider a symmetric, one-dimensional linear triatomic molecule, as shown in Figure 6.1. Let all masses be equal

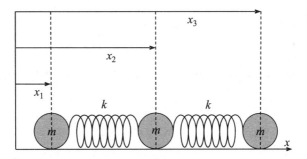

FIGURE 6.1   A linear triatomic molecule with three equal masses. The springs have equilibrium lengths L_0. The coordinates shown are defined with respect to an inertial frame.

and the motion be in one dimension. There are two oscillatory modes of longitudinal vibration and one cyclic mode corresponding to the translation of the center of mass. Assume small displacements from equilibrium. (This is not true for the cyclic mode, but this can be finessed.)

Define the small oscillations as deviations from the center of mass position R_0 at time $t = t_0$.

$$\begin{aligned}
\eta^1 &= x_1 - (R_0 - L_0), \\
\eta^2 &= x_2 - R_0, \\
\eta^3 &= x_3 - (R_0 + L_0),
\end{aligned} \tag{6.70}$$

where L_0 is the equilibrium length of the unstretched springs—for purists, a solution in center of mass frame can be chosen by substituting

$$R_0 \to R(t) = R_o + v_{cm}(t - t_0). \tag{6.71}$$

Note that the kinetic energy is diagonal in these coordinates

$$T = \frac{1}{2}m((\dot{\eta}^1)^2 + (\dot{\eta}^2)^2 + (\dot{\eta}^3)^2), \tag{6.72}$$

but the potential energy has off-diagonal contributions

$$V = \frac{1}{2}k(\eta^1 - \eta^2)^2 + \frac{1}{2}k(\eta^3 - \eta^2)^2. \tag{6.73}$$

Solving for the inertial and potential matrices, gives the matrices

$$\mathbf{M} = \begin{bmatrix} m & 0 & 0 \\ 0 & m & 0 \\ 0 & 0 & m \end{bmatrix}, \tag{6.74}$$

and

$$\mathbf{K} = \begin{bmatrix} k & -k & 0 \\ -k & 2k & -k \\ 0 & -k & k \end{bmatrix}. \tag{6.75}$$

The first step in solving this problem is solve the eigenvalue equation for the eigenvalues

$$(-\omega_j^2 \mathbf{M} + \mathbf{K}) \cdot \mathbf{a}_j = 0, \tag{6.76}$$

or

$$\begin{bmatrix} -\omega_j^2 m + k & -k & 0 \\ -k & -\omega_j^2 m + 2k & -k \\ 0 & -k & -\omega_j^2 m + k \end{bmatrix} \begin{bmatrix} a^1 \\ a^2 \\ a^3 \end{bmatrix}_j = 0. \tag{6.77}$$

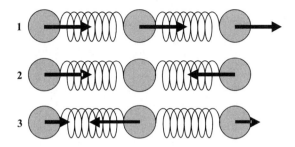

FIGURE 6.2 The normal modes of motion for the triatomic molecule. The first (top) mode is a cyclic translational mode. The other two modes are oscillatory in character.

The eigenvalues are determined by factoring the secular equation

$$\det\left(-\omega^2\mathbf{M}+\mathbf{K}\right)=(k-m\omega^2)(3k-m\omega^2)(-m\omega^2)=0. \qquad (6.78)$$

This yields three eigenvalues

$$\omega_1^2=0, \quad \omega_2^2=k/m, \quad \omega_3^2=3k/m. \qquad (6.79)$$

Once the eigenvalues are known, the eigenvectors can be determined and normalized. The zero frequency mode corresponds to the cyclic mode of center of mass translation, as shown in first diagram of Figure 6.2. The other two modes are stable oscillatory modes since $\omega^2 > 0$. The second mode, shown in the second diagram of Figure 6.2, has the middle mass fixed with the outer masses moving 180 degrees out of phase with each other. The remaining oscillatory mode (third diagram in Figure 6.2) has the middle mass oscillating against the common motion of the outer two masses.

The zero frequency mode is a solution to eigenvalue equation

$$\begin{bmatrix} k & -k & 0 \\ -k & 2k & -k \\ 0 & -k & k \end{bmatrix} \begin{bmatrix} a^1 \\ a^2 \\ a^3 \end{bmatrix}_1 = 0, \qquad (6.80)$$

with solution

$$\Rightarrow \begin{bmatrix} a^1 \\ a^2 \\ a^3 \end{bmatrix}_1 = \sqrt{N}\begin{bmatrix} 1 \\ 1 \\ 1 \end{bmatrix}. \qquad (6.81)$$

Applying the required normalization $\bar{\mathbf{a}}_i \cdot \mathbf{M} \cdot \mathbf{a}_j = \delta_{ij}$ gives

$$\begin{bmatrix} a^1 \\ a^2 \\ a^3 \end{bmatrix}_1 = \frac{1}{\sqrt{6m}}\begin{bmatrix} \sqrt{2} \\ \sqrt{2} \\ \sqrt{2} \end{bmatrix}. \qquad (6.82)$$

Likewise, solving for the second eigenvector, let $\omega_2 = \sqrt{k/m}$, giving the equation

$$\begin{bmatrix} 0 & -k & 0 \\ -k & k & -k \\ 0 & -k & 0 \end{bmatrix} \begin{bmatrix} a^1 \\ a^2 \\ a^3 \end{bmatrix}_2 = 0, \tag{6.83}$$

which has the normalized solution

$$\begin{bmatrix} a^1 \\ a^2 \\ a^3 \end{bmatrix}_2 = \frac{1}{\sqrt{6m}} \begin{bmatrix} \sqrt{3} \\ 0 \\ -\sqrt{3} \end{bmatrix}. \tag{6.84}$$

Finally, letting $\omega_3 = \sqrt{3k/m}$ gives the third eigenvalue equation

$$\begin{bmatrix} -2k & -k & 0 \\ -k & -k & -k \\ 0 & -k & -2k \end{bmatrix} \begin{bmatrix} a^1 \\ a^2 \\ a^3 \end{bmatrix}_3 = 0, \tag{6.85}$$

with normalized solution

$$\begin{bmatrix} a^1 \\ a^2 \\ a^3 \end{bmatrix}_3 = \frac{1}{\sqrt{6m}} \begin{bmatrix} 1 \\ -2 \\ 1 \end{bmatrix}. \tag{6.86}$$

Note that the center of mass is constant for the two oscillatory modes. Constructing the modal matrix from the normal mode vectors gives

$$\mathbf{A} = \frac{1}{\sqrt{6m}} \begin{bmatrix} \sqrt{2} & \sqrt{3} & 1 \\ \sqrt{2} & 0 & -2 \\ \sqrt{2} & -\sqrt{3} & 1 \end{bmatrix}, \tag{6.87}$$

with reciprocal

$$\mathbf{A}^{-1} = m\bar{\mathbf{A}} = \frac{\sqrt{m}}{\sqrt{6}} \begin{bmatrix} \sqrt{2} & \sqrt{2} & \sqrt{2} \\ \sqrt{3} & 0 & -\sqrt{3} \\ 1 & -2 & 1 \end{bmatrix}. \tag{6.88}$$

Therefore, the normal coordinates are given by

$$\begin{aligned} \varsigma^1 &= \sqrt{m/3}(\eta^1 + \eta^2 + \eta^3), \\ \varsigma^2 &= \sqrt{m/2}(\eta^1 - \eta^3), \\ \varsigma^3 &= \sqrt{m/6}(\eta^1 - 2\eta^2 + \eta^3), \end{aligned} \tag{6.89}$$

which agree with the sketch of the normal modes presented in Figure 6.2.

The Lagrangian, when rewritten in terms of the normal coordinates, separates into three independent pieces

$$L = \left(\frac{1}{2}(\dot{\varsigma}^1)^2\right) + \left(\frac{1}{2}(\dot{\varsigma}^2)^2 - \frac{k}{2m}(\varsigma^2)^2\right) + \left(\frac{1}{2}(\dot{\varsigma}^3)^2 - \frac{3k}{2m}(\varsigma^3)^2\right). \quad (6.90)$$

Clearly, ς^1 represents the cyclic center of mass coordinate. Ideally, this mode should have been eliminated apriori, and the reduced problem solved for the two true oscillatory modes of motion. One of the exercises at the end of the chapter approaches the problem from this point of view.

6.3.4 Double Pendulum

Another simple example is the double pendulum system. As shown in Figure 6.3, a double pendulum has a second mass hanging from a first pendulum. For large amplitudes, this system shows surprisingly complex and even chaotic behavior. Here, however, only the small oscillation limit is of interest. For simplicity, assume the motion lies in a plane. Then there are two degrees of freedom as shown in the figure. The equilibrium point has the two masses hanging collinearly along the vertical axis. Care must be used in constructing the Lagrangian since the coordinates of the bottom pendulum are given relative to an accelerating frame.

For specificity, let the masses and lengths of the two pendula be equal, $m_1 = m_2 = m$, $l_1 = l_2 = l$, and make the small angle expansion

$$\boldsymbol{\eta} = \mathbf{q} - \mathbf{q}_0 = \begin{bmatrix} \eta^1 \\ \eta^2 \end{bmatrix} = \begin{bmatrix} \theta_1 \\ \theta_2 \end{bmatrix}, \quad (6.91)$$

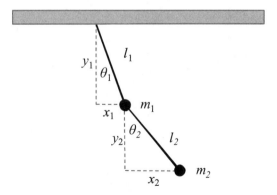

FIGURE 6.3 A possible coordinate system for the double pendulum system constrained to motion in a plane. Care must be used in constructing the Lagrangian since the coordinates of the bottom pendulum are given relative to an accelerating frame.

giving

$$x = l \sin \theta \approx l\theta, \quad l - y = l(1 - \cos \theta) \approx l\theta^2/2$$
$$\dot{x} = l\dot{\theta} \cos \theta = l\dot{\theta}, \quad \dot{y} = -l\dot{\theta}\sin\theta \approx 0 \tag{6.92}$$

The kinetic energy in this set of coordinates is given by

$$T = \frac{1}{2}m(\dot{x}_1^2 + \dot{y}_1^2) + \frac{1}{2}m((\dot{x}_1 + \dot{x}_2)^2 + (\dot{y}_1 + \dot{y}_2)^2)$$
$$\approx \frac{1}{2}ml^2\dot{\theta}_1^2 + \frac{1}{2}ml^2(\dot{\theta}_1 + \dot{\theta}_2)^2, \tag{6.93}$$

where the inertial matrix has off-diagonal terms due to the choice of coordinates. The potential energy, on the other hand, is diagonal in this coordinate system

$$V = -mgl(\cos\theta_1) - mgl(\cos\theta_1 + \cos\theta_2)$$
$$\approx V_0 + mgl\theta_1^2/2 + mgl(\theta_1^2 + \theta_2^2)/2. \tag{6.94}$$

The potential is a minimum when the pendula are hanging vertically, which occurs for $\theta_1 = \theta_2 = 0$. The small oscillation Lagrangian can be written as

$$L = \frac{1}{2}ml^2\dot{\theta}_1^2 - \frac{1}{2}ml^2(\dot{\theta}_1 + \dot{\theta}_2)^2 - mgl\theta_1^2/2 - mgl(\theta_1^2 + \theta_2^2)/2, \tag{6.95}$$

which can be written in matrix form as

$$L = \frac{1}{2}\dot{\boldsymbol{\eta}} \cdot \mathbf{M} \cdot \dot{\boldsymbol{\eta}} - \frac{1}{2}\boldsymbol{\eta} \cdot \mathbf{K} \cdot \boldsymbol{\eta}, \tag{6.96}$$

where

$$\mathbf{K}_{ij} = \left.\frac{\partial^2 V}{\partial q^i \partial q^j}\right|_{\eta=0} = mgl\begin{pmatrix} 2 & 0 \\ 0 & 1 \end{pmatrix}, \quad \mathbf{M}_{ij} = \left.\frac{\partial^2 T}{\partial \dot{q}^i \partial \dot{q}^j}\right|_{\eta=0} = ml^2\begin{pmatrix} 2 & 1 \\ 1 & 1 \end{pmatrix}. \tag{6.97}$$

Let $\omega_0^2 = g/l$ denote a scale factor with units of angular frequency. One gets the following linearized equations of motion

$$2\ddot{\theta}_1 + \ddot{\theta}_2 = -2\omega_0^2\theta_1,$$
$$\ddot{\theta}_1 + \ddot{\theta}_2 = -\omega_0^2\theta_2. \tag{6.98}$$

The normal mode decomposition results in the eigenvalue equation

$$(-\omega^2\mathbf{M} + \mathbf{K}) \cdot \mathbf{a} = 0, \tag{6.99}$$

or

$$ml^2 \begin{pmatrix} 2(-\omega^2 + \omega_0^2) & -\omega^2 \\ -\omega^2 & (-\omega^2 + \omega_0^2) \end{pmatrix} \cdot \mathbf{a} = 0. \qquad (6.100)$$

The determinant of the secular equation is

$$2(-\omega^2 + \omega_0^2)^2 - \omega^4 = \omega^4 - 4\omega^2\omega_0^2 + 2\omega_0^4 = 0, \qquad (6.101)$$

yielding the eigenvalues

$$\omega_\pm^2 = 2\omega_0^2 \left(1 \pm \frac{1}{\sqrt{2}} \right). \qquad (6.102)$$

The eigenvalue equations are

$$ml^2\omega_0^2 \begin{pmatrix} 2(1 - (\omega/\omega_0)^2) & -(\omega/\omega_0)^2 \\ -(\omega/\omega_0)^2 & (1 - (\omega/\omega_0)^2) \end{pmatrix} \cdot \mathbf{a} = 0, \qquad (6.103)$$

or

$$ml^2\omega_0^2 \begin{pmatrix} 2(-1 \mp \sqrt{2}) & -(2 \pm \sqrt{2}) \\ -(2 \pm \sqrt{2}) & (-1 \mp \sqrt{2}) \end{pmatrix} \cdot \mathbf{a}_\pm = 0. \qquad (6.104)$$

The resulting eigenvectors are

$$\mathbf{a}_\pm = \frac{1}{\sqrt{ml^2}} \begin{pmatrix} \dfrac{\mp 1/2}{\sqrt{1 \mp 1/\sqrt{2}}} \\ \dfrac{1/\sqrt{2}}{\sqrt{1 \mp 1/\sqrt{2}}} \end{pmatrix}, \qquad (6.105)$$

giving the modal matrix

$$\mathbf{A} = \frac{1}{\sqrt{ml^2}} \begin{pmatrix} \dfrac{1/2}{\sqrt{1 + 1/\sqrt{2}}} & \dfrac{-1/2}{\sqrt{1 - 1/\sqrt{2}}} \\ \dfrac{1/\sqrt{2}}{\sqrt{1 + 1/\sqrt{2}}} & \dfrac{1/\sqrt{2}}{\sqrt{1 - 1/\sqrt{2}}} \end{pmatrix}, \qquad (6.106)$$

with inverse

$$\mathbf{A}^{-1} = \sqrt{ml^2} \begin{pmatrix} \dfrac{1/\sqrt{2}}{\sqrt{1 - 1/\sqrt{2}}} & \dfrac{1/2}{\sqrt{1 - 1/\sqrt{2}}} \\ \dfrac{-1/\sqrt{2}}{\sqrt{1 + 1/\sqrt{2}}} & \dfrac{1/2}{\sqrt{1 + 1/\sqrt{2}}} \end{pmatrix}. \qquad (6.107)$$

One should verify that

$$\bar{\mathbf{A}} \cdot \mathbf{M} \cdot \mathbf{A} = \mathbf{1}. \tag{6.108}$$

Finally, the normal modes are given by $\varsigma = \mathbf{A}^{-1} \cdot \boldsymbol{\eta}$, or

$$\varsigma^1 = \varsigma_- = \sqrt{ml^2} \left\{ \frac{1/\sqrt{2}}{\sqrt{1 - 1/\sqrt{2}}} \theta_1 + \frac{1/2}{\sqrt{1 - 1/\sqrt{2}}} \theta_2 \right\},$$

$$\varsigma^2 = \varsigma_+ = \sqrt{ml^2} \left\{ \frac{-1/\sqrt{2}}{\sqrt{1 + 1/\sqrt{2}}} \theta_1 + \frac{1/2}{\sqrt{1 + 1/\sqrt{2}}} \theta_2 \right\}. \tag{6.109}$$

The normal modes of the double pendulum are sketched in Figure 6.4, where the smaller frequency mode has the two masses moving in phase and the higher frequency mode has the two masses moving in opposite directions.

The time behavior of the pendulum in the small amplitude approximation can be found in terms of the normal modes of oscillation

$$\begin{pmatrix} \theta_1(t) \\ \theta_2(t) \end{pmatrix} = \frac{1}{\sqrt{ml^2}} \begin{pmatrix} \dfrac{1/2}{\sqrt{1 + 1/\sqrt{2}}} & \dfrac{-1/2}{\sqrt{1 - 1/\sqrt{2}}} \\ \dfrac{1/\sqrt{2}}{\sqrt{1 + 1/\sqrt{2}}} & \dfrac{1/\sqrt{2}}{\sqrt{1 - 1/\sqrt{2}}} \end{pmatrix} \begin{pmatrix} c^1 \cos(\omega_- t - \phi_-) \\ c^2 \cos(\omega_+ t - \phi_+) \end{pmatrix}, \tag{6.110}$$

where the coefficients depend on the initial conditions. Assume the initial conditions illustrated by Figure 6.5, where the upper mass is initially displaced by some small angle Δ from the vertical and the second mass is initially

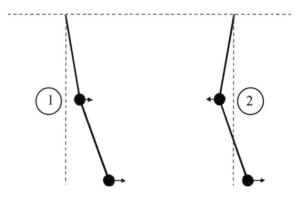

FIGURE 6.4   This sketch illustrates the normal modes of a double pendulum.

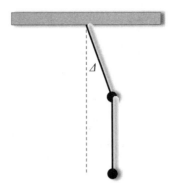

FIGURE 6.5   Initial conditions for the double pendulum example.

hanging vertically downward. At time $t = 0$, the system is released from rest. The initial conditions are

$$\theta_1(0) = \Delta, \quad \dot{\theta}_1(0) = 0,$$
$$\theta_2(0) = 0, \quad \dot{\theta}_2(0) = 0. \tag{6.111}$$

This implies that the phase angles are zero

$$\phi_+ = \phi_- = 0. \tag{6.112}$$

The motion of the normal modes for these conditions is given by

$$\varsigma^1 = \varsigma_- = \sqrt{ml^2} \left\{ \frac{1/\sqrt{2}}{\sqrt{1 - 1/\sqrt{2}}} \right\} \Delta \cos \omega_- t,$$

$$\varsigma^2 = \varsigma_+ = \sqrt{ml^2} \left\{ \frac{-1/\sqrt{2}}{\sqrt{1 + 1/\sqrt{2}}} \right\} \Delta \cos \omega_+ t, \tag{6.113}$$

and the time behavior of the solution is

$$\begin{pmatrix} \theta_1(t) \\ \theta_2(t) \end{pmatrix} = \frac{\Delta}{2} \begin{pmatrix} \cos(\omega_+ t) + \cos(\omega_- t) \\ \sqrt{2}(\cos(\omega_+ t) - \cos(\omega_- t)) \end{pmatrix}, \tag{6.114}$$

which can be rewritten as

$$\begin{pmatrix} \theta_1(t) \\ \theta_2(t) \end{pmatrix} = \Delta \begin{pmatrix} \cos \dfrac{(\omega_+ + \omega_-)t}{2} \cos \dfrac{(\omega_+ - \omega_-)t}{2} \\ \sqrt{2} \sin \dfrac{(\omega_+ + \omega_-)t}{2} \sin \dfrac{(\omega_+ - \omega_-)t}{2} \end{pmatrix}. \tag{6.115}$$

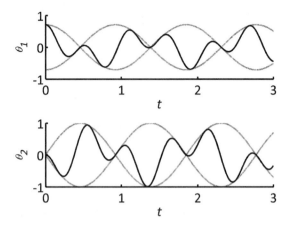

FIGURE 6.6 Solution for the double pendulum example showing how the amplitude of the motion is transferred from one pendulum to the other. The dotted lines indicate the envelope defined by the beat frequency.

Note that the motion of the laboratory coordinates has an average frequency of oscillation, at the mean of the two eigenfrequencies, with a smaller frequency envelope, depending on the difference in the two normal mode frequencies (This corresponds to a beat frequency). The results are sketched in Figure 6.6, which shows how the amplitude of oscillation is transferred back and forth between the two plumb bobs in time. This transfer of stored energy between coordinates is responsible for the phenomena of beats. The dotted lines indicate the envelope defined by the beat frequency. The mechanical energy stored in the normal modes is individually conserved, and the energy transfer between laboratory degrees of freedom must conserve the total energy.

6.4 FINITE LATTICE STRUCTURES

Consider now a case where a number of equally spaced lattice sites on a one-dimensional lattice structure are occupied with elements of equal mass. Assume that these elements are coupled to their nearest neighbors by uniform springs. Further assume that the end points of the linear structure are fixed. An example of such a structure would be a number of beads of mass Δm attached to a massless elastic string stretched out in the horizontal x-direction, clamped at the end points of the string and allowed to oscillate in the vertical y-direction as shown in Figure 6.7. The elements of mass on the lattice are restricted to a small region about their equilibrium points. Energy is

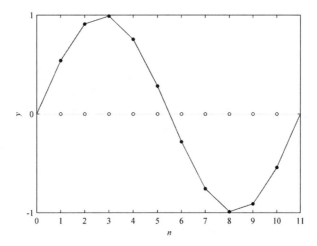

FIGURE 6.7   An example of a ten-element one-dimensional lattice structure that is clamped at both ends. The open circles represent the equilibrium positions of the lattice variables; the closed circles indicate the transverse deviations from equilibrium. The $n = 2$ normal mode of transverse vertical oscillations is illustrated.

transferred along the lattice structure by a wave-like coupling that arises from the nearest neighbor interactions.

Such linear lattice structures can, in principle, have longitudinal and transverse vibrational modes, but these modes decouple in the small amplitude limit [Fetter 1980, 108–110]. To simplify matters further, only one transverse mode is allowed. (If two transverse degrees of freedom are allowed, one would get both plane-polarized and circularly polarized solutions). In the figure, there are $n + 2$ lattice sites located at $\{x_i = i\Delta x : i = 0, \cdots, N + 1\}$ but the boundary conditions require that the amplitudes vanish at the end points of the lattice

$$y_0 = y_{N+1} = 0. \tag{6.116}$$

Therefore, there are N degrees of freedom corresponding to the N masses at $\{x_i = i\Delta x\}$ for $i = 1, \ldots, N$. Let each particle have an element of mass Δm and let the spring constant be $k = \Delta m \omega_0^2$, where ω_0 is some characteristic frequency that depends on the lattice spacing, string tension, and other physical properties of the system.

The Lagrangian for this system can be written as

$$L = \sum_{i=1}^{N} \Delta m \left\{ \frac{\dot{y}_i^2}{2} - \frac{\omega_0^2}{2}[(y_i - y_{i-1})^2 + (y_i - y_{i+1})^2] \right\} \tag{6.117}$$

which leads to the following coupled equations of motion

$$\ddot{y}_i = -\omega_0^2[2y_i - y_{i-1} - y_{i+1}].$$ (6.118)

Fetter and Walecka [Fetter 1980, pp. 108–110] obtain similar equations for coupled springs with a longitudinal mode of oscillation. Because the force couplings are usually different for longitudinal and transverse oscillations, the velocity of wave propagation will also be different. The general solution for transverse vertical lattice vibrations will be a sum over the normal modes

$$y_i(t) = \sum_n a_i{}^n c_n \cos(\omega_n t + \phi_n).$$ (6.119)

Substitution into the equations of motion results in the eigenvalue equations that the normal (standing wave) modes of motion must individually satisfy

$$\omega^2{}_n a_i{}^n = \omega_0^2[2a_i{}^n - a_{i-1}{}^n - a_{i+1}{}^n],$$ (6.120)

These modes are orthogonal, and one can impose the normalization condition

$$\mathbf{a}^n \cdot \mathbf{a}^{n'} = \delta^{nn'}.$$ (6.121)

One can try treating the eigenvectors of the lattice sites as discrete values of continuous functions

$$a_m{}^n = a^n(x_m) = a^n(m\Delta x),$$ (6.122)

subject to the boundary conditions

$$a^n(0) = a^n(L = (N+1)\Delta x) = 0.$$ (6.123)

In the continuum limit, the standing wave modes for a homogeneous string clamped at both ends is well known. They are given by

$$a^n(x) = A \sin\left(\frac{n\pi x}{L}\right).$$ (6.124)

Eigenvalue solutions to an n-element system are formidable, but the regularity of the structure and the limitations to nearest neighbor interactions come to the rescue here. One can use the continuum limit as a guide to attempt a discrete solution of the form

$$a_m{}^n = A \sin\left(\frac{n\pi m\Delta x}{(N+1)\Delta x}\right) = A \sin\left(\frac{nm\pi}{(N+1)}\right).$$ (6.125)

This solution meets the preliminary test of satisfying the required boundary conditions at the end points, so one can proceed by substituting the trial solution into the eigenvalue equation, yielding the relationship

$$\omega_n^2 \sin\left(\frac{nm\pi}{N+1}\right) = \omega_0^2 \left[2\sin\left(\frac{nm\pi}{N+1}\right) - \sin\left(\frac{n(m-1)\pi}{N+1}\right) \right. $$
$$\left. - \sin\left(\frac{n(m+1)\pi}{N+1}\right) \right]. \tag{6.126}$$

The right-hand side factors using standard trigonometric identities. The result is a valid solution to the eigenvalue problem with eigenfrequencies given by

$$\omega_n^2 = 4\omega_0^2 \sin^2\left(\frac{n\pi}{2(N+1)}\right). \tag{6.127}$$

The wave number associated with a normal mode is given by

$$k_n = \frac{n\pi}{L}. \tag{6.128}$$

Note that the discrete lattice result for the allowed wave numbers is identical to the continuous wave limit, except that the discrete system will have a finite number of distinct wave numbers. Equation 6.127 results in the following dispersion relationship relating frequency to wave number

$$\omega_n = 2\omega_0 \sin\left(\frac{k_n \Delta x}{2}\right). \tag{6.129}$$

A plot of the dispersion relation for a ten-element, one-dimensional lattice is shown in Figure 6.8. The straight line is the low-frequency (long wavelength)

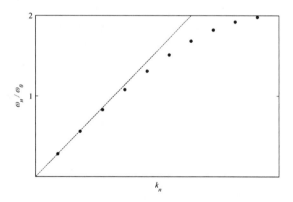

FIGURE 6.8 Dispersion relation for a ten-element, one-dimensional lattice. The straight line is the low-frequency (long wavelength) limit of the relation having a nearly constant phase velocity.

limit of the relation having a nearly constant *phase velocity* $v_n \sim \omega_n/k_n$. Note that there is a maximum cutoff frequency of $\omega_c < 2\omega_0$. There is also an associated minimum cutoff wavelength given by twice the lattice spacing $\lambda_c > 2\Delta x$.

The normalized, stationary eigenvector solutions are

$$a_m{}^n = \left(\frac{2}{N+1}\right)^{1/2} \sin\left(\frac{nm\pi}{N+1}\right). \tag{6.130}$$

The general motion of the lattice displacements is given by linear superposition over all normal mode excitations

$$\langle m \mid q^n(t)\rangle = q_m{}^n(t) = c^n \left(\frac{2}{N+1}\right)^{1/2} \sin\left(\frac{nm\pi}{N+1}\right) \cos\left(\omega_n t + \phi_n\right) \tag{6.131}$$

There are N linearly independent solutions for the normal mode corresponding to $1 \leq n \leq N$. Other choices of n outside of this range generate the same solutions up to a phase. For example, if $N = 3$, one gets the normal modes shown in Figure 6.9. There are three linearly independent vibrational modes corresponding to $n = 1, 2, 3$. The $n = 0$ mode is identically zero, and corresponds to maintaining the equilibrium positions of the particles. For $n > 3$, one repeats the previously obtained solutions up to a phase (sign) and a reordering of the sequence of terms. The dotted lines in the figures show the continuous function limit, giving the standing wave modes $a^n(x)$ associated with these discrete modes of oscillation.

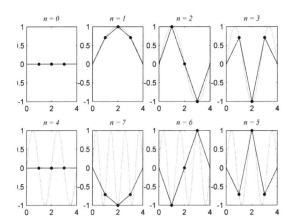

FIGURE 6.9 There are three linearly independent vibrational modes corresponding to $n = 1, 2, 3$ for the three-element, one-dimensional lattice. The $n = 0$ mode is identically zero, which corresponds to the equilibrium positions of the particles. For $n > 3$, the vibrational modes repeat up to a phase and a reordering of the terms.

6.4.1 Continuum Limit

Because the relationship between frequency and wave number is not a constant, a travelling wave with multiple frequencies will disperse for a system with a finite number of degrees of freedom. The phase velocity in the continuum limit, however, is a constant given by the limit

$$v = \lim_{\Delta x \to 0} \frac{\omega_n}{k_n} = \lim_{\Delta x \to 0} \omega_0 \Delta x \tag{6.132}$$

Note that for a finite phase velocity, the cutoff frequency becomes infinite in the continuum limit. In this limit, the finite dimensional vector space becomes a Hilbert space with an infinite number of degrees of freedom.

If one rewrites the equations of motion as a set of difference equations, one gets

$$\ddot{y}_i = -\omega_0^2 \Delta x^2 \left[\frac{2y_i - y_{i-1} - y_{i+1}}{\Delta x^2} \right] = -v^2 \left[\frac{2y_i - y_{i-1} - y_{i+1}}{\Delta x^2} \right]$$

$$= v^2 \frac{1}{\Delta x} \left\{ \left(\frac{y_{i+1} - y_i}{\Delta x} \right) - \left(\frac{y_i - y_{i-1}}{\Delta x} \right) \right\}. \tag{6.133}$$

In the limit that $y_i(t) = y(x_i, t) \to y(x, t)$ becomes continuous, the difference equation becomes a differential equation using

$$\frac{\partial^2 y(x,t)}{\partial t^2} \bigg|_x = \lim_{\Delta x \to 0} \frac{d^2 y_i(t)}{dt^2} \tag{6.134}$$

and

$$\frac{\partial y(x,t)}{\partial x} = \lim_{\Delta x \to 0} \frac{y(x + \Delta x, t) - y(x,t)}{\Delta x}. \tag{6.135}$$

Substituting into Equation (6.133) results in the wave equation

$$\frac{1}{v^2} \frac{\partial^2 y(x,t)}{\partial t^2} = \frac{\partial^2 y(x,t)}{\partial x^2}. \tag{6.136}$$

The general solution for a continuous string clamped at the limits of the interval $[0, L]$ can be written in terms of a sum over its standing wave normal modes

$$y(x,t) = \sum_{n=1}^{\infty} c_n \cos\left(\omega_n t + \phi_n\right) \sqrt{\frac{2}{L}} \sin\left(\frac{n\pi x}{L}\right). \tag{6.137}$$

In the continuous medium limit, there are an infinite number of normal modes, with the frequencies of oscillation given by the dispersion relation

$$\omega_n^2 = v^2 \left(\frac{n\pi}{L}\right)^2. \tag{6.138}$$

In the limit where the lattice is replaced by a continuous media, action at a distance, which was limited to nearest neighbor couplings in the model, is now replaced by a local derivative coupling

$$\frac{\partial^2 y(x,t)}{\partial x^2} \underset{\Delta x \to 0}{\leftrightarrow} \frac{1}{\Delta x}\left\{\left(\frac{y_{i+1} - y_i}{\Delta x}\right) - \left(\frac{y_i - y_{i-1}}{\Delta x}\right)\right\}. \tag{6.139}$$

Therefore, by going to a continuous description of the interactions of particles with each other and their fields, the non-local action-at-a-distance dilemma can be resolved. Fields couple locally to themselves in an infinitesimal neighborhood surrounding a point via a derivative coupling. This coupling propagates the disturbance continuously to more distance points.

The equivalence relationship of continuous and discrete systems in the small cell size limit given by Equation (6.139) can be carried out in either direction. Given a partial differential equation, for example, a continuous system can be converted to a discrete lattice with cell size Δx. In effect, this truncation of the continuum imposes a minimum wavelength to the accepted solutions, and results in a maximum frequency of oscillation. If one is not interested in resolving the dynamics at finer distance scales, this may prove to be an acceptable approximation to the partial differential equations of motion for a continuous system.

In the continuum limit, the Lagrangian summation over discrete degrees of freedom given by Equation 6.117 becomes an integral over a Lagrangian density $\mathcal{L}$

$$L = \int dx \mathcal{L}\left(\frac{\partial y}{\partial t}, \frac{\partial y}{\partial x}\right), \tag{6.140}$$

$$\mathcal{L}\left(\frac{\partial y}{\partial t}, \frac{\partial y}{\partial x}\right) = \frac{\rho}{2}\left[\left(\frac{\partial y}{\partial t}\right)^2 - v^2\left(\frac{\partial y}{\partial x}\right)^2\right]. \tag{6.141}$$

And the action involves a double integral over space and time

$$S = \iint dt dx \mathcal{L}\left(\frac{\partial y}{\partial t}, \frac{\partial y}{\partial x}\right). \tag{6.142}$$

This Lagrangian density can be used to define an independent starting point for the mechanics of continuous media. In Chapter 12, variations of the action defined as the variation of the integral of a continuous Lagrangian density

$$\delta S = \iint dt d\vec{x} \, \delta \mathcal{L} \left(\frac{\partial y}{\partial t}, \frac{\partial y}{\partial x^i}, y, \vec{x}, t \right) = 0, \tag{6.143}$$

leads to the Euler-Lagrange equations of motion for a continuous system of fields $y = \{y^j\}$. For the Lagrangian density of Equation 6.141, application of the continuum action principle results in the one-dimensional wave equation (6.136).

6.4.2 Periodic Lattice Structures

Periodic lattice structures, such as the one shown in Figure 6.10, can be analyzed in a similar manner. Consider a number N of equal mass elements Δm coupled by springs of spring constants k to their nearest neighbors, arranged in a circular configuration so that the first and last elements in the sequence are coupled to each other. If the springs are unstretched when the elements are in their equilibrium positions, $\theta_i = 2\pi i/N$ up to an overall phase, there is a longitudinal mode of excitation given by the coupled equations

$$\ddot{\eta}_i = -\omega_0^2 [2\eta_i - \eta_{i-1} - \eta_{i+1}], \tag{6.144}$$

where $\eta_i = \theta_i - 2\pi i/N$, $i = 1 \dots N$, $\omega_0^2 = k/\Delta m$, and the equations are subject to the periodic boundary condition

$$\eta_{i+N} = \eta_i \rightarrow \eta_0 = \eta_N, \eta_{N+1} = \eta_1. \tag{6.145}$$

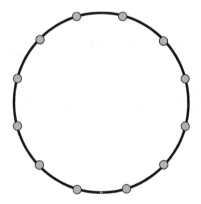

FIGURE 6.10 A periodic lattice structure consisting of equally spaced beads sliding freely on a ring of constant radius, coupled by uniform springs to their nearest neighbors.

The solutions are given by a discrete Fourier series in terms of cosine and sine normal modes. For periodic boundary conditions, there is also a zero-frequency cyclic mode corresponding to a collective rotation of all mass elements about the circle.

To use a different approach, assume that the eigenvalue equations have complex traveling wave solutions of the form

$$\eta_{j(n)}(t) = c_n e^{i(jn2\pi/N - \omega_n t)}$$
$$= c_n(\cos(jn2\pi/N - \omega_n t) + i\sin(jn2\pi/N - \omega_n t)) \qquad (6.146)$$

where the actual solutions correspond to the real part of the above expression. Plugging this expression into Equation (6.144) gives the dispersion relation

$$\omega_n^2 = \omega_0^2[2 - e^{-i2n\pi/N} - e^{i2n\pi/N}]$$
$$= 2\omega_0^2[1 - \cos 2n\pi/N] = 4\omega_0^2 \sin^2 \frac{n\pi}{N}, \qquad (6.147)$$

subject to the periodic boundary condition

$$e^{in2\pi} = 1. \qquad (6.148)$$

Therefore, the linearly independent values of n requires integers in the range $n = 0, \ldots, N - 1$, where $n = 0$ is the cyclic mode. The dispersion relation becomes

$$\omega_n = \pm 2\omega_0 \sin \frac{n\pi}{N}. \qquad (6.149)$$

Note that the standing wave solutions are superpositions of traveling waves moving in opposite directions.

6.5 FORCED AND DAMPED OSCILLATIONS

One can augment the small oscillation analysis by adding a time dependent driving term to the Lagrangian, giving a small oscillation Lagrangian of the form

$$L = \frac{1}{2}\dot{\boldsymbol{\eta}} \cdot \mathbf{M} \cdot \dot{\boldsymbol{\eta}} - \frac{1}{2}\boldsymbol{\eta} \cdot \mathbf{K} \cdot \boldsymbol{\eta} + \mathbf{F}(t) \cdot \boldsymbol{\eta}. \qquad (6.150)$$

Making the substitution

$$\boldsymbol{\eta} = \mathbf{A} \cdot \boldsymbol{\varsigma},$$
$$\mathbf{F} = \mathbf{Q} \cdot \mathbf{A}^{-1}, \qquad (6.151)$$

the Lagrangian completely separates

$$L = \sum_i \left(\frac{1}{2} \dot\varsigma_i \dot\varsigma^i - \frac{\omega_i^2}{2} \varsigma_i \varsigma^i + Q_i(t) \varsigma^i \right).$$ (6.152)

The result is n independent driven oscillators,

$$\ddot\varsigma^i + \omega_i^2 \varsigma^i = Q_i(t).$$ (6.153)

The solution, for any single mode, was given in Chapter 1 as a sum of homogeneous and inhomogeneous pieces

$$\varsigma^i = c^i \cos\left(\omega_i t - \phi_i\right) + \frac{1}{\sqrt{2\pi}} \int_{-\infty}^{\infty} \frac{Q(\omega) e^{i\omega t}}{\omega_i^2 - \omega^2} d\omega,$$ (6.154)

where $Q(\omega)$ is the Fourier transform of $Q(t)$.

A word of caution here: the denominator of the integral in Equation 6.154 goes to zero as $\omega^2 \to \omega_i^2$. Therefore, the small amplitude approximation is violated if frequencies near the resonance frequencies are present. If one is going to include driving forces, one probably should include damping terms as well, to avoid singularities in the response.

Including dissipative terms poses a new problem, however. In general, one cannot simultaneously diagonalize three symmetric matrices. One can introduce a linear damping term by including a dissipation function that is quadratic in the velocities

$$F_D = \frac{1}{2} \dot{\boldsymbol\eta} \cdot \mathbf{D}(q) \cdot \dot{\boldsymbol\eta}.$$ (6.155)

Evaluating $\mathbf{D}$ at the local minimum $\mathbf{D}_0 = \mathbf{D}(q_0)$, the linearized equations of motion become

$$\mathbf{M} \cdot \ddot{\boldsymbol\eta} + \mathbf{D}_0 \cdot \dot{\boldsymbol\eta} + \mathbf{K} \cdot \boldsymbol\eta = \mathbf{F}(t).$$ (6.156)

Here one has a velocity dependent damping term. Because the dissipation matrix cannot be diagonalized simultaneously with the inertia and potential matrices, the results are coupled differential equations of the form

$$\ddot\varsigma_i + \omega_i^2 \varsigma_i = Q_i(t) - \tilde{D}_{ij} \dot\varsigma^j$$ (6.157)

There are exceptional cases where the three matrices are simultaneously diagonalizable. In such cases, $\tilde{D}_{ij} = \gamma_i \delta_{ij}$, and the equations of motion decouple into n-independent damped, driven, linear oscillators

$$\ddot\varsigma_i + \gamma_i \dot\varsigma_i + \omega_i^2 \varsigma_i = Q_i(t).$$ (6.158)

6.5.1 DRIVEN TRIATOMIC MOLECULE

As an example of a coupled driven system, reconsider the linear triatomic atom by adding charges to the atomic sites as shown in Figure 6.11 [Morii 2003]. For simplicity, assume that the triatomic molecule has atoms of equal mass and zero net charge. One can simplify the problem by stipulating that the damping matrix diagonalizes simultaneously with the inertial and potential matrices. Further assume that the damping is small, so that the underdamped solution can be chosen.

Now, introduce an electromagnetic coupling by letting the middle atom have net charge $2q$ and the outer two atoms have net charge $-q$ each. If one further assumes that the wavelength of an externally applied electromagnetic field is long compared to molecular dimensions, one gets a dipole excitation of the molecule. The electric force in the long wavelength limit is given by

$$\mathbf{F}(t) = q_i \mathbf{E}_i(x, t) \approx \frac{q \langle \mathbf{E_0} \rangle}{\sqrt{6}} e^{-i\omega' t} \begin{pmatrix} 1 \\ -2 \\ 1 \end{pmatrix} \qquad (6.159)$$

$$\varsigma_3(t) = \varsigma_3^{(h)}(t) + \frac{q \langle E_0 \rangle e^{-i\omega' t}}{\omega_3^2 - \omega'^2 - i\omega' \gamma_3}. \qquad (6.160)$$

Only one normal mode (the third mode shown in Figure 6.2) is excited by this driving term. The homogeneous response is quickly damped, leaving only the dipole driving term to contribute. The result is resonance response at the normal mode frequency of the molecule. This resonance energy transfer approximates what happens when one heats food in a microwave oven, where one excites regions of the molecular spectrum with strong vibrational absorption bands. Water, sugars, and fat, for example, all have resonances at excitation frequencies of around 2.5 GHz, where the microwave radiation peaks in microwave ovens. The result is resonant excitation of molecules that

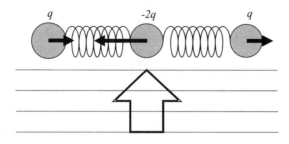

FIGURE 6.11   A triatomic charged molecule being driven by a long wavelength external electric field.

heats up the food as the energy is absorbed and thermalized by molecular collisions with their neighbors [Morii 2003].

6.5.2 Damped Double Pendulum

As an example of a simple damping mechanism, consider the double pendulum problem with the addition of a dissipation function due to air resistance that is proportional to the kinetic energy

$$F_D = \gamma T. \tag{6.161}$$

The dissipation matrix is proportional to the mass matrix giving the linear matrix equation

$$\mathbf{M} \cdot \ddot{\boldsymbol{\eta}} + \gamma \mathbf{M} \cdot \dot{\boldsymbol{\eta}} + \mathbf{K} \cdot \boldsymbol{\eta} = 0. \tag{6.162}$$

The modal matrix transformation given in Equation (6.106) continues to work, since $\mathbf{M}$ and $\mathbf{K}$ can be simultaneously diagonalized. One gets the same normal modes of motion, but with a modified time dependence given by

$$\ddot{\varsigma}_i + \gamma \dot{\varsigma}_i + \omega_i^2 \varsigma_i = 0 \tag{6.163}$$

where the frequencies ω_i^2 represent the eigenvalues given in Equation (6.102) for the transformed spring potential matrix

$$\tilde{\mathbf{K}} = \bar{\mathbf{A}} \cdot \mathbf{K} \cdot \mathbf{A} = 2\omega_0^2 \begin{bmatrix} 1 - \dfrac{1}{\sqrt{2}} & 0 \\ 0 & 1 + \dfrac{1}{\sqrt{2}} \end{bmatrix}. \tag{6.164}$$

Depending on the relative strength of the damping coefficient, the individual eigenvector solutions can be under-damped, over-damped, or critically-damped, as was the case for in the one dimensional problem studied in Chapter 1. Assuming weak damping, the time dependent solution to the double pendulum problem is given by

$$\begin{pmatrix} \theta_1(t) \\ \theta_2(t) \end{pmatrix} = \frac{e^{-\gamma t/2}}{\sqrt{ml^2}} \begin{pmatrix} \dfrac{1/2}{\sqrt{1 + 1/\sqrt{2}}} & \dfrac{-1/2}{\sqrt{1 - 1/\sqrt{2}}} \\ \dfrac{1/\sqrt{2}}{\sqrt{1 + 1/\sqrt{2}}} & \dfrac{1/\sqrt{2}}{\sqrt{1 - 1/\sqrt{2}}} \end{pmatrix} \begin{pmatrix} c^1 \cos(\omega_1' t - \phi_1) \\ c^2 \cos(\omega_2' t - \phi_2) \end{pmatrix},$$
$$\tag{6.165}$$

where

$$\omega_i' = \sqrt{\omega_i^2 - \gamma^2/4}. \tag{6.166}$$

Note that, in this model, both normal modes are damped at the same rate.

6.6 SUMMARY

The stationary points of a conservative Lagrangian system are those points for which the system is in equilibrium with respect to its noncyclic coordinates. Expansion of the Lagrangian about a stationary point $\mathbf{q}_0$, letting $\boldsymbol{\eta} = \mathbf{q} - \mathbf{q}_0$ denote small deviations from equilibrium, and ignoring a leading constant term, gives the quadratic Lagrangian

$$L = \frac{1}{2}\dot{\boldsymbol{\eta}} \cdot \mathbf{M} \cdot \dot{\boldsymbol{\eta}} - \frac{1}{2}\boldsymbol{\eta} \cdot \mathbf{K} \cdot \boldsymbol{\eta}, \tag{6.167}$$

where

$$\mathbf{K}_{ij} = \left.\frac{\partial^2 V}{\partial q^i \partial q^j}\right|_{\eta=0}, \tag{6.168}$$

and

$$\mathbf{M}_{ij} = \left.\frac{\partial^2 T}{\partial \dot{q}^i \partial \dot{q}^j}\right|_{\eta=0}. \tag{6.169}$$

This results in linearized equations of motion given by

$$\mathbf{M} \cdot \ddot{\boldsymbol{\eta}} = -\mathbf{K} \cdot \boldsymbol{\eta}. \tag{6.170}$$

The time-dependent solution can be found as a superposition over the normal modes of oscillation

$$\boldsymbol{\eta}(t) = \mathbf{A} \cdot \boldsymbol{\varsigma} = \mathbf{a}_j \varsigma^j(t) = \mathbf{a}_j c^j \cos(\omega_j t + f_j), \tag{6.171}$$

where

$$\mathbf{A} = \mathbf{a}_j \mathbf{n}'^j \tag{6.172}$$

is the modal matrix that simultaneously diagonalizes $\mathbf{M}$ and $\mathbf{K}$. The eigenvectors satisfy the eigenvalue equation

$$(-\omega_i^2 \mathbf{M} + \mathbf{K}) \cdot \mathbf{a}_i = 0. \tag{6.173}$$

The eigenvalues are determined by solving the secular equation

$$\det(-\omega^2 \mathbf{M} + \mathbf{K}) = 0. \tag{6.174}$$

The eigenvectors can then be evaluated from Equation 6.173 and are normalized such that

$$\bar{\mathbf{a}}_j \cdot \mathbf{M} \cdot \mathbf{a}_i = \delta_{ji}. \tag{6.175}$$

The equilibrium is stable if $\omega_i^2 > 0$, and unstable if $\omega_i^2 < 0$. Cyclic coordinates are in neutral equilibrium with $\omega_i^2 = 0$.

The modal matrix can be inverted to express the normal modes in terms of the physical coordinates

$$\varsigma = \mathbf{A}^{-1} \cdot \boldsymbol{\eta}. \tag{6.176}$$

The Lagrangian separates in the normal coordinate representation

$$L = \frac{1}{2}\dot{\varsigma} \cdot \dot{\varsigma} - \frac{1}{2}\varsigma \cdot \tilde{\mathbf{K}} \cdot \varsigma = \sum_i \left(\frac{1}{2}\dot{\varsigma}_i \dot{\varsigma}^i - \frac{\omega_i^2}{2}\varsigma_i \varsigma^i \right), \tag{6.177}$$

Lattice structures with nearest neighbor couplings and many degrees of freedom give the wave equation in the continuum limit.

6.7   EXERCISES

1. Using relative coordinates $y_1 = x_1 - R + L, y_2 = x_2 - R, y_3 = x_3 - R - L)$, and assuming that the center of mass R is at rest, find the eigenfrequencies for small oscillations about equilibrium for the linear triatomic atom shown in Figure 6.12, where the middle mass has mass M and the outer two masses have mass m. The atoms are equally spaced, with $y_1 = y_2 = y_3 = 0$ at equilibrium, and the effective spring constants are equal and given by k.

2. Consider a double pendulum shown in Figure 6.3 with $l_1 = l_2 = l$ and $m_1 \gg m_2$. Find the eigenfrequencies. Show that they are nearly degenerate. Find the modal matrix and the normal modes of oscillation. If the initial conditions are specified to be

$$\theta_1(0) = \Delta, \quad \theta_2(0) = 0, \quad \dot{\theta}_1(0) = \dot{\theta}_2(0) = 0,$$

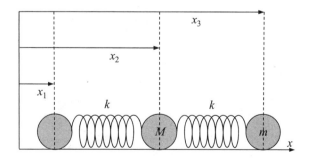

FIGURE 6.12   A triatomic molecule with $M \neq m$.

show that the solutions are given by

$$\theta_1(t) = \frac{\Delta}{2}(\cos\omega_+ t + \cos\omega_- t) = \Delta\cos\bar\omega t \cos\delta\bar\omega t,$$

$$\theta_2(t) = \frac{\Delta}{2\delta}(\cos\omega_+ t - \cos\omega_- t) = \frac{\Delta}{\delta}\sin\bar\omega t \sin\delta\bar\omega t,$$

$$\omega_\pm^2 = \bar\omega^2(1\pm\delta),$$

where $\delta\bar\omega$ is a small beat frequency. Show that $\delta = \sqrt{m_2/(m_1+m_2)}$, and therefore that the amplitude of the lighter mass is much greater than the heavier mass.

3. A charged particle of mass m and charge $+e$ is restricted to move in the $\hat{i}\wedge\hat{j}$ plane. It moves under the influence of two fixed charges of $+e$ each located at $\pm a\,\hat{i}$ respectively and two other fixed charges of $+3e$ located at $\pm a\,\hat{j}$ respectively. Show that the origin is a point of stable equilibrium and find the frequencies and normal modes of small oscillation.

4. A thin hoop of radius R and mass M oscillates in its own plane with one point of the hoop fixed, as shown in Figure 6.13. Attached to the hoop is a point mass M constrained to slide without friction along the hoop. The system is hanging in a uniform gravitational field $\mathbf{g}$. Consider only small oscillations about equilibrium.

a. Show that the normal mode frequencies are

$$\omega_1 = \frac{1}{2}\left(\frac{2g}{R}\right)^{1/2}, \quad \omega_2 = \left(\frac{2g}{R}\right)^{1/2}.$$

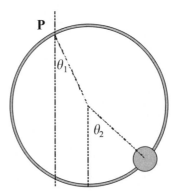

FIGURE 6.13 A bead of mass M, sliding freely on a thin hoop of mass M, which rotates in a plane about a fixed pivot point P.

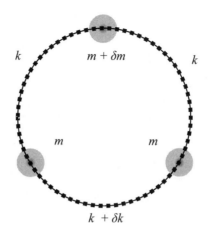

k $m + \delta m$ k

m m

$k + \delta k$

FIGURE 6.14   Three masses coupled by springs constrained to move on a circle.

b. Construct the modal matrix.

c. Find the normal coordinates, and show that they diagonalize the Lagrangian.

5. Consider three equally space beads on a ring as shown in Figure 6.14 with $m_1 = m_2 = m$, and the third mass is slightly heavier $m_3 = m(1 + \delta_m)$. Assume the masses are subject to 2-body linear restoring forces proportional to the change of the angle of separation with spring constants $k_1 = k_2 = k, k_3 = k(1 + \delta_k)$. The potential energy is given by

$$V = \frac{ka^2}{2}(\eta_3 - \eta_2)^2 + \frac{ka^2}{2}(\eta_3 - \eta_1)^2 + \frac{k(1 + \delta_k)a^2}{2}(\eta_2 - \eta_1)^2$$

where

$$\eta_1 = \theta_1, \quad \eta_1 = \theta_2 - \pi/3, \quad \eta_3 = \theta_3 - 2\pi/3,$$

Find the frequencies of oscillation, sketch the normal modes, and construct the modal matrix.

6. Consider the LC circuit shown in Figure 6.15. Kirkoff's loop equations can be derived from the Lagrangian

$$L = \frac{1}{2}L(I_1^2 + I_2^2) - \frac{1}{2}C(Q_1^2 + Q_2^2 + (Q_1 + Q_2)^2),$$

where $I = dQ/dt$ is the current flowing in a loop of the circuit, L is the inductance of an inductor, and C is the capacitance of a capacitor. By analogy to mechanical oscillators, the inductance plays the role of an

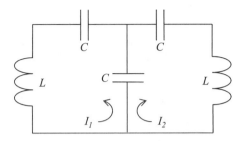

FIGURE 6.15   A coupled LC circuit.

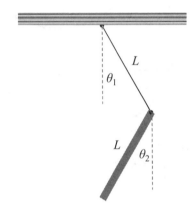

FIGURE 6.16   A rod suspended from the ceiling by a string.

inertial mass, and the capacitance plays the role of a spring constant. Treating the Lagrangian as a small oscillation problem, what are the normal modes and the frequencies of oscillation for the circuit?

7. Consider a thin rod of mass m and length l hung the ceiling by a massless string of length l attached to one end of the rod, as shown in Figure 6.16. Assume the motion lies in a plane. Find the frequencies of small oscillation and the normal modes of oscillation. The moment of inertia of a thin rod rotating about its center of mass is $ml^2/12$.

8. Two beads of mass m are coupled to three springs, as shown in Figure 6.17. The springs have their unstretched lengths when the masses are at their equilibrium positions. The outer two springs have spring constants k while the middle spring has a small spring constant δk. As $\delta k \to 0$ the two springs decouple. Find the eigenfrequencies for the modal matrix and the normal modes for this problem.

9. Find the eigenvectors and eigenfrequencies of the two coupled springs problem shown in Figure 6.17 by explicitly solving the eigenvalue problem when all the springs have the same spring constant k. Assume

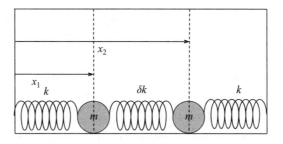

FIGURE 6.17 Two spring-coupled masses. When $\delta k \ll k$, one has the weak coupling limit. If $\delta k = k$, one has a uniform lattice structure, clamped at both ends, with two mass elements.

that springs are unstretched when the masses are in their equilibrium positions. Compare your results to the values obtained by treating the problem as a regular lattice structure, clamped at both ends, with longitudinal modes of oscillation.

10. A thin rigid rod of length L and mass m is hung from the ceiling by two massless, but also rigid, rods of length a each attached to the ends of the first rod. When the first rod is at rest, it is level with the floor, oriented in the x-direction, and the two other rods hang vertically in the z-direction. Find the normal modes and frequencies of small oscillations of the system. The first rod can move freely in the x-, y-, and z-directions, subject only to the fixed length constraints of the other rods. The momentum of inertia of a thin rod about its center of mass is $mL^2/12$. It has zero moment of inertia about its long axis.

11. Consider two mass points of mass m suspended from a ceiling by identical massless springs in a uniform gravitational field as shown in Figure 6.18. The springs have spring constants k and unstretched

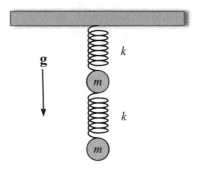

FIGURE 6.18 Two masses suspended from a ceiling by springs in a uniform gravitational field.

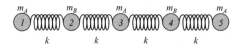

FIGURE 6.19   Linear spring-coupled system with five masses.

lengths l_0. Find the equilibrium positions of the mass points from the ceiling and solve for the eigenfrequencies and normalized eigenvectors of small vertical oscillations.

12. Consider the linear five-mass spring-coupled system shown in Figure 6.19. This 5-by-5 dimensional matrix system decouples into two smaller matrices if the small displacement coordinates are arranged in symmetric and antisymmetric pairs as follows:

$$a_1 = (\eta_1 - \eta_5)/\sqrt{2},$$
$$a_2 = (\eta_2 - \eta_4)/\sqrt{2},$$
$$s_1 = (\eta_1 + \eta_5)/\sqrt{2},$$
$$s_2 = (\eta_2 + \eta_4)/\sqrt{2},$$
$$s_3 = \eta_3.$$

Find the eigenfrequencies and normalized eigenvectors for this system.

13. Two pendula of equal lengths and equal masses are coupled together by a spring with spring constant k, as shown in Figure 6.20. The spring is unstretched when both pendula are hanging vertically in their equilibrium positions. Assume small angle motion throughout and that the motion is in the plane.

   a. Find the natural frequencies of oscillation and the normal modes of the system.

   b. For arbitrary, but small, initial angles and angular velocities, find the equations of motion for the two pendula.

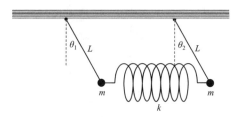

FIGURE 6.20   Two pendula coupled by a spring.

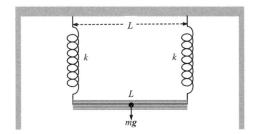

FIGURE 6.21   A beam of length L suspended from the ceiling by two equal massless springs of unstretched length a.

    c. Suppose the left pendulum is released from rest with an angle Δ, while the other is motionless and vertical. How long will it take before the right pendulum is moving and the left pendulum is at rest at $\theta_1 = 0$?

14. For the two pendula problem of Figure 6.20, assume that the pendula see damping forces due to a dissipation function

$$F_D = \frac{1}{2}\gamma_a(\dot{\theta}_1^2 + \dot{\theta}_2^2) + \frac{1}{2}\gamma_b(\dot{\theta}_1 - \dot{\theta}_2)^2,$$

where the first term can be attributed to air resistance acting on the pendula and the second term to dissipation in the spring. Show that the normal modes of small oscillation still decouple. Find the effect on the time behavior of the normal modes.

15. A thin beam of length L and mass m is suspended from the ceiling by two identical massless springs of spring constants k and unstretched lengths a, as shown in Figure 6.21. Find the eigenvectors and eigenfrequencies of small oscillations in the plane of the figure.

16. Three equal masses m in a plane coupled by identical springs of spring constant k form an equilateral triangle with sides of length L when in equilibrium, as shown in Figure 6.22. If the motion is restricted to a plane, there are 6 degrees of freedom, three of which are zero-frequency modes and three of which are vibrational in character. Use a computer program to find the eigenfrequencies and eigenvectors of vibrational motion. Identify and describe the three cyclic degrees of freedom.

17. Derive the dispersion relation

$$\omega_n^2 = 4\omega_0^2 \sin^2\left(\frac{n\pi}{2(N+1)}\right),$$

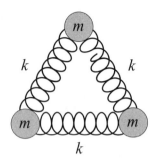

FIGURE 6.22   **Three masses arranged in an equilateral triangle configuration.**

for the finite oscillator lattice structure given by the coupled equations

$$\ddot{y}_i = -\omega_0^2[2y_i - y_{i-1} - y_{i+1}],$$
$$y_0 = y_{N+1} = 0.$$

18. Assuming that the dissipation function is simultaneously diagonalizable with the kinetic energy and potential energy terms in the small oscillation limit, the equations of motion for the damped driven oscillators in terms of the normal modes can be written as

$$\ddot{\xi}_i + \gamma_i\dot{\xi}_i + \omega_i^2\xi_i = Q_i(t).$$

These solutions can be further separated into damped homogenous solutions plus particular driven solutions: $\xi_i(t) = \xi_i^{(h)}(t) + \xi_i^{(p)}(t)$. Show that a particular solution to the inhomogeneous part can be written as

$$\xi_i^{(p)}(t) = \frac{1}{\sqrt{2\pi}}\int_{-\infty}^{\infty}d\omega\frac{\tilde{Q}_i(\omega)(-\omega^2 + i\omega\gamma_i + \omega_i^2)}{(\omega_i^2 - \omega^2)^2 + \omega^2\gamma_i^2}e^{-i\omega t},$$

where $\tilde{Q}_i(\omega)$ is the Fourier transform of the normal mode driving forces given by

$$Q_i(t) = \frac{1}{\sqrt{2\pi}}\int_{-\infty}^{\infty}d\omega\tilde{Q}_i(\omega)e^{-i\omega t}.$$

7

ROTATIONAL GEOMETRY AND KINEMATICS

I n this chapter, the structure of space is constructed using the geometric concepts of contraction and extension. These concepts combine to form a unified geometric algebra $GA(3)$, completely specifying the geometric properties of Euclidean space. Pauli Spin Algebra is developed as the minimal matrix representation of $GA(3)$. The $Spin(3)$ group of continuous orthogonal similarity transformations in three space dimensions is defined and found to be isomorphic to the $SU(2)$ rotation group. The connection to the special orthogonal group in three space dimensions $SO(3)$ is made, and the general properties of the three-dimensional orthogonal rotation matrices are developed. Newton's laws in a rotating frame of reference are then derived.

7.1 ROTATIONAL GEOMETRY OF EUCLIDEAN SPACE

A half-century before Newton, the French philosopher René Descartes (1569–1650) had already succeeded in reducing Euclidean geometry to algebraic methods by introducing his Cartesian system of coordinates. The two most obvious features of Cartesian space are its three-dimensional character and its manifest invariances. Cartesian space is a *metric space* (that is, a space defining a measure of separation) obeying the rules of Euclidean geometry. Newton's *Principia* asserts that the structure of this Euclidean space is translationally and rotationally invariant. The motion of physical objects in this space can be analyzed in terms of the continuous, differential motion of constituent particle points, consistent with the underlying symmetries of the spacetime structure.

Before directly addressing the kinematics of rotation, it is necessary to first develop a fuller understanding of the geometric structure of space. This is, in part, because of deficiencies in the standard notation, which obscure important geometric distinctions.

More to the point, the standard vector algebra notation has little direct connection with the way rotations are handled in quantum mechanics and field theories. In the larger scheme of things, it is important to understand how the various group structures are related. Moreover, much of this advanced notation is becoming increasing relevant for high-speed computer algorithms. Thus, concepts such as quaternions that were once fashionable, but have fallen into obscurity, are, once again, becoming essential tools for working physicists.

The procedure will be to first develop the Pauli spin algebra for Euclidean space, then use these properties to develop the orthogonal spin group. The spin group provides an elegant means to construct the special orthogonal group. Once the properties of the special orthogonal group are understood, the standard vector algebra notion will be resumed. Much of the properties of rigid body rotations are most easily analyzed using the traditional dyadic tensor notation. This notation, in any event, is the dominant one found in the existing literature.

7.1.1 Historical Context

As Lasenby and Doran point out in their review of geometric algebra, the study of rotational motion underwent revolutionary changes from the mid-nineteenth century to the early part of the twentieth century [Doran 2003]. Their analysis is expanded upon in the following discussion. A seminal development was the discovery of *quaternion algebra* by William Rowan Hamilton (1805–1865) in 1844 [Hamilton 1844]. Quaternions define the rotational transformation of vectors in terms of the fundamental *spin rotation group* $Spin(3) \equiv SU(2)$. Hamilton's *hyper-complex* operators (i, j, k) have the almost mystical algebraic properties

$$i^2 = j^2 = k^2 = ijk = -1. \tag{7.1}$$

The quaternion operator i is completely distinct from the complex number i. For example, the quaternion i anticommutes with the operators j and k. To avoid overloading the meaning of the symbol i (more than necessary) let $(i, j, k) \rightarrow (\sigma_1, \sigma_2, \sigma_3)$. Then, the *spin-rotation operators* σ_i satisfy the *Lie algebra* of the $SU(2)$ group

$$[\sigma_i, \sigma_j] = 2\varepsilon_{ijk}\sigma_k, \quad \{\sigma_i, \sigma_j\} = -2\delta_{ij}. \tag{7.2}$$

The sigma rotation operators are antiadjoint under vector reversion $\bar{\sigma}_i = -\sigma_i$. The quaternion numbers

$$q = q^0 + q^1\sigma_1 + q^2\sigma_2 + q^3\sigma_3 = |q|R \in \mathbb{H}, \tag{7.3}$$

have a positive-definite norm for all nonzero elements

$$\bar{q}q = |q|^2 = (q^0)^2 + \sum_{i=1}^{3} (q^i)^2, \tag{7.4}$$

and thus form an associative *division algebra*.[1] Note that the spin-rotation operators satisfy the cyclic property

$$\sigma_i \sigma_j \sigma_k = -\varepsilon_{ijk}, \tag{7.5}$$

where the three-index-symbols ε_{ijk} denotes the cyclic *Levi-Civita permutation symbols* defined on an orthonormal basis as

$$\varepsilon_{ijk} = e_i \cdot (e_j \times e_k) = \pm 1. \tag{7.6}$$

Now, let the Euclidean vector basis $\{e_i\}$ define a *Pauli Spin Algebra*, incorporating the quaternion group as a subgroup. In a two-dimensional complex-matrix representation, one can define self-adjoint basis vectors $e_i = i\sigma_i$ (referred to in the physics literature by the misleading name, the *Pauli "spin" operators*[2]), which have the standard representation given by

$$e_1 = i\sigma_1 = \begin{bmatrix} 0 & 1 \\ 1 & 0 \end{bmatrix}, \quad e_2 = i\sigma_2 = \begin{bmatrix} 0 & -i \\ i & 0 \end{bmatrix}, \quad e_3 = i\sigma_3 = \begin{bmatrix} 1 & 0 \\ 0 & -1 \end{bmatrix}. \tag{7.7}$$

The Pauli matrices satisfy the matrix commutation relationships

$$\begin{aligned} [e_i, e_j] &= 2i\varepsilon_{ijk}e_k = -2\varepsilon_{ijk}\sigma_k, \\ [\sigma_i, \sigma_j] &= 2\varepsilon_{ijk}\sigma_k, \\ e_1 e_2 e_3 &= i. \end{aligned} \tag{7.8}$$

From Equation (7.8), it can be seen that the quaternion generators of rotation can be expressed as

$$\sigma_i = -\frac{1}{2}\varepsilon_{ijk}e_j e_k \tag{7.9}$$

---

[1]A division algebra is a multiplicative group having a unique inverse for all nonzero elements. There are four known division algebras, three of which are associative. Unlike the real $\mathbb{R}$ and complex $\mathbb{C}$ division algebras, quaternion algebra $\mathbb{H}$ is non-commutative. Matrix algebras and Hilbert spaces can be based on any of the three associative division algebras: $\mathbb{R}, \mathbb{C}$, or $\mathbb{H}$.

[2]The Pauli matrices e_i are commonly called the Pauli "sigma" or "spin" matrices in the physics literature, but this usage conflicts with this book's convention that the sigma symbol be reserved for the actual, anti-adjoint (and anti-Hermitian), quaternion generators of orthogonal spin transformation. The Pauli matrices actually denote the self-adjoint (and Hermitian) basis vectors generating the three-dimensional "spin" space. In the present text, the two symbols are related by $e_i = i\sigma_i$.

and rotations built from them are part of the even-ranked subalgebra of the geometry. The generators are now unambiguously identifiable with the antisymmetric bivector product of two vectors.

Quaternions of unit magnitude R, called *spin rotators* or *rotors,* form a continuous multiplicative group with three parameters, isomorphic to the spin rotation group $Spin(3)$ in three space dimensions. A rotation of a vector in Euclidean space is given by the quaternion similarity transformations

$$V' = RV\bar{R}, \quad R\bar{R} = 1,$$
$$R(\hat{\sigma}, \theta) = e^{\hat{\sigma}\theta/2} \in Spin(3). \tag{7.10}$$

where $R(\hat{\sigma}, \theta)$ generates a rotation of a vector V about a rotational axis $\hat{\sigma}$ by an angle θ. Any rotation in the geometric algebra of three space dimensions $GA(3)$ can be decomposed into a product of three elementary rotations about three orthogonal space axes. It can be shown that the group of quaternions with unit normalization is isomorphic to the group of special-unimodular-unitary-matrices in two-complex dimensions $U \in SU(2),$[3] such that

$$UU^{\dagger} = 1, \quad \det(U) = +1. \tag{7.11}$$

Quaternion algebra is often used to represent vectors as well as rotations. The $(\hat{i}, \hat{j}, \hat{k})$ notion for Euclidean vector bases comes from this. This resulted in initial confusion as to how one should interpret the new quaternion numbers. Because of a duality relationship specific to three-dimensional space, a one-to-one mapping exists between the three rotational elements and the three directions of Euclidean space. In three-dimensional vector analysis, an overloaded notation was adopted to represent both vectors (displacement elements) and axial-vectors (rotational elements). Essentially the pseudoscalar element $i = i_3$, required for dimensional reasons, is conveniently suppressed. This blurring of the distinction between vectors and axial-vectors persists to the present-day in elementary vector analysis courses. In three space dimensions, the three linearly-independent bivector operators can be mapped to the three linearly-independent axial-vector orientation pseudovectors $\sigma^k = -ie^k$. This unique dual relationship between axial-vectors and space vectors in $GA(3)$ is illustrated in Figure 7.1.

---

[3]Note that $Spin(3) = SU(2) = Sp(1)$, where $SU(n) = SU(n, \mathbb{C})$ denotes the special group of unimodular, unitary, n-dimensional, complex matrices with determinant $+1$. The symplectic group $Sp(n) = U(n, \mathbb{H})$. is the group of unitary, n-dimensional, quaternion matrices. Despite the intricacies of group nomenclature, the three groups are algebraically equivalent. These equivalences are accidental however. For example, $SU(n)$, in general, is not a orthogonal spin group (unitarity is not equivalent to orthogonality).

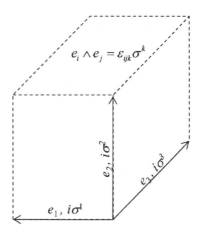

FIGURE 7.1 **The three space directions in GA(3) have a one-to-one dual relationship to the three linearly independent surface elements of rotation, a result that is unique to GA(3).**

At nearly the same time that Hamilton introduced quaternion algebra, Hermann Günter Grassmann (1809–1877) introduced the *Grassmann exterior product* [Grassmann 1844]. The exterior product (often referred to as the wedge product) defines a set of totally antisymmetric irreducible tensor classes generated from some linearly independent vector basis. The results can be interpreted as forming the tensor bases for higher geometrical constructs, such as surfaces and volumes. Given a vector basis $\{e_i\}$, the antisymmetric *bivector* products $e_i \wedge e_j = \sigma_{ij} = \varepsilon_{ijk}\sigma^k$ define a two-dimensional tensor basis for surface elements, while the totally antisymmetric *trivector* products $e_i \wedge e_j \wedge e_k = \tau_{ijk}$ define a tensor basis for volume elements. Unlike the vector cross product, the Grassmann product is associative, and it can be extended to spaces of any dimensionality. Grassmann's work would later inspire Elie Cartan's (1869–1951) work on exterior differential forms and Lie algebras.[4]

In 1878, William Kingdon Clifford (1845–1879) developed *Clifford algebra* by combining the inner product (contraction) and exterior product (extension) into a single geometric product [Clifford 1878]. Clifford algebra provides a complete representation of the geometric structure of space. Clifford algebras defined over the real number field are commonly

---

[4]Marius Sophus Lie (1842–1899) created the theory of continuous symmetry groups called Lie groups, and applied it to the study of geometry and differential equations. The infinitesimal generators of a Lie group define a Lie algebra.

referred to as *geometric algebras*. A geometric algebra defined on a space with n_s space-like components and m_t time-like components is designated as $GA(n_s, m_t) = Cl(n_s, m_t, \mathbb{R})$.

Geometric Clifford algebras are constructed by relating the symmetric product of two vectors with its inner dot product

$$\{a, b\} = 2a \cdot b = 2a^i b_i. \tag{7.12}$$

This relationship is an identity if the basis vectors obey the basic Clifford Algebra anti-commutation relationships

$$\{\lambda_i, \lambda_j\} = 2g_{ij}. \tag{7.13}$$

For example, the defining relationship for the Pauli Spin Algebra in three-dimensional Euclidean space is given by the Clifford Algebra

$$\{e_i, e_j\} = 2\delta_{ij}. \tag{7.14}$$

From this definition, all the other properties of the Pauli Spin Algebra listed in Equation (7.8) follow. Appendix C presents a more complete development of Geometric Algebras in spaces of arbitrary dimension.

Clifford, unfortunately, died young, before his algebra could be fully appreciated. Although Clifford Algebras resurface for use in defining internal spin spaces in quantum mechanics, it was left to David Hestenes [Hestenes 1966, 1984, 1986] to re-introduce their ordinary geometric significance. Meanwhile, Josiah Willard Gibbs (1839–1903) had successfully introduced his Vector Calculus [Gibbs 1906], which almost completely eclipsed the earlier work of Clifford and Grassmann.

After defining rotations from both the point of view of the Geometric Algebra and the Gibbs Vector Analysis frameworks, the kinematical properties of rotational motion will be employed to develop Newton's Laws of motion in a rotating frame of reference. This will prepare the stage for the following chapter describing the rotational motion of rigid bodies.

7.1.2 Representations of the Rotation Group

Depending on the representational scheme adopted, space vectors can be represented mathematically using either column vectors or square matrices. The former notation is used in the traditional *Gibbs Vector Analysis* taught in elementary mechanics textbooks, while the latter is found in abstract *Clifford Algebras* and in quantum mechanics. To avoid confusion between the column vector and square matrix representation of vector quantities, boldface **vector-dyadic** or $\langle bra \mid ket \rangle$ notation will continue to be used to represent column vectors and dyadics tensors constructed from these column vectors. Ordinary *math italics* notation will be used to represent the square matrix formulation of vectors and geometric tensors. The two notations differ significantly in their treatment of the operations of vector and tensor multiplication.

In the column vector notation, one defines a vector **a** in terms of its contravariant components with respect to a constant set of basis operators as

$$\mathbf{a} = a^i \boldsymbol{\lambda}_i \leftrightarrow |a\rangle = a^i |\lambda_i\rangle, \tag{7.15}$$

In the square matrix notation, the same vector is defined, using standard math italic notation, as

$$a = a^i e_i. \tag{7.16}$$

where the rank-1 tensor operators e_i can be represented as traceless square matrices and the scalar product is defined as the symmetric product of two vectors

$$a \cdot b = \frac{1}{2}(ab + ba) = \frac{1}{2}a^i b^j \{\lambda_i, \lambda_j\} = g_{ij} a^i b^j = a^i b_i,$$

provided that the basis satisfies the fundamental Clifford algebra relationship $\{\lambda_i, \lambda_j\} = 2g_{ij}$.

Mathematically, translational invariance is equivalent to asserting that space is *affine* in character, i.e., that it has no special origin. This will be the case if only relative displacements and relative motions are physically relevant for the description of the motion. An infinitesimal translation of a space point to a nearby space point is given by the *affine connection*, which defines the concept of *parallel transport* in curved spaces. In flat spaces, finite space translations are associative and commutative. They are, nearly universally, represented by the operation of vector addition, which involves component by component addition of the vector elements.

The description of rotational motion is significantly more complicated than that of translational motion. Unlike translations, finite rotations in Euclidean space are not commutative. They do, however, preserve the inner products of vectors. They are orthogonal transformations. Depending on the representation chosen, the operation of rotation can be expressed as either an *orthogonal transformation* of column vectors or as an *orthogonal similarity transformation* of an equivalent square-matrix representation of the vector basis. In the following, the "spin group" will be used, restrictively, to denote the special group of proper orthogonal similarity transformations of vectors into each other.[5] By "special," the proper subgroup of continuous transformation from the identity transformation is implied. For example, a rotation of a vector in three dimensions can be represented either as a *special orthogonal transformation* using the column vector representation

$$\mathbf{a}' = \mathbf{A}_R \cdot \mathbf{a}, \quad \bar{\mathbf{a}}' = \bar{\mathbf{a}} \cdot \bar{\mathbf{A}}_R, \tag{7.17}$$

where $\bar{\mathbf{A}}_R \cdot \mathbf{A}_R = 1$ and $\det(\mathbf{A}) = +1$, or as a *spin transformation of tensors* in the square matrix representation

$$a' = Ra\bar{R}, \tag{7.18}$$

where the *spin rotor R* is an element of the *spin group* in three dimensions with $R\bar{R} = 1$ and $\det(R) = +1$. *Spin*(3) is a double-valued representation of the rotation group, since the substitution $R \to -R$ in the similarity formula of Equation 7.18 results in the same physical rotation.[6]

Both transformational schemes do an equally satisfactory job in expressing transformations that map irreducible geometric tensors (such as scalars, vectors, axial-vectors, and pseudo-scalars) into themselves. But the Clifford algebra representation is more general in that it allows the transformation of objects of mixed tensor rank, such as spinors, and therefore generalizes more easily to the treatment of quantum mechanical systems. By definition, scalars and pseudoscalars transform by invariance under rotations, while vectors and pseudovectors transform by covariance under rotations.

Although spinors are not needed for the traditional treatment of classical mechanics, their existence should at least be acknowledged. *Spinors* and their adjuncts form left and right ideals (spin projection operators), which obey the following single-sided transformations under spin rotations

$$\psi' = R\psi, \quad \bar{\psi}' = \bar{\psi}\bar{R}, \quad \bar{\psi}'\psi' = \bar{\psi}\bar{R}R\psi = \bar{\psi}\psi, \tag{7.19}$$

Spinors come in pairs, defining *density matrices*, which transform as tensors. Spinors play a critical role in quantum mechanics, where they encode information about the internal spin polarization state of a quantum mechanical particle. Since only half of an orthogonal similarity transformation is applied

---

[5]More precisely, by the spin group, what is signified here is shorthand notation for what mathematicians would more properly call the singly connected spin subgroup of continuous transformations from the identity element. Improper, noncontinuous, transformations are explicitly excluded.

[6]In general, the singly connected spin subgroup $Spin^+(p,q)$, associated with a given geometric algebra $GA(p,q)$, defines a double-valued cover for the singly connected special orthogonal subgroup $SO^+(p,q)$.

to a spinor, spinors are allowed to be double-valued functions of the spacetime coordinates. Their density matrices are required to behave as single-valued tensors, however. Élie Joseph Cartan discovered spinors in 1913, and Clifford algebra regained prominence in the 1920s with the development of quantum spin. The *Pauli and Dirac spin algebras* have since become essential features of Quantum Mechanics.

The approach taken here is to first develop the more fundamental geometrical representation of a metric space, as this approach can be more easily related to quantum mechanics and transformation theory in general. The results will then be reduced to the more commonly taught vector algebra representation.

7.1.3 Algebraic Structure of Space

Mathematically, a geometric algebra is a matrix algebra that is capable of describing the complete geometric structure of a metric vector space. In geometric algebras, vectors, and all other geometric tensor objects, are represented as square matrices. It follows that the geometric product is *invertible*, since nonsingular matrices are invertible under group multiplication. The group product ab of vector elements a and b in a geometric algebra is generally denoted by an invisible multiplication symbol. The group product is closed, that is, the product of two geometric tensors is another geometric tensor. Like all matrix algebras, geometric algebras are associative under multiplication. All geometric tensors can be constructed from sums of inner and external products of the vector basis. The resulting algebra is finite.

In addition to the geometric product, two secondary vector products are defined in terms of the group product:

1. A symmetric vector dot product

$$a \cdot b = \frac{1}{2}(ab + ba) = \frac{1}{2}\{a, b\} \tag{7.20}$$

that denotes the scalar contraction of vectors, and

2. An antisymmetric vector wedge product

$$a \wedge b = \frac{1}{2}(ab - ba) = \frac{1}{2}[a, b] \tag{7.21}$$

that denotes the geometric extension of vectors, generating oriented elements of surface. Two vectors a and b are *linearly-independent*, if (and only if) their exterior or wedge products are nonzero ($a \wedge b \neq 0$).

The geometric product of two vectors can be expressed as a sum of the dot and wedge products

$$ab = (a \cdot b)1 + a \wedge b, \tag{7.22}$$

where the scalar element 1 represents the identity element of the algebra. The wedge product can be generalized to define the operation of maximally extending the geometric rank of a product of irreducible geometric tensors, while the dot product can be generalized to maximally contract the tensor rank of the product.

The geometric structure of space can be elucidated by starting from the basic postulates of Geometric Algebra [Hestenes 1984, 1986]. These are

 − Geometric algebras describe metric spaces which are defined over the field of real numbers.
 − Geometric vectors are self-dual under the operation of vector reversion, $\bar{v} = v$.
 − The square of any geometric vector is a real-valued scalar, $v^2 = v \cdot v$.

It is easy to demonstrate that the symmetric dot product of any two vectors is a scalar. Let $c = a + b$ denote the sum of two vectors; then

$$c^2 = (a + b)^2 = a^2 + ab + ba + b^2, \tag{7.23}$$

or,

$$2a \cdot b = \{a, b\} = ab + ba = c^2 - a^2 - b^2. \tag{7.24}$$

Therefore, the dot product can be written as a sum of squares of vectors which are scalars by the third postulate.

The above postulates are satisfied if one defines a Clifford Algebra over the field of real numbers. Using a linearly-independent covariant basis λ_i, the basis vector operators satisfy the Clifford algebra anticommutator relations

$$\{\lambda_i, \lambda_j\} = 2g_{ij}, \tag{7.25}$$

where g_{ij} is the metric of the space. In the following sections, the geometric properties of a metric space are described in terms of two fundamental operations, namely, *projection* and *extension*.

7.1.3.1 Projection

The angle bracket notation $\langle T \rangle$ will be used to denote the projection of the scalar part of a geometric tensor T. *Scalar projection*, or *contraction*, of a displacement vector involves the extraction or "measurement" of the projected length of a vector in the direction of some other vector, often using a set of mutually orthogonal unit vectors. The result of a scalar projection yields a potentially measurable, real-valued number or field.

In Clifford algebras, basis vectors and their extensions can always be represented by real-valued traceless matrices. In general, the scalar projection of a geometric tensor of mixed rank is defined as the extraction of the scalar part of the geometric tensor. For the N-dimensional real matrix representation of the algebra, one gets

$$\langle T \rangle \equiv \text{Re} \frac{1}{N} \text{Trace}(T) = \frac{1}{N} \text{Trace}(T). \tag{7.26}$$

For example,

$$\langle 1 \rangle = 1 = \frac{1}{N} \text{Trace}(1). \tag{7.27}$$

$$\langle e_i e_j \rangle = e_i \cdot e_j = g_{ij}. \tag{7.28}$$

With the preceding definitions, the measure of a vector invariant $a = a^i \lambda_i$ is given by projecting the inner product of a vector with its adjunct under reversion. Recall that real-valued vectors are self-adjoint under reversion. The scalar projection of the product of any two vectors is defined to be identical to the vector dot product since the wedge product, signifying extension, has no scalar part

$$\langle ab \rangle = \langle a \mid b \rangle = a \cdot b = g_{ij} a^i b^j. \tag{7.29}$$

As shown in Figure 7.2, the contraction of two vectors is a symmetric operation. That is, the projection of vector a on vector b is the same as that of vector b on vector a. If one of the two vectors is defined as having unit length, the

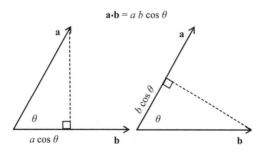

FIGURE 7.2 **The projection of vector _a_ on _b_ leads to the same result as that of vector _b_ on _a_.**

projection of a second vector on the normalized vector is given as a multiple of the length of the normalized vector. Likewise, extraction of a vector's components is given by the projection of the contravariant and covariant components of a vector, using contraction with corresponding vector basis

$$\langle e^i a \rangle = \langle e^i \mid a \rangle = a^i, \quad \langle e_i a \rangle = \langle e_i \mid a \rangle = a_i. \tag{7.30}$$

Since the symmetric dot product of two vectors contracts to a scalar, the only way to extend the geometry is through repeated application of the Grassmann totally antisymmetric product of vectors. All geometric tensors created by extension have zero scalar projection, therefore

$$\langle e_a \rangle = 0, \quad \langle e_a \wedge e_b \rangle = 0, \quad \langle e_a \wedge e_b \wedge e_c \cdots \rangle = 0. \tag{7.31}$$

The cyclic property of the scalar projection operator follows from the cyclic properties of the trace of its matrix representation

$$\langle ABC \rangle = \langle CAB \rangle = \langle BCA \rangle. \tag{7.32}$$

It follows that the scalar projection of a tensor is invariant under a similarity transformation using spin rotors R

$$\langle T' \rangle = \langle RT\bar{R} \rangle = \langle T\bar{R}R \rangle = \langle T \rangle. \tag{7.33}$$

Therefore, scalar projection is a rotationally invariant concept.

7.1.3.2 Extension

The exterior or wedge product of vectors is antisymmetric by definition. The wedge product encapsulates the concept of extension. The wedge product of two vectors generates an antisymmetric rank-two tensor (called a *bivector* or 2-blade) that can be identified with the surface element created by taking the exterior product of Grassmann. The sigma–matrices define a general rank-2 tensor basis spanning the space of bivectors

$$\sigma_{ij} = \lambda_i \wedge \lambda_j. \tag{7.34}$$

Bivectors play an important role in transformation theory. All Lie algebras can be constructed as bivector algebras defined with respect to some underlying parent geometric algebra of appropriate dimensions [Doran 1993].

The exterior product of Grassmann is equivalent to the geometrical construction of sweeping out the area generated by translating the tail of the first vector along the length of the second vector as shown in Figure 7.3. The orientation (or circulative sense) of the associated axial-vector in $GA(3)$ is

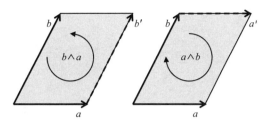

FIGURE 7.3   Generating elements of surface using the exterior product, the tail of the first vector is translated along the length of the second vector. Since the exterior product is anticommutative, $a \wedge b = -b \wedge a$.

given by rotating the second vector into the first vector using the right hand rule. Therefore, the Grassmann product produces oriented elements of area, having a distinct rotational signature.

The most general rank-2 geometric tensor can be written as a sum over all possible 2-blades

$$F = F_{ab} \frac{\sigma^{ab}}{2!}, \tag{7.35}$$

where the factor of 2! is needed to avoid double counting. The components of a 2-blade can be extracted by scalar projection with a bivector basis

$$F_{ab} = \langle \bar{\sigma}_{ab} F \rangle. \tag{7.36}$$

Likewise, a projection of the 2-blade part of a mixed rank tensor can be composed from products of these scalar projections with the rank-2 bivector basis operators

$$\langle F \rangle_2 = \langle \bar{\sigma}_{ab} F \rangle \frac{\sigma^{ab}}{2!}. \tag{7.37}$$

For example, the magnetic field generates a rotation for a moving charge about the magnetic field axis. Therefore magnetic fields are most appropriated described in terms of bivectors.[7]

One can repeat the process of extension, generating rank-3 totally antisymmetric tensors, by sweeping out a bivector area in the direction of a

---

[7]In four spacetime directions, Maxwell's field tensor is a six component bivector field, consisting of three rotational components (the magnetic field part) and three spacetime-dilation components (the electric field part), which accelerate a particle along a given space direction.

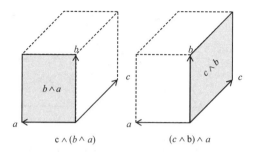

FIGURE 7.4   Generating an element of three-volume $c \wedge b \wedge a$ using the exterior product. The left hand side shows the grouping $c \wedge (b \wedge a)$, while the right hand side shows the grouping $(c \wedge b) \wedge a$. Since the exterior product is associative, both orderings generate the same volume element.

third vector, as shown in Figure 7.4. If three vectors $\{a,b,c\}$ are linearly independent, one gets a volume element of non-zero weight, provided $a \wedge b \wedge c \neq 0$.

The most general rank-3 geometric tensor can be written as a sum over all possible 3-blades

$$\langle T \rangle_3 = T_{abc} \frac{e^a \wedge e^b \wedge e^c}{3!}, \tag{7.38}$$

$$T_{abc} = \langle e_a \wedge e_b \wedge e_c T \rangle. \tag{7.39}$$

The process of extension can be continued, creating totally antisymmetric tensors of increasing rank, until one runs out of linearly-independent vectors.

The exterior product is associative, anticommutative, and distributive over addition. Given three vectors (a,b,c), the following algebraic rules apply

$$(a \wedge b) \wedge c = a \wedge (b \wedge c), \tag{7.40}$$

$$a \wedge b = -b \wedge a. \tag{7.41}$$

$$a \wedge (b+c) = a \wedge b + a \wedge c. \tag{7.42}$$

In algebraic expressions, the wedge product has higher priority than the dot product, and both have higher priority than the geometric product. This yields the following grouping rules

$$a \cdot b \wedge c = a \cdot (b \wedge c), \tag{7.43}$$

$$a \cdot bc = (a \cdot b)c, \tag{7.44}$$

$$ab \wedge c = a(b \wedge c). \tag{7.45}$$

7.2   PAULI SPIN ALGEBRA

The geometric algebra for three-dimensional Euclidean space has the e_i Cartesian basis representation, with metric $g_{ij} \to \delta_{ij}$, resulting in the Geometric Algebra

$$\{e_i, e_j\} = 2\delta_{ij}. \tag{7.46}$$

Working in a Euclidean basis has the advantages that the covariant and contravariant bases are identical $e_i = e^i$, etc. This simplifies the use of matrix notation, since raised and lowered matrix indices are identical ($A^{ij} = A^i_{\ j} = A_i^{\ j} = A_{ij}$). Furthermore, the determinant of the metric tensor has unit norm $\|g\| = +1$ and can be suppressed. The minimal representation of this algebra is given by the complex two-dimensional Pauli spin algebra defined by Equation 7.7.

The rank-3 pseudoscalar element $i_3 = i = e_1 e_2 e_3$ is the highest rank geometric tensor of $GA(3)$. The geometric tensor i has properties similar to the imaginary number i, but represents the pseudoscalar element of three-dimensional Euclidean space. It denotes a volume element of unit norm, and it is antiadjoint under reversion

$$i^2 = -1, \quad \bar{i} = e_3 e_2 e_1 = -i, \quad \bar{i}i = 1. \tag{7.47}$$

In three-dimensional space, i commutes with all other elements of the algebra, satisfying the condition

$$[i, e_j] = 0. \tag{7.48}$$

Using the unit volume element, one can define the *vector cross product* and the *triple scalar product* of vectors in terms of wedge products

$$a \wedge b = ia \times b, \tag{7.49}$$

$$a \wedge b \wedge c = i(a \times b \cdot c) = i(a \cdot b \times c). \tag{7.50}$$

Components of vectors and axial vectors cannot be summed since they have different transformation properties. They represent different irreducible classes of geometric objects.

Three-dimensional Euclidean space $GA(3)$ has eight linearly-independent geometric elements that can be classified into four irreducible tensors of rank $0 \le m \le 3$ as summarized in Table 7.1. The table illustrates how a general geometric tensor in $GA(3)$ can be decomposed into four reducible geometric tensors

$$T = S + iP + V^i e_i + ie^k A_k \tag{7.51}$$

TABLE 7.1   Classification of the irreducible geometric structure of $GA(3)$

Geometric Class	Tensor Basis	Linear Independent Elements
Scalar	$\Gamma^{(0)} = 1$	1 identity element: 1
Line	$\Gamma_i^{(1)} = e_i$	3 line elements: $e_i = i\sigma_i$
Surface	$\Gamma_{ij}^{(2)} = \sigma_{ij} = e_i \wedge e_j = -\varepsilon_{ijk}\sigma^k$	3 rotation elements: $\sigma^k = -ie^k$
Volume	$\Gamma_{ijk}^{(3)} = e_i \wedge e_j \wedge e_k = \varepsilon_{ijk}i$	1 pseudoscalar element: i

where $\{S, P, V, A\}$ denote scalar, pseudoscalar, vector and axial-vector components, respectively of a geometrical tensor in $GA(3)$.

The geometric product in $GA(3)$ defines a *Pauli Spin Algebra*

$$ab = a \cdot b + a \wedge b = a \cdot b + i\, a \times b, \tag{7.52}$$

where the Euclidean vector basis e_i plays the role of the *Pauli spin* matrices which satisfy

$$[e_i, e_j] = 2i\varepsilon_{ijk}e^k. \tag{7.53}$$

A position vector in this notation is given by the traceless matrix

$$r(x, y, z) = \begin{bmatrix} z & x - iy \\ x + iy & -z \end{bmatrix}. \tag{7.54}$$

7.2.1 Generating the Spin Rotation Group

Infinitesimal quaternion rotors are generators of rotations. To demonstrate this, first consider an arbitrarily-oriented, infinitesimal rotor

$$R(\hat{\sigma}, d\theta) = e^{\hat{\sigma}d\theta/2} = 1 + \hat{\sigma}d\theta/2 + \mathcal{O}(d\theta^2), \tag{7.55}$$

where $\hat{n} = i\hat{\sigma}$ is a unit pseudo-vector polar axis

$$\hat{\sigma}d\theta = -i\hat{n}d\theta = -ie_k d\theta^k, \tag{7.56}$$

Note that infinitesimal rotations commute to lowest order in the infinitesimals

$$R(\hat{\sigma}_1, d\theta^1)R(\hat{\sigma}_2, d\theta^2) = R(\hat{\sigma}_2, d\theta^2)R(\hat{\sigma}_1, d\theta^1)$$
$$= 1 + \hat{\sigma}_1 d\theta^1/2 + \hat{\sigma}_2 d\theta^2/2. \tag{7.57}$$

Applying an infinitesimal rotor to an arbitrary vector displacement r gives the rotated vector r'

$$r' = Rr\bar{R} = r + dr, \qquad (7.58)$$

Solving for this infinitesimal displacement gives

$$\begin{aligned} dr = r' - r &= Rr\bar{R} - r \\ &= (1 + \hat{\sigma} d\theta/2)r(1 - \hat{\sigma} d\theta/2) - r \\ &= \frac{1}{2}[\hat{\sigma}, r]d\theta = -i(\hat{n} \wedge r)d\theta, \end{aligned} \qquad (7.59)$$

or

$$dr = (\hat{n} \times r)d\theta, \qquad (7.60)$$

where the last term yields the expected geometric result of an infinitesimal rotation in terms of the vector cross product as demonstrated by the construction shown in Figure 7.5.

A finite rotation about a fixed axis can be generated continuously by taking by an infinite product of infinitesimal rotations, giving the well known result from elementary analysis

$$R(\hat{\sigma}, \theta) = \lim_{N \to \infty} \left(1 + \hat{\sigma} \frac{\theta/2}{N}\right)^N = e^{\hat{\sigma}\theta/2} \in Spin(3). \qquad (7.61)$$

Every element of $Spin(3)$ can be written in this form, so that the two groups are isomorphic. Every rotation is expressible as a unitary rotor. For positive angles $\theta > 0$, the rotor $R(\hat{\sigma}, \theta)$ rotates a vector in a counter-clockwise sense about the polar axis $\hat{\sigma}$ by the angle θ.

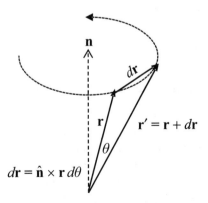

FIGURE 7.5   **An infinitesimal rotation of a vector in the counterclockwise sense about an axis.**

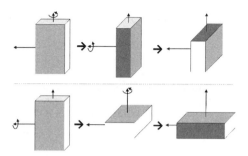

FIGURE 7.6 The noncommutativity of finite rotations: The ordering of two 90°
rotations affects the final orientation of a book.

Using Euler's identity, a finite rotation about a fixed axis $\hat{\sigma}$ is given in the
quaternion representation as

$$R(\hat{\sigma}, \theta) = e^{\hat{\sigma}\theta/2} = \cos\frac{\theta}{2} + \hat{\sigma}\sin\frac{\theta}{2}, \quad R \in Spin(3), \qquad (7.62)$$

where $\hat{\sigma} = n_i\sigma^i$, $\hat{\sigma}^2 = -1$, and θ is the angle of rotation in the counterclockwise sense, using the right hand rule. The angle and direction of rotation
of a $Spin(3)$ rotation can be explicitly evaluated by scalar projection

$$\cos(\theta/2) = \langle R \rangle, \qquad (7.63)$$

$$n^i = \frac{1}{\sin(\theta/2)}\langle \bar{\sigma}^i R \rangle. \qquad (7.64)$$

The direction of rotation is fixed using the right hand rule for positive rotation
angles in the domain $0 \le \theta < 2\pi$.

Finite rotations, unlike infinitesimal rotations, are noncommutative,
as illustrated by the effect of reversing the order of two 90° rotations,
demonstrated in Figure 7.6.

7.2.2 *Real-Valued Matrix Representation of Euclidean Space

The two-dimensional complex Pauli representation is commonly used in
elementary quantum mechanics textbooks. However, this complex matrix
representation obscures the geometric significance of the imaginary number
$i = i^{(3)}$ and unnecessarily conflates it with the complex phase algebra characteristic of quantum mechanics. Since the unit pseudoscalar $i^{(3)}$, is an element
of three volume, it changes sign under space inversion. Therefore it is not a
scalar and consequently must have a vanishing scalar projection. It is therefore more transparent to explicitly represent $i^{(3)}$ as a traceless real-valued

matrix in a higher-dimensional outer-product space. It can be shown that the pseudoscalar element in a space of even dimensions anticommutes with the vector basis and defines a space-inversion operator for that space.

One possible four-dimensional, real-valued, matrix representation of $GA(3)$ is as a proper subgroup of the four-dimensional space $GA(3,1)$, given by

$$e_0 = \sigma_3 \begin{bmatrix} 0 & 1 \\ 1 & 0 \end{bmatrix}, \quad e_1 = \tau_1 \begin{bmatrix} 1 & 0 \\ 0 & 1 \end{bmatrix}, \quad e_2 = \tau_2 \begin{bmatrix} 1 & 0 \\ 0 & 1 \end{bmatrix}, \quad e_3 = \sigma_3 \begin{bmatrix} 0 & -1 \\ 1 & 0 \end{bmatrix},$$
(7.65)

$$i^{(3)} = e_1 e_2 e_3 = \begin{bmatrix} 0 & -1 \\ 1 & 0 \end{bmatrix},$$
(7.66)

where the real, two-dimensional, submatrices $(\tau_1, \tau_2, \sigma_3)$ can be represented as

$$\tau_1 = \begin{bmatrix} 1 & 0 \\ 0 & -1 \end{bmatrix}, \quad \tau_2 = \begin{bmatrix} 0 & 1 \\ 1 & 0 \end{bmatrix}, \quad \sigma_3 = -\tau_1 \tau_2 = \begin{bmatrix} 0 & -1 \\ 1 & 0 \end{bmatrix}.$$
(7.67)

The four-dimensional, real-valued, matrix representation of $GA(3)$ given by Equation (7.65) in fact describes the pseudo-Euclidean Minkowski space labeled $GA(3,1)$, having three-space and one-time dimensions. In this representation, $GA(3) \subset GA(3,1)$ is the subspace of $GA(3,1)$ consisting of those elements of $GA(3,1)$ that commute with the three-dimensional pseudoscalar $i^{(3)}$. In other words, the time-like basis element e_0 is excluded from the $GA(3)$ subalgebra.

An immediate benefit of using a four-dimensional real matrix representation is that it adds a *parity* or *space inversion* operator, defined as $P = e_0$, to the algebra, where the parity operator inverts the three-dimensional space basis, leaving time basis invariant

$$P e_i P^{-1} = -e_i, \quad P e_0 P^{-1} = e_0, \quad P i_3 P^{-1} = -i_3.$$
(7.68)

In summary, the square-matrix representation of the geometric basis of a real metric space gives rise to a real-valued *Clifford algebra* in n dimensions $Cl(n, \mathbb{R})$, which describes the complete geometric structure of the space. $Cl(n, \mathbb{R})$ can also be referred to as the *Geometric Algebra GA(n)*. If the space has an indeterminate signature, the Geometric Algebra is denoted more precisely as $GA(n, m)$, where n denotes the number of space-like dimensions and m denotes the number of time-like dimensions. *Spin(n, m)* is defined, restrictively, as the continuous Lie Group formed from the Lie Algebra of surface elements (bivectors) in $GA(n, m)$. An interesting property of Lie algebras is that every continuous Lie group can be represented as a spin group acting on some Geometric Algebra [Doran 1993].

7.3 VECTOR ALGEBRA NOTATION

The basic properties of bra-ket and dyadic notations were developed in Chapter 2. Gibb's vector algebra is a subset of geometric algebra that describes how rank-one tensors, or vectors, map into other vectors via second-rank tensor (*dyadic*) transformations. In vector algebras, first-rank vectors are represented by column matrices, while second rank tensors are represented by square matrices. There is only one essential product, the dot product, denoting the contraction of adjacent tensors and vectors.

Vector algebras are dual algebras. Vectors $\mathbf{v}$ can only be multiplied from the left ($\mathbf{v}' = \mathbf{T} \cdot \mathbf{v}$), and their duals $\bar{\mathbf{v}}$ can only be multiplied from the right ($\bar{\mathbf{v}}' = \bar{\mathbf{v}} \cdot \bar{\mathbf{T}}$), by a second rank tensor transformation $\mathbf{T}$. Note that boldface type is used to denote vectors and tensors in this notation. Column vectors and their dual row vectors under the operation of reversion can also be represented as bra and ket vectors by making the association

$$\mathbf{v} \leftrightarrow |v\rangle, \tag{7.69}$$

$$\bar{\mathbf{v}} \leftrightarrow \langle\bar{v}|. \tag{7.70}$$

Real-valued vectors are defined to be self-adjoint under the operation of reversion, which is defined as the reversal of the order of all vectors in a product

$$\bar{\mathbf{v}} = \mathbf{v}. \tag{7.71}$$

Row and column matrices are constructed by projecting out the vector components.

Many treatments using dyadic or bra-ket index notation ignore the important distinction between covariant and contravariant vectors, so they are limited to working with orthonormal Cartesian bases. However, this is not a fundamental limitation. The *completeness relation*

$$1 = \lambda_i \bar{\lambda}^i; \tag{7.72}$$

can be used to facilitate projection with respect to a arbitrary tensor basis

$$\bar{\mathbf{b}} \cdot \mathbf{a} = \bar{\mathbf{b}} \cdot \lambda_i \bar{\lambda}^i \cdot \mathbf{a} = \begin{bmatrix} b_1 & b_2 & b_3 \end{bmatrix} \begin{bmatrix} a^1 \\ a^2 \\ a^3 \end{bmatrix} = b_i a^i. \tag{7.73}$$

Note that repeated (contracted) indices always involve sums over pairs of dummy index labels with one raised and one lowered index.

For consistency when extending real algebras into the complex domain, complex conjugation will be identified as the dual of a complex number $c = a + ib$ under reversion

$$\bar{c} = \overline{(a + ib)} = a - ib, \tag{7.74}$$

$$\bar{i} = i^* = -i. \tag{7.75}$$

This gives a positive definite norm for complex numbers

$$\bar{c}c = c^*c = (a - ib)(a + ib) = a^2 + b^2. \tag{7.76}$$

Therefore, when dealing with complex vectors,

$$\bar{\mathbf{v}} = \mathbf{v}^*, \tag{7.77}$$

and

$$\overline{|v\rangle} = \langle\bar{v}| = \langle v^*|. \tag{7.78}$$

This convention is consistent with the notion that the imaginary number i, which commutes with all other elements of the algebra, can always be reinterpreted as the generator of $Spin(2)$ rotations with respect to an internal, two-dimensional phase space. Bivector generators are always odd under reversion.

7.3.1 Orthogonal Transformations

Generalized orthogonal transformations on spaces of indefinite signature are transformations that leave the inner product of all vectors unchanged. Under orthogonal transformations, vectors and their duals transform as

$$\mathbf{a}' = \mathbf{O} \cdot \mathbf{a}, \quad \bar{\mathbf{a}}' = \bar{\mathbf{a}} \cdot \bar{\mathbf{O}}, \tag{7.79}$$

where $\bar{\mathbf{O}} \cdot \mathbf{O} = 1$, while second-rank tensors transform as

$$\mathbf{T}' = \mathbf{O} \cdot \mathbf{T} \cdot \bar{\mathbf{O}}. \tag{7.80}$$

Therefore all inner products are invariant under orthogonal transformations

$$\bar{\mathbf{a}}' \cdot \mathbf{b}' = \bar{\mathbf{a}} \cdot \mathbf{b}, \quad \bar{\mathbf{a}}' \cdot \mathbf{T}' \cdot \mathbf{b}' = \bar{\mathbf{a}} \cdot \mathbf{T} \cdot \mathbf{b}. \tag{7.81}$$

Note that the metric tensor $g_{ij} = \lambda_i \cdot \lambda_j$ is also invariant under orthogonal transformations.

For orthogonal matrices, reversion is equivalent to transposition of the indices,

$$\bar{O}^{ij} = O^{ji},\tag{7.82}$$

it follows that orthogonal transformations satisfy the *orthogonality condition*

$$O^{ki}O_{kj} = \delta^i_j.\tag{7.83}$$

Orthogonal matrices in $GA(3)$ are members of the orthogonal group $O(3)$. Calculate the determinant using

$$\det(1) = \det(\mathbf{O} \cdot \bar{\mathbf{O}}) = \det(\mathbf{O})\det(\bar{\mathbf{O}}) = \det(\mathbf{O})^2 = 1.\tag{7.84}$$

Since the determinant is real,

$$\det(\mathbf{O}) = \pm 1.\tag{7.85}$$

Orthogonal matrices with determinant -1 are said to be improper. The special orthogonal group $SO(3)$ adds the special constraint that restricts the determinant to $+1$. It represents the continuous three-parameter subgroup of orthogonal transformations that can be continuously evolved from the identity element.

7.3.2 Infinitesimal Rotations

Rotations in $GA(3)$ are continuous proper orthogonal transformations that map vectors into vectors while leaving the length of the vector unchanged. Finite rotations are built up from infinitesimal rotations.

Consider a rotation of a vector $\mathbf{r}$ by a counterclockwise rotation about an axial-vector rotation axis denoted by $\hat{\mathbf{n}}$, as shown in Figure 7.5. From Equation (7.60) an infinitesimal rotation about a fixed axis can be written in terms of the vector cross product as

$$d\mathbf{r} = d\theta\,\hat{\mathbf{n}} \times \mathbf{r} = d\theta^i \mathbf{e}_i \times \mathbf{r},\tag{7.86}$$

Since the cross product operator is peculiar to three space dimensions, it is better to replace the cross product by substituting

$$\mathbf{e}_k \times \mathbf{V} = -\varepsilon_{ijk}\mathbf{e}^i\mathbf{e}^j \cdot \mathbf{V},\tag{7.87}$$

Then an element of rotation of a basis vector can be written in dyadic form as

$$d\mathbf{e}'_i = d\theta^j \mathbf{e}_j \times \mathbf{e}'_i = -d\mathbf{\Theta} \cdot \mathbf{e}_i,\tag{7.88}$$

where $d\Theta$ is an infinitesimal antisymmetric dyadic operator

$$d\Theta = d\Theta_{ij}\mathbf{e}^i\mathbf{e}^j = \varepsilon_{ijk}\mathbf{e}^i\mathbf{e}^j d\theta^k \tag{7.89}$$

The time rate of change of a rotating basis can then be written in either of the two notations as

$$\frac{d\mathbf{e}'_i}{dt} = \boldsymbol{\omega} \times \mathbf{e}'_i = -\boldsymbol{\Omega} \cdot \mathbf{e}_i, \tag{7.90}$$

where $\boldsymbol{\omega}$ is the angular velocity axial-vector

$$\boldsymbol{\omega} = \omega^i\mathbf{e}_i = \frac{d\theta^i}{dt}\mathbf{e}_i, \tag{7.91}$$

and $\boldsymbol{\Omega}$ is the associated anti-symmetric dyadic derivative

$$\boldsymbol{\Omega} = \frac{d\Theta}{dt}. \tag{7.92}$$

with coefficients

$$\Omega_{ij} = \mathbf{e}_j \cdot \frac{d\mathbf{e}_i}{dt} = \varepsilon_{ijk}\omega^k = \begin{bmatrix} 0 & \omega^3 & -\omega^2 \\ -\omega^3 & 0 & \omega^1 \\ \omega^3 & -\omega^1 & 0 \end{bmatrix}. \tag{7.93}$$

7.3.3 Finite Rotations

To determine the $SO(3)$ representation of a vector rotation, first consider the rotation represented by the cover group $Spin(3)$

$$v' = R(\hat{\sigma},\theta)v\bar{R}(\hat{\sigma},\theta) = e^{\hat{\sigma}\theta/2}ve^{-\hat{\sigma}\theta/2}. \tag{7.94}$$

One needs to compare this to the equivalent orthogonal transformation

$$\mathbf{v}' = \mathbf{A}(\hat{\sigma},\theta) \cdot \mathbf{v}. \tag{7.95}$$

Equating the vector components in the two notations, one gets

$$v'^i = A^i_j v^j = \langle e^i R e_j \bar{R}\rangle v^j. \tag{7.96}$$

In column vector notation, a rotation applied to an arbitrary fixed axis $\hat{n}$ can be written as the general orthogonal transformation

$$\mathbf{V}' = \mathbf{A}(\sigma_{\hat{n}},\theta) \cdot \mathbf{V} = \mathbf{A}_{\hat{n}}(\theta) \cdot \mathbf{V}, \tag{7.97}$$

where the rotation matrices about the three coordinate axes are given by

$$\{A_1(\theta)^i{}_j\} = \begin{bmatrix} 1 & 0 & 0 \\ 0 & \cos\theta & -\sin\theta \\ 0 & \sin\theta & \cos\theta \end{bmatrix},$$

$$\{A_2(\theta)^i{}_j\} = \begin{bmatrix} \cos\theta & 0 & \sin\theta \\ 0 & 1 & 0 \\ -\sin\theta & 0 & \cos\theta \end{bmatrix}, \tag{7.98}$$

$$\{A_3(\theta)^i{}_j\} = \begin{bmatrix} \cos\theta & -\sin\theta & 0 \\ \sin\theta & \cos\theta & 0 \\ 0 & 0 & 1 \end{bmatrix}.$$

For infinitesimal rotations, the $SO(3)$ rotation matrices can be written in terms of infinitesimal generators $\mathfrak{M}_i$

$$\mathbf{A}_1(d\theta^1) = 1 + \mathfrak{M}_1 d\theta^1$$
$$\mathbf{A}_2(d\theta^2) = 1 + \mathfrak{M}_2 d\theta^2 \tag{7.99}$$
$$\mathbf{A}_3(d\theta^3) = 1 + \mathfrak{M}_3 d\theta^3$$

where

$$\mathfrak{M}_1 = \begin{bmatrix} 0 & 0 & 0 \\ 0 & 0 & -1 \\ 0 & 1 & 0 \end{bmatrix}, \quad \mathfrak{M}_2 = \begin{bmatrix} 0 & 0 & 1 \\ 0 & 0 & 0 \\ -1 & 0 & 0 \end{bmatrix}, \quad \mathfrak{M}_3 = \begin{bmatrix} 0 & -1 & 0 \\ 1 & 0 & 0 \\ 0 & 0 & 0 \end{bmatrix}.$$

$$\tag{7.100}$$

It is left as an exercise to demonstrate that these generators satisfy the Lie algebra

$$[\mathfrak{M}_i, \mathfrak{M}_j] = \varepsilon_{ij}{}^k \mathfrak{M}_k. \tag{7.101}$$

Most rotation formulas assume that vectors are rotated in a counterclockwise sense. One very important exception to this rule involves the definition of the sequence of Euler angle transformations, which are often, but not universally, defined in terms of counterclockwise rotations of the body basis. Erroneous results occur if the defined sign conventions and the prescribed sequence of rotations are not strictly observed.

7.3.4 Active and Passive Interpretations of a Transformation

Consider a contravariant transformation of a vector by a counterclockwise rotation about the $\hat{3}$-axis. Substituting

$$A(\sigma_3, \theta)^i{}_j = \langle e^i e^{\sigma_3\theta/2} e_j e^{-\sigma_3\theta/2} \rangle, \tag{7.102}$$

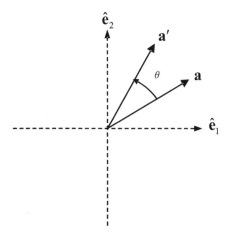

FIGURE 7.7 An active interpretation of a rotation about the $\hat{k}$ axis, resulting in a contravariant counterclockwise rotation of a vector $a = a^i e_i \rightarrow a' = a'^i e_i$ in the $(\hat{i}, \hat{j})$ plane, with respect to a fixed basis.

the orthogonal matrix representation of the transformation is given by

$$\begin{bmatrix} V'^i \\ V'^2 \\ V'^3 \end{bmatrix} = A(\sigma_3, \theta) \begin{bmatrix} V^i \\ V^2 \\ V^3 \end{bmatrix} = \begin{bmatrix} \cos\theta & -\sin\theta & 0 \\ \sin\theta & \cos\theta & 0 \\ 0 & 0 & 1 \end{bmatrix} \begin{bmatrix} V^i \\ V^2 \\ V^3 \end{bmatrix}. \qquad (7.103)$$

The active interpretation of this transformation, illustrated in Figure 7.7, involves the transformation of the vector, given by

$$\mathbf{V}' = \mathbf{A} \cdot \mathbf{V}. \qquad (7.104)$$

The transformation is applied to the components, holding the basis constant.

However the same contravariant components would be obtained, if one assumes a passive transformation of the coordinate basis, leaving the vector unchanged

$$\mathbf{V}' = V'^a e'_a = V^a e_a = \mathbf{V}. \qquad (7.105)$$

To maintain invariance, the basis vectors would have to rotate by the opposite rule of transformation from their vector components. Consider a vector invariant written in row-column notation as

$$a = a^i \mathbf{e}_i = [\mathbf{e}_1, \mathbf{e}_2, \mathbf{e}_3] \begin{bmatrix} a^1 \\ a^2 \\ a^3 \end{bmatrix} = [\mathbf{e}_1, \mathbf{e}_2, \mathbf{e}_3] \bar{A} A \begin{bmatrix} a^1 \\ a^2 \\ a^3 \end{bmatrix} = [\mathbf{e}'_1, \mathbf{e}'_2, \mathbf{e}'_3] \begin{bmatrix} a'^1 \\ a'^2 \\ a'^3 \end{bmatrix}.$$
$$(7.106)$$

where A and $\bar{A}$ denote the matrix coefficients of the transformation. In the passive interpretation, if the vector components rotate in a counter-clockwise sense, then the basis rotates in a clockwise sense, leaving the vector invariant. The basis therefore transforms by the reciprocal transformation

$$[\mathbf{e}'_1, \mathbf{e}'_2, \mathbf{e}'_3] = [\mathbf{e}_1, \mathbf{e}_2, \mathbf{e}_3]\bar{A}(\sigma_3, \theta). \tag{7.107}$$

Taking the transpose gives

$$\begin{bmatrix} \mathbf{e}'_1 \\ \mathbf{e}'_2 \\ \mathbf{e}'_3 \end{bmatrix} = A(\sigma_3, \theta) \begin{bmatrix} \mathbf{e}_1 \\ \mathbf{e}_2 \\ \mathbf{e}_3 \end{bmatrix} = \begin{bmatrix} \cos\theta & -\sin\theta & 0 \\ \sin\theta & \cos\theta & 0 \\ 0 & 0 & 1 \end{bmatrix} \begin{bmatrix} \mathbf{e}_1 \\ \mathbf{e}_2 \\ \mathbf{e}_3 \end{bmatrix}, \tag{7.108}$$

The result of this passive interpretation of the transformation is illustrated in Figure 7.8. In the geometric algebra notation, a passive transformation can be understood as sequentially applying two transformations, where the second transformation undoes the first.

$$V' = \bar{R}(RV\bar{R})R = V. \tag{7.109}$$

The first transformation is applied to the basis; the second, to the components.

Comparing Figure 7.7 with Figure 7.8 leads to the following observation: the contravariant rotational transformation of a vector $v'^i = a^i_j v^j$ can be interpreted as the result of either the *passive rotation* of a basis about an axis

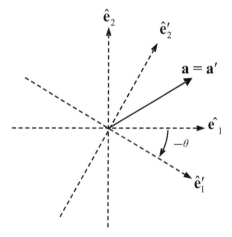

FIGURE 7.8   Passive interpretation of the rotation shown in Figure 7.7: the vector $a = a^i \mathbf{e}_i = a'^i \mathbf{e}'_i$ is unchanged. A covariant clockwise rotation of the $(\hat{i}, \hat{j})$ basis about the $\hat{k}$-axis is preformed, requiring a corresponding contravariant counter-clockwise change in the vector components.

in the clockwise sense, holding the vector constant, or as the *active rotation* of a vector in the counterclockwise sense, holding the basis constant.

7.3.5 Discrete Symmetries

There are a number of important discrete symmetries in nature. Many of these can be categorized as improper orthogonal transformations with $\det(A) = -1$. For example, systems are often invariant under space inversion, also known as parity, in which a right-handed basis transforms into a left-handed basis by a reflection of all space axes

$$\mathbf{e}_i \to -\mathbf{e}_i. \tag{7.110}$$

The parity transformation results in an inversion of the volume element

$$\mathbf{e}_1 \cdot (\mathbf{e}_2 \times \mathbf{e}_3) \to -\mathbf{e}_1 \cdot (\mathbf{e}_2 \times \mathbf{e}_3). \tag{7.111}$$

Discrete symmetries of a system, unlike continuous symmetries, do not give rise to new conservation laws.

7.3.5.1 Space Inversion

Space inversion can be written as an improper orthogonal matrix transformation with determinant -1. The orthogonal parity operator $\mathbf{P}$ is a member of $O(3)$ but not of $SO(3)$. In $GA(3)$, using bra-ket notation, the parity operator is given by discrete dyadic transformation

$$\mathbf{P} = -\mathbf{e}_i\mathbf{e}^i = -1, \quad \mathbf{P} \cdot \bar{\mathbf{P}} = 1, \quad \det(\mathbf{P}) = -1. \tag{7.112}$$

$$\mathbf{P} \cdot \mathbf{e}_i = -\mathbf{e}_i. \tag{7.113}$$

7.3.5.2 Mirror Reflection

Mirror reflection is a second example of an improper orthogonal transformation with determinant -1. Reflection about the $\hat{3}$-axis of Euclidean space inverts the 3-space axis. The transformation can be explicitly represented as

$$\mathbf{M}_{e_3} = 1 - 2\mathbf{e}_3\mathbf{e}^3 = \begin{bmatrix} 1 & 0 & 0 \\ 0 & 1 & 0 \\ 0 & 0 & -1 \end{bmatrix}, \quad \det(\mathbf{M}_{e_3}) = -1. \tag{7.114}$$

$$\mathbf{M}_{\mathbf{e}_3} \cdot \mathbf{e}_1 = \mathbf{e}_1, \quad \mathbf{M}_{\mathbf{e}_3} \cdot \mathbf{e}_2 = \mathbf{e}_2, \quad \mathbf{M}_{\mathbf{e}_3} \cdot \mathbf{e}_3 = -\mathbf{e}_3. \tag{7.115}$$

To describe a reflection about an arbitrary axis, one needs only to make an arbitrary rotation to a new arbitrary axis $\mathbf{A}_R \cdot \mathbf{e}_3 = \hat{\mathbf{n}}$, then reflection about any axis $\hat{n}$ can be written as

$$\mathbf{A}_{\hat{n}} = \mathbf{A}_R \cdot \mathbf{M}_{e_3} \cdot \bar{\mathbf{A}}_R = 1 - 2\hat{\mathbf{n}}\,\hat{\mathbf{n}}, \tag{7.116}$$

where the resulting reflection transformation also has determinant -1, since the determinant of a matrix is preserved under rotations.

The parity operator is equivalent to the application of three mutually orthogonal reflections. Application of an even number of reflections is equivalent to a rotation, since the product has determinant $+1$.

7.4 ROTATING FRAMES OF MOTION

A rigid body is an extended body where all mass points $\mathbf{r}'$ in the body maintain constant relative angles and distances with respect to an internal, fixed, *body frame*, which rotates with respect to a fixed space frame. As shown in Figure 7.9, the relative angles and separation distances between all points in the rigid body are fixed relative to each other and with respect to the body frame. The rigidity condition is satisfied if, and only if, all inner products are constants in the body frame

$$\mathbf{r}'_a \cdot \mathbf{r}'_b = \mathbf{r}_a \cdot \bar{\mathbf{A}} \cdot \mathbf{A} \cdot \mathbf{r}_b = \mathbf{r}_a \cdot \mathbf{r}_b. \tag{7.117}$$

A mapping of vectors into vectors that preserves all inner products requires that the transformation matrix satisfies $\mathbf{A} \cdot \bar{\mathbf{A}} = 1$. $\mathbf{A}$ is therefore an element of the orthogonal group $O(3)$.

But the motion of a rigid body about a fixed point also evolves continuously in time. Let the body frame instantaneously coincide with a fixed space frame at some time $t = 0$, then the body frame at some latter time is represented by

$$\mathbf{e}'_i(t) = \mathbf{A}(t) \cdot \mathbf{e}'_i(0) = \mathbf{A}(t) \cdot \mathbf{e}_i. \tag{7.118}$$

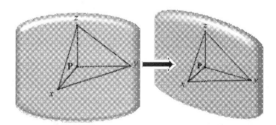

FIGURE 7.9 **Every point on a rigid body is fixed with respect to an embedded body frame, defined with respect to a body frame origin *P*. When the body is rotating about a fixed point, the origin is usually chosen to be coincident with the pivot point. When there is no fixed point with respect to the space frame, then the origin is usually taken to be the center of mass point of the rigid body.**

where $\mathbf{A}(0) = \mathbf{1}$, since the space and body basis vectors coincide initially, and

$$\mathbf{e}'_i(0) = \mathbf{e}_i. \tag{7.119}$$

Note that the body basis $\mathbf{e}'_i(t)$ is not a Riemann tensor basis since the transformation is nonholonomic (it is velocity dependent) in general

$$\frac{d\mathbf{e}'_i(t)}{dt} \neq 0 \tag{7.120}$$

whereas $d\mathbf{e}_i/dt = 0$ for the inertial Euclidean basis e_i of the space frame.

Since $\det(\mathbf{A} \cdot \bar{\mathbf{A}}) = 1$ for orthogonal transformations, one concludes that $(\det(\mathbf{A}_R))^2 = 1$, But there is no way to go continuously from a determinant $+1$ to -1 while keeping the square of the determinant constant. Therefore, the requirement that the motion be continuous imposes the special constraint on the determinant

$$\det(\mathbf{A}(t)) = \det(\mathbf{A}(0)) = +1 \tag{7.121}$$

Under continuous rotations, the body frame basis $\mathbf{e}'_i(t)$ must satisfy both the orthogonality condition

$$\mathbf{e}'^i(t) \cdot \mathbf{e}'_j(t) = \mathbf{e}'^i(0) \cdot \mathbf{e}'_j(0) = \delta^i_j, \tag{7.122}$$

implying the orthogonality condition

$$(A)^i_k (A)^k_j = \delta^i_j, \tag{7.123}$$

and the special continuity constraint

$$\det(\mathbf{A}) = +1. \tag{7.124}$$

It follows that a rotation $\mathbf{A}$ is an element of the "proper" or special group of continuous orthogonal transformations $SO(3)$.

7.4.1 Euler's Theorem

The special orthogonal group is defined by the set of vector transformations $\mathbf{A} \in SO(3)$ such that $\mathbf{A} \cdot \bar{\mathbf{A}} = 1$ and $\det(\mathbf{A}) = +1$. The 3×3 square matrices $\{A^i_j\}$ have nine real coefficients, but they must also satisfy six orthogonality constraints

$$(A)^i{}_k (A)_j{}^k = \delta^i{}_j, \tag{7.125}$$

leaving only three free parameters. Therefore, $SO(3)$ is a three-parameter continuous group.

It can be shown that every element of $SO(3)$ corresponds to a rotation about some axis. This is a consequence of *Euler's theorem*, which states that the general displacement of a rigid body with one point fixed is a rotation about some axis.

It has already been established that a rotation in E^3 is an element of the special orthogonal group. To prove the theorem, one only needs to show that every member of $SO(3)$ has a one-to-one mapping to a member of the rotation group. This can be done by explicitly finding the specific rotation axis and angle associated with any element $A \in SO(3)$.

The proof results from solving the eigenvector problem

$$\mathbf{A} \cdot \mathbf{n} = \lambda_n \mathbf{n}. \tag{7.126}$$

The transformation matrix $\mathbf{A}$ is unitary, but not symmetric (nor Hermitian), so its eigenvalues are complex in general. Therefore, to fully diagonalize the matrix, it is necessary to continue the solution into the complex field. A basic result of linear analysis is that one can always diagonalize a square matrix $\mathbf{A}$ by applying a complex unitary transformation

$$\mathbf{A}' = \mathbf{U} \cdot \mathbf{A} \cdot \bar{\mathbf{U}}, \tag{7.127}$$

where

$$\bar{\mathbf{U}} = \mathbf{U}^\dagger. \tag{7.128}$$

For complex vectors, the dual is given by

$$\bar{\mathbf{n}} = \mathbf{n}^\dagger = \mathbf{n}^* \tag{7.129}$$

The transformation that diagonalizes the matrix A therefore has the following properties

$$\mathbf{n}' = \mathbf{U} \cdot \mathbf{n}, \tag{7.130}$$

$$\bar{\mathbf{n}}' = \bar{\mathbf{n}} \cdot \bar{\mathbf{U}}, \tag{7.131}$$

where $\mathbf{A}'$ has the diagonal matrix form

$$A' = \begin{bmatrix} \lambda_1 & 0 & 0 \\ 0 & \lambda_2 & 0 \\ 0 & 0 & \lambda_3 \end{bmatrix}, \quad \bar{A}' = \begin{bmatrix} \lambda_1^* & 0 & 0 \\ 0 & \lambda_2^* & 0 \\ 0 & 0 & \lambda_3^* \end{bmatrix}, \tag{7.132}$$

Formally, one needs to find a solution to the eigenvalue equation with complex eigenvalues

$$(\mathbf{A} - \lambda 1) \cdot \mathbf{n} = 0, \tag{7.133}$$

$$\mathbf{n}^* \cdot (\bar{\mathbf{A}} - \lambda^* 1) = 0, \tag{7.134}$$

From linear analysis, the eigenvalue problem has a solution if and only if the following secular equation is satisfied

$$\det(\mathbf{A} - \lambda \mathbf{1}) = 0. \tag{7.135}$$

In three dimensions, this results in a cubic equation with real coefficients. From the fundamental theorem of algebra, this polynomial equation can be completely factored into three complex roots

$$\det(\mathbf{A} - \lambda \mathbf{1}) = (\lambda_1 - \lambda)(\lambda_2 - \lambda)(\lambda_3 - \lambda) = 0. \tag{7.136}$$

But every cubic equation with real coefficients crosses the origin at least once. Therefore, there is at least one real root to this eigenvalue problem. Moreover, orthogonality requires that $\mathbf{A}' \cdot \bar{\mathbf{A}}' = 1$. It follows that all the eigenvalues have unit modulus

$$|\lambda_a|^2 = 1, \tag{7.137}$$

so that all the eigenvalues λ_a lie on the unit circle

$$\lambda = e^{i\theta}. \tag{7.138}$$

In addition, the determinant of a matrix is preserved under unitary transformations. This implies that the product of the eigenvalues satisfies

$$\det(\mathbf{A}') = \lambda_1 \lambda_2 \lambda_3 = +1. \tag{7.139}$$

But a cubic equation with real coefficients has at least one real root, label it as λ_3, and the other two roots, if not real, must be complex conjugates of each other. It follows that one can always find that at least one eigenvector having the eigenvalue $\lambda_3 = +1$. The remaining two roots will always satisfy the criteria $\lambda_1 = \lambda_2^* = e^{i\phi}$.

If all three roots are real, the only two possibilities for the remaining two eigenvalues are $(1,1)$ and $(-1,-1)$, yielding the special diagonal cases

$$A'(\sigma_3, 0°) = \begin{bmatrix} 1 & 0 & 0 \\ 0 & 1 & 0 \\ 0 & 0 & 1 \end{bmatrix}, \quad A'(\sigma_3, 180°) = \begin{bmatrix} -1 & 0 & 0 \\ 0 & -1 & 0 \\ 0 & 0 & 1 \end{bmatrix}, \tag{7.140}$$

where the first matrix corresponds to the identity transform and second matrix to a $180°$ rotation about the $\hat{3}$-axis, respectively. All other diagonal solutions have complex eigenvalues of the form

$$A'(\sigma_3, \theta) = \begin{bmatrix} e^{i\theta} & 0 & 0 \\ 0 & e^{-i\theta} & 0 \\ 0 & 0 & 1 \end{bmatrix}, \tag{7.141}$$

Restricting the matrices $\mathbf{A}$ to real-valued members of $SO(3)$, the most general form of a rotation matrix, pre-diagonalized in the $\hat{3}$-direction, is given by the single parameter subgroup

$$A(\sigma_3, \theta) = \begin{bmatrix} \cos\theta & -\sin\theta & 0 \\ \sin\theta & \cos\theta & 0 \\ 0 & 0 & 1 \end{bmatrix}, \qquad (7.142)$$

where θ is the angle of rotation about the $\hat{3}$-axis.

But singling out the $\hat{3}$-axis for special treatment is solely a matter of convenience. Except for the identity transformation, the transformation matrix A has unique rotation axis $\hat{\mathbf{n}}$ (up to a sign which is correlated with the sign of the rotation angle) with an eigenvalue solution of $+1$

$$\mathbf{A}_{\hat{n}}(\theta) \cdot \hat{\mathbf{n}} = \hat{\mathbf{n}}. \qquad (7.143)$$

This uniquely defines the rotation axis. The rotation angle can be evaluated using the trace of the transformation matrix, which is invariant under orthogonal transformations

$$\text{Trace}(\mathbf{A}') = e^{i\theta} + e^{-i\theta} + 1 = \text{Trace}(\mathbf{A}) = 2\cos\theta + 1. \qquad (7.144)$$

This can be confirmed by evaluating the trace of Equation (7.142) for a specific representation where the axis of rotation is chosen to be the $\hat{3}$ axis. In general, a special orthogonal transformation can always be interpreted as a rotation about some fixed axis where the axis is a solution to the eigenvalue problem $\mathbf{A} \cdot \hat{\mathbf{n}} = \hat{\mathbf{n}}$ and the rotation angle is given by

$$\cos\theta = \frac{1}{2}(\text{Trace}(\mathbf{A}) - 1) \qquad (7.145)$$

If one restricts the domain to positive angles $0 \leq \theta \leq \pi$, the direction of the rotation axis is given by observing the effect of the rotation on a vector perpendicular to the rotation axis, using the right hand rule for positive rotation angles. Therefore, all rotations are elements of $SO(3)$, and all elements of $SO(3)$ are rotations. This completes the proof of Euler's theorem.

7.4.2 Pitch, Roll, and Yaw Angles

$SO(3)$ is a three-parameter continuous group. Any rotation can be decomposed into a product of no more than three elementary rotations about predefined axes. A common coordinate system used in flight dynamics and other engineering applications are the pitch α, roll β, and yaw γ rotations,

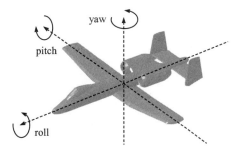

FIGURE 7.10 A flight dynamics model for three-dimensional rotations using roll, pitch, and yaw angles with respect to an aircraft body frame.

taken in that order with respect to three orthogonal axes labeled as $(\hat{1}, \hat{2}, \hat{3})$ or $(\hat{x}, \hat{y}, \hat{z})$, respectively. These are defined in the instantaneous body frame with respect to a pilot flying an aircraft as illustrated in Figure 7.10. The rotation angles for the coordinate basis are usually (but not universally) defined in a counterclockwise sense. Alternatively, vectors rotate in a clockwise sense. A general rotation can be expressed by the product of three elementary spinor rotations

$$R(\alpha, \beta, \gamma) = R_{3''}(\gamma)R_{2'}(\beta)R_1(\alpha) = e^{-\sigma_3'' \gamma/2}e^{-\sigma_2' \beta/2}e^{-\sigma_1 \alpha/2}, \tag{7.146}$$

or, equivalently, as three orthogonal transformations

$$\mathbf{A}(\alpha, \beta, \gamma) = \mathbf{A}_{3''}(\gamma) \cdot \mathbf{A}_{2'}(\beta) \cdot \mathbf{A}_1(\alpha). \tag{7.147}$$

Be aware that other conventions for the pitch, roll, and yaw angles can be found in the literature. It is important to pick a consistent notation and stick to it.

In the theoretical physics literature, the engineering pitch-roll-yaw convention is not as frequently encountered as the Euler angle convention to be described in the following section. Nevertheless, these engineering coordinates have an important practical advantage. They produce more stable numerical results, as there is no degeneracy in the coordinate axes.

Often one wants to express rotations with respect to a set of fixed space axes. In this case, a transformation of coordinates is required. The second mapping about the $2'$ body axis can be written in terms of the space basis as

$$\sigma_2' = R_1 \sigma_2 \bar{R}_1. \tag{7.148}$$

This results in effectively reversing the product of the two rotations

$$R_{2'}R_1 = e^{-\sigma_2' \beta/2}e^{-\sigma_1 \alpha/2} = (R_1 R_2 \bar{R}_1)R_1 = R_1 R_2 = e^{-\sigma_1 \alpha/2}e^{-\sigma_2 \beta/2}. \tag{7.149}$$

Likewise,

$$\sigma_3'' = R_1 R_2 \sigma_3 \bar{R}_2 \bar{R}_1, \tag{7.150}$$

leading to the result

$$R_{3''} R_{2'} R_1 = e^{-\sigma_3'' \gamma/2} e^{-\sigma_2' \beta/2} e^{-\sigma_1 \alpha/2} \big|_{\text{body}}$$
$$= R_1 R_2 R_3 = e^{-\sigma_1 \alpha/2} e^{-\sigma_2 \beta/2} e^{-\sigma_3 \gamma/2} \big|_{\text{space}}. \tag{7.151}$$

7.4.3 Euler Angles

Euler's angles, taken about the z, x', z'' axes, in that order, are commonly used in physics textbooks. Euler's prescription simplifies the analysis of some common physical problems. Euler's method is not recommended for critical engineering applications, since if the second rotation angle is zero, the first and third rotation axes become degenerate. Euler's convention is most useful for theoretical problems where the body frame has a symmetry axis, taken to be the body z-axis; then the kinetic energy of a rigid body becomes cyclic in two of the three generalized momenta.

Like the pitch, roll, and yaw angles, a number of conflicting conventions for Euler's angles can be found in the literature. Figure 7.11 shows the order of operations as defined by Goldstein's Classical Mechanics textbook [Goldstein 2002], where the second rotation, about the *line of nodes*, is taken to be about the $\hat{1}'$- (x'-) axis. In other standard textbooks, as, for example, Sakurai's Quantum Mechanics textbook [Sakurai 1967], the line of nodes is often defined to be along the $\hat{2}'$- (y'-) axis instead.

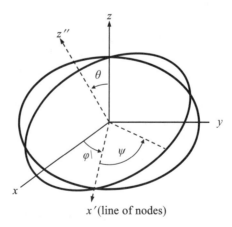

FIGURE 7.11 Definition of the Euler angles.

This text follows the notation used by Goldstein, which is also the convention used by many standard Mathematica and Matlab algorithms. Rotations of the body frame are defined to have a counterclockwise sense, with the rotations carried out in the following order

1. First, make a rotation by an angle ϕ about the initial z-axis.
2. Then, make a second rotation by an angle θ about the body x'-axis, called the line of nodes.
3. Finally, make a third rotation by an angle ψ about the new (and final) body z-axis.

This prescription gives a net rotation given by

$$R_{\text{Euler}}(\phi, \theta, \psi) = e^{-\sigma_3'' \psi/2} e^{-\sigma_1' \theta/2} e^{-\sigma_3 \phi/2}|_{\text{body}}$$
$$= e^{-\sigma_3 \phi/2} e^{-\sigma_1 \theta/2} e^{-\sigma_3 \psi/2}|_{\text{space}}, \tag{7.152}$$

or

$$\mathbf{A}_{\text{Euler}}(\phi, \theta, \psi) = \mathbf{A}_{3''}(\psi) \cdot \mathbf{A}_{1'}(\theta) \cdot \mathbf{A}_3(\phi)|_{\text{body}}$$
$$= \mathbf{A}_3(\phi) \cdot \mathbf{A}_1(\theta) \cdot \mathbf{A}_3(\psi)|_{\text{space}}. \tag{7.153}$$

The solution for the Euler rotation matrices can be found by making the explicit substitutions

$$A_3(\phi) = \begin{bmatrix} \cos\phi & \sin\phi & 0 \\ -\sin\phi & \cos\phi & 0 \\ 0 & 0 & 1 \end{bmatrix},$$

$$A_{1'}(\theta) = \begin{bmatrix} 1 & 0 & 0 \\ 0 & \cos\theta & \sin\theta \\ 0 & -\sin\theta & \cos\theta \end{bmatrix}, \tag{7.154}$$

$$A_{3''}(\psi) = \begin{bmatrix} \cos\psi & \sin\psi & 0 \\ -\sin\psi & \cos\psi & 0 \\ 0 & 0 & 1 \end{bmatrix},$$

The net result in the body frame is

$$A_{\text{body}} = \begin{bmatrix} \cos\psi\cos\phi - \cos\theta\sin\phi\sin\psi & \cos\psi\sin\phi + \cos\theta\cos\phi\sin\psi & \sin\theta\sin\psi \\ -\sin\psi\cos\phi - \cos\theta\sin\phi\cos\psi & -\sin\psi\sin\phi + \cos\theta\cos\phi\cos\psi & \sin\theta\sin\psi \\ \sin\theta\sin\phi & -\sin\theta\cos\phi & \cos\theta \end{bmatrix}.$$
$$\tag{7.155}$$

7.5 KINEMATICS OF ROTATING BODIES

The description of motion in a rotating frame of reference is complicated by the time dependence of the basis. Consider a vector $\mathbf{a}$ instantaneously described in two frames: a space frame s and a rotating body frame b

$$\mathbf{a} = a^i \mathbf{e}_{si} = a'^i \mathbf{e}'_{bi}. \tag{7.156}$$

From Equation (7.90), the rotating basis has a time-dependent behavior $\dot{\mathbf{e}}'_i = \boldsymbol{\omega} \times \mathbf{e}'_i = -\boldsymbol{\Omega} \cdot \mathbf{e}'_i$. The time derivative of the vector must include this contribution. Therefore, the time rate of change of a vector $\mathbf{a}$ in the two frames can be written as

$$\frac{d\mathbf{a}}{dt} = \frac{da^{si}}{dt}\mathbf{e}_{si} = \frac{da^{bi}}{dt}\mathbf{e}_{bi} + a^{bi}\frac{d\mathbf{e}_{bi}}{dt} = \frac{da^{bi}}{dt}\mathbf{e}_{bi} + \boldsymbol{\omega} \times \mathbf{a}. \tag{7.157}$$

or

$$\left.\frac{d\mathbf{a}}{dt}\right|_s = \left.\frac{d\mathbf{a}}{dt}\right|_b + \boldsymbol{\omega} \times \mathbf{a} = \left.\frac{d\mathbf{a}}{dt}\right|_b - \boldsymbol{\Omega} \cdot \mathbf{a}. \tag{7.158}$$

Equation (7.158) can be written in operator form as

$$\left.\frac{d}{dt}\right|_s = \left.\frac{d}{dt}\right|_b + \boldsymbol{\omega} \times = \left.\frac{d}{dt}\right|_b - \boldsymbol{\Omega}\cdot, \tag{7.159}$$

Equation (7.158) can be applied to the angular velocity vector itself giving

$$\left.\frac{d\boldsymbol{\omega}}{dt}\right|_s = \left.\frac{d\boldsymbol{\omega}}{dt}\right|_b. \tag{7.160}$$

Angular velocities involve infinitesimal rotations, so angular velocities are commutative. Contributions from rotations about all axes of rotation are additive. In terms of Euler angles, this gives an instantaneous angular velocity vector

$$\boldsymbol{\omega} = \mathbf{e}_3\dot{\phi} + \mathbf{e}_{1'}\dot{\theta} + \mathbf{e}_{3''}\dot{\psi}. \tag{7.161}$$

One can compute the components of ω by projecting components onto either the space or body frame. The body frame expression is generally more useful and will be evaluated here. The space frame representation is left as an exercise.

Note that the final rotation axis is already in the body frame

$$\mathbf{e}_{3''} = \mathbf{e}_{b3}. \tag{7.162}$$

Moreover, the projection of the line of nodes only requires evaluation in terms of the last transformation to express it in the body frame

$$\mathbf{e}_1' = \mathbf{e}_{b1} \cos \psi - \mathbf{e}_{b2} \sin \psi. \tag{7.163}$$

The first rotation is taken with respect to the space frame and requires the full transformation of all Euler angles to put it in the body frame

$$\mathbf{e}_3 = \mathbf{e}_{s3} = \mathbf{e}_{b1} \sin \theta \sin \psi + \mathbf{e}_{b2} \sin \theta \cos \psi + \mathbf{e}_{b3} \cos \theta. \tag{7.164}$$

Adding all contributions, the components of angular velocity with respect to the body frame are

$$\omega_{b1} = \mathbf{e}_{b1} \cdot \mathbf{e}_{s3} \dot{\phi} + \mathbf{e}_{b1} \cdot \mathbf{e}_{1'} \dot{\theta} = \dot{\phi} \sin \theta \sin \psi + \dot{\theta} \cos \psi,$$

$$\omega_{b2} = \mathbf{e}_{b2} \cdot \mathbf{e}_{s3} \dot{\phi} + \mathbf{e}_{b2} \cdot \mathbf{e}_{1'} \dot{\theta} = \dot{\phi} \sin \theta \cos \psi - \dot{\theta} \sin \psi, \tag{7.165}$$

$$\omega_{b3} = \mathbf{e}_{b3} \cdot \mathbf{e}_{s3} \dot{\phi} + \mathbf{e}_{b3} \cdot \mathbf{e}_{1'} \dot{\theta} + \mathbf{e}_{b3} \cdot \mathbf{e}_{b3} \dot{\psi} = \dot{\phi} \cos \theta + \dot{\psi},$$

or

$$\boldsymbol{\omega}|_b = (\mathbf{n_z} \dot{\phi} + \mathbf{n}_{x'} \dot{\theta} + \mathbf{n}_{z'} \dot{\psi})|_b = \begin{bmatrix} \dot{\phi} \sin \theta \sin \psi + \dot{\theta} \cos \psi \\ \dot{\phi} \sin \theta \cos \psi - \dot{\theta} \sin \psi \\ \dot{\phi} \cos \theta + \dot{\psi} \end{bmatrix}_b. \tag{7.166}$$

7.5.1 Newton's Laws in a Rotating Frame of Reference

A rotating frame is not an inertial frame, thereby requiring a correction to Newton's laws of motion. The time rate of change of the basis vectors in the body frame, given by Equation (7.90), must be taken into account. If one observes a particle from the viewpoint of the rotating frame, Equation (7.159) implies that the instantaneous velocity of the particle is given by

$$\mathbf{v}|_s = \mathbf{v}|_b + \boldsymbol{\omega} \times \mathbf{r} \tag{7.167}$$

Likewise, the acceleration of the particle is given by

$$\mathbf{a}_s = \left. \frac{d\mathbf{v}_s}{dt} \right|_b + \boldsymbol{\omega} \times \mathbf{v}_s = \mathbf{a}_b + 2\boldsymbol{\omega} \times \mathbf{v}_b + \boldsymbol{\omega} \times \boldsymbol{\omega} \times \mathbf{r} \tag{7.168}$$

But Newton's Second Law is valid in an inertial space frame, giving

$$\mathbf{F} = m\mathbf{a}_s = m(\mathbf{a}_b + 2\boldsymbol{\omega} \times \mathbf{v}_b + \boldsymbol{\omega} \times \boldsymbol{\omega} \times \mathbf{r}). \tag{7.169}$$

As observed in the rotating frame of the Earth, this result in a fictitious contribution to the apparent or effective force[8]

$$\mathbf{F}_{\text{eff}} = m\mathbf{a}_b = \mathbf{F} - m(2\boldsymbol{\omega} \times \mathbf{v}_b + \boldsymbol{\omega} \times \boldsymbol{\omega} \times \mathbf{r}), \tag{7.170}$$

---

[8]Derived by Gaspard-Gustave Coriolis (1792–1843) in 1835.

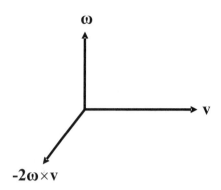

FIGURE 7.12   The direction of the fictitious Coriolis force in the rotating frame.

where $\mathbf{F}_{\text{eff}}$ is the apparent force in the accelerating frame, $\mathbf{F}$ is the physically applied force defined in an inertial frame, $-2m\,\boldsymbol{\omega} \times \mathbf{v}_b$ is the Coriolis force, and $-m\,\boldsymbol{\omega} \times \boldsymbol{\omega} \times \mathbf{r}$ is the centrifugal force. Figure 7.12 shows the direction of the Coriolis force in the rotating frame.

7.5.2 Equivalence of Gravitational and Inertial Mass

The angle made by a plumb bob, defining the local direction of the vertical, is a curious balance of centripetal acceleration, which depends on the inertial mass of an object, and the gravitational attraction of the Earth, which depends on the gravitational mass. The inertial mass establishes a relationship between the net force and acceleration, while the gravitational mass establishes a relationship to a specific gravitational force. Although Newton treated both masses as being equivalent, there is no a priori reason that they have to be. In fact this is the point of the Einstein's *Equivalence Principle* [Einstein 1915]. From Einstein's point of view, there is no gravitational mass. Inertial mass is all that exists.

Figure 7.13 shows a schematic of a plumb bob suspended above the surface of the Earth. The angle of the vertical with respect to the direction of the Earth's center is a function of the latitude. Assuming that the two masses are identical, the stationary plumb bob at the surface of the Earth will point in the direction of an effective gravitational field $\mathbf{g}_{\text{eff}}$.

$$m\mathbf{g}_{\text{eff}} = m\mathbf{g} - m\boldsymbol{\omega} \times (\boldsymbol{\omega} \times \mathbf{R}_{\text{E}}). \tag{7.171}$$

The biggest deviation is expected at 45° North or South Latitude. Consider the two extreme cases shown in the figure. In Case A, an object with gravitational mass, but no inertial mass, would point to the direct center of gravity of the Earth. Conversely, in Case B, an object with inertial mass, but

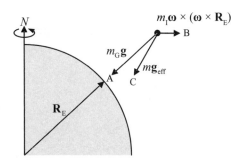

FIGURE 7.13 The angle of a plumb bob defining the direction of the vertical on the surface of the Earth is a function of the Latitude with the biggest effect expected at 45° North or South Latitude. The Eötvös experiment tested the angle of deflection for different materials to see if there was any difference between the inertial and gravitational masses.

no gravitational mass, would swing out to the perpendicular of the Earth's rotation axis. Case C shows, great exaggerated, the deflection expected if the two masses were the same.

The actual effect is small, about a tenth of a degree at most, but the relative deflection can be measured with great accuracy for different materials. Lóránd Eötvös (1848–1919) and his colleagues made a number of precision measurements from 1885 to 1909 at Budapest, Hungry (47°N Latitude). They found that all materials give the same deflection angle to an accuracy of 5×10^{-9}. Over the years a number of other tests of the equivalence principle have been made. None as yet have shown any discrepancies with the equivalence principle.

7.5.3 Coriolis Effect

The *Coriolis Effect* occurs when an object is moving in a rotating frame. It describes the apparent sideward deflection of particle in the rotating frame of reference due to its velocity. On the rotating Earth, this effect is responsible for the creation of cyclones in the southern hemisphere and anticyclones (or hurricanes) in the northern hemisphere. Hurricanes in the North Atlantic are strong storm systems with a pronounced rotational circulation and high sustained winds as shown in Figure 7.14. This satellite image of Hurricane Katrina was taken Aug. 28, 2005, when the storm was a Category 5 storm in the Gulf of Mexico. Hurricanes are formed about an eye when wind rushes to fill a low-pressure region caused by the rising of hot humid air. The circulation can build with time, creating potential devastating storms, provided that a sufficient heat source exists to sustain the low pressure region. The

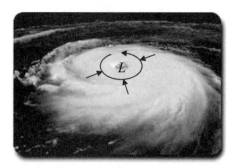

FIGURE 7.14   Hurricanes are formed in the North Atlantic Ocean about an "eye" as the result of a counterclockwise circulation of air about a low-pressure region due to the Coriolis Effect in the northern hemisphere. The arrows indicate pressure gradients. Shown is a Hurricane Katrina satellite image taken Aug. 28, 2005, when the storm was a Category 5 storm in the Gulf of Mexico (Satellite image courtesy of The National Oceanic and Atmospheric Administration, U.S. Department of Commerce[9]).

Coriolis Effect results in a counter-clockwise circulation of air when viewed from above the storm system in the northern hemisphere, and a clockwise circulation in the southern hemisphere.

7.5.4 Motion of a Falling Body on the Rotating Earth

When working with moving frames, one must account for the time dependence of the basis. This gives rise to fictional centrifugal and Coriolis forces. Ignoring air drag, the only physical force on a particle free falling in the atmosphere is that of gravity. The force equation can be written in the accelerating frame as

$$\mathbf{F}_{\text{eff}} = m\mathbf{a}_{\text{b}} = m\mathbf{g}_{\text{eff}} - 2m\boldsymbol{\omega} \times \mathbf{v}_{\text{b}}, \qquad (7.172)$$

where $\mathbf{g}_{\text{eff}}$ is the effective direction of the gravitational field corrected for the centripetal acceleration

$$\mathbf{g}_{\text{eff}} = \mathbf{g} - \boldsymbol{\omega} \times \boldsymbol{\omega} \times \mathbf{r} = -g\hat{\mathbf{r}} + \omega^2 \sin\theta\,\hat{\boldsymbol{\rho}}. \qquad (7.173)$$

The direction $\mathbf{g}_{\text{eff}}$ defines "vertically downward" relative to a suspended plumb bob.

The last term in Equation (7.173) is the centrifugal acceleration, which depends on the polar angle θ with respect of the Earth's rotation axis. The polar angle is also called the *co-latitude*. The rotational frequency of the Earth

---

[9]Source: *http://hurricanes.noaa.gov/*, downloaded June 2, 2007.

with respect to the fixed stars can be measured in terms of the ratio of the sidereal year to the calendar year, giving

$$\omega = \frac{2\pi}{1 \text{ day}} \left(\frac{366.25 \text{ days}}{365.25 \text{ days}} \right) = 7.292 \times 10^{-5} \text{ s}^{-1} \qquad (7.174)$$

At the Earth's surface, let R_E be the Earth's radius, then the centripetal acceleration is proportional to

$$\omega^2 R_E = 3.38 \text{ cm/s}^2 \simeq (0.3\%)g \qquad (7.175)$$

For example, consider dropping an object from rest at a height h above the surface at the equator, where $\mathbf{g}_{\text{eff}} \| \mathbf{g}$. The Coriolis force acts in the equatorial plane defined by the $\hat{x} \wedge \hat{z}$ plane with $\hat{z}$ being the vertical direction, $\hat{x}$ pointing to the east and $\hat{y}$ to the north. The equation of motion becomes

$$\mathbf{F}_{\text{eff}} = m\mathbf{a}_{\text{eff}} = m\mathbf{g}_{\text{eff}} - 2m\boldsymbol{\omega} \times \mathbf{v}_b, \qquad (7.176)$$

with initial conditions

$$x_0 = y_0 = 0, \quad z_0 = h, \quad \dot{x}_0 = \dot{y}_0 = \dot{z}_0 = 0. \qquad (7.177)$$

Make some approximations, letting $h \ll R_E$, and $g_{\text{eff}}(r) \simeq g = 9.8 \text{ m/s}^2$. Keeping only leading order contributions, and assuming the displacements are small compared to the Earth's radius, the equations of motion become

$$\ddot{y} = 0, \quad m\ddot{x} = -2m\omega_y \dot{z}, \quad m\ddot{z} = -mg_{\text{eff}} + 2m\omega_y \dot{x} \simeq -mg. \qquad (7.178)$$

In the vertical $\hat{z}$ direction, one gets approximately uniform acceleration. The time to fall a distance h is therefore $\sqrt{2h/g}$ and $\dot{z} = -gt$. Plugging this result into the equation for the x-component of acceleration gives

$$\ddot{x} = 2\omega_y \dot{z} = 2\omega_y gt, \qquad (7.179)$$

With solution

$$x = \omega_y gt^3/3 = \omega_y g(2h/g)^{3/2}/3 = 2\omega_y h(2h/g)^{1/2}/3 \qquad (7.180)$$

The falling particle deflects to the east. If one puts in some representative numbers, for example, dropping a particle from a height of 1 km gives a net deflection of only 69 cm, which is negligible compared to wind resistance effects.

The Coriolis Effect is usually small except in instances where one is dealing with large velocities (such as would be case in rocket motion), and/or where one is integrating over large distances (as in transcontinental travel) or times scales long compared to units of days (as would be the case for seasonal hurricane build up off of the east coast of Africa).

7.5.5 Foucault Pendulum

The Foucault pendulum, invented by engineer Jean Bernard Leon Foucault (1819–1876)[10], provides a classic demonstration of the proper rotational motion of the Earth about its polar axis. A Foucault pendulum is essentially a long pendulum that oscillates for a sufficient period of time for its rotation with respect to the Earth to become readily apparent. In most modern exhibits, frictional damping of the pendulum is compensated for by using an electromagnet to provide a small radial kick to the pendulum at maximum amplitude.

At the latitude of Paris, such a pendulum should precess by about 270° in 24 hours. At the third public demonstration of his device in 1851 at the Pantheon in Paris, Foucault created a public spectacle by using a 67 meter plumb bob. A working model of his pendulum can still be seen on display there, and similar devices are exhibited in many other museums worldwide.

Consider a Foucault pendulum of length l, initially set into motion in the $\hat{x} \wedge \hat{y}$ plane of the laboratory frame, where $\hat{z}$ is vertically upward as measured by a plumb bob. Denote the tension in the string by $\mathbf{T}$. Then, by Newton's equations of motion in a rotating frame, one gets

$$m\mathbf{a}_b = \mathbf{T} + m\mathbf{g}_{\text{eff}} - 2m\boldsymbol{\omega} \times \mathbf{v}_b. \tag{7.181}$$

With respect to this frame, $\omega_z = -\boldsymbol{\omega} \cdot \hat{\mathbf{g}}_{\text{eff}} \simeq \omega_E \cos\theta$, where θ is the polar angle, i.e., the colatitude. For a long pendulum, the small amplitude approximation for the motion can be made. Then, the bob of the pendulum moves in a horizontal plane $\hat{x} \wedge \hat{y}$, with $\hat{x}$ pointing to the east and $\hat{y}$ pointing to the north. The pendulum's oscillation frequency in the small amplitude approximation is $\omega_1 = \sqrt{g/l}$. The approximate planar equations of motion, including the Coriolis term, are given by

$$\ddot{x} = -\omega_1^2 x + 2\omega_2\dot{y}, \quad \ddot{y} = -\omega_1^2 y - 2\omega_2\dot{x}, \tag{7.182}$$

where $\omega_2 = \omega_E \cos\theta$ is the projection of the Earth's angular velocity on the pendulum's vertical axis. Let

$$z = x + iy, \tag{7.183}$$

---

[10]Jean Bernard Léon Foucault (1819–1868) was a physicist and inventor who showed that light travels faster in air than in water, and that its speed in a medium varies with the inverse of its refractive index. He measured the speed of light to better than 1% accuracy. Foucault also invented the Foucault pendulum and the gyroscope.

multiply the second equation in expression (7.182) by i, and add it to the first equation. This gives the complex equation

$$\ddot{z} + 2i\omega_2\dot{z} + \omega_1^2 z = 0. \tag{7.184}$$

If one tries solutions of form $z = e^{iut}$, one gets the quadratic solution

$$-u^2 - 2uw_2 + w_i^2 = 0, \tag{7.185}$$

where

$$u = -\omega_2 \pm \sqrt{\omega_1^2 + \omega_2^2} = -\omega_2 \pm \omega'. \tag{7.186}$$

The time-dependent solution for the amplitude of the swing is

$$x(t) = x_0 \cos\omega't \cos\omega_2 t, \quad y(t) = x_0 \cos\omega't \sin\omega_2 t, \tag{7.187}$$

where $\omega' = \sqrt{g/l + w_2^2}$ is the shifted oscillator frequency and $\omega_2 = \omega_E \cos\theta$ is the precession frequency. The precession frequency is one revolution per day at the poles. The rotation period increases to two days at 30° north latitude (60° polar angle). There is no precession at the equator. Note that the plane of the pendulum's motion is an inertial frame only at the poles or when the pendulum is set swinging east-west at the equator.

Figure 7.15 shows the precession of a Foucault pendulum in the horizontal plane of the Earth at 30° north latitude. The motion has been greatly

FIGURE 7.15 Precession of a Foucault pendulum in the horizontal plane of the Earth at a colatitude of 60°. The rotational period of the Earth has been increased to 1 hr and the length of the pendulum has been greatly exaggerated (90 km) to limit the number of pendulum swings per precession cycle.

exaggerated by increasing the rotational period of the Earth to one hour and the length of the pendulum to 90 km, thereby reducing the number of pendulum swings per precession cycle.

7.6 SUMMARY

A geometric algebra $GA(n)$ is a real Clifford algebra with n spacetime dimensions satisfying the anticommutator relations

$$\{\lambda_i, \lambda_j\} = 2g_{ij}, \tag{7.188}$$

where g_{ij} is the metric tensor of the space. A complete geometric algebra is generated from a vector basis by the process of extension. This generates a finite 2^n dimensional algebra. $Spin(n)$ is defined as the spin group formed from the Lie algebra of bivector surface elements $\sigma_i^j = e_i \wedge e^j$ in $GA(n)$ where $\bar{S}S = 1$ with $\det(S_j^i) = +1$ for all $S \in Spin(n)$.

$GA(3)$ with a constant diagonal metric $g_{ij} = dia(1, 1, 1)$ represents the geometric algebra of Euclidean space when using Cartesian coordinates. Its spin group of rotors $R \in Spin(3)$ is isomorphic to the $SU(2)$ group. Any rotor can be written as a quaternion rotation element of the form

$$R(n, \theta) = e^{\hat{\sigma}\theta/2} = e^{-i\hat{n}\theta/2}, \tag{7.189}$$

where the axial vector $\hat{n}$ is the axis of rotation, $i = e_1 e_2 e_3$ is the pseudoscalar element, and θ is the rotation angle. An infinitesimal rotation of a vector about a polar axis can be written as

$$d\mathbf{r} = d\theta\, \hat{\mathbf{n}} \times \mathbf{r}, \tag{7.190}$$

The connection between $Spin(3)$ and $SO(3)$ the two groups is given by the mapping

$$A^i_j = \langle e^i R e_j \bar{R} \rangle. \tag{7.191}$$

which has properties $\mathbf{A} \cdot \bar{\mathbf{A}} = 1$ and $\det(\mathbf{A}) = +1$. All elements $\mathbf{A} \in SO(3)$ are rotations with polar axis $\hat{\mathbf{n}}$ given by the eigenvalue equation

$$\mathbf{A}_{\hat{n}}(\theta) \cdot \hat{\mathbf{n}} = \hat{\mathbf{n}}, \tag{7.192}$$

where the rotation angle is given by

$$\text{Trace}(\mathbf{A}) = 2\cos\theta + 1. \tag{7.193}$$

Orthogonal transformations preserve the inner product of vectors. Euler's theorem states that the general motion of a rigid body with one point fixed is a rotation about some axis.

The continuous group $SO(3)$ is a three parameter group. A rotation of the body frame basis about an arbitrary axis can be written in terms of the three Euler angles (ϕ, θ, ψ).

$$R_{\text{Euler}}(\phi, \theta, \psi) = e^{-\sigma_3'' \psi/2} e^{-\sigma_1' \theta/2} e^{-\sigma_3 \phi/2}|_b = e^{-\sigma_3 \phi/2} e^{-\sigma_1 \theta/2} e^{-\sigma_3 \psi/2}|_s, \quad (7.194)$$

$$\mathbf{A}_{\text{Euler}}(\phi, \theta, \psi) = \mathbf{A}_{3''}(\psi) \cdot \mathbf{A}_{1'}(\theta) \cdot \mathbf{A}_3(\phi)|_b = \mathbf{A}_3(\phi) \cdot \mathbf{A}_1(\theta) \cdot \mathbf{A}_3(\psi)|_s. \quad (7.195)$$

Infinitesimal rotations commute, therefore angular velocities are additive. With respect to Euler's angles, the instantaneous axis of rotation is given by

$$\boldsymbol{\omega} = \mathbf{e}_3 \dot{\phi} + \mathbf{e}_{1'} \dot{\theta} + \mathbf{e}_{3''} \dot{\psi}. \quad (7.196)$$

The time rate of change of a vector in a rotating body frame can be written in terms of the inertial space frame as

$$\left. \frac{d\mathbf{a}}{dt} \right|_s = \left. \frac{d\mathbf{a}}{dt} \right|_b + \boldsymbol{\omega} \times \mathbf{a}. \quad (7.197)$$

Newton's Law of motion in a rotating coordinate system is given by

$$\mathbf{F}_{\text{eff}} = m\mathbf{a}_b = \mathbf{F} - m(2\boldsymbol{\omega} \times \mathbf{v}_b + \boldsymbol{\omega} \times \boldsymbol{\omega} \times \mathbf{r}) \quad (7.198)$$

where $\mathbf{F}_{\text{eff}}$ is the apparent force in the accelerating frame, $\mathbf{F}$ is the physical force defined in an inertial frame, $-2m\boldsymbol{\omega} \times \mathbf{v}_b$ is the Coriolis force, and $-m\boldsymbol{\omega} \times \boldsymbol{\omega} \times \mathbf{r}$ is centrifugal force.

7.7   EXERCISES

1. Find a real two-dimensional matrix representation

$$\{\tau_i, \tau_j\} = 2g_{ij},$$

of a geometric algebra with two space dimensions $GA(2,0)$, having the diagonal metric $g_{ij} = \text{dia}[1, 1]$. Find the normalized bivector element $i = i_2 = \tau_2 \wedge \tau_1$ for this algebra and show that the spin group of unimodular matrices generated is isomorphic to the group of complex numbers with unit norm. Note that $Spin(2,0) = SU(1) = U(1)$

defines a one-sphere (a one-parameter sphere of unit radius). Show that the spin group can be parameterized as $e^{i\theta/2} \in Spin(2,0)$ and that the special orthogonal group has the parameterization $e^{i\theta} \in SO(2,0)$, where θ is a counterclockwise rotation angle.

2. The special unitary group $SU(2)$ is defined as the three parameter group of complex 2 by 2 square matrices of the form

$$U = \begin{bmatrix} \alpha & \beta \\ -\beta^* & \alpha^* \end{bmatrix},$$

subject to the constraint

$$\alpha\alpha^* + \beta\beta^* = 1.$$

The $SU(2)$ group is unitary $U^\dagger U = 1$, and has determinant $\det(U) = +1$. Show that $SU(2)$ is isomorphic to the multiplicative group of quaternions $sp(1)$ with unit norm

$$q = q_0 + q_1\sigma^1 + q_2\sigma^2 + q_3\sigma^3 \ni \mathbb{H},$$

where

$$\bar{q} = q_0 - q_1\sigma^1 - q_2\sigma^2 - q_3\sigma^3.$$

and

$$\bar{q}q = q_0^2 + q_1^2 + q_2^2 + q_3^2 = 1$$

Note that $SU(2) = sp(1) = Spin(3)$ defines a *three-sphere* (a three-parameter sphere of unit radius).

3. Prove the following vector identities in $GA(3)$

$$a \wedge (i_3 b) = i_3(a \cdot b),$$
$$(a \wedge b) = -i_3(a \cdot (i_3 b)) = i_3(b \cdot (i_3 a)) = i_3(a \times b),$$
$$a \cdot (b \wedge c) = (a \cdot b)c - b(a \cdot c) = -a \times (b \times c),$$
$$(a \wedge b \wedge c) = \frac{1}{2}(abc - cba) = i_3 a \cdot (b \times c).$$

4. Prove that

$$(a \times b)^2 = -(a \wedge b)^2 = |a \wedge b|^2 = a^2 b^2 - (a \cdot b)^2 = (ab \sin\theta)^2.$$

5. Show that $GA(1,3)$—the four-dimensional spacetime algebra given by changing the overall sign of the metric of $GA(3,1)$—can be

represented by the two-dimensional square quaternion matrices of the form

$$\gamma_0 = \begin{bmatrix} 1 & 0 \\ 0 & -1 \end{bmatrix}, \quad \gamma_i = \begin{bmatrix} 0 & \sigma_i \\ \sigma_i & 0 \end{bmatrix},$$
$$\gamma_0^2 = -\gamma_i^2 = +1, \quad i = 1, 2, 3.$$

where the quaternion basis elements satisfy the conditions $\sigma_i^2 = -1$ and $\sigma_1 \sigma_2 \sigma_3 = -1$. Find the matrix representation for the four-dimensional pseudoscalar element $e_0 e_1 e_2 e_3$ in this representation. Show that this element is self-adjoint under reversion. This is the minimal matrix representation of $GA(1,3)$.

6. Show that the four-dimensional Euclidean space $GA(4,0)$ has the minimal quaternion representation

$$e_4 = \begin{bmatrix} 0 & 1 \\ 1 & 0 \end{bmatrix}, \quad e_i = \begin{bmatrix} 0 & \sigma_i \\ -\sigma_i & 0 \end{bmatrix}, \quad i = 1, 2, 3.$$

$\sigma_1 \sigma_2 \sigma_3 = -1$ and $\sigma_i^2 = -1$. Show that the Lie spin algebra $Spin(4,0)$ for this space is given by

$$[\sigma_i, \sigma_j] = 2\varepsilon_{ijk} \sigma_k,$$
$$[\sigma_i, d_j] = 2\varepsilon_{ijk} d_k,$$
$$[d_i, d_j] = 2\varepsilon_{ijk} \sigma_k,$$

where

$$d_i = e_5 \sigma_i,$$
$$e_5 = e_1 e_2 e_3 e_4 = \begin{bmatrix} 1 & 0 \\ 0 & -1 \end{bmatrix}.$$

7. How many linearly independent bivectors $\sigma^i{}_j = e^i \wedge e_j$ are there in an n-dimensional geometric space? Explicitly demonstrate that the set of bivectors form a Lie algebra given by

$$[\sigma^i{}_k, \sigma^k{}_j] = 2\sigma^i{}_j, \quad \text{(nsc)}$$

where the summation convention is not used for the intermediate index k.

8. Show that the nonrelativistic equations of motion of a charged particle with charge q and mass m in an Electric field $E = E^i e_i$ and magnetic field $B = B^i n_i$ can be written in $GA(3)$ notation as

$$m\frac{dv}{dt} = qE + qF_{\text{B}} \cdot v = qE + \frac{q}{2}[F_{\text{B}}, v],$$

where
$$F_{\text{B}ij} = \varepsilon_{ijk}B^k.$$

9. A matrix transforms as $M' = SMS^{-1}$ under a similarity transform:
   a. Show that the trace of a matrix is preserved under a similarity transformation.
   b. Show the antisymmetric or symmetric property of a matrix is preserved under an orthogonal similarity transform $S^{-1} = \tilde{S}$.

10. Show that three sequential $180°$ rotations about three mutually perpendicular space axis results in the identity transformation.

11. Show that a general $180°$ rotation about any axis can be written as $\mathbf{A}_{\hat{n}}(\pi) = (2\hat{\mathbf{n}}\,\hat{\mathbf{n}} - 1)$. Show also that $\mathbf{P}_{\pm} = \frac{1}{2}(1 \pm \mathbf{A})$ are projection operators satisfying $\mathbf{P}_{\pm}^2 = \mathbf{P}_{\pm}$, where $\mathbf{P}_{\pm}$ project the parallel and perpendicular components of a vector respectively.

12. Show that a rotation about an arbitrary axis by an angle θ can be represented as the product of two mirror reflections, where the rotation axis is given by the common line joining the two reflection planes, and where the angle between the two planes is $\theta/2$. Hint: use $\mathbf{M}_{\hat{n}} = 1 - 2\hat{\mathbf{n}}\,\hat{\mathbf{n}}$ for a general reflection about an axis $\hat{\mathbf{n}}$, and the following properties of a rotation about an axis $\hat{\mathbf{n}}$

$$\mathbf{A}_{\hat{n}}(\theta) \cdot \hat{\mathbf{n}} = \hat{\mathbf{n}},$$
$$\text{Trace}(\mathbf{A}_{\hat{n}}(\theta)) = 2\cos\theta + 1.$$

13. Demonstrate explicitly that the infinitesimal generators of orthogonal rotations satisfy the Lie Algebra formula given by

$$[\mathfrak{M}_i, \mathfrak{M}_j] = \varepsilon_{ij}{}^k \mathfrak{M}_k.$$

14. Use Euler's angles to calculate the angular velocity of a rotating rigid body in the space frame, showing that

$$\boldsymbol{\omega}|_s = (\mathbf{n}_z\dot{\phi} + \mathbf{n}_{x'}\dot{\theta} + \mathbf{n}_{z'}\dot{\psi})|_s = \begin{bmatrix} \dot{\theta}\cos\phi + \dot{\psi}\sin\theta\sin\phi \\ \dot{\theta}\sin\phi - \dot{\psi}\sin\theta\cos\phi \\ \dot{\psi}\cos\theta + \dot{\phi} \end{bmatrix}_s.$$

15. A bug holding on with a frictional force of magnitude $\leq \mu mg$ on a horizontally rotating phonograph record rotating at 33 1/3 rpm is observed to be crawling radially outward with a constant velocity of 3 cm/s in the rotating frame of the record. For what value of the coefficient of friction μ does the bug begin to slip just as it reaches the outer radius of 15 cm? Assume $g = 9.8$ m/s^2.

16. Calculate the horizontal deflection due to the Coriolis force at colatitude θ for a particle that is thrown vertical upwards at the surface of the Earth reaching a height $h \ll R_E$. Show that the answer is four times the value, with the opposite sign, than for the case where the particle is dropped from rest from the same height h at the same colatitude.

17. A spherical ball rolls without slipping on a flat horizontal plane. Express the constraint conditions in terms of the Euler angles and show that they are non-holonomic and therefore cannot be integrated.

18. Generate a plot of the motion in the horizontal $(\hat{x}, \hat{y})$ plane over two planetary-days for a long Foucault pendulum located at 30° North latitude on a planet with a fast rotational period of 1 hr. Assume that the planet has no atmosphere and that dissipation effects can be ignored. The pendulum is initially set swinging from rest with an amplitude of 10 m in the east direction, the length of the pendulum is 1 km, and the effective gravitation acceleration at the site is 1 m/s². The small amplitude approximation may be used. How many swings (cycles) does the pendulum go through?

8

RIGID BODY
MOTION

I n this chapter, the general motion of a rigid body is discussed. The rotational properties of a rigid body are defined in terms of its moment of inertia tensor. Euler's equations of rotational motion are derived. Lagrangian and Hamilton formulations of the rotating rigid body problem are developed. A number of examples are explored.

8.1 GENERAL MOTION OF A RIGID BODY

Charles's Theorem[1] states that the general displacement of a rigid body consists of a translation of a point plus a rotation of the body about that point. Figure 8.1 provides an illustration of Charles Theorem. The motion of a fixed wing aircraft consists of a translation of its center of mass point in space plus a rotation of its body frame about that point. The wings and fuselage of the plane delineate two of its three principal axes, with the third perpendicular to the first two. The translational motion of the rigid airframe is just the translation of the center of mass point under the net applied external force. The rotational motion of the airplane is given by the rotation of its body frame relative to a space frame of fixed orientation attached to the center of mass point.

In principle, one can pick any point in the body to be the origin of the body frame. However, if there is no fixed pivot, the center of mass point is a particularly good choice for the pivot, because the motion separates in this coordinate system. From Chapter 1's review of Newtonian dynamics, Newton's laws of motion for the center of mass point $\mathbf{r}_{cm}$ is

$$\mathbf{F}_a = m\ddot{\mathbf{r}}_{cm}, \tag{8.1}$$

where $\mathbf{F}_a$ is the net externally applied force, m is the total mass of the rigid body, and $\ddot{\mathbf{r}}_{cm}$ is the acceleration of the center of mass.

---

[1] Attributed to Michel Charles (1793–1880). A proof of Charles theorem is given in [Whittaker 1988].

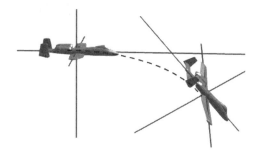

FIGURE 8.1   An illustration of Charles Theorem: the motion of a fixed wing aircraft consists of a translation of its center of mass point plus a rotation of its body frame. The wings and fuselage of the plane delineate two of its three principal axes, with the third perpendicular to the first two.

The kinetic energy and angular momentum of a system of particles separate in the center of mass frame. From the Section 1.4.6 (center of mass coordinates) one gets

$$T = T_{cm} + T' = \frac{1}{2} \mathbf{v}_{cm} \cdot \mathbf{p} + \sum_i \frac{1}{2} \mathbf{v}'_i \cdot \mathbf{p}'_i, \tag{8.2}$$

$$\mathbf{L} = \mathbf{L}_{cm} + \mathbf{L}' = \mathbf{r}_{cm} \times \mathbf{p} + \sum_i \mathbf{r}'_1 \times \mathbf{p}'_i. \tag{8.3}$$

where $\mathbf{r}_{cm}$ and $\mathbf{v}_{cm}$ are the position and velocity of the center of mass, respectively, $\mathbf{p}$ is the total momentum of the system, and $(\mathbf{r}'_1, \mathbf{v}'_i, \mathbf{p}'_i)$ are the positions, velocities, and momenta of the constituent points of the body in the center of mass frame. In a rigid body, the motion of its constituent mass points is severely constrained by the requirement that all points in the body maintain constant relative orientations and displacements.

If there is a fixed point of rotation, forces of constraint act to hold the pivot point constant. In this case, the motion of a rigid body is purely rotational when analyzed about the fixed point. Euler's Theorem, derived in Chapter 7, states that the general displacement of a rigid body with one point fixed is a rotation about some axis.

Because rotations don't commute in general, rotational motion can be surprisingly complex and sometimes counterintuitive. For example, the spinning bicycle wheel held by one end of its axle shown in Figure 8.2 sees an unbalanced torque about its pivot point. If released from rest, the wheel would simply fall down under the influence of gravity. However, if the wheel is set spinning about its axis fast enough, it appears to defy gravity and instead precesses about the vertical axis of the pivot point.

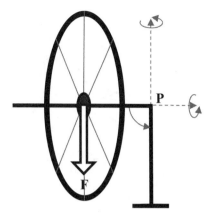

FIGURE 8.2 The rapidly spinning bicycle wheel, with one end of its axle balanced on a pivot point, appears to defy gravity by precessing about the vertical axis of the pivot rather than falling under gravity.

Since the translational motion of a rigid body is relatively trivial, this chapter concentrates on developing the description of the rotational motion of a rigid body about a fixed point. Consider two Cartesian coordinate systems both centered at the same origin P, rotating relative to each other as shown in Figure 8.3. One coordinate frame is fixed with respect to an inertial space frame s described by a constant tensor basis $\{e_1, e_2, e_3\}$ and the other is a rotating frame b fixed with respect to the body, having a rotating basis given by $\{e'_1, e'_2, e'_3\}$. The body frame basis is time dependent. The most common convention for the Euler angles treats rotation angles as positive if the body frame vectors are rotating in a counterclockwise sense. The motion of a point $\mathbf{r}'_i$ in the two frames is related by

$$\left.\frac{d\mathbf{r}'_i}{dt}\right|_s = \left.\frac{d\mathbf{r}'_i}{dt}\right|_b + \boldsymbol{\omega} \times \mathbf{r}'_i, \tag{8.4}$$

The body derivative, given by

$$\left.\frac{d\mathbf{r}'_i}{dt}\right|_b = \frac{dx_i^{j'}}{dt} e'_j, \tag{8.5}$$

is incomplete, since it ignores the variation of the rotating basis, which is accounted for in the $\boldsymbol{\omega} \times \mathbf{r}'_i$ term in Equation (8.4). All points on the rigid body are fixed in the body frame; therefore, the body derivative vanishes

$$\left.\frac{d\mathbf{r}'_i}{dt}\right|_b = 0. \tag{8.6}$$

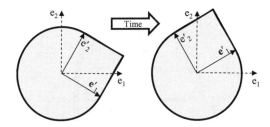

FIGURE 8.3   Rotation of a rigid body: the e_i space basis (dashed lines) is fixed in space, while the e_i' body basis (solid lines) is fixed in the rotating body. Points embedded in the rigid body are at rest with respect to the rotating body frame.

It follows that the equations of motion for mass points fixed in the rotating frame of a rigid body satisfy

$$\left.\frac{dr_i'}{dt}\right|_s = \boldsymbol{\omega} \times \mathbf{r}_i'. \tag{8.7}$$

This is the starting point for any in depth analysis of rotating rigid bodies. Figure 8.3 shows a possible evolution in time of a body frame relative to a space frame. At any given time, the rigid body has an angular velocity axial-vector $\boldsymbol{\omega}$ pointed along its instantaneous axis of rotation.

8.2 ROTATIONAL KINETIC ENERGY OF A RIGID BODY

If one point in the rigid body is fixed, the kinetic energy is due solely to the kinetic energy of rotation about this pivot point. Otherwise, the center of mass point $\mathbf{R}$ can be used as the effective pivot point. The kinetic energy separates in the center of mass frame giving

$$T = \frac{1}{2}M\dot{\mathbf{R}}^2 + \frac{1}{2}\sum_i m_i(\mathbf{v}_i')^2 = T_{cm} + T', \tag{8.8}$$

where the velocities in the center of mass frame are given by

$$\mathbf{v}_i' = \left.\frac{d\mathbf{r}_i'}{dt}\right|_s = \mathbf{v}_i - \dot{\mathbf{R}}. \tag{8.9}$$

The translational part of the kinetic energy is the kinetic energy of the center of mass point. The rotational kinetic energy T' depends only on the angular velocity since by Equation (8.7)

$$\mathbf{v}_i'|_s = \boldsymbol{\omega} \times \mathbf{r}_i'. \tag{8.10}$$

The kinetic energy of rotation is therefore a second order homogeneous function of the angular velocities. Working in the body frame, one can evaluate the rotational kinetic energy by expanding

$$T' = \frac{1}{2} \sum_i m_i \left(\frac{d\mathbf{r}'_i}{dt}\right)^2 = \frac{1}{2} \sum_i m_i (\boldsymbol{\omega} \times \mathbf{r}'_i) \cdot \frac{d\mathbf{r}'_i}{dt}$$

$$= \frac{1}{2} \sum_i \boldsymbol{\omega} \cdot \left(m\mathbf{r}' \times \frac{d\mathbf{r}'}{dt}\right)_i = \frac{1}{2} \sum_i \boldsymbol{\omega} \cdot \mathbf{L}_i. \tag{8.11}$$

Pulling the angular velocity vector out of the summation gives

$$T' = \frac{1}{2} \boldsymbol{\omega} \cdot \mathbf{L}'. \tag{8.12}$$

Next, the total angular momentum relative to the pivot point can also be expanded giving

$$\mathbf{L}' = \sum_i \left(m\mathbf{r}' \times \frac{d\mathbf{r}'}{dt}\right)_i = \sum_i m_i \mathbf{r}'_i \times (\boldsymbol{\omega} \times \mathbf{r}'_i)$$

$$= \sum_i m_i (\mathbf{r}'_i \cdot \mathbf{r}'_i)\boldsymbol{\omega} - m_i \mathbf{r}'_i (\mathbf{r}'_i \cdot \boldsymbol{\omega}). \tag{8.13}$$

The angular velocity can now be pulled out of this sum as well, giving the dyadic equation

$$\mathbf{L}' = \mathbf{I} \cdot \boldsymbol{\omega}, \tag{8.14}$$

where the moment of inertia tensor $\mathbf{I}$ is defined as

$$\mathbf{I} = \sum_n m_n (\mathbf{r}' \cdot \mathbf{r}' - \mathbf{r}'\mathbf{r}')_n, \tag{8.15}$$

The rotational kinetic energy can then be expressed as the fully contracted quadratic form

$$T' = \frac{1}{2} \boldsymbol{\omega} \cdot \mathbf{I} \cdot \boldsymbol{\omega}. \tag{8.16}$$

This expression is easiest to evaluate in the body frame, where the moment of inertia tensor of the rigid body is constant.

In the absence of torque, angular momentum is conserved, since

$$\mathbf{N} = \frac{d\mathbf{L}}{dt}. \tag{8.17}$$

Kinetic energy conservation in the absence of torque follows by contraction of Euler's equations with the angular velocity in the body frame, noting that

$$\frac{dT}{dt} = \boldsymbol{\omega} \cdot \mathbf{I} \cdot \frac{d\boldsymbol{\omega}}{dt} = \boldsymbol{\omega} \cdot \mathbf{N}. \tag{8.18}$$

8.3   MOMENT OF INERTIA TENSOR

The inertia tensor is a symmetric (self-adjoint) rank-two tensor, with components given by

$$I_{ij} = \mathbf{e}_i \cdot \mathbf{I} \cdot \mathbf{e}_j = \sum_n m_n \begin{bmatrix} (r_n^2 - x_n^2) & -x_n y_n & -x_n z_n \\ -y_n x_n & (r_n^2 - y_n^2) & -y_n z_n \\ -z_n x_n & -z_n y_n & (r_n^2 - z_n^2) \end{bmatrix}. \tag{8.19}$$

As illustrated in Figure 8.4, the on-diagonal elements are referred to as the *moments of inertia* defined as

$$I_{ii} = \sum_n m_n (r_{\perp n}^2), \quad r_{\perp n}^2 = (r^2 - r_i r_i)_n, \quad \text{(nsc)} \tag{8.20}$$

where $r_{\perp n}$ are the perpendicular displacements of the masses from an axis of rotation. Note that the moments of inertia are positive definite for extended three-dimensional matter distributions. The off-diagonal elements of the matrix are referred to the *products of inertia*, for example,

$$I_{xy} = -\sum_n m_n x_n y_n. \tag{8.21}$$

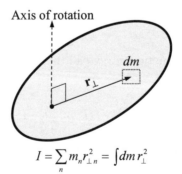

$$I = \sum_n m_n r_{\perp n}^2 = \int dm\, r_\perp^2$$

FIGURE 8.4   The moment of inertia of an element of mass rotating about an axis is given by $dI = r_\perp^2 dm$, whereas for a point mass it is given by $I = r_\perp^2 m$, where $r_\perp$ is the perpendicular displacement from the axis.

These products vanish if one is in a principal axis frame, where the moment of inertia tensor has been fully diagonalized. For continuous distributions of matter, the moment of inertia of a solid body with matter density $\rho(\mathbf{r})$, is given by the volume integral

$$I_{ij} = \int_V \rho(\mathbf{r})d^3\mathbf{r}(r^2\delta_{ij} - r_i r_j), \tag{8.22}$$

with a mass normalization of

$$m = \int_V \rho(\mathbf{r})d^3\mathbf{r}, \tag{8.23}$$

where r is the distance from the pivot to a point (not the perpendicular distance), δ_{ij} is the Kronecker delta, and m is the total mass of the body. Expanding in terms of Cartesian coordinates gives the matrix equation

$$\mathbf{I} = \int_V \rho(x,y,z)dxdydz \begin{bmatrix} y^2+z^2 & -xy & -xz \\ -xy & x^2+z^2 & -yz \\ -xz & -yz & x^2+y^2 \end{bmatrix}. \tag{8.24}$$

8.3.1 Principal Moments of Inertia

The moment of inertia tensor is a symmetric second-rank tensor. Such tensors can always be diagonalized by an orthogonal transformation into a principal axis frame. The principal moments are defined as the non-zero entries in the diagonalized moment of inertia matrix. One can make use of rotational invariance of the trace and determinant to find the constraints

$$\text{Trace}(\mathbf{I}) = I_1 + I_2 + I_3, \tag{8.25}$$

$$\det(\mathbf{I}) = I_1 I_2 I_3. \tag{8.26}$$

The principal axes of a rotating body are defined by finding eigenvectors of $\mathbf{I}$ such that

$$\mathbf{L} = \mathbf{I} \cdot \boldsymbol{\omega} = I\boldsymbol{\omega}. \tag{8.27}$$

In matrix notation this becomes

$$\begin{bmatrix} I_{11}-I & I_{12} & I_{13} \\ I_{21} & I_{22}-I & I_{23} \\ I_{31} & I_{32} & I_{33}-I \end{bmatrix} \begin{bmatrix} \omega_1 \\ \omega_2 \\ \omega_3 \end{bmatrix} = \begin{bmatrix} 0 \\ 0 \\ 0 \end{bmatrix}. \tag{8.28}$$

The above linear system of equations has a solution if, and only if, the determinant of the following equation vanishes

$$\det(\mathbf{I} - I_i\mathbf{1}) = f(I_i) = 0. \tag{8.29}$$

This is called the *secular equation*. It represents a cubic equation for the principal moments of inertia I_i. All three roots are real. Once the moments of inertia are found, the principal axes of revolution can be found by solving for the normalized eigenvectors

$$(\mathbf{I} - I_i\mathbf{1}) \cdot \hat{\mathbf{n}}_i = 0. \tag{8.30}$$

Usually the basis is constrained to form a right-handed triplet. Many common rotational problems involve the rotation of an object about a symmetry axis, where the equation is easily diagonalized, since axes of symmetry are principal axes. Once two principal axes are known, the third axis can be found by taking their cross-product.

If an object is set rotating about a principal axis, the angular momentum will point in the direction of that axis; then, because principal axes are solutions to the eigenvalue equation $\mathbf{L} = \mathbf{I} \cdot \boldsymbol{\omega} = I\boldsymbol{\omega}$, $\mathbf{L}$ and $\boldsymbol{\omega}$ point in the same direction; therefore, $\dot{\mathbf{L}} = I\dot{\boldsymbol{\omega}} = 0$ for the torque free motion about a principal axis. No torque is needed to maintain a uniform rotation rate about a principal axis.

For example, the balancing of an automobile tire requires both static and dynamical load balancing. *Static balancing* of a tire requires that the center of gravity lies along the axis of rotation. The externally applied gravitational force then produces no torque about the rotation axis. *Dynamic balancing* requires additionally that the axis of rotation be a principal axis. For example, if an automobile has a tire which is out of dynamical balance, as shown in the right side of Figure 8.5, the axle about which it is rotating is not a principal axis. Consequently, the tire will tend to wobble, since $\mathbf{L}$ and $\boldsymbol{\omega}$ are not collinear, and a periodic torque must be exerted by the axle of the car to keep it rolling straight. At certain speeds, this periodic torque may excite a resonant wobbling frequency, and the tire may begin to shake violently, vibrating the entire automobile. This causes excess wear on the tire and axle.

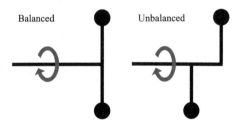

FIGURE 8.5 A rigid body rotating about a fixed axis is dynamically balanced if the axis of rotation is a principal axis.

8.3.2 Diagonalizing the Inertia Tensor

Unless one is clever enough to take advantages of obvious symmetries, the inertia tensor will not, a priori, be diagonal in the body frame. But since the tensor is transpose symmetric, it can always be diagonalized by an orthogonal transformation putting it into a principal axis frame. Let $\mathbf{A}$ be a orthogonal rotation of the current reference frame $\mathbf{e}_i$, into the new principal axis reference frame $\mathbf{n}_i$ then the transformation can be expressed as

$$\mathbf{A} = \mathbf{n}^i \mathbf{e}_i, \quad \bar{\mathbf{A}} = \mathbf{e}_i \mathbf{n}^i, \tag{8.31}$$

$$\mathbf{n}_i = \mathbf{A} \cdot \mathbf{e}_i, \quad \mathbf{e}_i = \bar{\mathbf{A}} \cdot \mathbf{n}_i, \tag{8.32}$$

where the principal axes $\mathbf{n}_i$ are solutions to the eigenvalue equations

$$\mathbf{I} \cdot \mathbf{n}_i = I \mathbf{n}_i \tag{8.33}$$

Applying the principal axis transformation, the eigenvalue equations becomes

$$\mathbf{I} \cdot \mathbf{A} \cdot \mathbf{e}_i = I \mathbf{A} \cdot \mathbf{e}_i. \tag{8.34}$$

Multiplying from the left by $\bar{\mathbf{A}}$ gives

$$\bar{\mathbf{A}} \cdot \mathbf{I} \cdot \mathbf{A} \cdot \mathbf{e}_i = \mathbf{I}' \cdot \mathbf{e}_i = I_i \mathbf{e}_i, \tag{8.35}$$

where $\mathbf{I}'$ has been diagonalized by the transformation $\mathbf{I}' = \bar{\mathbf{A}} \cdot \mathbf{I} \cdot \mathbf{A}$, giving

$$\mathbf{I}' = \begin{bmatrix} I_1 & 0 & 0 \\ 0 & I_2 & 0 \\ 0 & 0 & I_3 \end{bmatrix}. \tag{8.36}$$

Once the principal axes are known and normalized, one can construct the rotation matrix connecting the principal axes to body frame axes using Equation (8.31). The dynamical properties of a rigid body are therefore fully categorized by giving its mass, principal axes, and principal moments of inertia.

8.3.3 Diagonalization Procedure

To diagonalize the moment of inertia tensor, one first solves the secular equation to obtain the principal moments

$$\det \left(\mathbf{I} - I_i \mathbf{1} \right) = 0. \tag{8.37}$$

The above yields a cubic equation for the eigenvalues I_i. Since $\mathbf{I}$ is self-adjoint, the eigenvalues are real, and the eigenvectors are orthogonal if the eigenvalues are distinct. Then for each $I_j = \{I_1, I_2, I_3\}$ the solutions for the eigenvectors $\mathbf{n}_j$ are

$$(\mathbf{I} - I_j) \cdot \mathbf{n}_j = (\mathbf{I} - I_j) \cdot \mathbf{A} \cdot \mathbf{e}_j = 0, \tag{8.38}$$

the transformation matrix can then be constructed as

$$\mathbf{e}^i \cdot \mathbf{n}_j = \mathbf{e}^i \cdot \mathbf{A} \cdot \mathbf{e}_j = A^i{}_j, \tag{8.39}$$

subject the constraint that the eigenvectors have been previously orthonormalized

$$\mathbf{n}^i \cdot \mathbf{n}_j = \delta^i_j. \tag{8.40}$$

The orthogonal transformation $\mathbf{A}$ will be a rotation if $\det(\mathbf{A}) = +1$.

If the eigenvalues are distinct, the eigenvectors are orthogonal to each other as shown by the following analysis: Starting with the eigenvalue equation $\mathbf{I} \cdot \mathbf{n}_i = I_i \mathbf{n}_i$, construct its dual $\mathbf{n}_j \cdot \bar{\mathbf{I}} = \mathbf{n}_j I_j$, then, by associativity,

$$(\mathbf{n}_j \cdot \bar{\mathbf{I}}) \cdot \mathbf{n}_i - \mathbf{n}_j \cdot (\mathbf{I} \cdot \mathbf{n}_i) = (I_j - I_i)\mathbf{n}_j \cdot \mathbf{n}_i = 0. \tag{8.41}$$

If $(I_j - I_i) \neq 0$, the eigenvalues are distinct, implying that $\mathbf{n}_j \cdot \mathbf{n}_i = 0$, and the eigenvectors are orthogonal. If the eigenvalues are degenerate, on the other hand, any linear combination of the degenerate eigenvectors is also an eigenvector, and two degenerate eigenvectors need not be orthogonal. However, the Gramm-Schmidt orthonormalization procedure can always be applied, creating an orthonormal basis set from a set of degenerate, but linearly-independent, eigenvectors [Arfken 2001: pp. 506–599].

As an example of the diagonalization procedure, consider the following moment of inertia tensor in dimensionless units $ma^2 = 1$:

$$\mathbf{I} = ma^2 \begin{bmatrix} 2 & 1 & 0 \\ 1 & 2 & 0 \\ 0 & 0 & 4 \end{bmatrix} = \begin{bmatrix} 2 & 1 & 0 \\ 1 & 2 & 0 \\ 0 & 0 & 4 \end{bmatrix}. \tag{8.42}$$

The $\hat{3}$-axis $\hat{\mathbf{n}}_3 = [0, 0, 1]$ is clearly a principal axis, with a principal moment of inertia of $4ma^2$. The other two principal axes can be found by solving the eigenvalue problem

$$\det \begin{bmatrix} 2 - \lambda & 1 & 0 \\ 1 & 2 - \lambda & 0 \\ 0 & 0 & 4 - \lambda \end{bmatrix} = (1 - \lambda)(3 - \lambda)(4 - \lambda) = 0. \tag{8.43}$$

The principal moments are $I_i = ma^2\{1, 3, 4\}$. The eigenvalue equations are

$$
\begin{bmatrix} 2 - I_1 & 1 & 0 \\ 1 & 2 - I_1 & 0 \\ 0 & 0 & 4 - I_1 \end{bmatrix} \begin{bmatrix} \cos\phi_1 \\ \sin\phi_1 \\ 0 \end{bmatrix} = \begin{bmatrix} 1 & 1 & 0 \\ 1 & 1 & 0 \\ 0 & 0 & 3 \end{bmatrix} \begin{bmatrix} \cos\phi_1 \\ \sin\phi_1 \\ 0 \end{bmatrix}, \tag{8.44}
$$

$$
\begin{bmatrix} 2 - I_2 & 1 & 0 \\ 1 & 2 - I_2 & 0 \\ 0 & 0 & 4 - I_2 \end{bmatrix} \begin{bmatrix} \cos\phi_2 \\ \sin\phi_2 \\ 0 \end{bmatrix} = \begin{bmatrix} -1 & 1 & 0 \\ 1 & -1 & 0 \\ 0 & 0 & 1 \end{bmatrix} \begin{bmatrix} \cos\phi_2 \\ \sin\phi_2 \\ 0 \end{bmatrix}, \tag{8.45}
$$

which give the solutions

$$
\hat{\mathbf{n}}_1 = \begin{bmatrix} \cos\pi/4 \\ -\sin\pi/4 \\ 0 \end{bmatrix}, \quad \hat{\mathbf{n}}_2 = \begin{bmatrix} \cos\pi/4 \\ \sin\pi/4 \\ 0 \end{bmatrix}. \tag{8.46}
$$

The orthogonal transformation matrix that diagonalizes $\mathbf{I}$ is given by the rotation matrix

$$
\mathbf{A} = \begin{bmatrix} \cos\pi/4 & \sin\pi/4 & 0 \\ -\sin\pi/4 & \cos\pi/4 & 0 \\ 0 & 0 & 1 \end{bmatrix}, \tag{8.47}
$$

where

$$
\mathbf{n}_j = \mathbf{A} \cdot \mathbf{e}_j, \quad \tilde{\mathbf{A}} \cdot \mathbf{I} \cdot \mathbf{A} = \mathbf{I}' = ma^2 \begin{bmatrix} 1 & 0 & 0 \\ 0 & 3 & 0 \\ 0 & 0 & 4 \end{bmatrix}. \tag{8.48}
$$

8.3.4 Parallel Axis Theorem

The moment of inertia of any object about an axis through its center of mass is the minimum moment of inertia for an axis in that direction in space. The moment of inertia about any axis parallel to that axis through the center of mass can be found from

$$
I_\parallel = I_{cm} + mR_\perp^2, \tag{8.49}
$$

The expression added to the center of mass moment of inertia is the moment of inertia of a point of mass m located at a perpendicular offset $R_\perp$ from the axis of revolution. More generally, one gets the parallel axis theorem relating two sets of parallel axes with a relative displacement of origin. Let one axis be through the center of mass then

$$
x_i' = x_i - R_i, \tag{8.50}
$$

where the center of mass offset is given by

$$R_i = \frac{1}{m} \int x_i dm, \tag{8.51}$$

then

$$I_{ij} = \int dm(r^2 \delta_{ij} - x_i x_j) = I_{ij}^{cm} + m(R^2 \delta_{ij} - R_i R_j), \tag{8.52}$$

$$I_{ij}^{cm} = \int dm(r'^2 \delta_{ij} - x_i' x_j'). \tag{8.53}$$

where the cross terms in the products cancel since $\int dm x_i' = 0$ in center of mass coordinates.

This results in the *Parallel Axis Theorem*, which states that the moment of inertia for a set of axes translationally offset by a displacement **R** from a second set passing through the center of mass is the moment of inertia tensor of the center of mass frame plus the moment of inertia of a point mass placed at **R**

$$\mathbf{I} = \mathbf{I}_{cm} + m(R^2 \mathbf{1} - \mathbf{R}\,\mathbf{R}). \tag{8.54}$$

For example, as shown in Figure 8.6, a slender rod of length l and mass m has a moment of inertia of $ml^2/12$ when rotated about its midpoint. Its moment of inertia increase to $ml^2/3$ when rotated about a parallel axis through an end point of the rod.

8.3.5 Symmetry Considerations

If a body has rotational symmetry about an axis, the axis is a principal axis. Likewise, if the body has mirror symmetry when reflected about a plane,

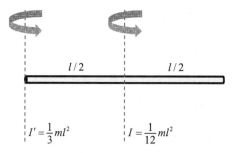

FIGURE 8.6   The moment of inertia a thin rod of length l and mass m is $I = ml^2/12$ when rotated about its midpoint. Its moment of inertia increases to $I' = ml^2/3$ when rotated about a parallel axis through an endpoint of the rod.

the axis of reflection is a principal axis. The proofs of both statements are straightforward.

Consider first a density distribution having mirror symmetry about the $z = 0$ plane. The density distribution satisfies the condition

$$\rho(x, y, z) = \rho(x, y, -z). \tag{8.55}$$

Here the density is an even function of z. Then the products of inertia linear in z vanish

$$I_{xz} = \int x \, dx \, dy \int_{-z_0}^{z_0} \rho(x, y, z) z \, dz = 0, \quad I_{yz} = \int y \, dx \, dy \int_{-z_0}^{z_0} \rho(x, y, z) z \, dz = 0, \tag{8.56}$$

Therefore the axis of mirror reflection, $\hat{z}$, must be a principal axis.

Likewise, if the $\hat{z}$ axis is an axis of rotational symmetry, the distribution is azimuthally independent about the symmetry axis

$$\rho(x, y, z) = \rho(r_\perp, z). \tag{8.57}$$

Therefore, the moments of inertia in the $\hat{x} \wedge \hat{y}$ plane are degenerate principal axes. Given two linearly-independent principal axes, a third axis can be generated by taking the cross product $\mathbf{n}_3 = \mathbf{n}_1 \times \mathbf{n}_2$. It follows that $\hat{z}$ is the third principal axis.

8.3.6 Perpendicular Axis Theorem

A thin *lamina* is a flat plate in a plane with negligible dimensions in a third space direction. If the z-axis is chosen as the thin dimension, the volume density distribution reduces to a surface density distribution

$$\rho(x, y, z) = \sigma(x, y)\delta(z). \tag{8.58}$$

It is clear that this axis forms a symmetry axis of mirror reflection, since the products of inertia with respect to z-displacements identically vanish

$$I_{xz} = I_{yz} = 0, \tag{8.59}$$

$$I_{ij} = \int \sigma(x, y) \, dx \, dy \begin{bmatrix} y^2 & xy & 0 \\ xy & x^2 & 0 \\ 0 & 0 & x^2 + y^2 \end{bmatrix} = \begin{bmatrix} I_{xx} & I_{xy} & 0 \\ I_{yx} & I_{yy} & 0 \\ 0 & 0 & I_{zz} \end{bmatrix}. \tag{8.60}$$

Note that integrating over the delta function at $z = 0$ gives

$$I_{zz} = \int \sigma(x, y)(x^2 + y^2) \, dx \, dy = \int \sigma(x, y) y^2 \, dx \, dy + \int \sigma(x, y) x^2 \, dx \, dy. \tag{8.61}$$

This result is the *perpendicular axis theorem*, which states that: the moment of inertia of a thin lamina about an axis perpendicular to the plane is the sum of the moments of inertia of two perpendicular axes through the same point in the plane of the object.

$$I_{zz} = I_{xx} + I_{yy}. \tag{8.62}$$

The utility of this theorem goes beyond that of calculating moments of strictly planar objects. The perpendicular axis theorem can be a valuable tool in the buildup of moments of inertia for three-dimensional objects (such as cylinders) by slicing them up into thin laminas and summing over the moments of inertia of each of the laminas.

8.3.6.1 Principal Axes of a Thin Lamina

The axis perpendicular to a planar lamina is automatically a principal axis $\mathbf{n}_3 = \mathbf{e}_3$. Therefore, one only needs to solve the eigenvalue problem in two dimensions

$$\begin{bmatrix} I_{11} - \lambda & I_{12} \\ I_{21} & I_{22} - \lambda \end{bmatrix} \begin{bmatrix} n_x \\ n_y \end{bmatrix} = 0. \tag{8.63}$$

The eigenvalues are given by solving the secular equation in two dimensions

$$(I_{11} - \lambda)(I_{22} - \lambda) - I_{12}^2 = 0, \tag{8.64}$$

with eigenvalues

$$I_{1,2} = \lambda_{\pm} = \left(\frac{I_{11} + I_{22}}{2} \right) \pm \sqrt{\left(\frac{I_{11} + I_{22}}{2} \right)^2 - (I_{11}I_{22} - I_{12}^2)}. \tag{8.65}$$

Therefore, the eigenfunctions are given by

$$\mathbf{n}_j = \mathbf{R}(e_3, \theta) \cdot \mathbf{e}_j = \begin{bmatrix} \cos\theta & -\sin\theta \\ \sin\theta & \cos\theta \end{bmatrix} \mathbf{e}_j, \quad \mathbf{n}_1 = \begin{bmatrix} \cos\theta \\ \sin\theta \end{bmatrix}, \quad \mathbf{n}_2 = \begin{bmatrix} -\sin\theta \\ \cos\theta \end{bmatrix}. \tag{8.66}$$

The eigenvectors satisfy

$$(I_{11} - I_j)n_{1j} + I_{12}n_{2j} = 0,$$
$$I_{12}n_{1j} + (I_{22} - I_j)n_{2j} = 0. \tag{8.67}$$

Since the desired solutions should be orthogonal to each other, only one eigenfunction is needed to get the other, using the third axis to create a right-handed triplet

$$\mathbf{n}_1 = \mathbf{n}_2 \times \mathbf{e}_3. \tag{8.68}$$

8.3.6.2 Thin Disk

The thin, homogenous, circular disk, shown in Figure 8.7, is a classic example where the perpendicular axis theorem can be applied. The surface density is constant. The rotational axis is a principal axis, with a moment of inertia given by integration

$$\mathbf{I}_{zz} = \int_0^{2\pi} d\phi \int_0^R \sigma r^2 r dr = \frac{m}{\pi R^2} \int_0^R 2\pi r^3 dr = \frac{1}{2} mR^2, \qquad (8.69)$$

using the normalization

$$m = \int_0^R \sigma dA = \sigma \pi R^2. \qquad (8.70)$$

By the rotational symmetry of the problem, all eigenvectors in the plane perpendicular to the symmetry axis are degenerate. By the perpendicular axis theorem, they must add up to the third moment of inertia. Therefore,

$$I_x = I_y = \frac{1}{2} I_z = \frac{1}{4} mR^2, \qquad (8.71)$$

In the center of mass frame for a disk of radius a and mass m, one gets

$$\mathbf{I} = \begin{bmatrix} \frac{1}{4} ma^2 & 0 & 0 \\ 0 & \frac{1}{4} ma^2 & 0 \\ 0 & 0 & \frac{1}{2} ma^2 \end{bmatrix}. \qquad (8.72)$$

One can now use the parallel axis theorem to find moments of inertia for a thin disk rotating about a parallel set of axes. For example, the

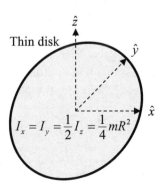

FIGURE 8.7 Applying the perpendicular axis theorem to a thin circular disk.

moment of inertia about an axis parallel to the symmetry axis located on the circumference of the disk at a displacement $\mathbf{R} = (a, 0, 0)$ is

$$I_{\parallel} = m \begin{bmatrix} a^2/4 & 0 & 0 \\ 0 & a^2/4 & 0 \\ 0 & 0 & a^2/2 \end{bmatrix} + m \begin{bmatrix} 0 & 0 & 0 \\ 0 & a^2 & 0 \\ 0 & 0 & a^2 \end{bmatrix}$$

$$= m \begin{bmatrix} a^2/4 & 0 & 0 \\ 0 & 5a^2/4 & 0 \\ 0 & 0 & 6a^2/4 \end{bmatrix}. \tag{8.73}$$

Note that the perpendicular axis theorem continues to apply in the new coordinate frame.

8.4 EULER'S EQUATION OF MOTION

Euler's equations of motion for a rigid body are obtained from evaluating the torque equation in the rotating frame

$$\mathbf{N} = \left.\frac{d\mathbf{L}}{dt}\right|_{space} = \left.\frac{d\mathbf{L}}{dt}\right|_{body} + \boldsymbol{\omega} \times \mathbf{L}. \tag{8.74}$$

Using a principal axis body frame, where the moments are diagonal and constants, gives Euler's equations for the rotation of a rigid body

$$\mathbf{N} = \mathbf{I} \cdot \frac{d\boldsymbol{\omega}}{dt} + \boldsymbol{\omega} \times \mathbf{L}, \tag{8.75}$$

which, in component form, becomes

$$\begin{aligned} N_i &= \frac{dL_i}{dt} + \varepsilon_i^{jk} \omega_j L_k \\ &= I_{(i)} \frac{d\omega_i}{dt} + \frac{1}{2} \varepsilon_i^{jk} \omega_j \omega_k (I_{(k)} - I_{(j)}), \end{aligned} \tag{8.76}$$

where no summation is taken over indices in parentheses. The result is cyclic over the three indices. Expanding the components gives

$$\begin{aligned} N_1 &= I_1 \frac{d\omega_1}{dt} + \omega_2 \omega_3 (I_3 - I_2), \\ N_2 &= I_2 \frac{d\omega_2}{dt} + \omega_3 \omega_1 (I_1 - I_3), \\ N_3 &= I_3 \frac{d\omega_3}{dt} + \omega_1 \omega_2 (I_2 - I_1). \end{aligned} \tag{8.77}$$

Note that Euler's equations are first order differential equations for the angular velocities and do not directly give the equations of motion for the coordinates. Moreover, they are expressed in the non-inertial frame of the rotating body. To evaluate Euler's equations, the torque must also be expressed in the body frame. Therefore, Euler's formalism is best adapted for the cases where the rotation axes are constrained or where the motion is torque-free. The more general case is best solved using Lagrange's equations of motion in terms of some appropriate set of angular coordinates.

8.4.1 Torque-Free Motion of a Rigid Body

As an example of Euler's equations, consider the torque free motion of a rigid body. Setting $\mathbf{N} = 0$ in Euler's equation of motion, gives

$$
\begin{aligned}
I_1 \frac{d\omega_1}{dt} &= (I_2 - I_3)\omega_2\omega_3, \\
I_2 \frac{d\omega_2}{dt} &= (I_3 - I_1)\omega_3\omega_1, \\
I_3 \frac{d\omega_3}{dt} &= (I_1 - I_2)\omega_1\omega_2.
\end{aligned}
\tag{8.78}
$$

In the absence of torque, angular momentum is conserved since

$$
\mathbf{N} = \frac{d\mathbf{L}}{dt} \to 0.
\tag{8.79}
$$

Rotational kinetic energy is also conserved, since

$$
\frac{dT}{dt} = \boldsymbol{\omega} \cdot \mathbf{N} = \boldsymbol{\omega} \cdot \mathbf{I} \cdot \frac{d\boldsymbol{\omega}}{dt} \to 0.
\tag{8.80}
$$

Under some conditions, even in the absence of torque, the motion may be neither simple nor intuitive, as the following cases studies illustrate.

8.4.1.1 Three Unequal Moments

When the three moments of inertia are distinct, the motion gets quite interesting. It is useful to begin by analyzing the rotational motion for stability under small deviations from a principal axis. Assume that the initial axis of rotation is close to one of three (unequal) principal axes, namely the $\hat{2}$ body axis with angular velocity Ω_2 for specificity. Expanding motion for small deviations from this principal axis gives

$$
\omega_1 = \eta_1(t), \quad \omega_2 = \Omega_2 + \eta_2(t), \quad \omega_3 = \eta_3(t),
\tag{8.81}
$$

If the axis of rotation is stable, the perturbations $\eta_i(t)$ will remain small for all subsequent times. To test this hypothesis, expand Euler's equations of motion to lowest order in $\eta(t)$

$$
\begin{aligned}
I_1 \frac{d\eta_1}{dt} &= (I_2 - I_3)\Omega_2\eta_3, \\
I_2 \frac{d\eta_2}{dt} &= 0, \\
I_3 \frac{d\eta_3}{dt} &= (I_1 - I_2)\Omega_2\eta_1.
\end{aligned}
\tag{8.82}
$$

This results in the following separated second order equations of motion

$$
\ddot{\eta}_1 = -\Omega_p^2\eta_1, \quad \ddot{\eta}_2 = 0, \quad \ddot{\eta}_3 = -\Omega_p^2\eta_3,
\tag{8.83}
$$

where, Ω_p is a precession frequency given by

$$
\Omega_p^2 = (I_2 - I_3)(I_2 - I_1)\Omega_2^2/I_1 I_3.
\tag{8.84}
$$

If $\Omega_p^2 > 0$, the motion is simple harmonic and restorative. However, the motion is unstable if $\Omega_p^2 < 0$. On closer examination, the solution proves to be unstable if I_2 is the intermediate moment of inertia and stable if I_2 is either the largest or smallest moment of inertia. This instability about the intermediate axis is the cause of the apparent complicated tumbling motion of torque-free asymmetric rigid bodies.

To facilitate visualization of the motion, a scale transformation can be made. This is a simplification of a method originally suggested by Louis Poinsot.[2] Given $\mathbf{L} = \ell\mathbf{n} = \mathbf{I} \cdot \boldsymbol{\omega}$, Let $\mathbf{S}$ be a scale transform that makes the maps the angular velocities to a unit vector $\mathbf{n}$ in the direction of the angular momentum; then

$$
\boldsymbol{\omega} = \mathbf{S} \cdot \mathbf{n} = \ell\mathbf{I}^{-1} \cdot \mathbf{n},
\tag{8.85}
$$

where the angular velocities scale as

$$
\omega_i = \frac{\ell n_i}{I_i}.
\tag{8.86}
$$

---

[2] Poinsot's construction is considerably more complicated, with the motion described by the point of contact of the inertia ellipsoid as it rolls without slipping on an invariable plane. For a description of Poinsot's method see [Goldstein 2002, 200–208].

The equations of motion for the trajectory angular momentum orientation unit vector in the body frame can now be written as

$$\frac{dn_i}{dt} = \varepsilon_{ijk}\frac{\ell}{I_k}n_j n_k. \tag{8.87}$$

The angular momentum orientation vector in the body frame is constrained to trace out a trajectory on the unit sphere, $\hat{\mathbf{n}} = (\sin\theta\cos\phi, \sin\theta\sin\phi, \cos\theta)$, $\mathbf{n}^2 = 1$. The precession of the angular momentum vector in the body frame moves along the constant energy contours, which for fixed angular momentum ℓ are given by

$$C(\theta,\phi) = \frac{E(\theta,\phi)}{\ell^2} = \sum_i \frac{n_i^2}{I_i} = \left(\frac{\cos^2\phi}{I_1} + \frac{\sin^2\phi}{I_2}\right)\sin^2\theta + \frac{\cos^2\theta}{I_3}, \tag{8.88}$$

where (θ,ϕ) represents the polar and azimuthal angles of the angular momentum orientation vector with respect to the body frame. Map $\hat{1},\hat{2},\hat{3}$ to $\hat{x},\hat{y},\hat{z}$ coordinates, where $(I_x < I_y < I_z)$.

As the system evolves in time, the unit vector $\hat{\mathbf{n}}$ moves around the unit sphere as shown by the trajectories shown in the sketch (Figure 8.8), which represents the projection of the scaled contours of Equation 8.88 as seen from above the unstable $\hat{y}$-axis. The two stable principal axes are shown in the $\hat{x} \wedge \hat{z}$ plane of the projection. Small perturbations of the orbits from the stable principal axes remain small. Angular momentum vectors nearly aligned

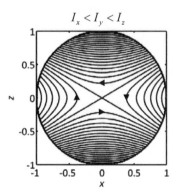

FIGURE 8.8 **Precession of the unit angular momentum orientation vector in the body frame: The figure shows the projection of the unit sphere onto the $\hat{x} \wedge \hat{z}$ plane as viewed from above the body $\hat{y}$-axis, where the $\hat{x}$ and $\hat{z}$ axes are stable and the intermediate $\hat{y}$-axis is unstable ($I_x < I_y < I_z$). The direction of precession is counterclockwise about the smallest moment of inertia and clockwise about the axis with the largest moment of inertia.**

with the intermediate $\hat{y}$-axis are unstable, however. The unstable trajectories wrap themselves about the sphere, causing the body frame of the object to appear to flip and tumble through space, even though its angular momentum is constant in the space frame.

The precession of angular momentum, as viewed from the body frame, is counterclockwise about the smallest moment (fastest spinning axis) and clockwise about the axis with the largest moment (slowest spinning axis). In the space frame, the body axes rotate with the opposite sense about the fixed angular momentum direction.

Therefore, even though the angular momentum vector is conserved, the observed motion can be far from simple, unless one is rotating about a principal axis. When the three principal axes are distinct, the body might be observed to tumble and even flip while moving in free space.

8.4.1.2 Axially Symmetry

Now, let two of the three principal axes have the same moment of inertia, $I_1 = I_2 \neq I_3$. Euler's equations become

$$I_0 \frac{d\omega_1}{dt} = (I_0 - I_3)\omega_2\omega_3$$
$$I_0 \frac{d\omega_2}{dt} = (I_3 - I_0)\omega_3\omega_1 \qquad (8.89)$$
$$I_3 \frac{d\omega_3}{dt} = 0$$

where the component of angular velocity along the symmetric $\hat{3}$ body axis ω_3 is conserved. The other two components precess about the $\hat{3}$ axis with a uniform rate of precession Ω given by the coupled linear equations

$$\frac{d\omega_1}{dt} = \Omega\omega_2, \quad \frac{d\omega_2}{dt} = -\Omega\omega_1, \qquad (8.90)$$

where

$$\Omega = \left(\frac{I_0 - I_3}{I_0}\right)\omega_3. \qquad (8.91)$$

The direction of precession depends on whether the inertia ellipsoid is prolate $I_0 < I_3$ or oblate $I_0 > I_3$. The solution for the precession frequency Ω describes uniform circular motion in the $\hat{1} \wedge \hat{2}$ plane with

$$\omega_1 = \omega_\perp \cos \Omega(t - t_0), \quad \omega_2 = \omega_\perp \sin \Omega(t - t_0), \qquad (8.92)$$

where the magnitude of the angular velocity vector is a constant

$$\omega^2 = \omega_\perp^2 + \omega_3^2. \tag{8.93}$$

From Euler's equations of motion, the magnitudes of L, ω_3, and $\omega_\perp$ are all constants of the motion. Consider these as providing given initial conditions. The angle between the space $\hat{z}$ direction and the body $\hat{z}'$ direction is given by

$$u = \cos\theta = \frac{\mathbf{L} \cdot \boldsymbol{\omega}_3}{|\mathbf{L}||\boldsymbol{\omega}_3|} \tag{8.94}$$

In the body frame, the angular momentum vector precesses about the body symmetry axis. In the space frame, pick the space $\hat{z}$ axis to be along the direction of angular momentum. Then the body $\hat{z}'$ axis precesses uniformly about the angular momentum axis (space $\hat{z}$ axis). Figure 8.9, shows the angular velocity and angular momentum diagrams for the free motion of the symmetric top. The total angular velocity vector rotates about the angular momentum vector in the space frame and about the body symmetry axis in the body frame. The left figure shows a prolate axis of symmetry and the right figure an oblate one. Geometrically, the angular velocity vector $\boldsymbol{\omega}$ forms the line of contact between the surfaces of two cones that roll in such a manner that the directions of $\boldsymbol{\omega}_3$, $\boldsymbol{\omega}$, and $\mathbf{L}$ lie in a common plane. The opening angle between the body $\hat{3}$-axis and the cone formed by the angular velocity vector is given by

$$\cos\theta_{\omega b} = \omega_3/\omega. \tag{8.95}$$

Using $T = \boldsymbol{\omega} \cdot \mathbf{L}/2$, the opening angle between the angular velocity vector and the cone formed by the angular velocity vector is given by

$$\cos\theta_{\omega s} = 2T/\omega L. \tag{8.96}$$

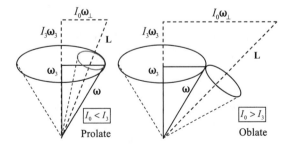

FIGURE 8.9 Free motion of the symmetric top: The angular velocity rotates about the angular momentum vector in the space frame and about the body symmetry axis in the body frame. The left figure shows a prolate axis of symmetry, and the right figure, an oblate one.

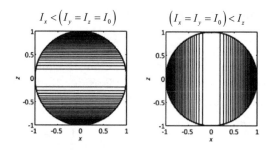

FIGURE 8.10 Constant energy contours for free motion of the symmetrical top with fixed angular momentum. Prolate inertia ellipsoids rotate about the smallest moment of inertia, and oblate ellipsoids rotate about the maximum moment. The angular momentum vector in the body frame precesses along contours of constant energy.

Constant energy contours for the precession of the unit angular momentum vector for free motion of a symmetrical rigid body are shown in Figure 8.10. The angular momentum vector precesses about the body symmetry axis tracing out scaled constant energy contour lines given by

$$C(\theta) = \frac{E(\theta)}{\ell^2} = (\sin^2\theta/I_0 + \cos^2\theta/I_3). \tag{8.97}$$

8.4.1.3 Earth's Wobble

Because the Earth bugles at the equator $I_0 \neq I_3$, it has a slight wobble (precession) of its rotational motion about the symmetry axis. The angular velocity vector points 10 meters from the pole and circles it. This leads to a precession period first estimated by Euler in 1749 to be about 300 days [Tong 2005].

$$\Omega = \left(\frac{I_0 - I_3}{I_0}\right)\omega_3 \approx \frac{1}{300}\frac{2\pi}{day} \tag{8.98}$$

The effect is very small and was not originally observed in the data. In 1891, astronomer Seth Carlo Chandler, Jr. (1846–1913) re-analyzed the data and found a precession with a period of 427 days, called *Chandler precession*. The discrepancy in the observed period with the rigid body prediction can be explained by the fact that the Earth is not a rigid body. It has a molten core, an ocean, and suffers tidal effects [Tong 2005].

8.4.1.4 Spherically Symmetry

If all three principal axes are degenerate $I_1 = I_2 = I_3 = I_0$. Then, the moment of inertia tensor is spherically symmetric $\mathbf{I} = I_0\mathbf{1}$. The result is stable, uniform, rotational motion of the freely rotating body about any axis,

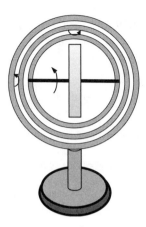

FIGURE 8.11   A sketch of a gyroscope. The central spinning mass is mounted on gimbals allowing it to freely rotate about any axis.

with the angular momentum collinear to the angular velocity $\mathbf{L} \parallel \boldsymbol{\omega}$, giving

$$I_0 \frac{d\omega}{dt} = 0. \tag{8.99}$$

8.4.2 Foucault's Gyroscope

Leon Foucault invented the *gyroscope* the year after demonstrating his famous pendulum. Foucault's gyroscope was a rapidly rotating disk with a heavy rim, mounted on low-friction gimbals. As the Earth rotates beneath the gyroscope, the rotor maintains its magnitude and direction in space in the absence of external torque. If a gyroscope is set spinning about a principal axis (see Figure 8.11), that axis will maintain a fixed orientation in an inertia space frame.

Sperry's *gyrocompass*, a practical adaptation of the gyroscope, is essentially a gyroscope set spinning in the direction of "true north." The gyrocompass is used in ship navigation and forms the basis for inertial guidance systems used by rockets.

8.5 LAGRANGIAN FORMULATION

Euler's formalism was developed in terms of angular velocities, and not directly in terms of a set of generalized coordinates. To develop Lagrange's equations of motion for a rotating body, these angular velocities have to be connected to a set of generalized coordinate degrees of freedom. The kinetic energy of a rotating body in Euler's formulation can be expressed in either the body or space frames as

$$\text{T} = \frac{1}{2}\boldsymbol{\omega} \cdot \mathbf{I} \cdot \boldsymbol{\omega}\bigg|_s = \frac{1}{2}\boldsymbol{\omega} \cdot \mathbf{I} \cdot \boldsymbol{\omega}\bigg|_b, \tag{8.100}$$

where the two frames are connected by a time-dependent rotation matrix

$$\boldsymbol{\omega}_b = \mathbf{A} \cdot \boldsymbol{\omega}_s, \quad \mathbf{I}_b = \mathbf{A} \cdot \mathbf{I}_s \cdot \bar{\mathbf{A}}. \tag{8.101}$$

In terms of Euler angles, a rotation from the space frame to the body frame given by orthogonal transformation

$$\mathbf{A}(\phi, \theta, \psi) = \mathbf{A}(\mathbf{n}_z'', \psi) \cdot \mathbf{A}(\mathbf{n}_x', \theta) \cdot \mathbf{A}(\mathbf{n}_z, \phi). \tag{8.102}$$

To develop a Lagrangian formulation, the kinetic energy must be rewritten as a functional of the coordinates, their generalized velocities, and the time. It is most convenient to evaluate the expression in the principal axis body frame, where the moment of inertia tensor is both constant and diagonal. As demonstrated in the previous chapter, in terms of Euler coordinates, the total angular velocity in the body frame is given by the functional relationship

$$\boldsymbol{\omega}_b = \mathbf{n}_z\dot{\phi} + \mathbf{n}_{x'}\dot{\theta} + \mathbf{n}_{z'}\dot{\psi}\big|_b = \begin{bmatrix} \dot{\phi}\sin\psi\sin\theta + \dot{\theta}\cos\psi \\ \dot{\phi}\cos\psi\sin\theta - \dot{\theta}\sin\psi \\ \dot{\phi}\cos\theta + \dot{\psi} \end{bmatrix}_b \tag{8.103}$$

One can now evaluate the kinetic energy functional in the principal axis body frame getting

$$T(\phi, \theta, \psi, \dot{\phi}, \dot{\theta}, \dot{\psi}, t) = \frac{1}{2}I_1(\dot{\phi}\sin\psi\sin\theta + \dot{\theta}\cos\psi)^2$$
$$+ \frac{1}{2}I_2(\dot{\phi}\cos\psi\sin\theta - \dot{\theta}\sin\psi)^2$$
$$+ \frac{1}{2}I_3(\dot{\phi}\cos\theta + \dot{\psi})^2. \tag{8.104}$$

Note that the result is cyclic in ϕ and t. Therefore, there are two first integrals of the motion immediately available for the free motion of a rigid body: The ϕ component of angular momentum p_ϕ is conserved, as is the kinetic energy T. But, otherwise, this expression remains messy. Nothing much else cancels.

To simplify the analysis of the dynamics, one can specialize the system to the heavy, axially symmetric rigid body $I_1 = I_2 = I_0 \neq I_3$. This gives a more tractable system of equations that are still pedagogically of interest. The kinetic energy simplifies to

$$T(\theta, \dot{\phi}, \dot{\theta}, \dot{\psi}) = \frac{1}{2}I_0(\dot{\theta}^2 + \dot{\phi}^2 \sin^2 \theta) + \frac{1}{2}I_3(\dot{\phi} \cos \theta + \dot{\psi})^2. \qquad (8.105)$$

There are now three first integrals of the motion available, $\{p_\phi, p_\psi, T\}$, and the motion can be reduced to an effective one-dimensional problem in a single angular coordinate θ. The canonical momenta for the free rotating symmetric top are given by

$$p_\theta = I_0\dot{\theta}, \quad p_\psi = I_3(\dot{\phi} \cos \theta + \dot{\psi}), \quad p_\phi = I_0\dot{\phi} \sin^2 \theta + p_\psi \cos \theta. \quad (8.106)$$

The free motion of the symmetric rigid body has previously been solved in Section 8.4.1.2 using Euler's equations of motion. The analytic results in the body frame involve only elementary circular functions. In this section the solution will be rederived from a Lagrangian point of view using Euler's coordinates. It is useful to set up the problem by letting the conserved angular momentum be in direction of the space $\hat{z}$-axis, as shown in Figure 8.9. In the body frame solution, given by Euler's equations, both the angular velocity and angular momentum vectors appear to rotate about the body $\hat{z}'$-axis, whereas, in the space frame, the body axes rotate about the fixed space $\hat{z}$-axis. The corotation of the body and space frames about each other is shown in Figure 8.9 for the free rotation of a rigid body with an axis of symmetry.

With all the momenta constant and the polar angle fixed, the equations for the canonical momenta become

$$p_\theta = I_0\dot{\theta} = 0, \qquad (8.107)$$

$$p_\psi = I_3(\dot{\phi} \cos \theta + \dot{\psi}) = I_3\omega_3, \qquad (8.108)$$

$$p_\phi = I_0\dot{\phi} \sin^2 \theta + p_\psi \cos \theta = I_0\omega_\perp \sin \theta + I_3\omega_3 \cos \theta. \qquad (8.109)$$

Solve for the constant angular velocities in terms of the canonical momenta and polar angle θ, using $u_0 = \cos \theta_0$, to get

$$\dot{\psi} = \left(\frac{p_\psi}{I_3} - \frac{(p_\psi - p_\phi u_0)u_0}{I_0(1 - u_0^2)}\right), \quad \dot{\phi} = \frac{(p_\psi - p_\phi u_0)}{I_0(1 - u_0^2)}, \qquad (8.110)$$

where, by direct integration, the angles are found to be either uniformly increasing or uniformly decreasing in time. The kinetic energy is a constant of the motion given by

$$T = \frac{(p_\phi - p_\psi u_0)^2}{2I_0(1 - u_0^2)} + \frac{p_\psi^2}{2I_3} = \frac{I_0\omega_\perp^2}{2} + \frac{I_3\omega_3^2}{2}. \qquad (8.111)$$

8.5.1 Spinning Top in a Uniform Gravitational Field

If one adds an interaction potential to the axially symmetric rigid body, the motion becomes significantly more complex. If the potential V depends only on the Euler angle θ, the previous first integrals of the motion are still available, and the motion reduces to an effective one-dimensional problem for the angle θ, which however, is no longer a constant of the motion. Figure 8.12 shows a symmetric top, with one point fixed in a horizontal plane, moving under a gravitational potential energy function $V(\theta) = mgl \cos \theta$. It experiences a torque due to gravity and starts to fall. If it is spinning with respect to ψ fast enough about its symmetric body axis, the angular range of polar motions in θ is restricted. The coupled equations of motion lead to a precession ϕ of the top about its vertical space axis. The top will also tend to nutate (bob up and down) as it precesses, causing $\dot{\theta}$ to speed up or slow down. Retrograde motion is possible. The motion of this top can be analyzed in terms of its Euler angle coordinates illustrated in Figure 8.12.

Using Euler's angles, the Lagrangian for the spinning symmetric top in a uniform gravitational field becomes

$$L = \frac{1}{2}I_1(\dot{\theta}^2 + \dot{\phi}^2 \sin^2 \theta) + \frac{1}{2}I_3(\dot{\phi} \cos \theta + \dot{\psi})^2 - mgl \cos \theta. \qquad (8.112)$$

Two of the three coordinates, (ϕ, ψ), are cyclic. Therefore, the problem should reduce to an effective one-body problem after eliminating the ignorable coordinates. The only non-conserved momentum is $p_\theta = I_0\dot{\theta}$, which

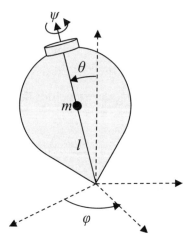

FIGURE 8.12 A sketch of the kinematic of a symmetric top. The top sees a torque due to gravity and starts to fall. The rotational spin causes the top to precess and nutate.

satisfies the torque equation

$$\dot{p}_\theta = -\frac{\partial}{\partial \theta} V_{eff}(p_\psi, p_\phi, \theta), \tag{8.113}$$

Here $V_{eff}(p_\psi, p_\phi, \theta)$ is an effective potential energy

$$V_{eff}(\theta, p_\phi, p_\psi) = \frac{(p_\phi - p_\psi \cos\theta)^2}{2I_1 \sin^2\theta} + \frac{p_\psi^2}{2I_3} + V(\cos\theta). \tag{8.114}$$

The cyclic momenta (p_ψ, p_ϕ) are constants of the motion given by

$$p_\psi = I_3(\dot{\phi}\cos\theta + \dot{\psi}) = I_3\omega_3, \tag{8.115}$$

$$p_\phi = I_3(\dot{\phi}\cos\theta + \dot{\psi})\cos\theta + I_1\dot{\phi}\sin^2\theta. \tag{8.116}$$

The conserved energy can be expressed as

$$E(\theta, \dot{\theta}, p_\phi, p_\psi) = T + V = \frac{1}{2}I_1\dot{\theta}^2 + \frac{(p_\phi - p_\psi \cos\theta)^2}{2I_1 \sin^2\theta} + \frac{p_\psi^2}{2I_3} + V(\cos\theta). \tag{8.117}$$

With several first integrals of the motion available, this is a good opportunity to employ the Hamiltonian approach. The Hamiltonian of the system is given by

$$H = T + V = \frac{p_\theta^2}{2I_1} + \frac{(p_\phi - p_\psi \cos\theta)^2}{2I_1 \sin^2\theta} + \frac{p_\psi^2}{2I_3} + V(\cos\theta) \tag{8.118}$$

Using the Hamiltonian method, the cyclic variables are eliminated yielding the noncyclic pair of first order equations

$$\dot{p}_\theta = -\frac{\partial H}{\partial \theta}, \quad \dot{\theta} = \frac{\partial H}{\partial p_\theta}, \tag{8.119}$$

which yields the second order equation

$$\dot{p}_\theta = I_1\ddot{\theta} = -\frac{\partial}{\partial \theta}H(\theta, p_\theta, p_\phi, p_\psi) = -\frac{\partial}{\partial \theta}V_{eff}(\theta, p_\phi, p_\psi). \tag{8.120}$$

This agrees with the Lagrangian result given by Equation (8.113).

Explicitly solving Equation (8.120) for the polar angle gives

$$\frac{I_1\dot{\theta}^2}{2} = E - V_{eff}(\theta) = E - \left[\frac{(p_\phi - p_\psi \cos\theta)^2}{2I_1 \sin^2\theta} + \frac{p_\psi^2}{2I_3} + V(\cos\theta)\right]. \tag{8.121}$$

This result also follows from energy conservation. Multiplying both sides by $\sin^2 \theta$ gives

$$\sin^2 \theta \left(\frac{d\theta}{dt}\right)^2 = \frac{2}{I_1}\left(E - \frac{p_\psi^2}{2I_3} - mg\ell \cos\theta\right)(1 - \cos^2\theta) - \frac{(p_\phi - p_\psi \cos\theta)^2}{I_1^2}.$$

(8.122)

Now make a change of coordinates $u = \cos\theta \rightarrow du = -\sin\theta d\theta$ to get a cubic equation for the effective interaction $f(u)$

$$\left(\frac{du}{dt}\right)^2 = f(u) = (1 - u^2)(\alpha - \beta u) - (b - au)^2 \geq 0,$$

(8.123)

where the parameters in term of physical constants are

$$\alpha = \frac{2E - I_3\omega_3^2}{I_1}, \quad \beta = \frac{2Mgl}{I_1},$$

$$a = \frac{p_\psi}{I_1}, \quad b = \frac{p_\phi}{I_1}.$$

(8.124)

The result is an elliptic integral of the form

$$t - t_0 = \int_{u_0}^{u} du/\sqrt{(1 - u^2)(\alpha - \beta u) - (b - au)^2}$$

(8.125)

where

$$f(u) = (1 - u^2)(\alpha - \beta u) - (b - au)^2$$
$$= \beta u^3 - (\alpha + a^2)u^2 + (2ab - \beta)u + (\alpha - b^2).$$

(8.126)

The physically allowed limits on the polar angle are constrained to

$$-1 \leq (u = \cos\theta) \leq 1$$

(8.127)

and the restriction to the allowed potential energy domain requires

$$f(u) \geq 0.$$

(8.128)

The turning points of the motion occur when $f(u) = 0$. In general, $f(u)$ is a cubic equation with $\beta = 2Mg\ell/I_1 > 0$, so the solution must go to positive infinity for $u \rightarrow \infty$, and to negative infinity for $u \rightarrow -\infty$. At the physical limits $(u \rightarrow \pm 1)$, the function is either negative or zero since

$$f(\pm 1) = -(b - au)^2 \leq 0.$$

(8.129)

By inspection, no root can be less than -1 ($f(u) < 0$ for all $u < -1$) and one root must always be greater than or equal to $+1$. These conditions constrain

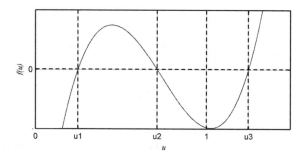

FIGURE 8.13 The effective "potential energy" diagram *f(u)* for the symmetric top. The assessable region is the region between the turning points u_1 and u_2 where *f(u)* > 0. The stable point is a local maximum of *f(u)*.

the functional form of the solution $f(u) = 0$ to the three roots

$$-1 \leq u_1 \leq u_2 \leq 1 \leq u_3 \tag{8.130}$$

The physical motion is bounded to the range $u_1 \leq u \leq u_2$, where $f(u) \geq 0$. $u = \cos\theta$ therefore oscillates (nutates or bobs) between u_1 and u_2 as shown in Figure 8.13.

After the time dependence $\theta(t)$ has been established, one can then solve for time behavior of cyclic coordinates, using

$$\phi(t) = \int \dot{\phi}(t)dt = \int \frac{\partial H}{\partial p_\phi}(\theta(t), p_\phi, p_\psi)dt,$$

$$\psi(t) = \int \dot{\psi}(t)dt = \int \frac{\partial H}{\partial p_\psi}(\theta(t), p_\phi, p_\psi)dt, \tag{8.131}$$

8.5.1.1 Precession with Nutation

Once $u(t)$ is known, one can integrate the remaining Hamilton equations of motion to find the time dependence of the cyclic variables

$$\psi(t) - \psi_0 = \int_0^t \left(\frac{I_1 a}{I_3} - \frac{(b-au)u}{(1-u^2)} \right) dt,$$

$$\phi(t) - \phi_0 = \int_0^t \frac{(b-au)}{(1-u^2)}dt, \tag{8.132}$$

where $a = p_\psi/I_1$ and $b = p_\phi/I_1$. Inspecting the equation for $\dot{\phi}$

$$\dot{\phi}(t) = \frac{b - au(t)}{1 - u^2(t)}, \tag{8.133}$$

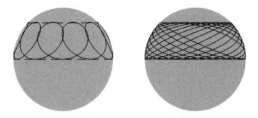

FIGURE 8.14 Precession with nutation (wobbly rotation) can be mono-directional (RHS) or there may be retrograde motion (LHS) depending on the initial conditions.

shows that the precession $\phi(t)$ reverses direction when $\dot{\phi}(t) = 0$. The corresponds to the turning points at

$$u' = b/a = p_\phi/p_\psi \tag{8.134}$$

Two cases are illustrated in Figure 8.14

1. $u' < u_1$ or $u' > u_2$ implies ϕ is monotonously increasing or decreasing, since the turning points are not in the physically allowed region (see the RHS of Figure 8.14).
2. $u_1 < u' < u_2$ implies ϕ reverses direction. This requires retrograde motion (see the LHS of Figure 8.14).

If the turning point is at an extremum $u' = u_1$ or $u' = u_2$, one gets a cusp (not shown in the figure).

Suppose the top is set spinning about its symmetry axis and released with zero initial precession and nutation

$$\dot{\theta}(0) = 0, \quad \dot{\phi}(0) = 0. \tag{8.135}$$

Then $u(0)$ is simultaneously a turning point in the polar angle

$$\dot{\theta}(0) = 0 \Rightarrow u(0) = u_2, \tag{8.136}$$

and of the precession

$$\dot{\phi}(0) = 0 \Rightarrow b - au(0) = 0 \Rightarrow u(0) = u', \tag{8.137}$$

which defines a cusp in the motion of the azimuthal angle. This is illustrated in Figure 8.15. Starting from the cusp, the figure axis initially falls. As it falls, it picks up speed and begins to precess in ϕ. The direction of precession can be determined from the conservation of the azimuthal contribution of angular momentum

$$p_\phi = I_3\omega_3 \cos\theta + I_1\dot{\phi}\sin^2\theta, \tag{8.138}$$

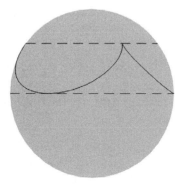

FIGURE 8.15   A cusp in the motion occurs when a turning point in the precession $\dot{\phi}(t) = 0$ occurs at an extremum of u ($\dot{u} = 0$).

where ω_3 is constant. As the figure axis falls, $\cos\theta$ decreases. Therefore, the magnitude of the ω_3 contribution to p_ϕ decreases. The $\dot{\phi}$ contribution must increase in the direction of ω_3 to make up the difference. Therefore, the direction of precession is the same as that of the spin about the symmetry axis.

8.5.1.2 Uniform Precession

To get uniform precession, (precession without bobbing) one needs to find a double root to $f(u) = 0$, giving an extremum in the effective potential diagram. First, consider the case where the double root is less than 1 ($u_1 = u_2 = u_0 < u_3$) as shown in Figure 8.16.

This gives the conditions

$$\dot{\theta} = 0, \quad \dot{\phi} = \text{constant}. \tag{8.139}$$

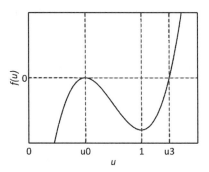

FIGURE 8.16   Conditions for uniform precession require a double root of $f(u)$, with $u_1 = u_2 = u_0 < 1$.

This requires $f'(u_0) = 0$. Solving the preceding equations simultaneously, gives a quadratic equation for $\dot{\phi}$

$$Mg\ell = \dot{\phi}(I_3\omega_3 - I_1\dot{\phi}\cos\theta_0). \tag{8.140}$$

This can also be derived using the Hamiltonian approach, by examining the stationary solutions to the noncyclic variables

$$\dot{p}_\theta = -\frac{\partial H}{\partial \theta} = \sin\theta\frac{\partial H}{\partial u} = \sin\theta(p_\psi\dot{\phi} - 2uI_1\dot{\phi}^2 + Mg\ell) = 0, \tag{8.141}$$

$$\dot{\theta} = \frac{p_\theta}{I_1} = 0. \tag{8.142}$$

Substituting $p_\psi = I_3\omega_3$ immediately gives the desired solution

$$Mg\ell = \dot{\phi}(I_3\omega_3 - I_1\dot{\phi}\cos\theta_0). \tag{8.143}$$

The solution to this quadratic equation gives two solutions: a fast precession solution and a slow precession solution

$$\dot{\phi} = \frac{I_3\omega_3}{2I_1\cos\theta_0}\left[1 \pm \sqrt{1 - \frac{4Mg\ell I_1\cos\theta_0}{I_3^2\omega_3^2}}\right] \tag{8.144}$$

$$\dot{\phi}|_{fast} \simeq \frac{I_3\omega_3}{I_1\cos\theta_0}, \quad \dot{\phi}|_{slow} \simeq \frac{Mg\ell}{I_3\omega_3} \tag{8.145}$$

For a stable solution to exist at all, the top has to have a minimum spin about its symmetry axis

$$p_\psi = I_3\omega_3 \geq \sqrt{4Mg\ell I_1\cos\theta_0}. \tag{8.146}$$

8.5.1.3 Sleeping Top

The sleeping top occurs when $f(u)$ has a double root at $u = 1$ as shown in Figure 8.17. The top spins without precession in an upright position ($\theta = \dot{\theta} = 0$). The two conserved canonical momenta are then degenerate, $p_\phi = p_\psi$. The solution is stable and the top remains upright so long as

$$I_3\omega_3 \geq \sqrt{4Mg\ell I_1}. \tag{8.147}$$

If $I_3\omega_3 < \sqrt{4Mg\ell I_1}$, The solution is unstable. Under a small perturbation, it will begin to wobble.

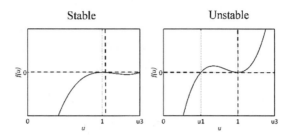

FIGURE 8.17 Energetics diagram for the sleeping top with a double root of $f(u)$ at $u_0 = 1$, showing stable (left-hand side) and unstable solutions (right-hand side).

8.5.2 Precession of the Equinoxes

If the Earth was a perfect sphere, it would experience no gravitational torque and it would act like a gyroscope, always pointing to the same direction of the celestial sphere. However, the bulging of the Earth gives rise to a gravitational quadrupole moment, which, in turn, causes a precession in the field of the sun-moon system. One can get a crude estimate of the size of the effect by noting that the Earth at the spring equinox had only recently entered the house of Pisces, the fish, at the time of Jesus of Nazareth's birth. After approximately 2000 years, it recently moved into the neighboring house of Aquarius the water carrier—each of the twelve houses of the zodiac is approximately $\frac{1}{12}$ of a full circuit about the celestial sphere. Thus, one can make a crude estimate of about 24,000 years for the precession of the equinoxes. Detailed calculations of this precession due to the torque of the sun and moon give a more precise period of 25,700 years [Tong 2005].

8.6   *PRECESSION OF MAGNETIC DIPOLES

Classically, a magnetic dipole with magnetic dipole moment $\boldsymbol{\mu}$ in a magnetic field $\mathbf{B}$ feels a torque given by [Jackson 1999; Morii 2003]

$$\mathbf{N} = \boldsymbol{\mu} \times \mathbf{B}. \tag{8.148}$$

This causes a particle with a magnetic moment to precess in a magnetic field, as shown in Figure 8.18.

The result can be parameterized in terms of a particle's *gyromagnetic ratio* $\tilde{\gamma}$, whereby a spinning charged rigid body has a magnetic moment given by

$$\boldsymbol{\mu} = \tilde{\gamma}\mathbf{L}. \tag{8.149}$$

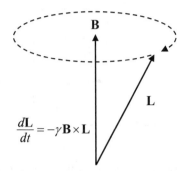

FIGURE 8.18 The precession of a magnetic dipole in a magnetic field.

For a classical body, it can be shown that as long as its charge and mass are distributed identically one gets

$$\tilde{\gamma} = q_e/2m, \tag{8.150}$$

where q_e is the charge of the particle and m is its mass. For more complicated distributions, an empirical *g-factor* is included giving

$$\tilde{\gamma} = g\left(\frac{q_e}{2m}\right). \tag{8.151}$$

Equations (8.148) and (8.149) give rise to the rotational equation of motion

$$\frac{d\mathbf{L}}{dt} = \tilde{\gamma}\mathbf{L} \times \mathbf{B} = \boldsymbol{\omega}_{\mathrm{p}} \times \mathbf{L}. \tag{8.152}$$

Note that the angular momentum $\mathbf{L}$ precesses around the magnetic field $\mathbf{B}$. The angular velocity of precession is called the *Larmor frequency*

$$\boldsymbol{\omega}_{\mathrm{p}} = -\frac{g}{2m}q_e\mathbf{B}. \tag{8.153}$$

Elementary particles such as electrons or protons have intrinsic spin angular momentum $\mathbf{S}$, with a magnetic moment $\boldsymbol{\mu}$ given by

$$\boldsymbol{\mu} = g\left(\frac{e\,\mathbf{S}}{2m}\right). \tag{8.154}$$

where e is the unit charge of an electron. The Dirac equation, for spin-1/2 charged leptons, predicts $g = 2$. This differs from classical charged objects by a factor of 2.[3] The value of g is modified slightly due to electroweak radiative

---

[3]Particle physicists predict $g = 2$ for leptons like the electron and muon, whereas $g = 2.8$ for protons and -1.9 for neutrons.

corrections, but the higher order corrections can be calculated to very high precision. However, new physics beyond the Standard Model of particle physics would modify the predicted results. Therefore, this quantity provides an excellent opportunity to test the Standard Model of particle physics.

What is usually measured is the spin vector $\hat{\mathbf{s}}$ of a particle in its instantaneous rest frame. But the direction of the laboratory velocity relative to this rest frame is changing in time due to the Lorentz Force Law, giving a precession of the coordinate basis as well. The solution for the time rate of change of the spin in slowly varying electromagnetic fields, including relativistic corrections, is given by the *Thomas Precession* formula [Thomas 1927][4]

$$\frac{d\hat{\mathbf{s}}}{dt} = \frac{e}{m}\hat{\mathbf{s}} \times \left(\left(\frac{g}{2} - 1 + \frac{1}{\gamma} \right) \mathbf{B} - \left(\frac{g}{2} - 1 \right) \frac{\gamma}{\gamma + 1} \frac{\mathbf{v} \cdot \mathbf{B}}{c^2} \mathbf{v} \right.$$
$$\left. - \left(\frac{g}{2} - \frac{\gamma}{\gamma + 1} \right) \frac{\mathbf{v}}{c^2} \times \mathbf{E} \right), \tag{8.155}$$

If $g = 2$, the spin, measured in the instantaneous rest frame of the particle, becomes a constant of the motion in the extreme relativistic limit $(\gamma = 1/\sqrt{1 - v^2/c^2} \to \infty)$, so that the result is sensitive to small deviations from $g = 2$.

The magnetic moment of leptons, such the electron and muon, can be calculated using Quantum Electrodynamics (QED) and the Standard Model couplings. Deviations from $g = 2$ occur because leptons are not structureless Dirac particles, but are effectively surrounded by a cloud of virtual particles due to quantum fluctuations. The deviation $a = (g - 2)/2$ is the called the *anomalous moment* of the leptons. To lowest order in the radiative correction QED predicts

$$a = \frac{g - 2}{2} = \frac{\alpha}{2\pi} \approx .0011614. \tag{8.156}$$

where α is the fine structure constant. The detailed QED predictions [Hagiwara 2002] of the gyromagnetic ratio based on including known standard model effects are

$$g_e = 2.002319304374 \pm 0.000000000008, \tag{8.157}$$
$$g_\mu = 2.002331832 \pm 0.0000000012. \tag{8.158}$$

The QED prediction for the anomalous magnetic moment of the electron agrees with the experimentally measured value to best than a part per

---

[4]A detailed derivation of the relativistic Thomas formula can be found in [Jackson 1999 : pp. 561–565].

trillion,[5] making it the most accurately verified prediction in the history of physics. A recent measurement [Brown 2001; Bennett 2004] of $(g_\mu - 2)$ at Brookhaven National Laboratory used long base line measurements of spin precession in a storage ring to determine the anomalous moment of the muon. The muon, being heavier, is more sensitive to hadronic corrections. A statistically significant deviation from the theoretical predictions would be an indication of the need to include new physics beyond the Standard Model. The experimental method requires precise knowledge of the fields and long muon storage times. The published results provide the most precise values yet obtained for the anomalous moment of the muon. A two sigma effect was observed, suggestive, but, not conclusive, of the possibility of new physics at the TeV mass scale. This experiment is one of a number of low energy tests currently underway to study the experimental limits of the Standard Model [Erler 2005].

8.7 SUMMARY

Charles's Theorem states that the general displacement of a rigid body consists of a translation of a point plus a rotation of the body about that point. The kinetic energy and angular momentum of a system of particles separate in the center of mass frame.

$$T = T_{cm} + T' = \frac{1}{2}\mathbf{v}_{cm} \cdot \mathbf{p} + \sum_i \frac{1}{2}\mathbf{v}'_i \cdot \mathbf{p}'_i, \tag{8.159}$$

$$\mathbf{L} = \mathbf{L}_{cm} + \mathbf{L}' = \mathbf{r}_{cm} \times \mathbf{p} + \sum_i \mathbf{r}'_i \times \mathbf{p}'_i. \tag{8.160}$$

If the rigid body is constrained to rotate about a fixed point, the equations of motion for mass points fixed in the rotating frame of a rigid body are given by

$$\left.\frac{d\mathbf{r}'_i}{dt}\right|_{space} = \boldsymbol{\omega} \times \mathbf{r}'_i. \tag{8.161}$$

The kinetic energy and angular momentum about the pivot can be written in terms of its momentum of inertia and angular velocity as

$$\mathbf{L} = \mathbf{I} \cdot \boldsymbol{\omega}, \tag{8.162}$$

$$T = \frac{1}{2}\boldsymbol{\omega} \cdot \mathbf{I} \cdot \boldsymbol{\omega}. \tag{8.163}$$

---

[5] $g/2 = 1.001\,159\,652\,180\,85\,(76)$ [Gabrielse 2006].

The momentum of inertia tensor can be evaluated relative to the body origin as

$$I_{ij} = \mathbf{e}_i \cdot \mathbf{I} \cdot \mathbf{e}_j = \sum_n m_n \begin{bmatrix} (r_n^2 - x_n^2) & -x_n y_n & -x_n z_n \\ -y_n x_n & (r_n^2 - y_n^2) & -y_n z_n \\ -z_n x_n & -z_n y_n & (r_n^2 - z_n^2) \end{bmatrix}. \qquad (8.164)$$

For a continuous distribution this becomes

$$\mathbf{I} = \int_V \rho(x, y, z) dx dy dz \begin{bmatrix} y^2 + z^2 & -xy & -xz \\ -xy & x^2 + z^2 & -yz \\ -xz & -yz & x^2 + y^2 \end{bmatrix}. \qquad (8.165)$$

The Parallel Axis Theorem states that the moment of inertia tensor for a set of axes translationally offset by a displacement $\mathbf{R}$ from a second set passing through the center of mass is the sum of the moment of inertia tensor in the center of mass frame plus the moment of inertia of a point mass placed at $\mathbf{R}$.

$$\mathbf{I} = \mathbf{I}_{\mathrm{cm}} + m(R^2 \mathbf{1} - \mathbf{R}\,\mathbf{R}). \qquad (8.166)$$

The moment of inertia tensor can be diagonalized by a rotation, where the diagonalized moments I_i have principal axes $\mathbf{n}_i$ which solutions to the eigenvalue equations

$$\mathbf{I} \cdot \mathbf{n}_i = I_i \mathbf{n}_i \qquad (8.167)$$

Euler's equations of motion for a rotating rigid body with one point fixed are

$$\mathbf{N} = \mathbf{I} \cdot \frac{d\boldsymbol{\omega}}{dt} + \boldsymbol{\omega} \times \mathbf{L}, \qquad (8.168)$$

$$\frac{dT}{dt} = \boldsymbol{\omega} \cdot \mathbf{N} \qquad (8.169)$$

In the principal axis body frame Euler's equations can be written as

$$\begin{aligned} N_1 &= I_1 \frac{d\omega_1}{dt} + \omega_2 \omega_3 (I_3 - I_2), \\ N_2 &= I_2 \frac{d\omega_2}{dt} + \omega_3 \omega_1 (I_1 - I_3), \\ N_3 &= I_3 \frac{d\omega_3}{dt} + \omega_1 \omega_2 (I_2 - I_1). \end{aligned} \qquad (8.170)$$

Angular velocities are additive. The angular velocity of rotation in the body frame can be expressed in terms of Euler angles by describing the angular velocity in terms of the body frame

$$\boldsymbol{\omega}_b = \mathbf{n}_z\dot{\phi} + \mathbf{n}_{x'}\dot{\theta} + \mathbf{n}_{z'}\dot{\psi}|_b = \begin{bmatrix} \dot{\phi}\sin\psi\sin\theta + \dot{\theta}\cos\psi \\ \dot{\phi}\cos\psi\sin\theta - \dot{\theta}\sin\psi \\ \dot{\phi}\cos\theta + \dot{\psi} \end{bmatrix}_b . \qquad (8.171)$$

The rotational kinetic energy of the rotating rigid body problem can therefore be expressed with respect to a principal axis body frame as

$$\begin{aligned} T(\phi,\theta,\psi,\dot{\phi},\dot{\theta},\dot{\psi},t) = & \frac{1}{2}I_1(\dot{\phi}\sin\psi\sin\theta + \dot{\theta}\cos\psi)^2 \\ & + \frac{1}{2}I_2(\dot{\phi}\cos\psi\sin\theta - \dot{\theta}\sin\psi)^2 \\ & + \frac{1}{2}I_3(\dot{\phi}\cos\theta + \dot{\psi})^2. \end{aligned} \qquad (8.172)$$

If there is an axis of symmetry this simplifies to

$$T(\theta,\dot{\phi},\dot{\theta},\dot{\psi}) = \frac{1}{2}I_0[\dot{\theta}^2 + \dot{\phi}^2\sin^2\theta] + \frac{1}{2}I_3(\dot{\phi}\cos\theta + \dot{\psi})^2. \qquad (8.173)$$

The Hamiltonian of an axially symmetric rigid body in a θ-dependent scalar potential can be written as

$$H = T + V = \frac{p_\theta^2}{2I_0} + \frac{(p_\phi - p_\psi\cos\theta)^2}{2I_0\sin^2\theta} + \frac{p_\psi^2}{2I_3} + V(\theta). \qquad (8.174)$$

The resulting motion reduces to an effective one-dimensional problem.

Rotational motion is complicated by the fact that rotations are not commutative and that the coupling is nonlinear in the angular velocities. Even the free rotation of a non-symmetric rigid body can exhibit complicated tumbling motion in space.

8.8 EXERCISES

1. What height to diameter aspect ratio would a uniform right circular cylinder need to have in order that its inertia ellipsoid be spherically degenerate?

2. If the $\hat{\mathbf{z}}$ axis is an axis of rotational symmetry, the density distribution satisfies

$$\rho(x,y,z) = \rho(r_\perp,z).$$

    a. By direct integration, show that $\hat{z}$ is a principal axis. (Hint: let $x = r_{\perp}\cos\phi$, $y = r_{\perp}\sin\phi$, and average over ϕ).

    b. Show that the moments of inertia about any two axes in the $\hat{x} \wedge \hat{y}$ plane are degenerate.

3. Three equal mass points of mass m are located at $(0,0,a)$, $(a,2a,0)$, $(2a,a,0)$, respectively. Find the principal moments of inertia about the origin and obtain a set of principal axes.

4. Find the principal moments of inertia for a thin lamina, shown in Figure 8.19, with uniform mass density, in the shape of a 45° right triangle with sides of length $\sqrt{2}a$ about its center of mass.

5. The mass quadrupole tensor for a continuous body is defined as

$$Q_{ij} = \int_V \rho(\mathbf{r})d^3\mathbf{r}(3r_i r_j - r^2\delta_{ij}),$$

Express **Q** in terms of the inertia tensor **I**.

6. Show that the moment of inertia with respect to the $\hat{z}$ axis of a solid uniform ellipsoid of mass M and surface $(x/a)^2 + (y/b)^2 + (z/c)^2 = 1$ is given by $M(a^2 + b^2)/5$.

7. Show that the principal moments of inertia about the center of mass of a solid rectangular prism of height h, width w, and depth d, and mass m are given by

$$I_h = \frac{1}{12}m(w^2 + d^2),$$

$$I_w = \frac{1}{12}m(h^2 + d^2),$$

$$I_d = \frac{1}{12}m(w^2 + h^2).$$

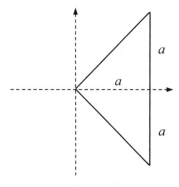

FIGURE 8.19 A thin lamina in the shape of a 45° right triangle.

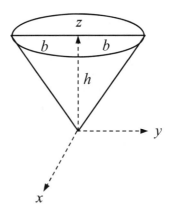

FIGURE 8.20   A right circular cone with height h and base radius b, aligned along its z-axis, with origin at its vertex.

8. Find the moments of inertia about the vertex point of the right circular cone, shown in Figure 8.20, with height h and base radius b, aligned along its z-axis, with origin at its vertex. Assume a uniform density distribution and an integrated mass m. Locate the center of mass of the cone in this coordinate system and use the Parallel Axis Theorem to find the moments of inertia about the center of mass point.

9. Find the principal moments of inertia about its center of a hollow sphere of mass m, with inner radius a and outer radius b. Assume a uniform density distribution. Take the limits $a \to 0$ and $a \to b$ to find the moments of inertia of a solid sphere and a thin walled sphere, respectively.

10. Find the principal moments of inertia about its center of mass of a hollow right circular cylinder of mass m, with inner radius a and outer radius b, and height h. Assume a uniform density distribution.

11. A thin lamina consists of a square metal plate of cross section $2a \times 2a$, with a circular hole of radius $a/4$ drilled through it at a location $(a/2, a/2)$ as shown in Figure 8.21. If the areal density of the material is σ, find the principal moments of inertia and their directions relative to the center of the plate.

12. A thin uniform hoop has radius R and mass M. It is suspended vertically from a point on its rim and allowed to oscillate in the plane of the hoop. Assume that it is in a uniform gravitational field g. Find the frequency of small oscillations.

13. A dumbbell (see Figure 8.22), consisting of two equal points of mass m, which are separated by a massless rigid rod of length l, is rotating

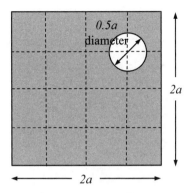

FIGURE 8.21   A square metallic plate of cross section $2a \times 2a$ with a hole of radius $a/4$ drilled through it at location $(a/2, a/2)$ relative to the center of the plate.

with fixed angular velocity ω about an axis that is at an angle θ with respect to the rod. Assume the axis passes through the center of mass of the dumbbell.

a. Find the components of the torque in the directions of the principal axes for the dumbbell.

b. Now find the components of the torque with respect to a fixed space frame.

14. Repeat Exercise 8.13, replacing the dumbbell by a thin rod of length l and mass m rotated with constant angular velocity ω about its midpoint at a fixed angle θ with respect to a rotation axis. Find the components of the torque in the body and space frames.

15. A car door initially at a right angle to a car will slam shut when a car accelerates, as shown in Figure 8.23. Assume that the door is a uniform rectangle, 1.2 m by 1.2 m square, and the acceleration is

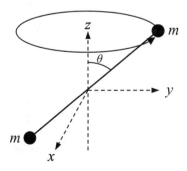

FIGURE 8.22   A rotating dumbbell.

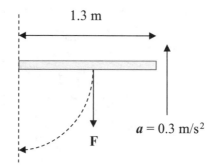

FIGURE 8.23 A rotating car door.

$0.3 \, \text{m/s}^2$. Find the time required for the door to close. The problem can be solved by numerical integration or as an elliptic integral.

16. Consider the torque free motion of an axially symmetric body that is uniformly precessing about its angular momentum axis. Solve for the analytic solutions for the Euler angles as a function of time.

17. A Foucault gyrocompass [Fetter 1980, problem 5.2] is a gyroscope in the form of a symmetric top that is mounted with no gravitational torque, and its symmetry axis is constrained to move in the horizontal plane parallel to the Earth's surface as shown in Figure 8.24. The gyroscope is set spinning about its symmetry axis with an angular velocity $\boldsymbol{\Omega}$.

   a. If $\Omega \gg \omega$ where $\boldsymbol{\omega}$ is the angular velocity of the Earth's rotation, show that the Coriolis force exerts the following torque on the

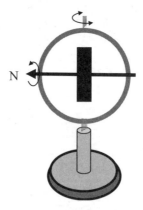

FIGURE 8.24 A Foucault Gyrocompass with its symmetry axis constrained to move in the horizontal plane parallel to the Earth's surface.

gyroscope relative the center of mass:

$$N = -2 \sum_n m_n \{ \mathbf{r}_n \times [\boldsymbol{\omega} \times (\boldsymbol{\Omega} \times \mathbf{r}_n)] \},$$

where the sum runs over all the particles in the gyrocompass with positions $\mathbf{r}_n$ relative to its center of mass.

b. The gyroscope is located at a polar angle (colatitude) θ. Consider its angular motion ϕ about the vertical (1) axis, Show that the symmetry (3) axis of the gyroscope will oscillate about the northerly direction according to

$$\ddot{\phi} = - \left(\frac{I_3}{I_1} \Omega \omega \sin \theta \right) \sin \phi.$$

It can therefore be used as a compass.

18. Given a symmetric top ($I_1 = I_2$) spinning in a uniform gravitational field, find the following condition for uniform precession without nutation using Euler's equations of motion:

$$Mgl = \dot{\phi}(I_3 \omega_3 - I_1 \dot{\phi} \cos \theta_0).$$

19. A pen is set spinning in an upright position. Assume that the pen is a uniform cylinder of length 15 cm and diameter 1 cm. How fast must it spin if it is to remain upright?

20. A symmetric top ($I_1 = I_2$) spinning with one point fixed in a gravitational field starts with the initial conditions [Fetter 1980, problem 5.10]:

$$\theta = \frac{\pi}{3}, \quad \dot{\theta} = 0,$$

$$\dot{\phi} = 2 \left(\frac{Mgl}{3I_1} \right)^{1/2}, \quad \dot{\psi} = \left(\frac{3I_1 - I_3}{I_3} \right) \left(\frac{Mgl}{3I_1} \right)^{1/2}.$$

a. Find the conserved moments p_ϕ and p_ψ, and the effective potential.

b. Show that the solution for θ can be written as

$$\dot{u}^2 = \frac{Mgl}{I_1}(1 - u)^2(2u - 1),$$

where $u = \cos \theta$.

c. Hence, derive the explicit solution

$$\sec\theta = 1 + \sec h\left(\left(\frac{Mgl}{I_1}\right)^{1/2}t\right),$$

d. What is the corresponding behavior of $\dot\phi$ and $\dot\psi$?

e. Plot θ, ϕ, ψ as a function of time. Discuss the qualitative behavior of the solution and its asymptotic limit as $t \to \infty$.

9

CANONICAL
TRANSFORMATION
THEORY

I n this chapter, the action principle will be generalized, allowing for an extended class of canonical transformations between sets of phase space observables. Poisson brackets form a new class of canonical invariants that describe the time evolution of physical operators. Generating functions provide the connection linking sets of canonical coordinates. The role between the generators of infinitesimal canonical transformations and the symmetries of the physical system are explored. The Lie algebra of the rotation group is developed further. The symplectic structure of the theory is seen to give rise to the conservation of phase space and Liouville's theorem.

9.1   GENERALIZING THE ACTION

In Chapter 4, The Euler-Lagrange equations of motion

$$\frac{d}{dt}\frac{\partial L(q,\dot{q},t)}{\partial \dot{q}^i} = \frac{\partial L(q,\dot{q},t)}{\partial q^i}, \tag{9.1}$$

were shown to be a direct result of minimizing the action functional

$$\delta \int_1^2 L(q,\dot{q},t)dt = 0, \tag{9.2}$$

subject to the constraints that the variations are time-independent ($\delta t = 0$) and that the variation of the coordinates vanish at the end points of the integral

$$\delta q_i|_1^2 = 0. \tag{9.3}$$

The equations of motion are unchanged by adding a total time derivative of a function of the coordinates and time to the Lagrangian.

$$L \to L' = L + \frac{dF(q,t)}{dt}. \tag{9.4}$$

When the Legendre transformation exists, the Lagrange and Hamiltonian formulations lead to equivalent descriptions of the physical system. The important role that symmetries of the motion play is readily apparent in either approach. The Hamiltonian approach, however, treats coordinates and momenta on a more equal footing, making the dynamical structure of the theory more apparent.

The real advantage of the Hamiltonian system comes in realizing that, by doubling the number of coordinates, the Hamiltonian method allows greater flexibility in the class of allowed transformations. Lagrangian dynamics allow transformations under all invertible point transformations of the form $q^{\prime i} = q^{\prime i}(q, t)$. It will be shown that this class of point transformations can be extended to the class of all *canonical transformations* of phase space coordinates $q^{\prime i}(q, p, t)$ and momenta $p_i^{\prime}(q, p, t)$ that preserve the invariance of the Hamiltonian form. Canonical transformations are therefore a superset of point transformations. All point transformations are canonical but the converse is not true.

The word "canonical" when used in this context means a standard accepted form. It describes Lagrangian actions that can be written in the standard form

$$\delta \int_1^2 L(q, \dot{q}, p, \dot{p}, t) dt = 0, \tag{9.5}$$

where

$$L(q, \dot{q}, p, \dot{p}, t) = p_i \dot{q}^i - H(q, p, t) + \frac{dF(q, p, t)}{dt}. \tag{9.6}$$

The total time derivative of a function of the coordinates and the time explicitly included in the Lagrangian is needed to allow for the more general class of transformations. Variations of the extended action principle treat generalized coordinates and momenta as independent degrees of freedom. As in the standard action principle, variations of the canonical coordinates are defined to vanish at the end points of integration

$$\delta q^i |_1^2 = 0, \quad \delta p_i |_1^2 = 0, \tag{9.7}$$

so that the additive total time derivative $dF(q, p, t)/dt$ has no effect on the action

$$\delta \int_1^2 \left(\frac{dF(q, p, t)}{dt} \right) dt = \delta F(q, p, t)|_1^2 \equiv 0. \tag{9.8}$$

The minimization procedure gives rise to the generalized Euler-Lagrange equations of motion

$$\frac{d}{dt}\frac{\partial L(q,\dot{q},p,\dot{p},t)}{\partial \dot{q}^i} = \frac{\partial L(q,\dot{q},p,\dot{p},t)}{\partial q^i},$$

$$\frac{d}{dt}\frac{\partial L(q,\dot{q},p,\dot{p},t)}{\partial \dot{p}_i} = \frac{\partial L(q,\dot{q},p,\dot{p},t)}{\partial p_i},$$

(9.9)

which reduce to Hamilton's equations of motion. This invariance of the Lagrangian under the addition of a total time derivative of the phase space coordinates in Equation (9.5) proves to be an essential point of departure for the study of canonical transformation theory that will be developed more fully in the present and next chapters.

As in the earlier version of the action principle, the variations are taken at constant time $\delta t \equiv 0$. The variations are required to be continuous differential functions of the time, but are otherwise arbitrary

$$\delta p_i(\alpha,t) = \alpha \xi_i(t), \quad \delta q^i(\alpha,t) = \alpha \eta^i(t),$$

(9.10)

where α is an expansion parameter that can be made as small as desired, and $\xi_i(t), \eta^i(t)$ are arbitrary differentiable functions of the time. This ensures that the order of differentiation can be reversed in the following terms

$$\delta \frac{d}{dt}p_i = \frac{d}{dt}\delta p_i = \alpha \dot{\xi}_i(t),$$

$$\delta \frac{d}{dt}q^i = \frac{d}{dt}\delta q^i = \alpha \dot{\eta}^i(t).$$

(9.11)

Variation of the action gives

$$\delta S = \delta \int_{t_1}^{t_2} L dt = \delta \int_{t_1}^{t_2} (p_i \dot{q}^i - H(q,p,t))dt$$

$$= \int_{t_1}^{t_2} ((\delta p_i)\dot{q}^i + p_i(\delta \dot{q}^i) - \delta H(q,p,t))dt,$$

(9.12)

where

$$\delta H(q,p,t) = \frac{\partial H}{\partial q^i}\delta q^i + \frac{\partial H}{\partial p_i}\delta p_i.$$

(9.13)

Exchanging the order of differentiations, $\delta \dot{q}^i = d(\delta q^i)/dt$, and integrating by parts gives

$$\int_{t_1}^{t_2}\left(\left(\dot{q}^i - \frac{\partial H}{\partial p_i}\right)\delta p_i - \left(\dot{p}_i + \frac{\partial H}{\partial q^i}\right)\delta q^i\right)dt = -p_i\delta q^i|_1^2 = 0.$$

(9.14)

Since the variations are all independent and vanish at the end points of integration, the above formula results in Hamilton's equations of motion:

$$\dot{p}_i = -\frac{\partial H(q,p,t)}{\partial q^i}, \quad \dot{q}^i = \frac{\partial H(q,p,t)}{\partial p_i}. \tag{9.15}$$

In the generalized action, the coordinates and momenta are independent variables, therefore one can consider general transformations connecting them.

9.2 POISSON BRACKETS

The *Poisson Bracket* plays a central role in the definition of the time evolution of a dynamical system in the Hamiltonian formulation. They are named after mathematician and physicist Siméon-Denis Poisson (1781–1840). Poisson is best known for his works on electricity and magnetism, which essentially created a new branch of mathematical physics, and for his work on celestial mechanics. Poisson brackets have algebraic properties similar to quantum mechanical commutator relations. Therefore, the Poisson bracket formulation is an important point of departure for studies of quantum mechanics and statistical mechanics.

Poisson brackets arise naturally from the Hamiltonian form of the time evolution of a physical operator. A *classical observable* $A(q,p,t)$ is defined as any real function of the phase space coordinates and the time. Using the chain rule to evaluate the time derivative of a classical observable gives

$$\frac{dA(q,p,t)}{dt} = \frac{\partial A}{\partial q^i}\dot{q}^i + \frac{\partial A}{\partial p_i}\dot{p}_i + \frac{\partial A}{\partial t}. \tag{9.16}$$

Substitution of Hamilton's equations of motion results in the regular form

$$\frac{dA(x,p,t)}{dt} = \left(\frac{\partial A}{\partial x_i}\frac{\partial H}{dp_i} - \frac{\partial A}{\partial p_i}\frac{\partial H}{dq_i}\right) + \frac{\partial A}{\partial t}, \tag{9.17}$$

which defines the time evolution of a classical observable, as measured along the particle trajectory, in terms of the Poisson bracket of the observable and the Hamiltonian

$$\frac{dA}{dt} = [A,H]_{qp} + \frac{\partial A}{\partial t}, \tag{9.18}$$

where the Poisson bracket of two observables is defined as the antisymmetric bilinear form

$$[A, B]_{q,p} = \frac{\partial A}{\partial q^i} \frac{\partial B}{\partial p_i} - \frac{\partial A}{\partial p_i} \frac{\partial B}{\partial q^i}. \tag{9.19}$$

9.2.1 Fundamental Properties

Poisson Bracket Algebra obeys rules similar to matrix commutation algebra. In will be shown in a latter section that Poisson Brackets are a type of *canonical invariant*; i.e., the Poisson Bracket of any two observables is the same for all choices of canonical coordinates, so that the subscripts with respect to the coordinate frame can be dropped

$$[A, B]_{qp} = [A, B]_{QP} = [A, B]. \tag{9.20}$$

Poisson Brackets have the following important mathematical properties:

1. Poisson Brackets are antisymmetric under exchange of observables

$$[A, B]_{qp} = \frac{\partial A}{\partial q^i} \frac{\partial B}{\partial p_i} - \frac{\partial A}{\partial p_i} \frac{\partial B}{\partial q^i} = -[B, A]_{qp}, \tag{9.21}$$

Therefore, the Poisson Bracket of an observable with respect to itself identically vanishes

$$[A, A]_{qp} \equiv 0. \tag{9.22}$$

2. Poisson Brackets are bilinearly distributive

$$[aA + bB, C] = a[A, C] + b[B, C], \tag{9.23}$$

where (A, B, C) are physical observables and (a, b) are c-numbers (constant parameters).

3. The Poisson Bracket of a product of two observables with a third observable satisfies Leibniz's Rule:

$$[AB, C] = A[B, C] + [A, C]B. \tag{9.24}$$

4. The Jacobi Identity is satisfied

$$[A, [B, C]] + [B, [C, A]] + [C, [A, B]] = 0. \tag{9.25}$$

The previous properties are similar to the commutator algebra of quantum mechanics. The proofs of all these statements are straightforward, see [Goldstein 2002], or work it out as one of the exercises in this text.

Dropping the coordinate subscripts, the time derivative of a physical observable can now be written in the compact form

$$\frac{dA(q,p,t)}{dt} = [A, H] + \frac{\partial A}{\partial t}. \tag{9.26}$$

From this, one can conclude that any physical observable that does not explicitly depend on the time is a constant of the motion, if, and only if, its Poisson bracket with the Hamiltonian of the system vanishes. Equation 9.26 defines an alternate definition of classical mechanics on a par with the Lagrangian and Hamiltonian prescriptions. Applying the formula to the coordinates and momenta simply recovers Hamilton's equations of motion

$$\dot{q}^i = [q^i, H] = \frac{\partial H}{\partial p_i},$$
$$\dot{p}_i = [p_i, H] = -\frac{\partial H}{\partial q^j}. \tag{9.27}$$

It is straightforward to show that the fundamental Poisson bracket relationships between phase space coordinates are similar to those from quantum mechanics. The classical bracket relationships are

$$[q_i, q_j]_{\text{PB}} = 0 \quad [p_i, p_j]_{\text{PB}} = 0 \quad [q_i, p_j]_{\text{PB}} = \delta_{ij}. \tag{9.28}$$

In Section 9.3.5, a transformation will be shown to be canonical if it preserves the fundamental Poisson bracket relationships

$$[Q_i, Q_j]_{\text{PB}} = 0 \quad [P_i, P_j]_{\text{PB}} = 0 \quad [Q_i, P_j]_{\text{PB}} = \delta_{ij}. \tag{9.29}$$

The Poisson bracket given by Equation 9.19 evaluates to the same result independent of the choice of canonical coordinates. Therefore, the Poisson bracket representation results in a coordinate-independent statement of physical law.

9.2.2 Series Solution to the Equations of Motion

Consider a one-dimensional problem, where the solution of the problem can be expressed as a Taylor expansion

$$p(q_0, p_0, t) = p_0 + \frac{dp}{dt}\bigg|_0 (t - t_0) + \cdots = \sum_{n=0}^{\infty} \frac{1}{n!} \left(\frac{d^n p}{dt^n}\right)_0 (t - t_0)^n, \tag{9.30}$$

$$q(q_0, p_0, t) = q_0 + \frac{dq}{dt}\bigg|_0 (t - t_0) + \cdots = \sum_{n=0}^{\infty} \frac{1}{n!} \left(\frac{d^n q}{dt^2}\right)_0 (t - t_0)^n. \tag{9.31}$$

This series solution is valid for all times within its radius of convergence. Substituting Poisson brackets for the total time derivatives evaluated at initial time gives, by iteration

$$\frac{dp}{dt}\bigg|_0 = [p, H]|_0, \quad \frac{d^2p}{dt^2}\bigg|_0 = [[p, H], H]|_0, \cdots,$$
$$\frac{dq}{dt}\bigg|_0 = [q, H]|_0, \quad \frac{d^2q}{dt^2}\bigg|_0 = [[q, H], H]|_0, \cdots. \tag{9.32}$$

Substitution into the series gives the Lie bracket expansion

$$q(q_0, p_0, t) = q|_0 + [q, H]_0 t + \frac{1}{2}[[q, H], H]_0 t^2 + \cdots,$$
$$p(q_0, p_0, t) = p|_0 + [p, H]_0 t + \frac{1}{2}[[p, H], H]_0 t^2 + \cdots. \tag{9.33}$$

For example, consider a particle moving in a uniform gravitational field. The Hamiltonian of the system is

$$H(q, p) = \frac{p^2}{2m} + mgq. \tag{9.34}$$

Using Poisson brackets, one can expand the solution as a power series in time. In this simple case, the series for the coordinate terminates after the first few terms

$$[q, H] = \left[q, \frac{p^2}{2m}\right] = p\left[q, \frac{p}{2m}\right] + \left[q, \frac{p}{2m}\right]p = \frac{p}{m},$$
$$[[q, H], H] = \left[\frac{p}{m}, H\right] = \left[\frac{p}{m}, mgq\right] = -g, \tag{9.35}$$
$$[[[q, H], H], H] = 0,$$

giving the series solution

$$q(q_0, p_0, t) = q|_0 + [q, H]_0(t - t_0) + \frac{1}{2}[[q, H], H]_0(t - t_0)^2$$
$$= q_0 + \frac{p_0}{m}(t - t_0) - \frac{g}{2}(t - t_0)^2. \tag{9.36}$$

Likewise, the series for the momentum also truncates

$$[p, H] = -mg, \quad [[p, H], H] = 0, \tag{9.37}$$

giving the series solution

$$p(q_0, p_0, t) = p|_0 + [p, H]_0 (t - t_0)$$
$$= p_0 - mg(t - t_0). \tag{9.38}$$

As a second example, the simple harmonic oscillator can be described by the Hamiltonian

$$H(p, q) = \frac{p^2}{2m} + \frac{kq^2}{2}. \tag{9.39}$$

The classical commutators of displacement q and momentum p with respect to this Hamiltonian are

$$\dot{q} = [q, H] = \left[q, \frac{p^2}{2m} \right] = \frac{p}{m},$$
$$\dot{p} = [p, H] = \left[p, \frac{kq^2}{2} \right] = -kq, \tag{9.40}$$

The above equations are simply a restatement of Hamilton's equations of motion. Solving for the time behavior of the displacement by substituting into the series expansion, then summing the series gives the well-known solution to the initial value problem

$$q(t) = q_0 \cos \omega t + \frac{p_0}{\omega m} \sin \omega t. \tag{9.41}$$

The details of the solution are left as an exercise. Note that no integration is required to solve the equations of motion.

9.2.3 Connection to Quantum Mechanics

Paul A. M. Dirac (1902–1984), a pioneer of quantum mechanics, was the first to realize the connection between classical Poisson Brackets and quantum mechanical commutators. In quantum mechanics, the canonical coordinates (q, p) are replaced by quantum mechanical operators $(\hat{q}, \hat{p})$ satisfying the fundamental quantum mechanical matrix commutation relations [Dirac 1981]

$$[\hat{q}^i, \hat{q}^j]_{QM} = 0, \quad [\hat{p}_i, \hat{p}_j]_{QM} = 0, \quad [\hat{q}^i, \hat{p}_j]_{QM} = i\hbar \delta^i_j, \tag{9.42}$$

Note that $\hbar$ is Plank's Constant divided by 2π,[1] and $[A, B]_{QM}$ is the quantum mechanical commutator between two quantum operators. The factor $\hbar$, with

---

[1] Max Karl Ernst Ludwig Planck (1858–1947) is considered to be the founder of quantum theory. He introduced the concept of the discrete quantum of energy to deal with problem of the ultraviolet catastrophe in black-body radiation.

units of angular momentum, is required on dimensional grounds. The factor i is needed to make the commutator product Hermitian (Hermitian operators have real eigenvalues).

The connection between classical mechanics and quantum mechanics is established by making the general substitution of the quantum mechanical brackets for the classical Poisson Brackets

$$[A, B]_{PB} \rightarrow \frac{1}{i\hbar}[\hat{A}, \hat{B}]_{QM}. \tag{9.43}$$

where the time evolution of a physical observable in the *Heisenberg Representation* of quantum mechanics is generated by the Hamiltonian operator

$$\frac{d\hat{A}}{dt} = \frac{1}{i\hbar}[\hat{A}, \hat{H}]_{QM} + \frac{\partial\hat{A}}{\partial t}. \tag{9.44}$$

This procedure is called *canonical quantization* and can be applied to any classical Hamiltonian system, using any set of canonical coordinates, provided some basic ordering rules are applied to guarantee the observables are Hermitian.

For example, the simple harmonic oscillator can be quantized by defining a Hamiltonian

$$H(\hat{p}, \hat{q}) = \frac{\hat{p}^2}{2m} + \frac{k\hat{q}^2}{2}. \tag{9.45}$$

where the classical observables are replaced by quantum mechanical operators. The commutators of displacement q and momentum p with respect to this Hamiltonian are

$$\begin{aligned}
\frac{d\hat{q}}{dt} &= \frac{1}{i\hbar}[\hat{q}, H] = \frac{1}{i\hbar}\left[\hat{q}, \frac{\hat{p}^2}{2m}\right] = \frac{\hat{p}}{m}, \\
\frac{d\hat{p}}{dt} &= \frac{1}{i\hbar}[p, H] = \frac{1}{i\hbar}\left[\hat{p}, \frac{k\hat{q}^2}{2}\right] = -k\hat{q},
\end{aligned} \tag{9.46}$$

A comparison of Equation (9.46) with the classical equivalent (9.40) shows that the quantum mechanical operators evolve in time in exactly the same way that the classical observables do.

The time evolution of classical and quantum observables are formally the same in the Poisson Bracket and Heisenberg models of the respective theories. What is different is that, in quantum mechanics, one cannot a priori know the value of a coordinate and its conjugate momentum simultaneously.

Because they don't commute, they represent incommensurate observables. The best one can hope to do is predict the expectation value of the trajectory with respect to some state vector, $|\psi\rangle$. The expectation value of the path is given by

$$\langle \hat{q}(t) \rangle = \langle \psi(t_0)| \bar{U}(t,t_0) \hat{q}(t_0) U(t,t_0) | \psi(t_0) \rangle. \tag{9.47}$$

where $U(t,t_0)$ is the *time evolution operator*.

The quantum state vector encodes additional information about the internal structure of the particle that is not present in the classical particle picture, including its internal spin polarization state and particle favor identification. As a wave-amplitude, its phase information can be coherently summed to reproduce the wave-like aspects of particle behavior.

One of the ontological problems with the standard interpretation of quantum mechanics is that it provides no satisfactory explanation for the collapse of the wave function at the moment of measurement. This "collapse", of course, attributes a reality to the wave function that Niels Bohr and many others would deny exists [Murdock 1987; Pais 1991].

In realistic interpretations of quantum mechanics, one can determine the previous history of a particle by a measurement of outcomes, but such a posteriori knowledge has no predictive value. For example, by measuring a charged particle's position and angle at both the entrance and exit of a magnetic spectrometer, one can deduce, after the fact of measurement, its classical trajectory and, by inference, the momentum of a particle. Blocking any part of the presumed classical path leads to a loss of the signal in the short wavelength limit, confirming that the particle's evolution, in the classical limit, is accurately predicted by the classical path. However, this is insufficient to predict its future path after it has been interfered with by the second measurement of position and angle.

If one repeatedly measures a particle's position, as in an ionizing bubble chamber, the trajectory of a particle, in between hard collisions, or its possible decay into other particles, appears to adhere closely to a classical trajectory. Since it is implausible to believe that a particle is better behaved when destructively inferred with, rather than left alone in free space, one might come to suspect that the "collapse of the wave function" in a measurement of position has nothing to do with the intrinsic properties of the particle, and everything to do with the reduced uncertainty in the knowledge of its location. Niels Bohr warned against attempting to attach more physical significance to the quantum wave function than it merits.

This inability to dynamically interpret the wave function bothered Einstein and others, who, without questioning the general correctness of the theory, questioned its completeness. In 1935, Einstein, Podolsky, and

Rosen proposed a thought experiment [Einstein 1935], referred to as the EPR paradox, suggesting that hidden variables could be used to sustain a realistic interpretation of quantum mechanics by completing the theory. In 1964, John Stewart Bell (1928–1990) demonstrated that any local hidden variable theory is incompatible with quantum mechanics [Bell 1964]. Today, the EPR paradox serves as a cautionary tale of how quantum mechanics (and, by inference, experimental reality) violates classically formed intuitions.

Bell's Theorem is often cited, mistakenly, as an argument against any possible realistic (dynamical) interpretations of quantum mechanics. Measurements of Bell's Inequality [Aspect 1982; Weihs 1998] are consistent with the predictions of quantum mechanics and inconsistent with a local dynamical theory involving hidden variables. However, this is a straw man argument. One can only measure a particle by measuring its interactions with its gauge fields. These gauge interactions modify both the particle itself and its environment. When one integrates out the gauge fields in an interacting quantum field theory, one finds that quantum mechanics represents both a non-local and non-linear theory. Bell's position, in fact, is that quantum mechanics is demonstrably and irreducibly nonlocal. In other words, what measurements of Bell's Inequality demonstrate is that any hypothetical, future, dynamical theory of particle motion, to be consistent with quantum mechanics and experiment, must conform to what is already known about the nature of physical reality: it is non-local and non-linear. Nonlocal theories can be expressed as local, Lorentz invariant, theories by the addition of auxiliary gauge fields. Quantum mechanics, in itself, provides an example that such a class of theories exist.

There is no doubt that any hypothetical "classical" path of a photon passing through a double slit interference experiment would have to be acutely sensitive to its boundary conditions. If one only knew why and how, one would be better able to understand nature. David Joseph Bohm (1917–1992) attempted to address some of these issues with his interpretation of quantum mechanics, which will be discussed in the context of the Hamilton-Jacobi Equation in Chapter 10.

9.3   GENERATING FUNCTIONS

The extended action principle defines a Lagrangian which can be written in the canonical form of Equation (9.6). The addition of an arbitrary function of time to the Lagrangian is not simply a footnote to mention in passing. It provides the mechanism for generating canonical transformations of the equations of motion which preserve the Hamiltonian form. In general, the transformation will individually transform both the Hamiltonian and the total derivative terms, while leaving the Lagrangian invariant.

$$p_i\dot{q}^i - H(p,q,t) + \frac{dF(q,p,t)}{dt} = P_i\dot{Q}^i - \tilde{H}(Q,P,t) + \frac{dF'(Q,P,t)}{dt}. \quad (9.48)$$

Although the functions F and F' are individually unknown and arbitrary, their difference is a property of the desired transformation. Defining $\tilde{F} = F' - F$, this mapping by invariance can be rewritten as

$$p_i\dot{q}^i - H(p,q,t) = P_i\dot{Q}^i - \tilde{H}(Q,P,t) + \frac{d\tilde{F}(q,p,Q,P,t)}{dt}. \quad (9.49)$$

Here $\tilde{F}$ is called the *generator of the canonical transformation* between the two coordinate systems. Hamilton's equations of motion are form-invariant under canonical transformations, giving the identical form to the equations of motion with respect to the new Hamiltonian system of equations

$$\dot{Q}^i = \frac{\partial\tilde{H}}{\partial P_i}, \quad \dot{P}_i = -\frac{\partial\tilde{H}}{\partial Q^i}. \quad (9.50)$$

Invariance of the Lagrangian provides a preliminary test for the "canonical" property: A transformation $(q,p) \rightarrow (Q,P)$ is *canonical* if a generating function exists such that

$$p_i\dot{q}^i - H(p,q,t) = P_i\dot{Q}^i - \tilde{H}(Q,P,t) + \frac{d\tilde{F}}{dt} \quad (9.51)$$

The function $\tilde{F}$ is referred to as the generator of a Legendre transformation between the two coordinate systems. In order to connect the coordinate systems, the function $\tilde{F}$ needs to be expressed as some mixture of initial and final coordinates. Half of the coordinates are to be chosen from the initial set of coordinates and the other half are to be chosen from the final set of coordinates. For example, consider a hypothetical transformation of canonical coordinates where the new Hamiltonian identically vanishes, resulting in

$$\dot{P}_i = \dot{Q}^i = 0. \quad (9.52)$$

Then $(P,Q) = (P_0, Q_0)$ are constants of the motion. The transformation

$$q(t) = q(Q_0, P_0, t), \quad p(t) = p(Q_0, P_0, t). \quad (9.53)$$

is, in fact, the desired solution of the initial value problem, where (Q_0, P_0) are some specified initial conditions. Therefore, the solution of Hamilton's equations of motion can be reduced to finding an appropriate generator $\tilde{F}$ for the time evolution of the dynamical system. This is the essential goal of the Hamilton-Jacobi method which will be developed in Chapter 10. A caveat: a complete solution to the Hamilton-Jacobi problem exists only if the equations of motion are completely integrable; this, in turn, depends on the existence of a sufficient number of mutually commuting observables.

Generators of canonical transformations are normally written in one of four standard forms, related to each other by Legendre transformations, provided that the following transforms exist

$$
\begin{aligned}
\tilde{F}_1 &= F_1(q, Q, t), \\
\tilde{F}_2 &= -P_i Q^i + F_2(q, P, t), \\
\tilde{F}_3 &= p_i q^i + F_3(p, Q, t), \\
\tilde{F}_4 &= p_i q^i - P_i Q^i + F_4(p, P, t).
\end{aligned}
\tag{9.54}
$$

The preceding classification of basic Legendre transformations give rise to the generator summary chart shown in Table 9.1, which relates the partial derivatives of the generating functions to the remaining coordinates.

Hamilton's equations of motion are also form-invariant under a multiplicative change of scale of the Lagrangian. The combination of a *scale*

TABLE 9.1 Generator summary table for the four basic types of generating functions

	Generator $\tilde{F}$	Transformation		Trivial Case
Type-1	$F_1(q, Q, t)$	$p_i = \dfrac{\partial F_1}{\partial q^i}$ $P_i = -\dfrac{\partial F_1}{\partial Q^i},$		$F_1 = q^i Q_i$
		$\tilde{H} - H = \dfrac{\partial F_1}{\partial t}$		
Type-2	$-P_i Q^i + F_2(q, P, t)$	$p_i = \dfrac{\partial F_2}{\partial q^i},$ $Q^i = \dfrac{\partial F_2}{\partial P_i},$		$F_2 = q^i P_i$
		$\tilde{H} - H = \dfrac{\partial F_2}{\partial t}$		
Type-3	$p_i q^i + F_3(p, Q, t)$	$q^i = -\dfrac{\partial F_3}{\partial p_i},$ $P_i = -\dfrac{\partial F_3}{\partial Q^i}$		$F_3 = Q^i p_i$
		$\tilde{H} - H = \dfrac{\partial F_3}{\partial t}$		
Type-4	$p_i q^i - P_i Q^i + F_4(p, P, t)$	$q^i = -\dfrac{\partial F_4}{\partial p_i},$ $Q^i = \dfrac{\partial F_4}{\partial P_i},$		$F_4 = p_i P^i$
		$\tilde{H} - H = \dfrac{\partial F_4}{\partial t}$		

transformation with a canonical transformation is sometimes called an *extended canonical transformation*. Extended canonical transformations have the general form

$$P_i \dot{Q}^i - H' + d\tilde{F}/dt = \lambda(p_i \dot{q}^i - H). \tag{9.55}$$

9.3.1 Type-2 and Type-3 Transforms

The Type-2 and Type-3 transforms are proper subgroups of all canonical transformations since they both contain the identity element. The F_2 transformation is the most commonly encountered and will be used to develop Hamilton-Jacobi theory in Chapter 10. The F_3 transform merely replaces the role played by the initial and final set of coordinates in the F_2 transform. Consider an F_2 transform of the form $F_2(q, P, t)$. Requiring that the Lagrangian transform by invariance gives the relation

$$p_i \dot{q}^i - H(p, q, t) = P_i \dot{Q}^i - \tilde{H}(Q, P, t) + \frac{d(dF_2(q, P, t) - P_i Q^i)}{dt}, \tag{9.56}$$

which results in

$$\frac{dF_2(q, P, t)}{dt} = p_i \dot{q}^i + Q^i \dot{P}_i - (H(q, p, t) - \tilde{H}(Q, P, t)). \tag{9.57}$$

Taking the partial derivatives of $dF_2(q, P, t)/dt$

$$\frac{\partial F_2}{\partial q^i} \dot{q}^i + \frac{\partial F_2}{\partial P_i} \dot{P}^i + \frac{\partial F_2}{\partial t} = p_i \dot{q}^i + Q^i \dot{P}_i + \tilde{H} - H, \tag{9.58}$$

and comparing terms, gives

$$\frac{\partial F_2(q, P, t)}{\partial q^i} = p_i, \quad \frac{\partial F_2(q, P, t)}{\partial P_i} = Q^i, \quad \frac{\partial F_2(q, P, t)}{\partial t} = \tilde{H} - H, \tag{9.59}$$

which agrees with the entries for a Type-2 transformation made in Table 9.1.

9.3.1.1 Identity Transform

The simplest example of an F_2 transformation is the identity transformation defined as

$$F_2(q, P, t) = q^i P_i \tag{9.60}$$

resulting in the equations

$$P_i = p_i = \frac{\partial F_2}{\partial q^i}, \quad q^i = Q^i = \frac{\partial F_2}{\partial P_i}, \quad \tilde{H} - H = \frac{\partial F_2}{\partial t} = 0. \tag{9.61}$$

Therefore, the function

$$\tilde{F}_2 = F_2(q, P) - P_i Q^i = P_i(q^i - Q^i) \qquad (9.62)$$

generates the identity transform

$$p_i = P_i, \quad q^i = Q^i, \quad H(q, p) = \tilde{H}(Q, P). \qquad (9.63)$$

9.3.1.2 Point Transformations

All invertible point transformations are canonical transformations as can be seen from the generator equation

$$F_2(q, P) = P_i Q^i(q, t) + \Lambda(q, t), \qquad (9.64)$$

resulting in

$$Q^i = \frac{\partial F_2}{\partial P_i} = Q^i(q, t), \quad p_i = \frac{\partial F_2}{\partial q^i} = P_j \frac{\partial Q^j(q, t)}{\partial q^i} + \frac{\partial \Lambda}{\partial q^i}, \qquad (9.65)$$

where $Q^i = Q^i(q, t)$ is an arbitrary, invertible, point transform of coordinates. Note that the transformation for the momentum is arbitrary up to a gauge transformation. Note also that velocities are contravariant vectors in configuration space, transforming as

$$\dot{Q}^i = \frac{\partial Q^i(q, t)}{\partial q^j} \dot{q}^j, \qquad (9.66)$$

while momenta are covariant vectors (up to a gauge term) transforming as

$$\left(p_i - \frac{\partial \Lambda}{\partial q^i} \right) = P_j \frac{\partial Q^j(q, t)}{\partial q^i}. \qquad (9.67)$$

9.3.2 Type-1 and Type-4 Transforms

The Type-1 and Type-4 transforms are not proper subgroups of the set of canonical transformations as they lack an identity element. However, they do include the *exchange transformation*, which interchanges coordinate and momentum degrees of freedom.

A Type-1 transformation can be written as

$$P_i \dot{Q}^i - H' + \frac{dF_1(q, Q, t)}{dt} = p_i \dot{q}^i - H. \qquad (9.68)$$

Applying the chain rule to the total time derivative gives

$$\frac{dF_1(q,Q,t)}{dt} = \frac{\partial F_1}{\partial t} + \frac{\partial F_1}{\partial q^i}\dot{q}^i + \frac{\partial F_1}{\partial Q^i}\dot{Q}^i. \qquad (9.69)$$

Substituting into the Lagrangian gives

$$P_i\dot{Q}^i - H'(Q,P,t) + \frac{\partial F_1}{\partial t} + \frac{\partial F_1}{\partial q^i}\dot{q}^i + \frac{\partial F_1}{\partial Q^i}\dot{Q}^i = p_i\dot{q}^i - H(q,p,t). \qquad (9.70)$$

Comparing terms gives the following set of partial differential equations

$$P_i = -\frac{\partial F_1(q,Q,t)}{\partial Q^i}, \quad p_i = \frac{\partial F_1(q,Q,t)}{\partial q^i}, \quad \tilde{H}(Q,P) - H(q,p) = \frac{\partial F_1(q,Q,t)}{\partial t}, \qquad (9.71)$$

which agrees with the entries for a Type-1 transformation given in Table 9.1. If a generating function is specified, these equations result in a mapping between the old and new coordinate systems. Conversely, if the coordinate transformation is known, one can attempt to integrate the equations to find a generating function, assuming that the equations are integrable using the F_1 transform. Not all canonical transformations can be written in any one of forms listed in Table 9.1. For example, the identity transform can be written using F_2 or F_3 forms but not using F_1 or F_4 forms.

9.3.2.1 Exchange Transformation

The trivial F_1 transformation, $F_1 = q^i Q_i$, generates an exchange of coordinates and momenta

$$P_i = -q_i = -g_{ij}q^j, \quad Q^i = p^i = g^{ij}p_j, \qquad (9.72)$$

while leaving the Hamiltonian unchanged $\tilde{H}(Q,P) = H(q,p)$. The trivial Type-4 transform provided in the generator table (Table 9.1) exhibits similar behavior. The sign flip under the exchange transformation is due to the *skew symmetry* of the underlying *symplectic*[2] structure of the theory. The origin of this skew symmetry can be seen most clearly by integrating the standard canonical form by parts

$$p_i\dot{q}^i - H = -q^i\dot{p}_i + \frac{d(p_i q^i)}{dt} - H, \qquad (9.73)$$

---

[2]Symplectic means skewed or twisted in Greek.

and rewriting the Lagrangian in the explicitly skew-symmetric or symplectic form

$$L = \frac{1}{2}(p_i \dot{q}^i - q^i \dot{p}_i) - H + \frac{dF(q,p,t)}{dt}. \tag{9.74}$$

9.3.2.2 F_1 Transformation of the Simple Harmonic Oscillator

As an example of the application of an F_1 transform, consider a one-dimensional harmonic oscillator with Hamiltonian

$$H(q,p) = \frac{p^2}{2m} + \frac{kq^2}{2} = \frac{1}{2m}(p^2 + m^2\omega^2 q^2) = \omega J. \tag{9.75}$$

The solution to this problem has an elegant solution in terms of action-angle coordinates (J, φ), where the Hamiltonian is cyclic in the angle coordinate φ. The detailed properties of action-angle coordinates will be discussed in detail in the following chapter. For the present purposes, consider the following coordinate mapping to a new set of coordinates which produces the known solution to the simple harmonic oscillator if the substitution $\varphi(t) = \omega t + \varphi_0$ is made

$$p(J,\varphi) = \sqrt{2m\omega J}\cos(\varphi), \quad q(J,\varphi) = \sqrt{2J/m\omega}\sin(\varphi). \tag{9.76}$$

The existence of an F_1 generator for this transformation will be demonstrated after exploring the properties of the transformation. The new coordinates are $Q = \varphi$, the phase angle of oscillation, and its canonical angular momentum $P = J$, also called the *action variable*. Since the transformation is time-independent, the Hamiltonian remains invariant under the canonical transformation. In the new system of coordinates, however, the Hamiltonian is cyclic in the angle coordinate

$$H(J,\varphi) = \omega J, \tag{9.77}$$

where $\omega = \sqrt{k/m}$ is a constant number (*c-number*) equal to the oscillator frequency. Hamilton's equations of motion in the new coordinate system are

$$\dot{J} = -\frac{\partial H}{\partial \varphi} = 0, \quad \dot{\varphi} = \frac{\partial H}{\partial J} = \omega. \tag{9.78}$$

The action variable with units of angular momentum is a constant of the motion. The angular velocity is given by the oscillator frequency, and the phase angle φ is a uniform linear function of time giving the action-angle solution

$$J = H/\omega. \tag{9.79}$$

$$\varphi(t) = \omega t + \varphi_0. \tag{9.80}$$

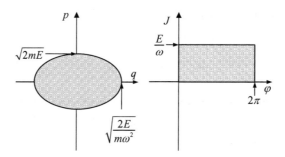

FIGURE 9.1 Phase space diagram of the Simple Harmonic Oscillator (left), and the mapping of the phase space into action-angle coordinates (right).

The phase space trajectories for constant energy in the original system of coordinates (p, q) are ellipses, as shown in the left side of Figure 9.1. The transformation to action-angle coordinates straightens out the lines of flow in phase space. In action-angle coordinates, the lines of flow are geodesics of constant action on the surface of a cylinder where the action represents the cylinder axis and φ is the cyclic angle, usually taken to be to be periodic on interval 2π. Its canonical momentum has units of action, hence the term action-angle variables.

The transformation to action-angle coordinates is clearly canonical since the equations of motion are form-invariant under the transformation. To close the argument, one must now show that an appropriate generating function exists that generates this transformation, thereby preserving the canonical form. Assuming an appropriate F_1 transform exists, then

$$p\dot{q} - H(pq) = J\dot{\varphi} - H(J, \varphi) + \frac{dF_1(q, \varphi)}{dt}, \tag{9.81}$$

where the generator satisfies the conditions

$$p = \frac{\partial F_1(q, \varphi)}{\partial q}, \quad J = -\frac{\partial F_1(q, \varphi)}{\partial \theta} \quad \tilde{H} - H = \frac{\partial F_1(q, \varphi)}{\partial t} = 0. \tag{9.82}$$

To carry out the integration of the first equation, one needs to first express the momentum as a function of the initial and final coordinates $p = p(q, Q)$. Taking the ratio

$$p/q = m\omega \cot \varphi, \tag{9.83}$$

gives the desired functional form

$$p = m\omega q \cot \varphi = \partial F_1(q, \varphi)/\partial q. \tag{9.84}$$

Integration gives

$$F_1(q, \varphi) = \frac{1}{2} m\omega q^2 \cot \varphi + f(\varphi). \tag{9.85}$$

where $f(\varphi)$ is an arbitrary function of the new angle coordinate, which can be set to zero in this case. One can verify the result by substitution into the second differential equation, giving

$$J = -\frac{\partial F_1(q, \varphi)}{\partial \varphi} = -\frac{1}{2} \frac{\partial m\omega q^2 \tan \varphi}{\partial \varphi} = \frac{m\omega q^2}{2 \sin^2 \varphi}, \tag{9.86}$$

or

$$q = \sqrt{2J/m\omega} \sin \varphi. \tag{9.87}$$

Since a F_1 transform exists, the action-angle transformation is canonical.

The mapping defined by Equation (9.76) can also be generated using a F_2 transformation. This will be demonstrated in Chapter 10. The generating function associated with a given canonical transformation is therefore not unique. There may be multiple generating functions, related by Legendre transformations, which produce the same canonical transformation equations. Moreover, one can always add an arbitrary function of time alone to a generating function, since the transformation

$$H \to H - \frac{dg(t)}{dt} \tag{9.88}$$

has no affect on the action integral. (It modifies the Hamiltonian without affecting the equations of motion.)

9.3.3 Active and Passive Interpretations

Assume the Hamiltonian is time-independent and transforms by invariance

$$H(q, p) = \tilde{H}(Q, P). \tag{9.89}$$

The mapping

$$Q^i = Q^i(q, p), \quad P_i = P_i(q, p) \tag{9.90}$$

can then be interpreted as a passive transformation at fixed time from one set of conjugate observables to another. The four basic generators can be written as

$$\frac{dF_1(q,Q)}{dt} = p_i \dot{q}^i - P_i \dot{Q}^i, \qquad \frac{dF_2(q,P)}{dt} = p_i \dot{q}^i + Q^i \dot{P}_i,$$

$$\frac{dF_3(p,Q)}{dt} = -q^i \dot{p}_i - P_i \dot{Q}^i, \qquad \frac{dF_4(p,P)}{dt} = -q^i \dot{p}_i + Q^i \dot{P}_i. \tag{9.91}$$

Note that the "canonical" character of a restricted, time-independent transformation is independent of the Hamiltonian of the system. It depends only on preserving the special contracted form relating the momenta to velocities required by the definition of the Lagrangian.

In the dynamical view of canonical transformations, the system evolves in time from some initial state to some final state in phase space via the continuous mapping

$$(q_0(t_0), p_0(t_0)) \rightarrow (q(t), p(t)). \tag{9.92}$$

The time evolution of a physical path in a Hamiltonian system describes an active canonical transformation. The proof will be given after exploring in detail the properties of infinitesimal canonical transformations. For now, it will suffice to illustrate that this is true for a specific system. Consider a particle falling in a uniform gravitational field, having a Hamiltonian given by

$$H(p,q) = \frac{p^2}{2m} + mgq. \tag{9.93}$$

Hamilton's equations of motion for this system are

$$\dot{p} = -\frac{\partial H}{\partial q} = -mg, \quad \dot{q} = p/m. \tag{9.94}$$

The solution is well known and can be expressed in terms of the initial coordinate and momentum

$$p = p_0 - mgt, \quad q = q_0 + p_0 t/m - 1gt^2/2. \tag{9.95}$$

One can try to find a generating function that eliminates the Hamiltonian (this is the essence of the Hamilton-Jacobi equation to be studied in the next chapter), then

$$\tilde{H} = H + \partial F_2(q,P)/\partial t = 0. \tag{9.96}$$

Since the new Hamiltonian identically vanishes, it follows that

$$\dot{P} = 0, \quad \dot{Q} = 0. \tag{9.97}$$

This implies that the new coordinates and momenta are constants of the motion. Pick these constants to be the initial conditions

$$P = p_0, \quad Q = q_0 \tag{9.98}$$

Now show that the transformation given by Equations (9.95) is indeed canonical. The generating function must satisfy

$$p = \frac{\partial F_2(p_0, q, t)}{\partial q} = p_0 - mgt,$$

$$q_0 = \frac{\partial F_2(p_0, q, t)}{\partial p_0} = q - p_0 t/m + 1gt^2/2, \tag{9.99}$$

$$H = -\frac{\partial F_2(p_0, q, t)}{\partial t}.$$

The required generating function that identically satisfies this equation is

$$F_2(q, p_0) = (p_0 - mgt)q - p_0^2 t/2m + p_0 gt^2/2. \tag{9.100}$$

This can be verified by substitution. Since a F_2 transformation exists, the mapping is indeed canonical. This will turn out to be true in general for all continuous time-dependent maps of the time evolution equations.

Continuous canonical transformations allow the representations of a system in a continuum of reference frames, such as the continuous rotation of the system with respect to a variational parameter. From the passive viewpoint, illustrated by the right-hand-side of Figure 9.2, the same system is being represented by a continuous mapping to different sets of phase space coordinates. In the active viewpoint, illustrated by the left-hand- side of Figure 9.2, the transformation represents the continuous mapping of a system point to a new system point. If the varied parameter is the time, the

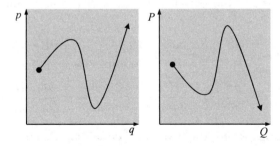

FIGURE 9.2 Active and passive interpretation of a canonical transformation. In the active interpretation, the system point evolves to a new system point (left-hand side). In the passive viewpoint (right-hand side), it is the coordinate system that changes.

canonical transformation describes the dynamic evolution of the system, and the trajectory describes the Hamiltonian flow of the system point.

9.3.4   Direct Conditions

Finding generating functions involves integration, which is often difficult. More direct ways of determining whether a transformation is canonical without integrating partial differential equations are available. The *direct conditions* define necessary and sufficient criteria that must be met in order that a transformation preserves the canonical character of the Lagrangian. The necessary and sufficient criteria for a given transformation to be canonical are given by the following direct conditions on the partial derivatives of the maps:

$$\left(\frac{\partial P_i}{\partial q^j}\right)_{qp} = -\left(\frac{\partial p_j}{\partial Q^i}\right)_{QP}, \quad \left(\frac{\partial P_i}{\partial p_j}\right)_{qp} = \left(\frac{\partial q^j}{\partial Q^i}\right)_{QP},$$
$$\left(\frac{\partial Q^i}{\partial q^j}\right)_{qp} = \left(\frac{\partial p_j}{\partial P_i}\right)_{QP}, \quad \left(\frac{\partial Q^i}{\partial p_j}\right)_{qp} = -\left(\frac{\partial q^j}{\partial P_i}\right)_{QP}.$$

(9.101)

For time independent transformations, leaving the Hamiltonian invariant, the direct conditions can easily be validated by direct differentiation. First, expand the time derivatives in the new frame using the chain rule and the initial set of Hamilton's equations of motion

$$\dot{Q}^i = \frac{\partial Q^i}{\partial q^j}\dot{q}^j + \frac{\partial Q^i}{\partial p_j}\dot{p}_j = \frac{\partial Q^i}{\partial q^j}\frac{\partial H}{\partial p_j} - \frac{\partial Q^i}{\partial p_j}\frac{\partial H}{\partial q^j},$$
$$\dot{P}_i = \frac{\partial P_i}{\partial q^j}\dot{q}^j + \frac{\partial P_i}{\partial p_j}\dot{p}_j = \frac{\partial P_i}{\partial q^j}\frac{\partial H}{\partial p_j} - \frac{\partial P_i}{\partial p_j}\frac{\partial H}{\partial q^j},$$

(9.102)

Next, expand the partial derivatives of the new Hamiltonian equations of motion

$$\dot{Q}^i = \frac{\partial H'}{\partial P_i} = \frac{\partial H}{\partial q^j}\frac{\partial q^j}{\partial P_i} + \frac{\partial H}{\partial p_j}\frac{\partial p_j}{\partial P_i} = \frac{\partial Q^i}{\partial q^j}\frac{\partial H}{\partial p_j} - \frac{\partial Q^i}{\partial p_j}\frac{\partial H}{\partial q^j},$$
$$\dot{P}_i = -\frac{\partial H'}{\partial Q^i} = -\frac{\partial H}{\partial q^j}\frac{\partial q^j}{\partial Q^i} - \frac{\partial H}{\partial p_j}\frac{\partial p_j}{\partial Q^i} = \frac{\partial P_i}{\partial q^j}\frac{\partial H}{\partial p_j} - \frac{\partial P_i}{\partial p_j}\frac{\partial H}{\partial q^j}.$$

(9.103)

Comparing terms, the two sets of equations are consistent if, and only if, the direct conditions of Equation (9.101) are satisfied.

These direct conditions are the conditions that any time-independent canonical transformation must satisfy. In a later section, it will be shown that time-dependent canonical transformations also satisfy the direct conditions.

Applying the direct conditions is, comparatively speaking, "a turn the crank procedure" which is more straightforward than trying to find a generating function. In essence, they provide the integrability criteria that must be met for a generator to exist.

As an example, the action-angle transformation used to solve the simple harmonic oscillator problem involves the invertible set of coordinate transformations given by

$$p = \sqrt{2m\omega J}\cos\varphi, \quad q = \sqrt{2J/m\omega}\sin\varphi,$$

$$\cot\varphi = p/m\omega q, \quad J = E/\omega = \frac{1}{2m\omega}(p^2 + m^2\omega^2 q^2). \tag{9.104}$$

Verifying that the direct conditions are met is a straightforward application of partial differentiation

$$\left(\frac{\partial J}{\partial q}\right)_{qp} = -\left(\frac{\partial p}{\partial \phi}\right)_{J\varphi} \quad \rightarrow \quad m\omega q = \sqrt{2m\omega J}\sin\varphi,$$

$$\left(\frac{\partial J}{\partial p}\right)_{qp} = +\left(\frac{\partial q}{\partial \varphi}\right)_{J\varphi} \quad \rightarrow \quad \frac{p}{m\omega} = \sqrt{2J/m\omega}\cos\varphi,$$

$$\left(\frac{\partial \varphi}{\partial q}\right)_{qp} = +\left(\frac{\partial p}{\partial J}\right)_{J\varphi} \quad \rightarrow \quad \frac{m\omega\cos^2\varphi}{p} = \sqrt{\frac{m\omega}{2J}}\cos\varphi, \tag{9.105}$$

$$\left(\frac{\partial \varphi}{\partial p}\right)_{qp} = -\left(\frac{\partial q}{\partial J}\right)_{J\varphi} \quad \rightarrow \quad -\frac{\sin^2\varphi}{m\omega q} = -\sqrt{\frac{1}{2m\omega J}}\sin\varphi.$$

A comparison of both sides of the four direct conditions shows that they are identically satisfied. Other simple tests of canonicality exist, such as the invariance of the fundamental Poisson Brackets, which will be introduced in the next section.

9.3.5 Canonical Invariance of Poisson Brackets

By use of the direct conditions, Poisson Brackets can be shown to be canonical invariants

$$[p_i, p_j]_{Q,P} = \frac{\partial p_i}{\partial Q^k}\frac{\partial p_j}{\partial P_k} - \frac{\partial p_j}{\partial Q^k}\frac{\partial p_i}{\partial P_k} = \frac{\partial p_i}{\partial Q^k}\frac{\partial Q^k}{\partial q^j} + \frac{\partial q^j}{\partial P_k}\frac{\partial P_k}{\partial q^j} = \frac{\partial p_i}{\partial q^j} = 0,$$

$$[q^i, q^j]_{Q,P} = \frac{\partial q^i}{\partial Q^k}\frac{\partial q^j}{\partial P_k} - \frac{\partial q^j}{\partial Q^k}\frac{\partial q^i}{\partial P_k} = -\frac{\partial q^i}{\partial Q^k}\frac{\partial Q^k}{\partial p_j} - \frac{\partial q^i}{\partial P_k}\frac{\partial P_k}{\partial p_j} = -\frac{\partial q^i}{\partial p_j} = 0,$$

$$[q^i, p_j]_{Q,P} = \frac{\partial q^i}{\partial Q^k}\frac{\partial p_j}{\partial P_k} - \frac{\partial p_j}{\partial Q^k}\frac{\partial q^i}{\partial P_k} = \frac{\partial q^i}{\partial Q^k}\frac{\partial Q^k}{\partial q^j} + \frac{\partial q^i}{\partial P_k}\frac{\partial P_k}{\partial q^j} = \frac{\partial q^i}{\partial q^j} = \delta^i_j. \tag{9.106}$$

Therefore, the fundamental Poisson Brackets are invariant under canonical transformations. The conserve is also true. The following theorem involving Canonical Invariance is stated without proof: A transformation is canonical if and only if the fundamental Poisson Brackets are preserved. Now, construct the Poisson Brackets of any two physical observables, considering (Q, P, t) as independent variables

$$
\begin{aligned}
[A, B]_{q,p} &= \frac{\partial A}{\partial q^i} \frac{\partial B}{\partial p_i} - \frac{\partial A}{\partial p_i} \frac{\partial B}{\partial q^i} \\
&= \left(\frac{\partial A}{\partial Q^j} \frac{\partial Q^j}{\partial q^i} + \frac{\partial A}{\partial P_j} \frac{\partial P_j}{\partial q^i} \right) \left(\frac{\partial B}{\partial Q^k} \frac{\partial Q^k}{\partial p_i} + \frac{\partial B}{\partial P_k} \frac{\partial P_k}{\partial p_i} \right) \\
&\quad - \left(\frac{\partial B}{\partial Q^k} \frac{\partial Q^k}{\partial q^i} + \frac{\partial B}{\partial P_k} \frac{\partial P_k}{\partial q^i} \right) \left(\frac{\partial A}{\partial Q^j} \frac{\partial Q^j}{\partial p_i} + \frac{\partial A}{\partial P_j} \frac{\partial P_j}{\partial p_i} \right) \\
&= \frac{\partial A}{\partial Q^j} \frac{\partial B}{\partial Q^k} [Q^j, Q^j] + \frac{\partial A}{\partial Q^j} \frac{\partial B}{\partial P_k} [Q^j, P_k] \\
&\quad - \frac{\partial B}{\partial Q^j} \frac{\partial A}{\partial P_k} [Q^j, P_k] + \frac{\partial A}{\partial P_j} \frac{\partial B}{\partial P_k} [P_j, P_k] \\
&= \left(\frac{\partial A}{\partial Q^j} \frac{\partial B}{\partial P_k} - \frac{\partial B}{\partial Q^j} \frac{\partial A}{\partial P_k} \right) \delta_j^k = [A, B]_{Q,P}.
\end{aligned} \tag{9.107}
$$

Therefore it follows that the Poisson Bracket, applied to any two physical observables, is a canonical invariant. The suffixes therefore can be dropped as being totally unnecessary, justifying the convention of Equation (9.20).

From the above discussion it should be clear that it is possible to demonstrate that a transformation is canonical in a number of equivalent ways, including:

1. demonstrating the existence of a generating function,
2. applying the direct conditions, or
3. showing that the fundamental Poisson Brackets are preserved.

Earlier, it was shown that the transformation

$$
p = \sqrt{2m\omega J} \cos \varphi, \quad q = \sqrt{2J/m\omega} \sin \varphi \tag{9.108}
$$

was canonical by use of the first two methods. The third method, using Poisson Brackets, is perhaps the easiest to demonstrate

$$
[q, q]_{\varphi J} = 0, \quad [p, p]_{\varphi J} = 0,
$$

$$
[q, p]_{\varphi J} = \frac{\partial q}{\partial \varphi} \frac{\partial p}{\partial J} - \frac{\partial p}{\partial \varphi} \frac{\partial q}{\partial J} = \cos^2 \varphi + \sin^2 \varphi = 1, \tag{9.109}
$$

In an earlier example, it was shown that the time evolution of the motion of a particle in a uniform gravitational field is given by the time dependent transformation

$$p = p_0 - mgt, \quad q = q_0 + p_0 t/m - 1gt^2/2. \tag{9.110}$$

It is easy to demonstrate that this time-dependent transformation preserves the fundamental bracket relationships

$$[q, q]_{q_0 p_0} = 0, \quad [p, p]_{q_0 p_0} = 0. \quad [q, p]_{q_0 p_0} = \frac{\partial q}{\partial q_0} \frac{\partial p}{\partial p_0} - \frac{\partial p}{\partial q_0} \frac{\partial q}{\partial p_0} = 1, \tag{9.111}$$

Poisson Brackets were introduced to write the total derivative of an observable in the canonical invariant form given by Equation (9.26). It follows that a physical observable in a Hamiltonian system is a constant of the motion along its physical trajectory if, and only if,

$$\frac{dA}{dt} = [A, H] + \frac{\partial A}{\partial t} = 0. \tag{9.112}$$

This implies that a time-independent observable (one that has no explicit dependence on time) is a constant of the motion if its Poisson Bracket with the Hamiltonian vanishes

$$\frac{dA(q, p)}{dt} = [A, H] = 0. \tag{9.113}$$

Likewise, the Hamiltonian is itself a constant of the motion if it is cyclic in the time, since

$$\frac{dH(q, p, t)}{dt} = [H, H] + \frac{\partial H(q, p, t)}{\partial t} = \frac{\partial H(q, p, t)}{\partial t}. \tag{9.114}$$

Moreover, if A and B are time-independent constants of the motion, then, by the Jacobi Identity, the Poisson Bracket of the two observables $[A, B]$ is also a constant of the motion

$$[[A, B], H] = -[[B, H], A] - [[H, A], B] = 0 \tag{9.115}$$

Some additional useful identities that are easy to demonstrate (see the exercises) are

$$\frac{\partial A(q, p, t)}{\partial p_i} = [q^i, A], \quad \frac{\partial A(q, p, t)}{\partial q^i} = -[p_i, A]. \tag{9.116}$$

These identities allow one to express partial derivatives with respect to the coordinates and momenta in the Poisson Bracket formalism.

9.4   INFINITESIMAL CANONICAL TRANSFORMATIONS

An *infinitesimal canonical transformation* (ICT) is a canonical transformation that differs from the identity transform by an infinitesimal amount

$$Q^i = q^i + \delta q^i, \quad P_i = p_i + \delta p_i \tag{9.117}$$

The generating function for this transformation can be written as a F_2 transform which deviates from the identity transformation by the addition of a physical observable G multiplied by small variational parameter $\delta\varepsilon$.

$$F_2(q, P, t) = q^i P_i + \delta\varepsilon G(q, P, t) \tag{9.118}$$

$$\tilde{F}_2 = F_2(q, P, t) - Q^i P_i = (q^i - Q^i)P_i + \delta\varepsilon G(q, P, t) \tag{9.119}$$

The first term on the right hand side of the previous equation is the identity transformation. The second term is a difference due to the infinitesimal generator $G(q, P, t)$. Note the somewhat inconsistent uses of the term "generator". The observable $G(q, P, t)$ is often called the *infinitesimal generator of the transformation*, or sometimes just the generator of the transformation. The generating function for the transformation $\tilde{F}_2$ (or F_2 in a short-handed fashion) is also often referred to as the generator of the transformation. One has to rely on context to know which is being referenced.

The generating function results in a transformation by a magnitude $\delta\varepsilon$ of its conjugate observable. From the generating function table (Table 9.1), one gets

$$p_i = \frac{\partial F_2}{\partial q^i} = P_i + \delta\varepsilon\frac{\partial G}{\partial q^i}, \quad Q^i = \frac{\partial F_2}{\partial P_i} = q^i + \delta\varepsilon\frac{\partial G}{\partial P_i}, \tag{9.120}$$

Therefore, the infinitesimal generator $G(q, P, t)$ results in the following relationships

$$\delta q^i = Q^i - q^i = \delta\varepsilon\frac{\partial G}{\partial p_i}, \quad \delta p_i = P_i - p_i = -\delta\varepsilon\frac{\partial G}{\partial q^i}. \tag{9.121}$$

Since the transformation is infinitesimal, however the initial and final momenta are the same to lowest order

$$G = G(q, P, t) \cong G(q, p, t). \tag{9.122}$$

9.4.1   Generating the Motion

If one replaces the arbitrary generator $G(q, p, t)$ with the Hamiltonian and applies Hamilton's equations, one gets

$$Q^i = q^i + \delta\varepsilon\frac{\partial H}{\partial p_i} = q^i + \delta\varepsilon\dot{q}^i, \quad P_i = p_i - \delta\varepsilon\frac{\partial H}{\partial q^i} = p_i + \delta\varepsilon\dot{p}_i \tag{9.123}$$

This becomes an identity, if the variational parameter is interpreted as an infinitesimal displacement in time, $\varepsilon = \delta t$, giving

$$\delta q^i = \frac{\partial H}{\partial p_i}\delta t = \dot{q}^i \delta t, \quad \delta p_i = -\frac{\partial H}{\partial q^i}\delta t = \dot{p}_i \delta t. \tag{9.124}$$

But, this is just the criteria for the time rate of change of the Hamiltonian flow. It follows that the Hamiltonian is the generator of the motion in time.

9.4.2 ICTs and the Direct Conditions

Make the expansion

$$q^i = Q^i_o + \delta\varepsilon\frac{\partial G}{\partial P_i} \simeq q^i_o + \delta\varepsilon\frac{\partial G}{\partial p_i}, \quad p_i = P_{io} - \delta\varepsilon\frac{\partial G}{\partial q^i} \simeq p_{io} - \delta\varepsilon\frac{\partial G}{\partial Q^i}. \tag{9.125}$$

Then, by direct differentiation and by comparing terms on the same line in the following formulas, one can verify that infinitesimal variations satisfy the direct conditions to lowest order in $\delta\varepsilon$

$$
\begin{aligned}
\frac{\partial q^j}{\partial P_i} &= \frac{\partial(Q^i - \delta q^i)}{\partial P_i} = -\delta\varepsilon\frac{\partial G}{\partial P_i \partial p_j}, \\[4pt]
-\frac{\partial Q^i}{\partial p_j} &= -\frac{\partial(q^i + \delta q^i)}{\partial p_j} = -\delta\varepsilon\frac{\partial G}{\partial p_j \partial P_i}, \\[4pt]
\frac{\partial p_j}{\partial P_i} &= \frac{\partial(P_j - \delta p_j)}{\partial P_i} = \delta^i_j + \delta\varepsilon\frac{\partial G}{\partial P_i \partial q^j}, \\[4pt]
\frac{\partial Q^i}{\partial q^j} &= \frac{\partial(q^i + \delta q^i)}{\partial q^j} = \delta^i_j + \delta\varepsilon\frac{\partial G}{\partial q^j \partial P_i}, \\[4pt]
\frac{\partial p_j}{\partial Q^i} &= \frac{\partial(P_j - \delta p_j)}{\partial Q^i} = \delta\varepsilon\frac{\partial G}{\partial Q^i \partial q^j}, \\[4pt]
-\frac{\partial P_i}{\partial q^j} &= -\frac{\partial(p_i + \delta p_i)}{\partial q^j} = \delta\varepsilon\frac{\partial G}{\partial q^j \partial Q^i}, \\[4pt]
\frac{\partial q^j}{\partial Q^i} &= \frac{\partial(Q^j - \delta q^j)}{\partial Q^i} = \delta^j_i - \delta\varepsilon\frac{\partial G}{\partial Q^i \partial p_j}, \\[4pt]
\frac{\partial P_i}{\partial p_j} &= \frac{\partial(p_i - \delta p_i)}{\partial p_j} = \delta^j_i - \delta\varepsilon\frac{\partial G}{\partial p_j \partial Q^i},
\end{aligned}
\tag{9.126}
$$

where the order of differentiation can be reversed. Therefore, all infinitesimal canonical transformations obey the direct conditions and are canonical. Finite transformations with respect to a continuous parameter are built up from successive ICTs and can be therefore also shown to be canonical by invoking group closure.

Consider a general finite, time-dependent canonical transformation:

$$Q^i = Q^i(q,p,t), \quad P_i = P_i(q,p,t), \quad H' = H + \partial F/\partial t. \tag{9.127}$$

This transformation can be separated conceptually into two steps:

1. Canonical time-independent transformations of the phase space coordinates at some initial time t_0

$$q,p \rightarrow Q(q,p,t_0), P(q,p,t_0). \tag{9.128}$$

   (This step clearly satisfies the direct conditions.)

2. A time-only mapping of the evolution of the new coordinates in time

$$Q(q,p,t_0), \quad P(q,p,t_0) \rightarrow Q(q,p,t), \quad P(q,p,t). \tag{9.129}$$

But the time variation is continuous and can be therefore divided into infinitesimal canonical transformations, which also obey the direct conditions:

$$Q(q,p,t_0), \quad P(q,p,t_0) \rightarrow Q(q,p,t_0+\delta t), \quad P(q,p,t_0+\delta t). \tag{9.130}$$

Integration over all infinitesimal steps gives

$$\begin{aligned}(Q(q,p,t_0), P(q,p,t_0)) &\rightarrow (Q(q,p,t_0+\delta t), P(q,p,t_0+\delta t)) \rightarrow \cdots \\ &\rightarrow (Q(q,p,t), P(q,p,t)).\end{aligned} \tag{9.131}$$

Therefore, one has *canonical invariance*: All continuous time-dependent canonical transformations satisfy the direct conditions. The proof uses the criteria that time evolution of the system is continuous and, therefore, can be built up from the repetition of infinitesimal transformations that individually satisfy the direct conditions.

9.4.3 Generators of Coordinate Transformations

Any physical observable can, in principle, be used to generate some transformation. Consider the use of a momentum observable as a generator. Let $G_j = P_j$ be a momentum component, then, from Equation (9.118),

$$Q^i = q^i + \delta\varepsilon\frac{\partial P_j}{\partial P_i} = q^i + \delta\varepsilon\delta_j^i, \quad P_i = p_i - \delta\varepsilon\frac{\partial P_j}{\partial q^i} = p_i, \tag{9.132}$$

The momentum P_j generates a displacement $q^j + \varepsilon$ in its associated canonical coordinate. Therefore, it follows that the canonical momenta are the generators of coordinate displacements. Similarly, use of a coordinate q^j as a generator leads to the following transformation

$$Q^i = q^i, \quad P_i = p_i - \delta \varepsilon \delta_i^j. \tag{9.133}$$

Therefore, the canonical coordinates are the generators of momentum displacements. This is similar to results obtained using Heisenberg's quantum mechanics.

9.4.4 Symmetry and the Constants of the Motion

Suppose a generator G has no explicit time dependence, then the variation of a physical observable A by the transformation generated by $\delta \varepsilon G$ is given by the Poisson bracket

$$\delta A = \frac{\partial A}{\partial q^i} \delta q^i + \frac{\partial A}{\partial p_i} \delta p_i = \delta \varepsilon \frac{\partial A}{\partial q^i} \frac{\partial G}{\partial p_i} - \delta \varepsilon \frac{\partial A}{\partial p_i} \frac{\partial G}{\partial q^i} = \delta \varepsilon [A, G], \tag{9.134}$$

$$\frac{dA}{d\varepsilon} = [A, G]. \tag{9.135}$$

In operator form, repeated differentiation gives

$$\frac{d}{d\varepsilon} = [, G], \quad \frac{d^2}{d\varepsilon^2} = [[, G], G], \quad \cdots \quad \frac{d^n}{d\varepsilon^n} = [[[\ldots], G], G]. \tag{9.136}$$

Note that the observable A is cyclic in the parameter ε if the Poisson Bracket $[A, G]$ vanishes. A physical observable that is a continuous differential function of a scalar parameter ε can be expanded in a Taylor series

$$A(\varepsilon) = \sum_n \frac{1}{n!} \frac{d^n A}{d\varepsilon^n}\bigg|_0 \varepsilon^n, \tag{9.137}$$

but this can be expressed as a series of terms involving repeated Poisson Brackets

$$A(\varepsilon) = A|_0 + [A, G]_0 \varepsilon + \frac{1}{2!} [[A, G], G]_0 \varepsilon^2 + \cdots. \tag{9.138}$$

Now consider how the Hamiltonian changes under such a continuous transformation

$$\delta H = \frac{\partial H}{\partial q^i} \delta q^i + \frac{\partial H}{\partial p_i} \delta p_i = \delta \varepsilon \left(\frac{\partial H}{\partial q^i} \frac{\partial G}{\partial p_i} - \frac{\partial H}{\partial p_i} \frac{\partial G}{\partial q^i} \right) = \delta \varepsilon [H, G]. \tag{9.139}$$

But if $\delta H = 0$ then the variation generated by G is a symmetry of the Hamiltonian. The Hamiltonian is invariant under the transformation. But $[H, G(q, p)] = 0$ also implies the infinitesimal generator G is a constant of the motion if $\partial G/\partial t = 0$. This produces a connection between the symmetries of the Hamiltonian and the constants of the motion: Any time-independent generator that leaves Hamiltonian invariant (i.e., any continuous symmetry of the Hamiltonian) is a constant of motion. In other words, for a given time-independent generator, the equation

$$\frac{\delta H}{\delta \varepsilon} = [H, G(q, p)] \tag{9.140}$$

implies that a symmetry of the Hamiltonian under that transformation can be related to a constant of motion satisfying

$$\frac{\delta H}{\delta \varepsilon} = [H, G(q, p)] = 0. \tag{9.141}$$

Often, the Hamiltonian can be associated with the energy of a mechanical system. Since the Hamiltonian's bracket with itself vanishes, it follows from Equation 9.114 that any Hamiltonian that does not explicitly depend on the time is a constant of the motion. If the Hamiltonian does not depend on a coordinate, the Poisson bracket with the associated conjugate momentum vanishes. It follows that if H is cyclic in a coordinate, the conjugate momentum is a constant of the motion

$$\delta H = \delta \varepsilon^j [H, p_j] = \delta \varepsilon^j \frac{\partial H}{\partial q^j} = -\delta \varepsilon^j \dot{p}_j = 0. \tag{9.142}$$

9.5 LIE ALGEBRA OF THE ROTATION GROUP

In Cartesian coordinates, the angular momentum observable of a single particle can be written as

$$\mathbf{L} = \mathbf{r} \times \mathbf{p} \rightarrow L_i = \varepsilon_{ijk} r^j p^k, \tag{9.143}$$

with Cartesian components

$$L_x = yp_z - zp_y, \quad L_y = zp_x - xp_z, \quad L_z = xp_y - yp_x. \tag{9.144}$$

The z-component of the angular momentum is the canonical momentum p_ϕ associated with the azimuthal angle ϕ in a spherical basis. Figure 9.3 shows an infinitesimal rotation of the particle about the z-axis. If the single particle Hamiltonian is cyclic in the azimuthal angle, then the z-component of angular momentum is conserved.

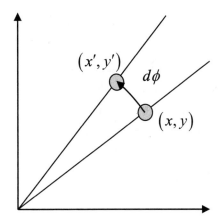

FIGURE 9.3 Sketch of an infinitesimal rotation.

The angular momentum operators form a closed Lie algebra under the Poisson bracket product

$$[L_i, L_j] = \varepsilon_{ijk} L_k. \tag{9.145}$$

This can be explicitly shown for the first of the three products using the fundamental Poisson bracket relationships

$$[L_x, L_y] = [yp_z - zp_y, zp_x - xp_z] = -yp_x + xp_y = L_z. \tag{9.146}$$

The other combinations are similar and follow from cyclic permutation of the indices.

Now extend this concept to the angular momentum of an N-particle system expressed in Cartesian coordinates $\mathbf{r}_n = (x_n, y_n, z_n)$. Consider a counterclockwise rotation of the system as a whole about the z-axis. The result is well known and can be written as

$$x'_n = x_n - y_n d\phi, \quad y'_n = y_n + x_n d\phi, \quad z'_n = z_n. \tag{9.147}$$

Momenta are rotated as well, so

$$p'_{xn} = p_{xn} - p_{yn} d\phi, \quad p'_{yn} = p_{yn} + p_{xn} d\phi, \quad p'_{zn} = p_{zn} \tag{9.148}$$

The generator for this rotation is the z-component of the angular momentum operator summed over all particles in the system

$$G = L_z = x_n p_{yn} - y_n p_{xn} = (\mathbf{r}_n \times \mathbf{p}_n)_z \tag{9.149}$$

By explicit substitution, the Poisson Brackets can be expanded to demonstrate the validity of the rotation formula given above

$$\delta x_n = d\phi[x_n, L_z] = d\phi[x_n, x_m p_{ym} - y_m p_{xm}] = -d\phi y_m \delta_{nm} = -d\phi y_n,$$
$$\delta y_n = d\phi[y_n, L_z] = d\phi[y_n, x_m p_{ym} - y_m p_{xm}] = d\phi x_m \delta_{nm} = d\phi x_n,$$
$$\delta p_{xn} = d\phi[p_{xn}, L_z] = d\phi[p_{xn}, x_m p_{ym} - y_m p_{xm}] = -d\phi p_{ym} \delta_{nm} = -d\phi p_{yn},$$
$$\delta p_{yn} = d\phi[p_{yn}, L_z] = d\phi[p_{yn}, x_m p_{ym} - y_m p_{xm}] = d\phi p_{xm} \delta_{nm} = d\phi p_{xn},$$
$$(9.150)$$

But if $\mathbf{L}_z = \mathbf{L} \cdot \mathbf{e}_z$ is the generator of a rotation about the z-axis, it follows that the infinitesimal generator of an arbitrary rotation about the n-axis is

$$G_n = \mathbf{L} \cdot \mathbf{n}. \qquad (9.151)$$

Total angular momentum generates a global rotation in space. If angular momentum about the $\mathbf{n}$ axis is conserved, then

$$[\mathbf{L} \cdot \mathbf{n}, H] = 0. \qquad (9.152)$$

In the chapter on rotational kinematics, a vector was shown to rotate under an infinitesimal rotation as follows

$$\mathbf{V}' = \mathbf{V} + d\phi\, \mathbf{n} \times \mathbf{V} \qquad (9.153)$$

Comparison with the angular momentum Poisson Bracket relationships validates this formula

$$d\mathbf{r}_m = d\phi\, \mathbf{n} \times \mathbf{r}_m = d\phi[\mathbf{r}_m, \mathbf{L} \cdot \mathbf{n}] \Rightarrow [\mathbf{r}_m, \mathbf{L} \cdot \mathbf{n}] = \mathbf{n} \times \mathbf{r}_m,$$
$$d\mathbf{p}_m = d\phi\, \mathbf{n} \times \mathbf{p}_m = d\phi[\mathbf{p}_m, \mathbf{L} \cdot \mathbf{n}] \Rightarrow [\mathbf{p}_m, \mathbf{L} \cdot \mathbf{n}] = \mathbf{n} \times \mathbf{p}_m. \qquad (9.154)$$

The fundamental angular momentum bracket relationships for an n-particle system are given by

$$[\mathbf{L}, \mathbf{L} \cdot \mathbf{n}] = \mathbf{n} \times \mathbf{L} \rightarrow [L_i, L_j] = \varepsilon_{ijk} L_k. \qquad (9.155)$$

An arbitrary vector observable, must be composed from the available vectors of the algebra, namely the phase space vectors $(\mathbf{r}_m, \mathbf{p}_n)$. Scalar products $\mathbf{A} \cdot \mathbf{B}$ of these vectors are rotationally invariant since

$$[\mathbf{A} \cdot \mathbf{B}, \mathbf{L} \cdot \mathbf{n}] = \mathbf{A} \cdot (\mathbf{n} \times \mathbf{B}) + (\mathbf{n} \times \mathbf{A}) \cdot \mathbf{B} \equiv 0. \qquad (9.156)$$

A Hamiltonian scalar observable composed of only such scalar products will be rotationally invariant under a global rotation of the system. Therefore, an isolated system (one with no external interaction fields) will be globally rotationally invariant.

9.5.1 Finite Rotations

For a rotation about an instantaneous axis, the rotation formula is

$$\frac{d\mathbf{r}_m}{d\phi} = [\mathbf{r}_m, \mathbf{L} \cdot \mathbf{n}] \tag{9.157}$$

If the axis of rotation is fixed, it can be labeled as the z-axis giving

$$\frac{dx_m}{d\phi} = [x_m, L_z] = -y_m, \quad \frac{dy_m}{d\phi} = [y_m, L_z] = x_m, \quad \frac{dz_m}{d\phi} = [z_m, L_z] = 0. \tag{9.158}$$

The integral over a finite angle can be expanded in a Taylor series giving

$$x_m(\phi) = \sum_i \frac{1}{n!} \frac{d^n x_m}{d\phi^n}\bigg|_0 \phi^n = x_m(0) + [x_m, L_z]_0 \phi + \frac{1}{2!}[[x_m, L_z], L_z]_0 \phi^2 + \cdots$$

$$= x_m(0) - y_m(0)\phi - \frac{1}{2!}x_m(0)\phi^2 + \frac{1}{3!}y_m(0)\phi^3 \cdots$$

$$= x_m(0)\cos\phi - y_m(0)\sin\phi \tag{9.159}$$

Likewise,

$$y_m(\phi) = \sum_i \frac{1}{n!} \frac{d^n y_m}{d\phi^n}\bigg|_0 \phi^n = y_m(0) + [y_m, L_z]_0 \phi + \frac{1}{2!}[y_m, L_z]_0 \phi^2 + \cdots$$

$$= y_m(0)\cos\phi + x_m(0)\sin\phi \tag{9.160}$$

The matrix representation of this rotation is a SO3 rotation matrix given by

$$\begin{bmatrix} x_m(\phi) \\ y_m(\phi) \\ z_m(\phi) \end{bmatrix} = \begin{bmatrix} \cos\phi & -\sin\phi & 0 \\ +\sin\phi & \cos\phi & 0 \\ 0 & 0 & 1 \end{bmatrix} \begin{bmatrix} x_m(0) \\ y_m(0) \\ z_m(0) \end{bmatrix}. \tag{9.161}$$

Products of rotations about different axes can be reduced to a single rotation about some space axis.

9.5.2 Mutually Commuting Observables

Recall, from the fundamental Poisson Brackets, that canonical momenta commute with each other

$$[p_j, p_j] = 0, \tag{9.162}$$

but angular momenta do not

$$[L_i, L_j] = \varepsilon_{ijk}L_k. \tag{9.163}$$

At most one component of angular momenta may be treated as a canonical momentum. On the other hand,

$$[L^2, L_i] = [\mathbf{L} \cdot \mathbf{L}, L_i] = 0, \tag{9.164}$$

so that the squared magnitude of angular momentum can be used as a second canonical momentum. In quantum mechanics, one can simultaneously measure only mutually commuting observables such as the set (L^2, L_z, H) for a rotationally invariant Hamiltonian.

9.5.3 *SO4 Symmetry of the Kepler Problem

In reduced coordinates, the central force problem can be described by the one-body Hamiltonian

$$H = \frac{\mathbf{p}^2}{2m} + V(r) = \frac{1}{2m}\left(p_r^2 + \frac{1}{r^2}p_\theta^2 + \frac{1}{r^2\sin\theta}p_\phi^2\right) + V(r) \tag{9.165}$$

Since scalars are rotational invariant,

$$[\mathbf{p}^2, \mathbf{L} \cdot \mathbf{n}] = 0, \quad [r, \mathbf{L} \cdot \mathbf{n}] = [\sqrt{\mathbf{r}^2}, \mathbf{L} \cdot \mathbf{n}] = 0. \tag{9.166}$$

Therefore, the angular momentum vector and its square is a constant of the motion.

$$[H, \mathbf{L} \cdot \mathbf{n}] = 0, \quad [H, L^2] = 0. \tag{9.167}$$

For the Kepler problem, the Runge-Lenz vector defines three additional constants of the motion

$$[\mathbf{A}, H] = 0, \tag{9.168}$$

where

$$\mathbf{A} = \mathbf{p} \times \mathbf{L} - mk\hat{\mathbf{r}}, \quad \hat{\mathbf{r}} = \mathbf{r}/r \quad \mathbf{A} \cdot \mathbf{L} = 0. \tag{9.169}$$

Note that the constants of motion $(\mathbf{L}, \mathbf{A}, H)$ form a closed algebra under the Poisson Bracket product. Defining

$$\mathbf{D} = \mathbf{A}/\sqrt{-2mE} \tag{9.170}$$

allows one to generate the Poisson Bracket algebra

$$[L_i, L_j] = \varepsilon_{ijk}L^k, \quad [D_i, L_j] = \varepsilon_{ijk}D^k, \quad [D_i, D_j] = \varepsilon_{ijk}L^k. \tag{9.171}$$

with the following constants of the motion.

$$[\mathbf{D}, H] = 0, \quad [\mathbf{L}, H] = 0, \tag{9.172}$$

The proof of these relationships is straightforward, if tedious, and will be left as an exercise.

Note that $SO(3)$ rotational symmetry is associated with the conservation of angular momentum when $[\mathbf{L}, H] = 0$, which is valid for any spherically symmetric potential. For the special case of a $1/r$ potential, the symmetry group is the larger transformation group including the Runge-Lenz vector. The enlarged symmetry group given by conservation of both angular momentum and Runge-Lenz vectors turns out to be isomorphic with the six generators of a $SO(4)$ rotational group in a four dimensional space. In quantum mechanics, the presence of this special symmetry reveals itself by the greater degeneracy observed in the energy spectrum of the Hydrogen atom.

9.6 SYMPLECTIC STRUCTURE OF THE THEORY

In the Hamiltonian formulation, coordinates and momentum come in coupled pairs. (q^i, p_i). This is reminiscent of complex numbers, which can be thought of as pairs of real numbers $z = x + iy$. Consider a 2n-dimensional phase space of coordinates and their momenta. A phase space vector in this space can be written as[3]

$$\mathbf{x} = (p_1, \ldots p_n, q^1, \ldots q^n) = \begin{bmatrix} \mathbf{p} \\ \mathbf{q} \end{bmatrix} = (\mathbf{q} + \mathbf{j}\mathbf{p}) \begin{bmatrix} 0 \\ 1 \end{bmatrix}. \tag{9.173}$$

The Lagrangian can be expressed in the symplectic form as

$$\begin{aligned} L &= \frac{1}{2}\bar{\mathbf{x}} \cdot \mathbf{j} \cdot \dot{\mathbf{x}} - H(x,t) + \frac{dF(x,t)}{dt} \\ &= \frac{1}{2}(p_i \dot{q}^i - q^i \dot{p}_i) - H(p,q,t) + \frac{dF(p,q,t)}{dt} \\ &= p_i \dot{q}^i - H(p,q,t) + \frac{dF(p,q,t)}{dt} - \frac{1}{2}\frac{d}{dt}(p_i q^i), \end{aligned} \tag{9.174}$$

where

$$\mathbf{j} = \begin{bmatrix} 0 & 1 \\ -1 & 0 \end{bmatrix} \tag{9.175}$$

is the *symplectic operator*. $\mathbf{j}$ is a 2n by 2n dimensional matrix with properties that commute with the phase space variables and satisfy the conditions

$$\bar{\mathbf{j}} = -\mathbf{j}, \quad \mathbf{j}^2 = -1. \tag{9.176}$$

---

[3]Note that [Goldstein 2002] defines p and q in the opposite order from the convention used here: $\mathbf{x}_{\text{Goldstien}} = [\mathbf{q} \quad \mathbf{p}]^T$. This affects the overall sign of the symplectic operator in the resulting equations.

Hamilton's equations of motion can now be written more compactly in terms of the gradient of a function in a 2n-dimensional space

$$\dot{x} = -\mathbf{j} \cdot \nabla_x H. \tag{9.177}$$

Expansion gives

$$\begin{pmatrix} \dot{\mathbf{p}} \\ \dot{\mathbf{q}} \end{pmatrix} = \begin{bmatrix} 0 & -1 \\ +1 & 0 \end{bmatrix} \begin{bmatrix} \nabla_p H \\ \nabla_q H \end{bmatrix}. \tag{9.178}$$

where the $2n$ dimensional gradient can be written as the column matrix

$$\nabla_x = [\nabla_p, \nabla_q]^T = \left[\frac{\partial}{\partial p_1}, \cdots \frac{\partial}{\partial p_n}, \frac{\partial}{\partial q^1}, \cdots \frac{\partial}{\partial q^n} \right]^T. \tag{9.179}$$

Note that the derivative $\dot{\mathbf{x}}$ gives the direction of the Hamiltonian flow in phase space, as defined by the continuous distribution $-\mathbf{j} \cdot \nabla_x H(x, t)$. As shown by previous analysis, lines of Hamiltonian flow can never cross since the gradient is a single-valued function. If the flow lines crossed, there would be multiple solutions to the time evolution problem; but, the time evolution equations are single valued functions of time.

It follows that the Hamiltonian flow is incompressible if the Hamiltonian is a real valued scalar function. Taking the divergence of the flow gives

$$\mathbf{Div}(\dot{\mathbf{x}}) = -\nabla \cdot \mathbf{j} \cdot \nabla H(x, t) = -\begin{bmatrix} \nabla_p \\ \nabla_q \end{bmatrix}^T \cdot \begin{bmatrix} 0 & 1 \\ -1 & 0 \end{bmatrix} \cdot \begin{bmatrix} \nabla_p \\ \nabla_q \end{bmatrix} H$$

$$= -(\nabla_p \cdot \nabla_q - \nabla_q \cdot \nabla_p) H \equiv 0. \tag{9.180}$$

Dissipation is required to get the phase space to decay to an invariant subspace called the *attractor*.

Consider now a canonical mapping to a new set of coordinates, where for simplicity, the transformation does not explicitly involve the time. It can be written as a matrix transformation

$$\dot{\mathbf{x}}' = \begin{bmatrix} \dot{\mathbf{p}}' \\ \dot{\mathbf{q}}' \end{bmatrix} = \mathbf{M} \begin{bmatrix} \dot{\mathbf{p}} \\ \dot{\mathbf{q}} \end{bmatrix}, \tag{9.181}$$

where $M^i{}_j = \partial x^i / \partial x'^j$. Expansion of the transformation gives

$$\dot{\mathbf{x}}' = \mathbf{M} \cdot \dot{\mathbf{x}} = -\mathbf{M} \cdot \mathbf{j} \cdot \nabla H = -\mathbf{M} \cdot \mathbf{j} \cdot \bar{\mathbf{M}} \cdot \nabla' H = -\mathbf{M} \cdot \mathbf{j} \cdot \bar{\mathbf{M}} \cdot \mathbf{j} \cdot \dot{\mathbf{x}}'. \tag{9.182}$$

This results in the symplectic condition for canonical transformations[4]

$$\mathbf{M} \cdot \mathbf{j} \cdot \bar{\mathbf{M}} = \mathbf{j}, \qquad (9.183)$$

or

$$\mathbf{M} \cdot \mathbf{j} = \mathbf{j} \cdot \bar{\mathbf{M}}^{-1}, \qquad (9.184)$$

which is equivalent to requiring that the direct conditions be satisfied. Note that

$$\det\left(M^2\right) = 1 \rightarrow \det\left(M\right) = \pm 1 \qquad (9.185)$$

where, for transformations that evolve continuously from the identity element, the Jacobian of the transformation is identically $+1$, therefore

$$\det\left(M\right) = \left| \frac{\partial x^i}{\partial x'^j} \right| = 1. \qquad (9.186)$$

9.6.1 Conservation of Phase Space Volume

The symplectic condition of Equation (9.183) results in the conservation of phase space for Hamiltonian systems. Define an element of phase space volume

$$dV = d\mathbf{x} = \prod_i dq^i dp_i = |M| d\mathbf{x}' = |\partial x^i / \partial x'^j| \prod_i dq'^i dp'_i = |M| dV'. \quad (9.187)$$

But $|M^i_j| = |\partial x^i / \partial x'^j|$ is just the Jacobian of the transformation from one set of coordinates to another. Setting $|M| = 1$ for canonical transformations implies that the volume of phase space is conserved as shown in Figure 9.4. In other words, volume elements of phase space are invariant under canonical transformations. Conservation of phase space volume is easy to directly demonstrate for a two-dimensional phase space consisting of a single coordinate and its conjugate momentum

$$dV = dq dp = \begin{vmatrix} \dfrac{\partial q}{\partial Q} & \dfrac{\partial p}{\partial Q} \\[2mm] \dfrac{\partial q}{\partial P} & \dfrac{\partial p}{\partial P} \end{vmatrix} dQ dP = |M| dV'. \qquad (9.188)$$

---

[4]The space of smooth functions on a symplectic manifold defines an infinite-dimensional Lie algebra with the Poisson bracket acting as the Lie bracket. The corresponding Lie group is the group of canonical transformations.

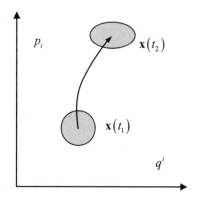

FIGURE 9.4 Motion of a volume element of phase space conserves its volume but not necessarily its shape.

Expanding the Jacobian gives

$$dqdp = \{q,p\}_{QP} dQdP = dQdP. \tag{9.189}$$

where the fundamental Poisson Bracket definition $\{q^i, p_j\}_{QP} = \delta^i_j$ is used to get the final result.

9.6.2 Liouville's Theorem

Next consider a number of particles confined within a region of phase space as illustrated in Figure 9.5, whose boundary surface is evolving in time.

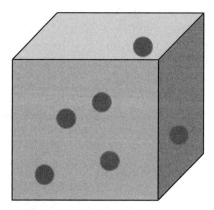

FIGURE 9.5 Since lines of Hamiltonian flow never intersect, particles can never enter or leave a volume of phase space, whose boundary surface is evolving in time; therefore the number of particles in a given volume of phase space is conserved.

Since lines of Hamiltonian flow never cross, particles within the volume must remain within the volume, while particles outside the volume must remain outside the boundary. Since particles cannot cross the boundary, the number density is a constant of the motion. *Liouville's theorem*[5] states that the number density of particles in a volume of phase space is conserved under a general canonical transformation.

If the statistical ensemble is large, one can take the continuum limit and define the density of states in phase space as a continuous distribution

$$\rho(q, p, t) = \lim_{\Delta V \to 0} \frac{\Delta N}{\Delta V} = \frac{dN}{dV}. \tag{9.190}$$

But the number density is constant in time, This gives rise to Liouville's Equation for the density of a Hamiltonian system

$$\frac{d\rho}{dt} = [\rho, H] + \frac{\partial \rho}{\partial t} = 0. \tag{9.191}$$

If the particle density is replaced by an incompressible fluid, the system evolves in such a way as to preserve the density as shown in Figure 9.6.

The normalization of the density function depends on how the distribution is interpreted. One may have a single particle system but not know the exact state very well. Then ρ is understood as a probability distribution, and the distribution is normalized to a single particle, corresponding to a 100% probability that the particle will be in one of the allowed states:

$$\int \rho \prod_i dq^i dp_i = 1. \tag{9.192}$$

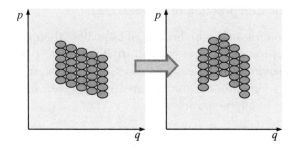

FIGURE 9.6 The time evolution of particles in an incompressible fluid, moving in such a way as to preserve the volume.

---

[5]Developed by Joseph Liouville (1809–1882) in 1838.

One the other hand, the system in question could consist of a large number of identical, non-interacting particles. For example, there are Avogadro's number ($\sim 6.028 \times 10^{23}$) of gas molecules in one gm-mole of gas. In this case, the statistical distribution would be normalized to total number of particles in the system:

$$\int \rho \prod_i dq^i dp_i = N. \tag{9.193}$$

If particles (or probability) are neither created nor destroyed, then $dN/dt = 0$.

9.6.3 Stationary Ensembles

A stationary ensemble is one that is not evolving in time, therefore

$$\partial \rho / \partial t = 0, \tag{9.194}$$

implying

$$[\rho, H] = 0. \tag{9.195}$$

Any time-independent statistical ensemble can be written as a function of the time independent constants of the motion which have vanishing Poisson Brackets the Hamiltonian. A famous example of a stationary ensemble is the *Boltzmann Distribution* of thermodynamics given by

$$\rho(H) = e^{-H/kT}, \tag{9.196}$$

where k is Boltzmann's constant and the parameter T is the absolute temperature in Kelvin.

9.6.4 Adiabatic Compression of an Ideal Gas

As an example of the application of Liouville's theorem, consider the adiabatic compression of an ideal gas [Morii 2003]. Figure 9.7 shows a gas localized within some volume being compressed by a piston.

The Hamiltonian for an ideal gas is given by the sum of the kinetic energies of all the particles

$$H = \sum_i \frac{\mathbf{P}_i^2}{2m}. \tag{9.197}$$

Liouville's theorem applies to statistical ensembles of Hamiltonian systems with nearly continuous distributions in phase space. If one adiabatically compresses an ideal gas, no heat flows into or out of the system. Since phase space

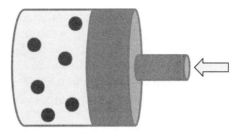

FIGURE 9.7 Adiabatic compression of a gas by pushing in a piston slowly.

is conserved, a reduced volume in coordinate space increases the volume in momentum space. Therefore, an ideal gas gets hotter when it is compressed. The effect on the phase space is illustrated in Figure 9.8. Phase space is conserved for Hamiltonian systems.

9.7   *ELEVATING TIME TO A COORDINATE

If the Hamiltonian is cyclic in a coordinate, the associated momentum is a constant of the motion since

$$\dot{p} = \left.\frac{\partial L(q,\dot{q},t)}{\partial q^i}\right|_{\dot{q},t} = -\left.\frac{\partial H(q,p,t)}{\partial q^i}\right|_{p,t} \tag{9.198}$$

Note that the Hamiltonian is cyclic in a coordinate if, and only if, the equivalent Lagrangian is also cyclic in the coordinate. Calculating the total derivative of the Hamiltonian by the chain rule gives

$$\frac{dH(q,p,t)}{dt} = \frac{\partial H}{\partial p_i}\dot{p}_i + \frac{\partial H}{\partial q^i}\dot{q}^i + \frac{\partial H}{\partial t}. \tag{9.199}$$

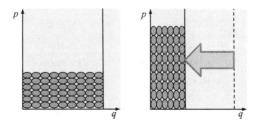

FIGURE 9.8 In the adiabatic compression of an ideal gas, as the coordinate volume decreases, the momentum volume increases to conserve phase space. The result is that the gas gets hotter.

Substituting Hamilton's equations of motion, the above equation reduces to

$$\frac{dH}{dt} = \dot{q}^i \dot{p}_i - \dot{p}_i \dot{q}^i + \frac{\partial H}{\partial t} = \frac{\partial H}{\partial t}. \tag{9.200}$$

Therefore, the Hamiltonian of a monogenic system is a constant of the motion if, and only if, it cyclic in the time. Comparing with the Lagrangian formulation, one has

$$\frac{dH}{dt} = \left.\frac{\partial H(q,p,t)}{\partial t}\right|_{q,p} = -\left.\frac{\partial L(q,\dot{q},t)}{\partial t}\right|_{q,\dot{q}}. \tag{9.201}$$

This suggests that the Hamiltonian can be formally treated as the conjugate momentum associated with the time dependence of the physical system. In nonrelativistic mechanics, time is usually treated as a parameter and not a coordinate. To elevate time to a coordinate, one needs to introduce a new scalar parameter τ, which parameterizes the motion of a particle in space and time. Consider, for simplicity, a single particle system. The path of a particle in space and time consists of a single parameter family of curves called its trajectory, parameterized as

$$
\begin{aligned}
q^i &= q^i(\tau), \quad i = 1, 2, 3 \\
q^0 &= t(\tau).
\end{aligned} \tag{9.202}
$$

where the common four-dimensional indexing notion is used that coordinate indices start at zero with the time coordinate labeled by the zero index $i = 0$. The three space coordinates remain labeled as $i = 1, 2, 3$. Greek symbols are often used to represent indices in this four-dimensional space-time. At this point, the notion is just a convenience, as space-time is still explicitly nonrelativistic and time still transforms like a scalar.

The action can now be written in a four-component form in terms the parameter τ

$$\delta S = \delta \int_1^2 p_\alpha dq^\alpha = \delta \int_1^2 p_\alpha u^\alpha d\tau, \tag{9.203}$$

where $u^\alpha = \dot{q}^\alpha = dq^\alpha/d\tau$ are components of a four-velocity, and the dot derivative denotes the derivative with respect to the independent scalar parameter τ. With respect to the four-velocity, a Lagrangian functional can be defined as

$$L = p_\alpha \dot{q}^\alpha = p_i \dot{q}^i + p_0 \frac{dt}{d\tau} = p_i \dot{q}^i - H(q,p,t)\frac{dt}{d\tau}. \tag{9.204}$$

where the formalism is not yet symmetric in the treatment of the time and the other coordinates, since the momentum component p_0 associated with the time has been replaced by the holonomic constraint condition $p_0 = -H$, or

$$C(q^\alpha, p_\alpha) = p_0 + H(q^i, p_i, t) = 0. \tag{9.205}$$

Instead of directly applying the constraint, one can keep p_0 as an independent degree of freedom and use the method of undetermined multipliers to enforce the constraint condition

$$\int_1^2 \delta(p_\alpha \dot{q}^\alpha) - \gamma \left(\frac{\partial C}{\partial q^\alpha} \delta q^\alpha + \frac{\partial C}{\partial p_\alpha} \delta p_\alpha \right) d\tau = 0, \tag{9.206}$$

$$\int_1^2 \left(\dot{q}^\alpha \delta p_\alpha - \dot{p}_\alpha \delta q^\alpha - \gamma \left(\frac{\partial C}{\partial q^\alpha} \delta q^\alpha + \frac{\partial C}{\partial p_\alpha} \delta p_\alpha \right) \right) d\tau = -p_\alpha \delta q^\alpha |_1^2 = 0, \tag{9.207}$$

where γ is an unknown multiplier, yet to be determined.

In this four-dimensional treatment, Hamilton's equations of motion are

$$\dot{q}^\alpha = \gamma \frac{\partial C}{\partial p_\alpha},$$
$$\dot{p}_\alpha = -\gamma \frac{\partial C}{\partial q^\alpha}. \tag{9.208}$$

Separation of the time and space coordinates gives the following sets of equations

$$\dot{q}^i = \frac{dq^i}{dt} \frac{dt}{d\tau} = \gamma \frac{\partial H}{\partial p_i}, \quad \alpha = i = 1, 2, 3,$$

$$\dot{q}^0 = \frac{dq^0}{dt} \frac{dt}{d\tau} = \frac{dt}{d\tau} = \gamma \quad \alpha = 0,$$

$$\dot{p}_i = \frac{dp_i}{dt} \frac{dt}{d\tau} = -\gamma \frac{\partial H}{\partial q^i} \quad \alpha = i = 1, 2, 3, \tag{9.209}$$

$$\dot{p}_0 = \frac{dp_0}{dt} \frac{dt}{d\tau} = -\gamma \frac{\partial H}{\partial t} \quad \alpha = 0.$$

The undetermined multiplier is therefore found to be $\gamma = dt/d\tau$. Eliminating the multiplier and applying the constraint $p_0 = -H$ reproduces Hamilton's equations of motion

$$\frac{dq^i}{dt} = \frac{\partial H}{\partial p_i}, \quad \frac{dp_i}{dt} = -\frac{\partial H}{\partial q^i}, \quad \frac{dH}{dt} = \frac{\partial H}{\partial t}. \tag{9.210}$$

The preceding analysis, although derived in a nonrelativistic framework, is reminiscent of special relativity, where $\gamma = 1/\sqrt{1 - v^2/c^2}$. All necessary clues are hidden, but hinted at, in the classical framework. In fact, by simply replacing the standard nonrelativistic Hamiltonian for a particle in an electromagnetic field

$$H_{NR} = (p - eA)^2/2m + e\Phi, \tag{9.211}$$

with the relativistic Hamiltonian

$$H_R = c\sqrt{m^2c^2 + (p - eA)^2} + e\Phi \tag{9.212}$$
$$\approx mc^2 + (p - eA)^2/2m + e\Phi,$$

where c is the speed of light in vacuum, one obtains the correct relativistic form of the Lorentz force equation. The holonomic constraint condition is equivalent to the statement that there is a maximum three-velocity given by the speed of light, which is constant in all frames of reference. This will be demonstrated in Chapter 11, where a Lorentz-invariant, relativistic action principle will be developed.

9.8 SUMMARY

The extended action principle defines a canonical form for the Lagrangian given by

$$L(q, \dot{q}, p, \dot{p}, t) = p_i \dot{q}^i - H(q, p, t) + \frac{dF(q, p, t)}{dt}. \tag{9.213}$$

Varying this action with respect to the independent phase space coordinates (p, q), while holding the end points of the variation fixed, yields Hamilton's equations of motion. Canonical transformations are transformations that maintain this canonical form. The initial and final sets of canonical observables are related by generating functions $\tilde{F}$ such that the invertible mappings $q, p \to Q, P$ are given by

$$p_i \dot{q}^i - H(p, q, t) = P_i \dot{Q}^i - \tilde{H}(Q, P, t) + \frac{d\tilde{F}(q, p, Q, P, t)}{dt}, \tag{9.214}$$

where half of the coordinates in $\tilde{F}$ come from the initial set of coordinates and half from the final set. The direct conditions for a mapping to be canonical are

$$\left(\frac{\partial P_i}{\partial q^j}\right)_{qp} = -\left(\frac{\partial p_j}{\partial Q^i}\right)_{QP}, \quad \left(\frac{\partial P_i}{\partial p_j}\right)_{qp} = \left(\frac{\partial q^j}{\partial Q^i}\right)_{QP},$$

$$\left(\frac{\partial Q^i}{\partial q^j}\right)_{qp} = \left(\frac{\partial p_j}{\partial P_i}\right)_{QP}, \quad \left(\frac{\partial Q^i}{\partial p_j}\right)_{qp} = -\left(\frac{\partial q^j}{\partial P_i}\right)_{QP}. \tag{9.215}$$

In the Poisson bracket formulation of Hamiltonian dynamics, the time evolution of a physical observable is given by

$$\frac{dA(q,p,t)}{dt} = [A, H] + \frac{\partial A}{\partial t}, \tag{9.216}$$

where

$$[A, B]_{q,p} = \frac{\partial A}{\partial q^i}\frac{\partial B}{\partial p_i} - \frac{\partial A}{\partial p_i}\frac{\partial B}{\partial q^i}, \tag{9.217}$$

denotes the Poisson Bracket between any two physical observables. Poisson Brackets are invariant under canonical transformations. General brackets can be generated from the properties of the fundamental bracket relations

$$[p_i, p_j] = 0, \quad [q^i, q^j] = 0, \quad [q^i, p_j] = \delta^i_j. \tag{9.218}$$

The fundamental angular momentum bracket relationship for the rotation symmetry group is given by

$$[L_i, L_j] = \varepsilon_{ijk} L_k \tag{9.219}$$

Canonical quantitization comes from substituting quantum mechanical operators for the classical brackets

$$[A, B]_{PB} \rightarrow \frac{1}{i\hbar}[\hat{A}, \hat{B}]_{QM}. \tag{9.220}$$

An infinitesimal canonical transformation (ICT) is a canonical transformation that differs from the identity transform by an infinitesimal amount

$$Q^i = q^i + \delta q^i, \quad P_i = p_i + \delta p_i \tag{9.221}$$

The generating function can be written as a F_2 transform which deviates from the identity transformation by the addition of a physical observable multiplied by small variational parameter $\delta\varepsilon$

$$F_2(q, P, t) = q^i P_i + \delta\varepsilon G(q, P, t) \tag{9.222}$$

where the observable $G(q, P, t)$ generates an infinitesimal transformation. Finite continuous transformations are built from infinitesimal transformations. Any time-independent generator that leaves Hamiltonian invariant (i.e., any continuous symmetry of the Hamiltonian) is a constant of motion.

The canonical Lagrangian can be expressed in the symplectic form as

$$L = \frac{1}{2}\bar{\mathbf{x}} \cdot \mathbf{j} \cdot \dot{\mathbf{x}} - H(x, t) + \frac{dF(x, t)}{dt}, \tag{9.223}$$

where $\mathbf{x} = [\mathbf{p}, \mathbf{q}]^T$ is a phase space vector and $\mathbf{j}$ is the symplectic operator satisfying $\bar{\mathbf{j}} = -\mathbf{j}$, $\mathbf{j}^2 = -1$, with representation

$$\mathbf{j} = \begin{bmatrix} 0 & 1 \\ -1 & 0 \end{bmatrix}. \tag{9.224}$$

In this representation, Hamilton's equations become

$$\dot{\mathbf{x}} = -\mathbf{j} \cdot \nabla_x H. \tag{9.225}$$

It is easy to show that phase space is conserved since the Hamiltonian flow is divergentless

$$\mathbf{Div}(\dot{\mathbf{x}}) = -\nabla \cdot \mathbf{j} \cdot \nabla H(x, t) = 0. \tag{9.226}$$

Liouville's theorem states that the number density of particles in a volume of phase space is conserved under a general canonical transformation. Since the time evolution of a Hamiltonian system is canonical, this implies that particle density is conserved

$$\frac{d\rho}{dt} = [\rho, H] + \frac{\partial \rho}{\partial t} = 0, \tag{9.227}$$

where ρ denotes the density of states of a statistical ensemble in the continuum limit. Stationary ensembles satisfy

$$[\rho, H] = 0, \tag{9.228}$$

where the stationary ensemble can be expressed as a function of those physics observables that commute with the Hamiltonian.

9.9 EXERCISES

1. Prove the following mathematical identities of Poisson Brackets
   a. $[A, B] = -[B, A]$,
   b. Leibniz's Rule: $[AB, C] = A[B, C] + [A, C]B$.
   c. Jacobi Identity: $[A, [B, C]] + [B, [C, A]] + [C, [A, B]] = 0$.
2. Prove the following two identities

$$\frac{\partial A(q, p, t)}{\partial p_i} = [q^i, A], \qquad \frac{\partial A(q, p, t)}{\partial q^i} = -[p_i, A].$$

3. Obtain the motion in time $x(t)$ of a linear harmonic oscillator of mass m and spring constant k from the time series expansion of the Poisson Bracket equations of motion

$$\frac{dA(p,q)}{dt} = [A, H]$$

Assume the particle's initial position and momentum are given by

$$x(0) = x_0, \quad p(0) = p_0.$$

4. Show by any of the methods listed in Section 9.3.5 that the transformation

$$Q = \ln\left(\frac{1}{q}\sin p\right), \quad P = q\cot p$$

is canonical.

5. Show that the following transformations are canonical
   a. $P = \frac{1}{2}(p^2 + q^2)$, $Q = \tan^{-1}(q/p)$,
   b. $P = q^{-1}$, $Q = pq^2$.

6. Show that the following transformation is canonical for any value of the parameter α

$$Q_1 = q_1\cos\alpha + p_2\sin\alpha, \quad Q_2 = q_2\cos\alpha + p_1\sin\alpha,$$
$$P_1 = -q_2\sin\alpha + p_1\cos\alpha, \quad P_2 = -q_1\sin\alpha + p_2\cos\alpha.$$

7. By requiring that the Lagrangian be of the form $p_i\dot{q}^i - H = P_i\dot{Q}^i - H + dF/dt$, verify the following relationships between the various Legendre transforms

$$F = F_1(q, Q) = F_2(q, P) - Q_iP^i = F_3 + q^ip_i = F_4(p, P) + q^ip_i - Q^iP_i,$$

provided the indicated transformation exists.

8. Show that the F_3 transformation can generate an identity transform, and the F_4 transformation can generate an exchange transform.

9. Show that the transformation

$$Q = p + i\alpha q, \quad P = \frac{p - i\alpha q}{2i\alpha}$$

is canonical and find a generating function. Use the transformation to solve the linear harmonic oscillator problem.

10. Show that for an orthogonal transformation of a vector $\mathbf{q}$ in configuration space $\mathbf{q}' = \mathbf{A} \cdot \mathbf{q}$, where $\bar{\mathbf{A}} \cdot \mathbf{A} = 1$, that the new momenta are

given by the reciprocal orthogonal transformation plus the gradient of a scalar function of the coordinates. [Hint: Try the following F_2 transform: $F_2(q, P) = \mathbf{P} \cdot \mathbf{A} \cdot \mathbf{q} + \Lambda(q)$].

11. Find the Hamiltonian for the spherical pendulum in spherical coordinates. Show explicitly, in these coordinates, that the fundamental angular momentum commutator relationships are satisfied

$$[L_i, L_j] = \varepsilon_{ijk} L_k.$$

12. Show that the Runge-Lenz vector is a constant of the motion for the Kepler problem by explicitly evaluating the Poisson Bracket $[\mathbf{A}, H]$.

13. Show that the Runge-Lenz vector $\mathbf{A}$ and the angular momentum vector $\mathbf{L}$ have mutually commuting Poisson Brackets.

14. Defining $\mathbf{D} = \mathbf{A}/\sqrt{-2mE}$, where $\mathbf{A}$ is the Runge-Lenz vector, verify the following $SO(4)$ Poisson Bracket algebra for the Hydrogen atom:

$$[\mathbf{D}, H] = 0, \quad [\mathbf{L}, H] = 0,$$

$$[L_i, L_j] = \varepsilon_{ijk} L^k, \quad [D_i, L_j] = \varepsilon_{ijk} D^k, \quad [D_i, D_j] = \varepsilon_{ijk} L^k.$$

15. Show that the symplectic Lagrangian

$$L = \bar{z} \cdot \mathbf{j} \cdot \dot{z} - \omega_0 \bar{z} \cdot z,$$

where

$$\omega_0^2 = k/m,$$

$$z = \frac{1}{\sqrt{2}} \left(\frac{p}{\sqrt{m\omega_0}} - \mathbf{j}\sqrt{m\omega_0}q \right) \begin{bmatrix} 1 \\ 0 \end{bmatrix}$$

$$= \frac{1}{\sqrt{2}} \begin{bmatrix} \dfrac{p}{\sqrt{m\omega_0}} \\ \sqrt{m\omega_0}q \end{bmatrix}, \quad \mathbf{j} = \begin{bmatrix} 0 & 1 \\ -1 & 0 \end{bmatrix}$$

gives the equations of motion for a simple harmonic oscillator with spring constant k.

HAMILTON-JACOBI THEORY

$\mathbf{I}$n this chapter, the Hamilton-Jacobi equation will be developed. The relationship between the Hamilton-Jacobi equation and the geometric optics limit of quantum mechanics will then be explored. The Bohm interpretation of quantum mechanics is discussed. Action-angle variables will be introduced and applied to the solution of quasi-periodic systems. The role played by the action variable as an adiabatic invariant will be introduced.

10.1 HAMILTON-JACOBI EQUATION

The Hamilton-Jacobi equation is a unique reformulation of classical mechanics in which the motion of a particle can be represented as a wave. The Hamilton-Jacobi equation was originally developed by William Hamilton in relation to geometric optics. Carl Gustav Jacob Jacobi (1804–1851) developed its mechanical interpretation. The Hamilton-Jacobi equation is particularly useful in identifying the conserved quantities for mechanical systems, even when the system cannot be fully solved.

The analytic solution of Hamilton's equations of motion is closely related to the question of whether a system is integrable. Every continuous symmetry of a mechanical system corresponds to a constant of the motion, which results in a first integral of the equations of motion. But, every completely integrable Hamiltonian system has $2n$ constants of the motion. This raises the question of whether one cannot directly transform the dynamical variables of the system into these constants of the motion. This is the goal of the Hamilton-Jacobi transformation.

For example, the time evolution of a Hamiltonian system is given by the canonical transformation to a set of constant initial coordinates $(q_0, p_0) = (q(t_0), p(t_0))$ such that

$$q^i = q^i(q_0, p_0, t), \quad p_i = p_i(q_0, p_0, t), \tag{10.1}$$

where the Hamiltonian has been annihilated by the transformation, giving

$$\dot{\mathbf{q}}_0 = 0, \quad \dot{\mathbf{p}}_0 = 0, \quad \tilde{H}(q_0, p_0, t) = 0. \tag{10.2}$$

This solution can be generated as a continuous evolution from the identity transform by finding an appropriate Type 2 generating function satisfying the conditions

$$p_i = \frac{\partial F_2(q, p_0, t)}{\partial q^i}, \quad q_0^i = \frac{\partial F_2(q, p_0, t)}{\partial p_{0i}}, \quad H = -\frac{\partial F_2(q, p_0, t)}{\partial t}. \quad (10.3)$$

Clearly, the transformation F_2 can be deduced after the dynamical solution is known, by solving the above partial differential equations for F_2. The converse is also true, the time dependence of the motion can derived by first finding a suitable generator function.

10.1.1 Hamilton's Principal Function

One can generalize the problem slightly, by defining the transformed phase space coordinates (α, β) to be arbitrary constants of integration, perhaps more intuitive than the set of initial coordinates, where

$$Q^i = \beta^i = \beta^i(q_0, p_0), \quad P_i = \alpha_i = \alpha_i(q_0, p_0). \quad (10.4)$$

In the Hamilton-Jacobi formalism, it is conventional to use the functional notation $S(q, \alpha, t)$ to denote the generating function called *Hamilton's principal function*

$$S(q, \alpha, t) = F_2(q, P, t). \quad (10.5)$$

The transformation satisfies the conditions

$$H(q, p, t) = -\frac{\partial S(q, \alpha, t)}{\partial t},$$

$$p_i = \frac{\partial S(q, \alpha, t)}{\partial q^i}, \quad (10.6)$$

$$\beta^i = \frac{\partial S(q, \alpha, t)}{\partial \alpha_i}.$$

The Hamilton-Jacobi equation results from substituting the functional derivatives for momentum $p_i = \partial S(q, \alpha, t)/\partial q^i$ back into the Hamiltonian, yielding a first-order, nonlinear, partial differential equation that can be solved to obtain Hamilton's principal function

$$H\left(q, \frac{\partial S}{\partial q^i}, t\right) = -\frac{\partial S}{\partial t}. \quad (10.7)$$

The Hamilton-Jacobi equation has a solution provided that the following determinant does not vanish

$$\left| \frac{\partial^2 S}{\partial q^i \partial \alpha_j} \right| \neq 0. \tag{10.8}$$

Hamilton's equations in their original form consisted of $2n$ first-order differential equations or n second-order ordinary differential equations. By incorporating n constants of integration into the principal function, one is left with one partial differential equation in n variables. When Hamilton's principal function is separable in the coordinates, the problem decouples into a series of ordinary first-order integral equations where the n generalized momentum α_i can be chosen to be their constants of integration.

Once Hamilton's principal function $S(q, \alpha, t)$ is found, the time evolution of the phase space coordinates are given by straightforward partial differentiation

$$p_i = \frac{\partial S(q, \alpha, t)}{\partial q^i}, \quad \beta^i = \frac{\partial S(q, \alpha, t)}{\partial \alpha_i}. \tag{10.9}$$

The Hamilton-Jacobi equation is enormously useful in solving the equations of motion for classical particles. It yields all constants of motion automatically, and the solution itself becomes formulated in terms of those constants of motion. Even when the physical problem cannot be fully reduced to quadrature, the Hamilton-Jacobi equation can often be used as a starting point for a perturbative solution to the problem.

Not all problems are completely separable, and thereby reducible to quadrature. Reduction to quadrature is always possible in coordinate systems where there are n constants of the motion that have vanishing Poisson brackets with the Hamiltonian. These constants can then be used to define the new generalized momenta. Generally speaking, classical systems are separable in the same set of coordinates as their corresponding quantum mechanical (Schrödinger equation) analogs.

Hamilton's principal function can be related to the classical action by calculating the change in the action along a physical trajectory, where the αs are constants of the motion

$$\int_1^2 L \, dt = \int_1^2 (p_i \dot{q}^i - H) dt = \int_1^2 (p_i dq^i - H dt)$$

$$= \int_1^2 \left(\frac{\partial S}{\partial q^i} dq^i + \frac{\partial S}{\partial t} dt \right) = \int_1^2 (dS(q, \alpha, t))$$

$$= S_2 - S_1 = \Delta S(\alpha, t). \tag{10.10}$$

10.1.2 Hamilton's Characteristic Function

A special, but important, case is where the Hamiltonian is time-independent. The Hamiltonian itself is then one of the constants of the motion. It is a function only of the new canonical momenta of the system, since it is cyclic in the new coordinates. Using separation of variables, one can attempt a separable solution of the form

$$S(q, \alpha, t) = W(q, \alpha) - E(\alpha)t, \tag{10.11}$$

where $W(q, \alpha)$ is referred to as *Hamilton's characteristic function*. In terms of Hamilton's characteristic function, the Hamilton-Jacobi equation then takes on the time-independent form

$$H\left(q, \frac{\partial W(q, \alpha)}{\partial q^i}\right) = E(\alpha) \tag{10.12}$$

where the solution is subject to the conditions

$$p_i = \frac{\partial W(q, \alpha)}{\partial q^i}, \tag{10.13}$$

$$\beta^i = \frac{\partial W(q, \alpha)}{\partial \alpha_i} - \frac{\partial E(\alpha)}{\partial \alpha_i}(t - t_0). \tag{10.14}$$

The first equation is used to set up the Hamilton-Jacobi equation. After the characteristic function is reduced to quadrature, the second equation can be inverted to find the dependence of the coordinates on the time.

An interesting feature of the Hamilton-Jacobi equation is that constant contours of the characteristic function behave like the lines of constant phase of a wave front, in that particle trajectories cross surfaces of constant W perpendicularly. Figure 10.1, for example, illustrates lines of constant W being bisected by a dashed line which represents the parabolic trajectory of a projectile moving in a uniform gravitational field.

For time-independent Hamiltonians, Hamilton's characteristic function can be treated as a separate independent transformation of the physical system, distinct from that generated by Hamilton's principal function. Consider a time-independent mapping from one set of phase space coordinates to another $(q, p) \rightarrow (Q, P)$ that leaves the Lagrangian invariant and the Hamilton unchanged

$$L = P_i \dot{Q}^i - \tilde{H}(P, Q) + \frac{d\tilde{F}}{dt} = p_i \dot{q}^i - H(p, q), \tag{10.15}$$

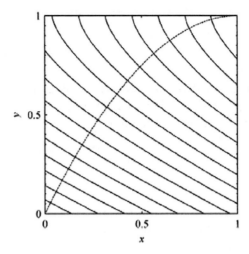

FIGURE 10.1 The dashed line shows a parabolic trajectory of a projectile in a uniform gravitational field cutting perpendicularly through lines of constant *W*.

where

$$\tilde{F} = Q^i P_i + F_2(q, P). \tag{10.16}$$

Let $(Q, P) \to (Q, \alpha)$ and $F_2(q, P) = W(q, \alpha)$. The transformation makes all the new coordinates cyclic and results in the partial derivatives

$$Q^i = \frac{\partial W(q, \alpha)}{\partial \alpha_i}, \quad p_i = \frac{\partial W(q, \alpha)}{\partial q^i}, \quad \tilde{H} = H = E(\alpha), \tag{10.17}$$

where $W(q, \alpha)$ is Hamilton's characteristic function and $\alpha = \{\alpha_1, \ldots, \alpha_n\}$ denotes a complete set of constants of integration that have vanishing Poisson brackets with each other and with the Hamiltonian.

$$[\alpha_i, \alpha_j] = 0, \quad [\alpha_i, H] = 0. \tag{10.18}$$

In particular, assume that the problem is completely separable. The transformation can then be written as a sum over functions of single coordinates.

$$W(q, \alpha) = \sum_i W_i(q^i, \alpha). \tag{10.19}$$

If a coordinate q^k is already cyclic in the original system, separation of variables gives the trivial generating function

$$W_k|_{cyclic} = \alpha_k q^k, \quad (\text{nsc}), \tag{10.20}$$

which leaves the associated momentum unchanged

$$p_k = \frac{\partial(\alpha_k q^k)}{\partial q^k} = \alpha_k. \quad \text{(nsc)}. \tag{10.21}$$

When separation of variables is possible, the momenta are functions of single separated variables, converting the partial differentials to ordinary differential equations of motion

$$p_i(q^i, \alpha) = \frac{\partial W(q, \alpha)}{\partial q^i} = \frac{\partial \sum_j W_j(q^j, \alpha)}{\partial q^i} = \frac{\partial W_i(q^i, \alpha)}{\partial q^i}. \tag{10.22}$$

The method does not specify how the integration constants α are to be chosen. But given a coordinate system in which the problem completely separates, there are n first integrals of the motion whose Poisson brackets mutually vanish. In addition to the energy itself, any cyclic coordinate would be a good candidate to chose as one of the αs.

Given a complete set of integration constants α, the energy can be written as a function of these constants

$$E(\alpha) = \tilde{H}(\alpha) = H(p, q). \tag{10.23}$$

Hamilton's equations of motion are therefore cyclic in terms of the new coordinates. Note that under the transformation, the new coordinates Q^i are not constants of the motion, since the Hamiltonian has not been annihilated, but are reduced to uniform parameterizations of the time

$$\dot{\alpha}_i = -\frac{\partial E(\alpha)}{\partial Q^i} \equiv 0, \quad \dot{Q}^i = \frac{\partial E(\alpha)}{\partial \alpha_i} = \frac{\partial E(\alpha)}{\partial \alpha_i} = v^i(\alpha), \tag{10.24}$$

with solutions

$$Q^i(t) = \beta_0^i + v^i(\alpha)(t - t_0). \tag{10.25}$$

The characteristic function transformation is similar to the principal function transformation, except that the Hamiltonian has not been removed.

Compare, for example, the principal function transformation $(q, p) \rightarrow (\beta, \alpha)$

$$F_2(q, \alpha, t) = S(q, \alpha, t) = W(q, \alpha) - E(\alpha)t, \quad H' = 0 = H - E, \tag{10.26}$$

with an equivalent characteristic function transformation $(q, p) \rightarrow (Q, \alpha)$

$$F_2(q, \alpha) = W(q, \alpha), \quad H' = H = E. \tag{10.27}$$

The two approaches are related by the coordinate substitution

$$Q^i(t) = \frac{\partial W(q, \alpha)}{\partial \alpha_i} = \beta^i + v^i(t - t_0). \tag{10.28}$$

When the Hamiltonian is a constant of the motion, it is sometimes useful to select it as the first constant of integration.

$$P_i = \alpha_i, \quad \text{where } P_1 = E = \alpha_1. \tag{10.29}$$

Then only the first coordinate depends on the time, and the remaining coordinates are constants of the motion

$$Q^i(t) = \begin{cases} \beta^1 + (t - t_0) & i = 1 \\ \beta^i & i \neq 1 \end{cases}. \tag{10.30}$$

10.1.3 Solution to the Simple Harmonic Oscillator

The Hamiltonian for the simple harmonic oscillator is

$$H(q, p) = \frac{1}{2m}(p^2 + m^2\omega^2 q^2) = E, \tag{10.31}$$

where the Hamilton is a constant of the motion and may be identified as the canonical momentum of the new coordinate system, setting $\alpha = E$. Use separation of variables to separate the time behavior

$$S = W(q, E) - Et. \tag{10.32}$$

By substituting, $p = \partial W/\partial q$, the Hamilton-Jacobi equation for Hamilton's principal function becomes

$$\frac{1}{2m}\left(\frac{\partial W(q, E)}{\partial q}\right)^2 + \frac{m\omega^2}{2}q^2 = E, \tag{10.33}$$

which can now be solve for $p(q, E)$

$$p(q, E) = \frac{\partial W(q, E)}{\partial q} = \pm\sqrt{2m(E - m\omega^2 q^2/2)}, \tag{10.34}$$

where the two branches are related by time reversal invariance. If one picks the positive branch, this yields the formal solution for the characteristic function

$$W(q, E) = \int \sqrt{2m(E - m\omega^2 q^2/2)}dq. \tag{10.35}$$

The principal function now has the formal solution

$$S(q, E, t) = \int \sqrt{2m(E - m\omega^2 q^2/2)}\, dq - Et. \tag{10.36}$$

Letting $\beta = -t_0$ when $q_0 = 0$, the time dependence of the motion $q(t)$ for the positive branch can be found by evaluating

$$-t_0 = \frac{\partial S}{\partial E} = \sqrt{m/2} \int_0^q \frac{dq}{\sqrt{(E - m\omega^2 q^2/2)}} - t, \tag{10.37}$$

which has an analytic solution

$$\omega(t - t_0) = \sin^{-1}(\omega q/\sqrt{2E/m}), \tag{10.38}$$

which can be inverted to get the single valued function

$$q(E, t_0, t) = \sqrt{2E/m\omega^2} \sin(\omega(t - t_0)) = A \sin\phi. \tag{10.39}$$

Here, A is the amplitude of the oscillation given by

$$E = \frac{1}{2} m\omega^2 A^2, \tag{10.40}$$

and ϕ is a phase angle

$$\phi = (\omega t + \beta). \tag{10.41}$$

The simplest way to calculate the momentum is to differentiate the solution in Equation (10.39) to get the coordinate's velocity

$$p = m\dot{q} = \sqrt{2mE} \cos(\omega t + \beta) = m\omega A \cos\phi. \tag{10.42}$$

However, one can calculate it, in principle, directly from the transformation equations

$$p = \frac{\partial S}{\partial q} = \frac{\partial W}{\partial q} = \int \frac{-m^2\omega^2 q\, dq}{\sqrt{2m(E - m\omega^2 q^2/2)}} = \int \frac{-m^2\omega^2 dq^2/2}{\sqrt{2m(E - m\omega^2 q^2/2)}}$$
$$= \sqrt{2mE} \cos(\omega(t - t_0)) = m\omega A \cos(\phi), \tag{10.43}$$

Note that one never needs to completely integrate the principal function itself in Equation (10.36). It is sufficient to reduce it to formal quadrature, since only its partial derivatives contribute to the equations of motion. One can use a symbolic integrator, or look it up the solution in a book of integral tables, if a closed form is desired.

10.1.4 Kepler Problem in Plane Polar Coordinates

The effective one-particle Hamiltonian for the Kepler problem can be reduced to motion in a plane with a Hamiltonian given by

$$H(r, p_r, p_\phi) = \frac{p_r^2}{2m} + \frac{p_\phi^2}{2mr^2} - \frac{mk}{r}. \tag{10.44}$$

Time and angle are cyclic variables, giving two first integrals the motion, namely, the energy E and the angular momentum ℓ. Chose these as the new momenta, giving

$$\alpha_1 = H = E, \quad \alpha_2 = p_\theta = \ell. \tag{10.45}$$

Hamilton's principal function separates in these coordinates, giving

$$S(r, \phi, E, \ell, t) = W_r(r, E, \ell) + W_\phi(\phi, \ell) - Et. \tag{10.46}$$

The Hamilton-Jacobi equation can now be written as

$$H = \frac{1}{2m} \left(\frac{\partial W_r}{\partial r} \right)^2 + \frac{1}{2mr^2} \left(\frac{\partial W_\phi}{\partial \phi} \right)^2 - \frac{k}{r} = E \tag{10.47}$$

The solution for the angular separation is trivial since the coordinate is cyclic

$$\frac{\partial W_\phi}{\partial \phi} = p_\theta = \ell \rightarrow W_\phi = \phi \ell. \tag{10.48}$$

The separation of the radial dependence gives

$$\frac{\partial W_r}{\partial r} = p_r = \pm \sqrt{2m \left(E + \frac{k}{r} - \frac{\ell^2}{2mr^2} \right)}. \tag{10.49}$$

Note that there are two branches that are time-reversed images of each other. The principal function of the transformation can be written as

$$S(r, \phi, E, \ell, t) = \pm \int dr \sqrt{2m \left(E + \frac{k}{r} - \frac{\ell^2}{2mr^2} \right)} + \phi \ell - Et. \tag{10.50}$$

The problem has the formal solutions given by

$$\beta_r = -t_0 = \frac{\partial S}{\partial E} = -t \pm \int_{r_0}^{r} dr \, m \Big/ \sqrt{2m \left(E + \frac{k}{r} - \frac{\ell^2}{2mr^2} \right)}, \tag{10.51}$$

$$\beta_\phi = \frac{\partial S}{\partial l} = \phi_0 = \phi - \int_{r_0}^{r} dr (\ell)/r^2 \Big/ \sqrt{2m \left(E + \frac{k}{r} - \frac{\ell^2}{2mr^2} \right)}. \tag{10.52}$$

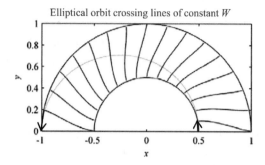

FIGURE 10.2 Lines of constant W for an elliptical Kepler orbit with a trajectory given by $r_0 = r_p = 1/2$, $\phi_0 = 0$, and $e = 1/3$. The physically allowed region is $0.5 \leq r \leq 1$.

The first equation can be solved for $r(t)$, and the second for $r(\phi)$. A comparison with the results found in Chapter 5 on central force motion using elementary techniques reveals that the solutions are identical.

From Equation 10.50, Hamilton's characteristic function for the Kepler problem can be written as

$$W(r, \phi, E, \ell) = \pm \int_b^r \sqrt{2mE + \frac{2mk}{r} - \frac{\ell^2}{r^2}} dr + \phi \ell, \qquad (10.53)$$

where b is the distance of closest approach (perihelion). Figure 10.2 shows that a typical elliptical orbit traverses lines of constant W perpendicularly

$$\mathbf{p} = \nabla W. \qquad (10.54)$$

10.1.5 Simple Pendulum Revisited

The simple pendulum was solved previously using elementary techniques in Chapter 1. Nevertheless, it remains a fascinating problem, demonstrating both librational and rotational modes of periodic motion. The Hamiltonian for the simple pendulum is

$$H = T + V = \frac{p_\theta^2}{2ml^2} + mgl(1 - \cos\theta) = E. \qquad (10.55)$$

To recap the results of the first chapter: the phase space map of the pendulum (shown in Figure 10.3) is a cylinder with the momentum direction oriented along the cylinder's axis. This phase space diagram has the following features:

- There is an elliptical stable point at $\theta = 0, E = 0$. This corresponds to the minimum of energy.
- There is an unstable hyperbolic branch point at $\theta = \pm\pi, E = 2mgl$.

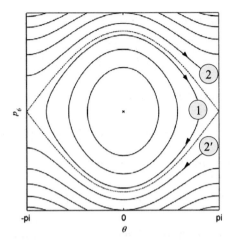

FIGURE 10.3 Phase space diagram for the Pendulum problem. The separatrix between librational and rotational modes of motion is shown as the broken line. At the separatrix, a single librational mode labeled 1 splits into two rotational modes labeled 2 and 2' respectively.

- There are two kinematical regions, separated by the energy $E_0 = 2mgl$.
- For $E > 2mgl$, the motion is rotational (clockwise or counterclockwise) about the pivot.
- For $0 < E < 2mgl$, the motion is librational (oscillatory) about the pivot point.
- At the separatrix between the two modes $E_0 = 2mgl$, a pendulum takes an infinite time to reach the point of unstable equilibrium. If already there, it remains there indefinitely.

One can solve this equation using the energy $E = \alpha$ as the constant of integration. In the new coordinate system

$$\dot{\alpha} = -\frac{\partial E}{\partial \beta} = 0, \quad \dot{Q} = \frac{\partial E}{\partial E} = 1 \tag{10.56}$$

This implies that $Q(t)$ is simply the time

$$Q = t - t_0 \tag{10.57}$$

The characteristic function transformation that gives us this result is given by

$$p_\theta = \frac{\partial W(\theta, E)}{\partial \theta}, \quad Q = \frac{\partial W(\theta, E)}{\partial E} \tag{10.58}$$

Replacing p_θ in the time-independent Hamilton-Jacobi equation gives

$$\frac{1}{2ml^2}\left(\frac{\partial W}{\partial \theta}\right)^2 + mgl(1 - \cos \theta) = E. \tag{10.59}$$

Solving for the momentum

$$p_\theta = \frac{\partial W}{\partial \theta} = \pm\sqrt{2ml^2(E + mgl(\cos \theta - 1))}. \tag{10.60}$$

and reducing to formal quadrature, gives

$$W(E, \theta) = \pm\int d\theta\sqrt{2ml^2(E + mgl(\cos \theta - 1))}. \tag{10.61}$$

As usual, one only needs to completely evaluate the partial derivatives of $W(E, \theta)$. The second transformation condition in Equation (10.56) gives

$$Q = t - t_0 = \frac{\partial W}{\partial E} = \int_0^\theta \frac{\sqrt{ml^2/2}d\theta}{\sqrt{(E + mgl(\cos \theta - 1))}}. \tag{10.62}$$

The time evolution of the system can now be solved by inverting this last equation for $\theta(t)$.

The simple pendulum demonstrates two distinct kinds of periodic motion. The separatrix (shown as the dotted line in Figure 10.3) between librational and rotational modes occurs when $E = 2mgl$. As one crosses the separatrix from lower energies, a single librational mode splits into two rotational modes, one rotating in a clockwise sense and the second in a counterclockwise sense. This splitting is called *bifurcation*.

For the librational solution, there is a maximum angle of oscillation and the motion is vibrational in character (bounded) about the minimum of potential

$$mgl(1 - \cos \theta_{\max}) = E < 2mgl. \tag{10.63}$$

The integral to be evaluated becomes

$$t - t_0 = \frac{1}{\sqrt{2}\omega_0}\int_0^\theta \frac{d\theta}{\sqrt{(\cos \theta - \cos \theta_{\max})}}, \quad \omega_0 = \sqrt{g/l}. \tag{10.64}$$

From a book of integral tables or using a symbolic integrator, the analytic solution is found to be an *elliptic integral of the first kind* $F(\phi, k^2)$,

$$\omega_0 t = F(\phi, k) = \int_0^\phi \left(\sqrt{1 - k^2\sin^2\phi}\right)^{-1} d\phi = \text{sn}^{-1}(\sin \phi, k^2), \tag{10.65}$$

where $k = \sin\theta_{max}/2$. Inverting the solutions gives the result in terms of a *Jacobi elliptic* function $\mathrm{sn}(u, k^2)$

$$\sin\phi = \frac{\sin\theta/2}{\sin\theta_{max}/2} = \mathrm{sn}(\omega_0 t, k^2). \tag{10.66}$$

The period for oscillatory orbits is given by completing the integral over the allowed domain of motion

$$T_{Osc} = \frac{4}{\omega_0} F(\pi/2, k) = \frac{4}{\omega_0} \int_0^{\pi/2} \left(1/\sqrt{1 - k^2 \sin^2\phi} \right) d\phi$$

$$= \frac{2T_0}{\pi} K(k^2), \tag{10.67}$$

where $K(k) = F(\pi/2, k^2)$ is the *complete elliptic integral of the first kind*, with $K(0) = \pi/2$ and $K(1) = \infty$. This result is identical to the results found by using elementary methods in Chapter 1. The analytic solution of the rotational case $E > 2mgl$ is left as an exercise.

10.2   BOHM INTERPRETATION OF QUANTUM MECHANICS

The Hamilton-Jacobi equation can be derived as the geometric optics limit of the Schrödinger[1] equation of quantum mechanics, by representing the Schrödinger wave function in the polar notational form

$$\Psi(\mathbf{r}, t) = A(\mathbf{r}, t)e^{-S(\mathbf{r},t)/i\hbar}, \tag{10.68}$$

where the amplitude A and the phase S are real-valued functions of the coordinates. The Schrödinger equation for a single quantum particle is given by

$$H\Psi = i\hbar\frac{\partial}{\partial t}\Psi, \tag{10.69}$$

where $\hbar = h/2\pi$ in terms of Plank's constant. For a particle in a potential well,

$$H(p, q) = \frac{\mathbf{p}^2}{2m} + V(\mathbf{r}) \rightarrow \frac{-\hbar^2\nabla^2}{2m} + V(\mathbf{r}). \tag{10.70}$$

The second expression results from the quantum mechanical operator substitution $\mathbf{p} = -i\hbar\nabla$, which is consistent with the canonical quantitization

---

[1] Erwin Rudolf Josef Alexander Schrödinger (1887–1961).

condition $[q^i, p_j] = i\hbar\delta^i_j$. The Schrödinger equation then reduces to

$$\left[\frac{-\hbar^2\nabla^2}{2m} + V(\mathbf{r})\right]\Psi = i\hbar\frac{\partial}{\partial t}\Psi. \tag{10.71}$$

Substituting $\Psi(\mathbf{r}, t) = A(\mathbf{r}, t)e^{-S(\mathbf{r},t)/i\hbar}$ gives the following expansions:

$$i\hbar\frac{\partial}{\partial t}\Psi = \left(-\frac{\partial S}{\partial t} + \frac{i\hbar}{A}\frac{\partial A}{\partial t}\right)\Psi, \tag{10.72}$$

$$\mathbf{p}\Psi = -i\hbar\nabla\Psi = \left(\nabla S - \frac{i\hbar\nabla A}{A}\right)\Psi, \tag{10.73}$$

$$\mathbf{p}\cdot\mathbf{p}\Psi = -i\hbar\nabla\cdot\left(\left(\nabla S - \frac{i\hbar\nabla A}{A}\right)\Psi\right)$$

$$= -i\hbar\nabla\cdot\left(\nabla S - \frac{i\hbar\nabla A}{A}\right)\Psi + \left(\nabla S - \frac{i\hbar\nabla A}{A}\right)^2\Psi, \tag{10.74}$$

$$H\Psi = \left[\frac{(\nabla S)^2}{2m} + V(r) - \frac{\hbar^2\nabla^2 A}{2mA}\right]\Psi - \frac{i\hbar}{2m}\left[\frac{2\nabla A\cdot\nabla S + A\nabla^2 S}{A}\right]\Psi. \tag{10.75}$$

Putting the preceding intermediary results back into the Schrödinger equation gives

$$\left(-\frac{\partial S}{\partial t} + \frac{i\hbar}{A}\frac{\partial A}{\partial t}\right)\Psi$$

$$= \left[\frac{(\nabla S)^2}{2m} + V(r) - \frac{\hbar^2\nabla^2 A}{2mA}\right]\Psi - \frac{i\hbar}{2m}\left[\frac{2\nabla A\cdot\nabla S + A\nabla^2 S}{A}\right]\Psi. \tag{10.76}$$

Separating real and imaginary parts of the preceding equation gives a real part

$$\frac{(\nabla S)^2}{2m} + V(r) - \frac{\hbar^2\nabla^2 A}{2mA} = -\frac{\partial S}{\partial t}, \tag{10.77}$$

and an imaginary part

$$\frac{\partial A}{\partial t} = -\frac{2\nabla A\cdot\nabla S + A\nabla^2 S}{2m}. \tag{10.78}$$

This latter equation can be rewritten as

$$\frac{\partial A^2}{\partial t} = -\nabla\cdot\left(\frac{A^2\nabla S}{m}\right). \tag{10.79}$$

Letting $A^2 = \rho$ denote the density of the wave, the imaginary part of the equation can be interpreted as the *continuity equation* for a conserved particle. Defining the velocity as

$$\mathbf{v} = \frac{\nabla S}{m}, \qquad (10.80)$$

The continuity equation becomes

$$\frac{\partial \rho}{\partial t} = -\nabla \cdot (\rho \mathbf{v}), \qquad (10.81)$$

The real part of the Schrödinger equation can be reordered to give

$$\frac{(\nabla S)^2}{2m} + V(r) + \frac{\partial S}{\partial t} = \frac{\hbar^2 \nabla^2 A}{2mA}. \qquad (10.82)$$

In the geometric ray limit, as $\hbar \to 0$, the left-hand-side of this equation is reduces to the Hamilton-Jacobi equation for the imaginary phase S. Therefore, the statement is often made that classical particle mechanics represents the geometric ray limit of the Schrödinger equation.

In the mid Twentieth Century, David Joseph Bohm (1917–1992) pointed out that the Schrödinger equation can be treated exactly as a Hamilton-Jacobi equation for the particle trajectories if one introduces a "quantum potential" of the form [Bohm 1951; 1952;1953]

$$V_Q = -\hbar^2 \nabla^2 A / 2mA, \qquad (10.83)$$

yielding the non-local Hamilton-Jacobi equation

$$\frac{(\nabla S)^2}{2m} + V(\mathbf{r}) + V_Q(\mathbf{r}) = -\frac{\partial S}{\partial t}. \qquad (10.84)$$

In Bohm's interpretation, particles are viewed as having definite positions and velocities, with a probability distribution that may be calculated from the wave function. The quantum potential does not vanish as one moves away from a quantum particle. This is the source of the nonlocality observed in quantum mechanics. The wave function effectively guides the particles by means of the quantum potential that is responsible for the observed interference and diffraction effects. The potential itself is an integral aspect of a quantum particle's nature, and is both the cause and the limit to its spread. Letting $\rho = A^2$, the particle's density distribution is given by the imaginary part of the Schrödinger equation

$$\frac{\partial \rho}{\partial t} = -\nabla \cdot (\rho \mathbf{v}) = -\nabla \cdot (\mathbf{J}), \qquad (10.85)$$

where $\mathbf{J} = \rho \nabla S / m$ can be interpreted as a current density.

Hamilton's principal function can be variously interpreted as the forming function of a canonical transformation that annihilates the Hamiltonian of the system, or, as the imaginary phase of a quantum mechanical wave function representing a quantum particle.

Bohm's interpretation of quantum mechanics, based as it is on the philosophy of physical realism, has strong affinities to classical mechanics. Particles, according to Bohm, exist independently of measurement and have well-defined positions and momenta. A particle going through a double-slit interference measurement apparatus, for example, goes through only one of the two slits, but the nonlocal response of the pilot wave gives rise to an observed interference pattern, nonetheless. Both a particle's wave and trajectory can be determined, although only after the fact, from where it hits the screen.

There are numerous objections to Bohm's hypothesis, not the least of which is that it does not (yet) reproduce quantum field theory. Thus far, a Lorentz-invariant expression of Bohm's ideas has proved challenging. To some, Bohm's construct seems to be contrived. To such as these, the quantum potential appears to be either deeply mysterious or poorly motivated. What actually happens when quantum particles collide is still a matter of active speculation. Particles and fields are only conceptual partitions of the short- and long-range behavior manifested by a single unified phenomenon. No current theory is able to properly account for the self-interaction of a particle with its self-consistent gauge fields, which presumably is at the origin of particle condensation (although renormalization theory allows one to sweep the mess under the rug). What Bohm's analysis shows is that quantum phenomena are subject to multiple interpretations, some more exotic than others. It provides a proof, in principle, that a dynamical interpretation of quantum mechanics remains possible, if elusive.

10.3   ACTION-ANGLE VARIABLES

Action-angle variables are a special set of canonical coordinates that can be used to solve the Hamilton-Jacobi equation. They are valuable as a starting point for a perturbative analysis of the motion. The action-angle method uses the separated actions integrated over a closed path as its canonical momenta. Action-angle variables are particularly useful in finding the frequencies of librational and rotational modes of motion without completely solving the problem. The action variables can be shown to be adiabatic invariants. The *Bohr-Sommerfeld quantization conditions* of the old quantum

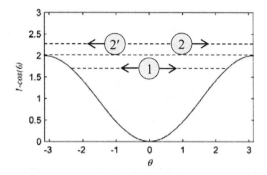

FIGURE 10.4   **Energy diagram of the periodic cosine potential: Region 1 shows a bounded librational mode. Region 2 shows a cyclic rotational mode. In both cases the phase space is bounded.**

theory required the action integral to be an integral multiple of Planck's constant.

The action-angle method can be applied when the following conditions are met

- The problem is completely integrable.
- The Hamiltonian is a constant of the motion.
- The motion is bounded to a finite region of phase space as indicated in Figure 10.4.

The last condition is satisfied if the particle is either bound in a potential well, where the motion oscillates, and is referred to as libration, or where the coordinate is cyclical in character, such as rotational motion on a cylinder, which is periodic in interval 2π. Action-angle variables map a closed path in phase space to a rotation angle that is moving uniformly in time. The mapping takes the form

$$(p, q) \rightarrow (J(\alpha), \varphi), \tag{10.86}$$

where the new constant momenta J_i have the units of action and the new coordinates φ^i are periodic on the standard interval 2π. Let the principal value of the angles be $-\pi < \varphi^i \leq \pi$. In effect, the mapping straightens out the lines of flow in a periodic potential, as shown in Figure 10.5. In the figure, the librational phase space of the simple harmonic oscillator is mapped onto uniform rotational modes on the surface of a cylinder. In effect, action-angle coordinates define a set of invariant tori. The coordinates of the surface are described by the angles, and the motion is characterized by the action variables. The angles are uniform parameterizations of the time about a closed path.

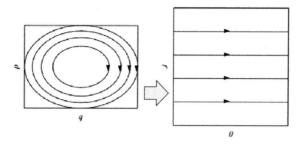

FIGURE 10.5   The librational phase space of the simple harmonic oscillator is mapped into a cylinder by use of action-angle variables.

Action-Angle variables are a special case of using Hamilton's characteristic function to find a solution to the Hamilton-Jacobi partial differential equations

$$p_i = \partial W_i(q^i, J)/\partial q^i, \tag{10.87}$$

$$\varphi^i = \sum_i \partial W_i(q^i, J)/\partial J_i, \tag{10.88}$$

$$H(p, q) = E(J_i), \tag{10.89}$$

where Hamilton equations of motion are cyclic in all the coordinates giving

$$\dot{J}_i = -\partial H'(J)/\partial \varphi^i = 0, \tag{10.90}$$

$$\dot{\varphi}^i = \partial H'(J)/J_i = \omega^i(J). \tag{10.91}$$

The angular velocity of a given mode of oscillation is related to the period by

$$\dot{\varphi}^i = \omega^i = 2\pi/T_i, \tag{10.92}$$

since

$$\oint d\varphi^i = 2\pi = \omega^i T_i, \quad (\text{nsc}), \tag{10.93}$$

where the units of angle are in radians. The action variables are normalized to preserve the integral of the momentum taken over a closed path[2]

$$2\pi J_i(\alpha) = \oint J_i d\varphi^i = \oint p_i(q^i, \alpha) dq^i. \tag{10.94}$$

---

[2]The normalization of action follows the convention of [Jose 1998], where J is quantized in units of $\hbar$ and angles are in radian measure. This convention differs by a factor of 2π from that given in [Goldstein 2002], where the action variable $J = \oint p dq$ is quantitized in units of h and angles are in units of cycles.

This last expression provides the relationship between any convenient set of separation constants α and the desired action observables. Once $\alpha_i(J)$ are known, the energy can be expressed as a function of the action variables

$$E(\alpha) \rightarrow E(J). \tag{10.95}$$

The frequencies of the motion can be obtained by differentiation using Equation 10.91. Finally, Equations 10.87 and 10.88 can be inverted to solve for the time behavior of the coordinates

$$\varphi_0^i + \omega^i(t - t_0) = \sum_i \partial W_i(q^i, J)/\partial J_i$$

$$= \sum_i \frac{\partial}{\partial J_i} \int_{q_0^i}^{q^i} p_i(q^i, J) dq^i. \tag{10.96}$$

By self consistency, the integral of the angle around a closed path is

$$\Delta\varphi = \oint d\varphi = \oint \frac{\partial\varphi}{\partial q^k} dq^k$$

$$= \oint \frac{\partial^2}{\partial q^k \partial J_k} W dq^k = \oint \frac{\partial^2}{\partial J_k \partial q^k} W dq^k$$

$$= \oint \frac{\partial}{\partial J_k} p_k dq^k = \frac{\partial}{\partial J_k} \oint p_k dq^k$$

$$= \frac{\partial}{\partial J_k}(2\pi J_k) = 2\pi. \tag{10.97}$$

10.3.1 Simple Harmonic Oscillator in Action Angle Coordinates

As an example of the action-angle method, consider the one-dimensional harmonic oscillator

$$H(q,p) = \frac{p^2}{2m} + \frac{m\omega^2 q^2}{2}. \tag{10.98}$$

where $\omega = \sqrt{k/m}$ and k is the spring constant. Substituting the first transformation equation $p = \partial W/\partial q$, the time-independent Hamilton-Jacobi equation can be written as

$$\frac{1}{2m}\left(\frac{\partial W(q,J)}{\partial q}\right)^2 + \frac{m\omega^2 q^2}{2} = E(J) \tag{10.99}$$

Now generate the solution using action-angle variables. The transformation will map the lines of flow from a plane to a cylinder as shown in

Figure 10.5. Solving for the momenta

$$p = \frac{\partial W(q,J)}{\partial q} = \pm\sqrt{2m(E - m\omega^2 q^2/2)}, \tag{10.100}$$

and integrating gives

$$W(q,J) = \pm\int_{q_0}^{q} dq\sqrt{2m(E - m\omega^2 q^2/2)}, \tag{10.101}$$

The action variable is formed by evaluating this integral over a complete period. This eliminates the dependence on the coordinate q

$$2\pi J(E) = 4\int_{0}^{q_{max}} dq\sqrt{2m(E - m\omega^2 q^2/2)}. \tag{10.102}$$

The integral can be solved using elementary techniques giving

$$2\pi J = 2\pi \frac{E}{\omega}. \tag{10.103}$$

Therefore, the energy is related to the action by the angular frequency c-number

$$E(J) = \omega J. \tag{10.104}$$

The second of Hamilton's equations in the action-angle frame immediately yields the angular frequency

$$\dot{\varphi} = \frac{\partial E(J)}{\partial J} = \omega = \frac{2\pi}{T_0}, \tag{10.105}$$

Therefore, in the new frame the angle variable is a uniform parameterization of the time

$$\varphi = \varphi_0 + \omega t, \tag{10.106}$$

Finally, solve the second transformation equation

$$\varphi = \partial W/\partial J, \tag{10.107}$$

using Equation (10.101) to find the coordinate dependence on time to get

$$\varphi(q,J) = \frac{\partial W(q,J)}{\partial J} = \frac{\partial}{\partial J}\int_{0}^{q} dq\sqrt{2m(\omega J - m\omega^2 q^2/2)},$$

$$= \int_{0}^{q} dq \frac{m\omega J}{\sqrt{2m(\omega J - m\omega^2 q^2/2)}}$$

$$= \sin^{-1}(m\omega q/2J). \tag{10.108}$$

where the positive root of Equation (10.101) is taken for specificity. This can be combined with Equation (10.100) to complete the solution. The canonical mapping into action-angle coordinates is given by the invertible formulas

$$p(J, \varphi) = \sqrt{2m\omega J} \cos(\varphi), \quad q(J, \varphi) = \sqrt{2J/m\omega} \sin(\varphi). \tag{10.109}$$

$$\cot \varphi = p/m\omega q, \quad J = E/\omega = \frac{1}{2m\omega}(p^2 + m^2\omega^2 q^2). \tag{10.110}$$

It is legitimate to question whether the solution of such a simple problem merits the introduction of the heavy duty machinery of the Hamilton-Jacobi method. It appears to be the proverbial sledge hammer applied to the cracking of the peanut.

10.3.2 Anisotropic Oscillator

Continuing the above approach, n-dimensional anisotropic oscillator separates in Cartesian coordinates. The Hamiltonian is

$$H = \sum_{i=1}^{n} \left(\frac{p_i^2}{2m} + \frac{m\omega_i^2 x_i^2}{2} \right) = \sum_{i=1}^{n} \omega_i J_i, \tag{10.111}$$

where the frequencies are given by

$$\omega_i = \sqrt{k_i/m} = \frac{2\pi}{T_i}. \tag{10.112}$$

If the frequencies are commensurate (rational fractions of each other), the orbits are closed with period given by

$$T_{closed} = n_1 T_1 = n_2 T_2 = n_3 T_3. \tag{10.113}$$

This results in familiar Lissajous patterns. If the periods are incommensurate, the phase portrait eventually fills the allowed phase space, and the motion is called quasiperiodic.

Since the rational numbers are a vanishing small subset of the real numbers, in most cases the orbits never close, but fill the available surface. For example, two Lissajous figures for the anisotropic oscillator calculated over 50 cycles for the vertical axis period are shown in Figure 10.6. The ratio of the two frequencies selected are nearly the same to about a percent, but the figure on the left closes with a $7/5 = 1.4$ ratio of frequencies, while frequency ratio for the figure to the right is irrational, with ratio $\sqrt{2}/1 \sim 1.41$. The irrational ratio indicates that the path is space filling. If the figure on the right were extended for a longer time interval the figure becomes black as the curve sweeps the available phase space.

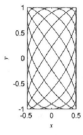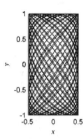

FIGURE 10.6   Lissajous figures for the anisotropic oscillator are calculated over 50 vertical axis periods. The ratio of the two frequencies selected are nearly the same to about a percent, but the figure on the left closes with a $5/7 = 1.4$ ratio of frequencies, while frequency ratio for the figure to the right is irrational, with ratio$1/\sqrt{2} \sim 1.41$, and the orbit is space filling.

Figure 10.7 shows the orbits of the same two Lissajous patterns plotted in action-angle coordinates. The action variables define an invariant manifold in the form of a torus, and the angles define the coordinates for surface of the invariant torus. The orbit to the left is closed with a frequency ratio of 7:5. The orbit is said to be periodic. The frequencies to the right are incommensurate with ratio $\sqrt{2} : 1$, and eventually the trajectory fills the space. The orbit is said to be quasiperiodic due to the irrational ratio of periods.

10.3.3 Kepler Problem Revisited

Up to now, the Kepler problem has been treated in plane polar coordinates by restricting the solution to the plane defined by the angular momentum vector. The more general case of spherical coordinates is sometimes needed, but the integrals are more difficult to evaluate. The reduced Hamiltonian for

FIGURE 10.7   The orbits of the Lissajous figures of Figure 10.6 are shown plotted in action-angle coordinates. The action values define an invariant manifold in the form of a torus, and the angles define the coordinates for the invariant torus. The orbit to the left is closed with a frequency ratio of 7:5. The frequencies to the right are incommensurate with ratio $\sqrt{2}$:1. Eventually, the trajectory fills the available phase space.

the Kepler problem in spherical polar coordinates is

$$H = \frac{g^{ij}p_ip_j}{2m} - \frac{k}{r}$$

$$= \frac{1}{2m}\left(p_r^2 + \frac{1}{r^2}\left(p_\theta^2 + \frac{p_\varphi^2}{\sin^2\theta}\right)\right) - \frac{k}{r}, \tag{10.114}$$

where $m = m_1 m_2/(m_1 + m_2)$ is the reduced mass. The problem separates in these coordinates $W = W_r + W_\theta + W_\phi$, yielding the equations

$$p_\phi = \frac{\partial W_\phi}{\partial \phi} = \ell_z,$$

$$\left(\frac{\partial W_\theta}{\partial \theta}\right)^2 + \frac{\ell_z^2}{\sin^2\theta} = \ell^2, \tag{10.115}$$

$$\frac{1}{2m}\left(\left(\frac{\partial W_r}{\partial r}\right)^2 + \frac{\ell^2}{r^2} - \frac{k}{r}\right) = E.$$

Note that the constants of integration are taken to be the $\hat{z}$-component of angular momentum $L_z = \ell_z$, the square of the angular momentum $L^2 = \ell^2$, and the energy $H = E$. The equations all reduce to quadrature, giving

$$W_\phi = \ell_z \phi = J_\phi \phi,$$

$$W_\theta = \int p_\theta d\theta = \int d\theta \sqrt{\ell^2 - \frac{\ell_z^2}{\sin^2\theta}} = \int d\theta \sqrt{(J_\theta + J_\phi)^2 - \frac{J_\phi^2}{\sin^2\theta}},$$

$$W_r = \int p_r dr = \int dr \sqrt{2m\left(E + \frac{k}{r} - \frac{\ell^2}{r^2}\right)} \tag{10.116}$$

$$= \int dr \sqrt{2m\left(E + \frac{k}{r} - \frac{(J_\theta + J_\phi)^2}{r^2}\right)}.$$

The evaluation of the integrals is left as an exercise;[3] the results for the action variables are

$$J_\phi = \frac{1}{2\pi}\oint p_\phi d\phi = \ell_z,$$

$$J_\theta = \frac{1}{2\pi}\oint p_\theta d\theta = \ell - \ell_z, \tag{10.117}$$

$$J_r = \frac{1}{2\pi}\oint p_r dr = -\ell + k\sqrt{-m/2E},$$

---

[3]See [Goldstein 2002] for the solution.

The constants of integration are

$$\ell_z = J_\phi, \quad \ell = J_\theta + J_\phi, \quad J_r + J_\theta + J_\phi = k\sqrt{-m/2E}, \tag{10.118}$$

therefore,

$$E = -k/2a = -\frac{mk^2}{2(J_r + J_\theta + J_\phi)^2}, \quad mka = (J_r + J_\theta + J_\phi)^2. \tag{10.119}$$

The periods are all degenerate with a common period; therefore, the bound orbits all close.

$$\omega^i = \frac{\partial E}{\partial J_i} = \frac{mk^2}{(J_r + J_\theta + J_\phi)^3} = \frac{1}{k\sqrt{m}}(2E)^{2/3} = \frac{2\pi}{T}. \tag{10.120}$$

This gives Kepler's third law, since using $E = -k/2a$, one finds

$$T^2 = (2\pi)^2 ma^3/k = (2\pi)^2 a^3/G(m_1 + m_2). \tag{10.121}$$

Finally, Hamilton's principal function is given by

$$S(q, J, t) = W - Et = W_r + W_\theta + W_\phi + \frac{mk^2}{2(J_r + J_\theta + J_\phi)^2}t, \tag{10.122}$$

The time dependence of the coordinates can be found by inverting the following equations

$$\varphi^i = \beta^i + \omega^i t = \partial W(q, J)/\partial J_i. \tag{10.123}$$

The solutions appear to be more complicated than they truly are, in part because the simplifying assumption of motion in a plane was not imposed ab initio. To verify the calculation, examine the radial solution, replacing $\ell^2 = (J_\theta + J_\varphi)^2$,

$$m\dot{r} = p_r = \frac{\partial W_r}{\partial r} = \sqrt{2m\left(\frac{k}{r} - \frac{mk^2}{2(J_r + J_\theta + J_\phi)^2} - \frac{(J_\theta + J_\varphi)^2}{2mr^2}\right)}, \tag{10.124}$$

$$\dot{r} = \sqrt{\frac{2}{m}\left(E + \frac{k}{r} - \frac{\ell^2}{2mr^2}\right)}, \tag{10.125}$$

$$\int dt = \int \left[\frac{2}{m}\left(E + \frac{k}{r} - \frac{\ell^2}{2mr^2}\right)\right]^{-1/2} dr = \int \frac{dr}{\sqrt{2(E - V_{eff}(r))/m}}. \tag{10.126}$$

This is the solution that was initially found when studying central force motion in Chapter 5. Further simplification occurs if the total angular momentum vector is aligned with the z-axis. Then, $J_\theta = 0$, and the motion is in the equatorial plane $\theta = \pi/2$, reducing the problem to one in plane polar coordinates.

Action-angle variables are useful for describing completely integrable systems that are localized in phase space (bound systems). Note that most systems are not completely integrable, so this is a special, although important, case. The method offers a straightforward way of calculating the periods of various motions of the system, which may be what one is primarily interested in evaluating. The action-angle method is also a useful starting point to begin studying the stability of systems under small perturbations, where the phase orbits take on the form of invariant tori.

10.3.4 Bohr Atom and the Rydberg Spectrum

Building on Louis de Broglie's (1892–1987) concept of standing waves, action-angle variables were invoked in early attempts to quantize classical physics. The general features of the *old quantum theory* [Bohr 1913] propounded by Niels Henrik David Bohr (1885–1962) and extended by Arnold Johannes Wilhelm Sommerfeld (1868–1951) are as follows:

1. A bound atom can exist without radiating, but only for certain discrete stationary states.
2. These states require the action variable to be quantized in units of $\hbar$

$$J_i = \frac{1}{2\pi} \oint p_i dq^i = n_i \hbar, \tag{10.127}$$

   where $h = 2\pi\hbar$ is Plank's constant.
3. When an atom transitions between two states, it emits a photon of energy

$$\Delta E = h\nu = hc/\lambda = E_2 - E_1. \tag{10.128}$$

4. The energy levels of hydrogen, according to this model, are

$$E_n = -\frac{me^4/(4\pi)^2}{2(J_r + J_\theta + J_\phi)^2} = -\frac{m\alpha^2(\hbar c)^2}{2(J_r + J_\theta + J_\phi)^2} = -\frac{mc^2\alpha^2}{2n^2} = -\frac{E_0}{n^2}, \tag{10.129}$$

   where α is the fine structure constant, and one quantizes the action in units of $\hbar$

$$(J_r + J_\theta + J_\phi) = n\hbar. \tag{10.130}$$

5. Using $\lambda \nu = c$, the wavelengths of emitted light are given by

$$\frac{1}{\lambda} = R \left(\frac{1}{n_2^2} - \frac{1}{n_1^2} \right), \tag{10.131}$$

where $R = E_0/\hbar c$ is Rydberg's[4] constant, and $E_0 = 13.6\,\text{eV}$. This result is in good agreement with the gross features of the observed *Rydberg spectrum* of hydrogen.

Although the Bohr-Sommerfeld model achieved some early success, in the end it could not explain adequately the details of atomic spectra. This theory was eventually replaced by the matrix mechanics approach using canonical quantization of the coordinates and momenta described in the previous chapter on canonical transformations.

10.3.5 Degeneracy

When the frequencies are degenerate, as is the case in the hydrogen atom, the phase space collapses, and the orbits close. It such cases, only one angle is needed to parameterize the time-dependence of the trajectory, since the other angles differ only in their relative phases. In quantum mechanics, this is handled by assigning a principal quantum number to the spectrum. In classical physics one defines a principal action variable, by replacing $J_r \rightarrow J_n$

$$J_n = J_r + J_\theta + J_\phi, \tag{10.132}$$

The energy now depends only on this action

$$E(J_n) = -\frac{mk^2}{2J_n^2}, \tag{10.133}$$

the angular frequencies are degenerate

$$\dot{\varphi}_i = \frac{\partial E(J_n)}{\partial J_i} = \frac{\partial E(J_n)}{\partial J_n} \frac{\partial J_n}{\partial J_i} = \frac{\partial E(J_n)}{\partial J_n} = \omega \tag{10.134}$$

and the angle coordinates differ only in phase

$$\varphi_i = \varphi + \beta_i = \omega(t - t_0) + \beta_i. \tag{10.135}$$

The Bohr-Sommerfeld quantization condition becomes

$$\oint J_n d\varphi = nh = (n_r + n_\theta + n_\phi)h. \tag{10.136}$$

---

[4]Johannes Robert Rydberg (1854–1919).

Degenerate systems are separable in more than one set of coordinates. For example, the Kepler problem is separable in spherical polar and plane polar coordinates.

10.4   ADIABATIC INVARIANTS

At the first international physics conference at Solvay in 1911, a deceptively simple problem was poised. If one had a simple pendulum of variable length l and changed the length slowly enough, how would the energy change? See Figure 10.8 for a schematic of this thought experiment. After much debate, the consensus arrived at was that the action $J = E/\omega$ would remain nearly invariant under these conditions. Therefore, the energy should scale linearly with the angular frequency $E \propto \omega = \sqrt{g/l}$.

This invariance of the action is a general feature of adiabatic changes. When the Hamiltonian is a constant and the problem separable, the action variables are absolute constants of the motion. However, even if the Hamiltonian is time-dependent, the action belongs to a class of observables known as *adiabatic invariants*.

To understand why this would be so, consider a one-dimensional system, with a time dependent Hamiltonian, $H_0(q, p; \lambda(t))$ where the time dependence is due to a parameter λ that slowly varies in time. The change in λ is said to be adiabatic if the change is very slow compared to the period of the motion

$$|\Delta\lambda| \approx |\dot{\lambda}|\frac{2\pi}{\omega} \ll \lambda \tag{10.137}$$

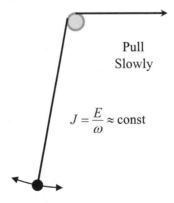

FIGURE 10.8   The action J of a simple pendulum remains nearly constant under an adiabatic change of its length. Therefore, the energy would scale linearly with the frequency $E(t) \propto \omega(t)$.

This is called the *adiabatic condition*. One performs a canonical transformation into action-angle variables using the same generating function $F_2(q,J)$ as would have been be used in the case of constant λ. Of course, since λ is a function of time, the generating function will be also a function of time through its dependence on λ.

$$H'(\phi,J) = H_0(q,p;\lambda) + \frac{\partial F_2(q,J;\lambda)}{\partial t} = H_0(J;\lambda) + \frac{\partial F_2}{\partial \lambda}\dot{\lambda}$$

$$= H_0(J;\lambda) + f(\varphi,J;\lambda)\dot{\lambda}. \tag{10.138}$$

Here, f is an auxiliary function defined as

$$f(\varphi,J;\lambda) = \left.\frac{\partial F(q,J;\lambda)}{\partial \lambda}\right|_{q\to q(\varphi,J)}. \tag{10.139}$$

The change to action-angle coordinates is made after taking the partial derivative. Note that the unperturbed Hamiltonian $H_0(J;\lambda)$ is cyclic in the angle variable. Because φ is periodic on interval 2π, $f(\varphi,J,\lambda)$ can be expanded in a Fourier series

$$f(\varphi,J,\lambda) = \frac{a_0}{2} + \sum_{n=1}^{\infty} a_n(J,\lambda)\cos n\varphi + \sum_{n=1}^{\infty} b_n(J,\lambda)\sin n\varphi, \tag{10.140}$$

implying that $\partial f(\varphi,J,\lambda)/\partial\varphi$ is oscillatory

$$\frac{\partial f(\varphi,J,\lambda)}{\partial\varphi} = -\sum_{n=1}^{\infty} na_n(J,\lambda)\sin n\varphi + \sum_{n=1}^{\infty} nb_n(J,\lambda)\cos n\varphi. \tag{10.141}$$

Hamilton's equations of motion are

$$\dot{J} = -\frac{\partial H'}{\partial\varphi} = -\frac{\partial f(\varphi,J,\lambda)}{\partial\varphi}\dot{\lambda}, \tag{10.142}$$

$$\dot{\varphi} = \frac{\partial H'}{\partial J} = \frac{\partial H_0(J,\lambda)}{\partial J} + \frac{\partial f(\varphi,J,\lambda)}{\partial J}\dot{\lambda}$$

$$= \omega(\lambda) + \frac{\partial f(\varphi,J,\lambda)}{\partial J}\dot{\lambda}. \tag{10.143}$$

It should be clear that the action is no longer constant, although its time dependence is small due to the smallness of $\dot{\lambda}$. Nor is the frequency $\dot{\varphi}$ constant. It is slow-changing due to the dependence of $H_0(J,\lambda)$ on λ. The equation for $\dot{J}$ contains a product of a small, slowly changing, factor $\dot{\lambda}(t)$ multiplied by a periodically oscillating term $\partial f(\varphi,J,\lambda)/\partial\varphi$. Qualitatively, if $\dot{\lambda}(t)$ is slowly changing over a cycle, one can expect that the oscillations of $\dot{\lambda}(\partial f/\partial\varphi)$ averaged over a period to cancel its contribution to some higher order. Therefore, $J(t)$ remains almost constant even when tracked over long periods of time, during which the parameter $\lambda(t)$ may have changed considerably. Quantities that remain nearly constant under an adiabatic change of parameters are said to be adiabatically invariant.

Consider for example, a time-dependent oscillator, where the oscillator frequency is adiabatically changing in time. The unperturbed Hamiltonian is

$$H_0(q,p;\omega(t)) = \frac{p^2}{2m} + \frac{m\omega^2}{2}q^2 = \omega J. \qquad (10.144)$$

Then, by the adiabatic assumption, the action remains almost constant even when the oscillatory frequency changes, if the change is slow enough. The time-dependence of the energy is given approximately by

$$E(t) = \omega(t)J. \qquad (10.145)$$

$E(t)$ and $\omega(t)$ are slowly changing functions, although the net change in each may be large over a long time interval. However, if the adiabatic condition is valid, the action will remain nearly unchanged. In many cases, analysis shows that the change in the action will be exponentially smaller than the change of the angular frequency.

As a second example, consider a gyrating charged particle in a slowly varying magnetic field. Assuming a constant magnetic field, the unperturbed Lagrangian separates into components parallel and perpendicular to the magnetic field, which can be taken to be in the $\hat{z}$-direction. The magnetic field can be written as $\mathbf{B} = \nabla \times \mathbf{A} = B\,\hat{z}$, where the vector potential in plane polar coordinates is

$$\mathbf{A} = A_\phi \hat{\phi} = \frac{1}{2}Br\,\hat{\phi}. \qquad (10.146)$$

The Lagrangian for this interaction is

$$L = \frac{1}{2}m\mathbf{v}^2 + q_e\mathbf{A}\cdot\mathbf{v}, \qquad (10.147)$$

or

$$L = \frac{1}{2}m(r^2\dot{\phi}^2 + \dot{r}^2 + \dot{z}^2) + \frac{1}{2}q_eBr^2\dot{\phi}. \qquad (10.148)$$

The canonical momentum about the azimuthal axis is conserved and can be used as an action variable

$$J_\phi = p_\phi = mr^2\left(\dot{\phi} + \frac{q_e}{2m}B\right). \qquad (10.149)$$

In the gyroscopic frame $\dot{\phi} = \omega = -q_e B/m$ and $\dot{r} = 0$. In the transverse direction to the magnetic field, the motion is circular with a radius of gyration $r = m v_\perp / q_e B$. If q_e/m is constant, the magnetic moment of the gyrating charge is an adiabatic variable given by

$$\mu = \frac{q_e}{m} J_\phi = -\frac{1}{2} q_e r^2 \dot{\phi} = -\frac{q_e^2}{2m} r^2 B. \tag{10.150}$$

The magnetic moment is known as the *first adiabatic invariant* of the charged plasma. Note that $J \sim \pi r^2 B = \Phi_B$, so that this effect can also be considered as a conservation of magnetic flux passing through the orbit of the particle. Calculating the transverse component of kinetic energy gives

$$T_\perp = \frac{1}{2} m v_\perp^2 = \frac{1}{2} m r^2 \omega^2 = -\frac{1}{2} m r^2 \omega \frac{q_e}{m} B = \mu B. \tag{10.151}$$

In a slowly changing magnetic field, the particle will move in such a way as to keep its magnetic moment constant. The total kinetic energy of the particle in a time independent magnetic field is a constant of the motion

$$E = \frac{m}{2} (v_\parallel^2 + v_\perp^2), \tag{10.152}$$

implying that

$$v_\parallel = \pm \sqrt{2m \left(E - \frac{m}{2} v_\perp^2 \right)}, \tag{10.153}$$

In a slowly changing magnetic field, the particles spiral along the magnetic field lines, changing direction when $v_\parallel = 0$. In the adiabatic approximation, the magnetic moment is conserved, and the particles reflect from a "magnetic mirror" when the field reaches a critical magnitude

$$v_\parallel = \pm \sqrt{2m(E - \mu B)} = 0. \tag{10.154}$$

Shaped magnetic fields (magnetic bottles) are often used to trap charged particles in a plasma.

10.5 SUMMARY

The Hamilton-Jacobi equation is a partial differential equation of the form

$$H \left(q, \frac{\partial S}{\partial q^i}, t \right) = -\frac{\partial S}{\partial t}, \tag{10.155}$$

where $S(q, \alpha, t)$ is Hamilton's principal function and $p_i = \partial S / \partial q^i$ are the canonical momentum observables of a classical system. It is the result of making a canonical transformation $(q,p) \rightarrow (\beta,\alpha)$ to a set of constant canonical momenta α and constant coordinates β that annihilates the Hamiltonian. The transformation equations are

$$H(q,p,t) = -\frac{\partial S(q,\alpha,t)}{\partial t}, \quad p_i = \frac{\partial S(q,\alpha,t)}{\partial q^i}, \quad \beta^i = \frac{\partial S(q,\alpha,t)}{\partial \alpha_i}. \quad (10.156)$$

The solution is usually found by separation of variables, where the αs are constants of integration.

If the Hamiltonian is a constant of the motion, the time is cyclic. Then, assuming that the principal function separates in the coordinate system chosen, one gets

$$S(q,\alpha,t) = W(q,\alpha) - E(\alpha)t = \sum_i W_i(q^i,\alpha) - E(\alpha)t, \quad (10.157)$$

where $W(q,\alpha)$ is Hamilton's characteristic function. The characteristic function can also be generated as an independent transformation $(q,p) \rightarrow (Q,\alpha)$ that leaves the Hamiltonian unchanged, using

$$\tilde{H} = H = E(\alpha), \quad p_i = \frac{\partial W(q,\alpha)}{\partial q^i}, \quad Q^i = \frac{\partial W(q,\alpha)}{\partial \alpha_i} = \beta^i + v^i(t-t_0),$$

$$(10.158)$$

where $\dot{Q}^i = \partial E(\alpha)/\partial \alpha^i = v^i(\alpha)$. Since $\mathbf{p} = \nabla_q W(q,\alpha)$, the motion is orthogonal to lines of constant W. Therefore, the characteristic function acts like the phase of a wave, and classical mechanics can be thought of as the classical limit of a wave equation.

For quasi-periodic systems, the action-angle transformation $(q,p) \rightarrow (\varphi,J)$ is particularly useful. This is a variant of the Hamilton's characteristic function transformation. The generating function is

$$S(q,J,t) = \sum_i W_i(q^i,J) - E(J)t, \quad (10.159)$$

where the action variables are defined in terms of the arbitrary constants of integration as

$$J_i(\alpha) = \frac{1}{2\pi} \oint p_i dq^i = \frac{1}{2\pi} \oint \frac{\partial W_i(q,\alpha)}{\partial q^i} dq^i. \quad (10.160)$$

Once these integrals have been obtained, they can be inverted to eliminate the integration constants αs as functions of the actions J. Hamilton's equations of motion in action-angle variables are

$$\dot{J}_i = -\frac{\partial E(J)}{\partial \phi^i} = 0, \quad \dot{\varphi}^i = \frac{\partial E(J)}{\partial J_i} = \omega^i(J). \tag{10.161}$$

The periods of oscillation are given by

$$T_i = \frac{2\pi}{\omega^i}. \tag{10.162}$$

Finally, the time-dependence of the motion is given as a solution to the transformation equations

$$\varphi^i = \varphi_0^i + \omega^i(t - t_0) = \frac{\partial W(q,J)}{\partial J_i}. \tag{10.163}$$

Under time-dependent perturbations, action variables are examples of adiabatic invariants, i.e., observables that remain nearly constant, provided that the time evolution of the parameters is slow compared to the characteristic periods of the quasi-periodic motion.

The Bohr-Sommerfeld quantitization rule of the old quantum theory quantizes the action variable in units of Planck's constant

$$J_i = \frac{1}{2\pi} \oint p_i dq^i = n_i \hbar. \tag{10.164}$$

10.6 EXERCISES

1. Show that the function

$$S(q, p_0, t) = q p_0 \sec \omega t - \frac{1}{\omega} \left(\frac{m\omega^2}{2} q^2 + \frac{p_0^2}{2m} \right) \tan \omega t$$

is a solution to the Hamilton-Jacobi equation for the simple harmonic oscillator with Hamiltonian

$$H = \frac{1}{2m}(p^2 + m^2 \omega^2 q^2).$$

Then show that the principal function S generates the following initial value solution to the equations of motion

$$q(t) = q_0 \cos \omega t + \frac{p_0}{m\omega} \sin \omega t, \quad p(t) = p_0 \cos \omega t - m\omega q_0 \sin \omega t.$$

2. A particle moving in one dimension sees a force $F(t) = \lambda t$ that is linearly increasing with time. The time dependent Hamiltonian for this system is given by

$$H(p, x, t) = \frac{p^2}{2m} - \lambda t x$$

where λ is a constant. Solve the equations of motion using Hamilton's principal function for the initial conditions $x_0 = t_0 = 0$ and $p_0 = mv_0$.

3. Find Hamilton's characteristic function for a projectile moving in a vertical plane under a constant force of gravity, with Hamiltonian

$$H = \frac{p_x^2}{2m} + \frac{p_z^2}{2m} + mgz,$$

Show that application of this function yields the correct equations of motion. The initial conditions are $t = x = z = 0$, and the projectile has an initial velocity v_0 at an angle α with respect to the horizontal. (Figure 10.1 shows what the constant contours of the characteristic function might look like for some set of numerical input.)

4. A symmetrical top with one point fixed in a uniform gravitational field has a Hamiltonian given by

$$H = \frac{p_\theta^2}{2I_0} + \frac{(p_\phi - p_\psi \cos \theta)^2}{2I_0 \sin^2 \theta} + \frac{p_\psi^2}{2I_3} + mgl \cos \theta.$$

Use Hamilton's principal function to reduce the problem to quadrature.

5. A particle of charge q and mass m is constrained to move in a plane acted on by a central force harmonic oscillator potential

$$V(r) = kr^2/2.$$

In addition, there is a constant magnetic field $\mathbf{B}$ perpendicular to the plane, so that the particle is also influenced by a vector potential

$$\mathbf{A} = \frac{1}{2} \mathbf{B} \times \mathbf{r}.$$

Find the Hamiltonian for this system in plane polar coordinates (r, ϕ). Set up the Hamilton-Jacobi equation for this system. Separate the equation and reduce it to quadrature. Discuss the motion if the canonical momentum p_ϕ is zero at initial time $t = 0$.

6. Use the Hamilton-Jacobi equation to reduce the problem of the energetic pendulum with energy $E > E_s = 2mgl$ to quadrature.

7. The Hamiltonian in elliptic cylindrical coordinates can be written as

$$H = \frac{p_\mu^2 + p_\nu^2}{2ma^2(\sinh^2 \mu + \sin^2 \nu)} + \frac{p_z^2}{2m} + V(\mu, \nu, z),$$

where the foci of the ellipses are located at $\pm a$ on the x-axis. Show that the Hamilton-Jacobi equation is completely separable into ordinary differential equations in these coordinates, provided that V has a similar structure, namely

$$V(\mu, \nu, z) = \frac{V_\mu(\mu) + V_\nu(\nu)}{2ma^2(\sinh^2 \mu + \sin^2 \nu)} + V_z(z),$$

where $V_\mu(\mu), V_\nu(\nu), V_z(z)$ are arbitrary functions.

8. The Hamiltonian in parabolic cylindrical coordinates can be written as

$$H = \frac{p_\sigma^2 + p_\tau^2}{2m(\sigma^2 + \tau^2)} + \frac{p_z^2}{2m} + V(\sigma, \tau, z),$$

Show that the Hamilton-Jacobi equation is completely separable into ordinary differential equations in these coordinates provided, that V has a similar structure, namely,

$$V(\sigma, \tau, z) = \frac{V_\sigma(\sigma) + V_\tau(\tau)}{\sigma^2 + \tau^2} + V_z(z)$$

where $V_\sigma(\sigma), V_\tau(\tau), V_z(z)$ are arbitrary functions.

9. Integrate Hamilton's characteristic function for the Kepler problem. Plot $W(r, \theta)$ for the case of a parabolic orbit $E = 0$. Use dimensionless units $b = m = k = 1$. Draw a parabolic trajectory with initial conditions $(r_0 = b, \phi_0 = 0, t_0 = 0)$ and show that the trajectory is perpendicular to lines of constant W. The results should be similar to the plot shown in Figure 10.9.

10. Find the radial and angular periods for a particle with energy $E < V_0$ trapped in a two dimensional potential well given by

$$V(r) = \begin{cases} 0 & r < a \\ V_0 & r > a \end{cases}.$$

11. Use action-angle variables to show that the librational period of the energetic pendulum with energy $E < E_s = 2mgl$ is given by

$$T_{Osc} = \frac{2T_0}{\pi} K(k^2),$$

where $T_0 = 2\pi \sqrt{l/g}$ is the small amplitude period and $K(k) = F(\pi/2, k^2)$ is the complete elliptic integral of the first kind.

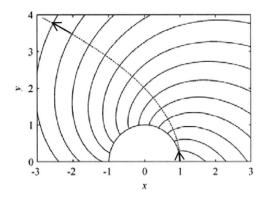

FIGURE 10.9 Lines of constant W for a parabolic Kepler orbit with an impact parameter $b = 1$. The physically allowed region is $r \geq b$.

12. Use action-angle variables to show that the rotational period of the energetic pendulum with energy $E > E_s = 2mgl$ is given by

$$T_{Rot} = \frac{T_0}{\pi \sqrt{E/E_s}} K \left(\frac{1}{E/E_s} \right).$$

where $T_0 = 2\pi \sqrt{l/g}$ is the small amplitude period and $K(k) = F(\pi/2, k^2)$ is the complete elliptic integral of the first kind.

13. A particle moves in periodic motion in one dimension under the influence of a potential $V(x) = k|x|$. Use action-angle variables to find the period of the motion and express the period as a function of the particle energy E.

14. Consider a particle of charge q with mass m moving in a plane while gyrating under the influence of a slowly changing uniform magnetic field $B(t)$. Find the adiabatic change of kinetic energy with time.

15. A particle moving in the vertical $\hat{x} \wedge \hat{z}$ plane under the constant force of gravity is constrained to move on a cycloid parameterized by

$$x = a(\theta + \sin \theta), \quad z = a(1 - \cos \theta).$$

a. Find a suitable Hamiltonian for this problem.
b. Show, using action-angle variables, that the period of the motion is independent of its amplitude for $\theta_{\max} < \pi$.

16. Consider the Hamiltonian

$$H = \frac{p^2}{2m} + \lambda q^{2n}$$

where λ is a positive constant and n is a positive integer.

a. Show that the energy is related to the action variable J by

$$E = \frac{1}{\lambda^{n+1}} (\pi^2 J^2 / 8m I_n^2)^{n/(n+1)},$$

where $I_n = \int_0^1 dx \sqrt{1 - x^{2n}}$.

b. Find the period of the motion.

c. If λ is slowly changing in time, what is the adiabatic change in the energy?

11

SPECIAL RELATIVITY

I n this chapter, Einstein's Special Theory of Relativity will be examined. Spacetime will be treated as a four-dimensional flat Minkowski space consisting of three space-like and one time-like dimension. The spin group $Spin(3, 1)$ for this space is the cover group for the proper homogeneous Lorentz group $SO(3, 1)$, which consists of space rotations and velocity boosts. Inertial frames in Minkowski space are connected by the Poincaré group of transformations. Four-momentum is conserved in the absence of a four-force. Particle motion reduces to Newtonian dynamics in the limit of small velocities. Particle kinematics in the impulse approximation and relativistic single-particle Lagrangians will be explored. The Lorentz force law is derived.

11.1 UNIFYING SPACE AND TIME

Newton viewed space and time as existing independently of matter. Together, they parameterize the motion of particles, providing an invariant field of view in which matter evolves. Newton assumed that three-dimensional space is isotropic, homogenous, and Euclidean. The Euclidean (flatness) property implies that finite displacements can be treated as vector elements of a tangent space.

Denote the displacement vector of a point from the origin in E^3 by ΔP. The Newtonian laws of physics preserve the global, ten-parameter, Galilean group of point transformations given by

$$\Delta P' = R\Delta P\bar{R} + \Delta X_0 + V_0\Delta t. \tag{11.1}$$

where R is a space rotation, ΔX_0 is a constant displacement, and $V_0\Delta t$ is time dependent displacement due to a Galilean velocity transformation.

Galilean invariance is preserved if one derives the laws of motion from a scalar Lagrangian that depends on scalar products of vectors (relative coordinates) defined in some inertial frame. Additional postulates of Newtonian mechanics include:

1. Matter consists of point-like particles with definite mass.
2. The momentum of a particle is the product of its mass and velocity.

451

3. The applied force acting on a particle is equal to its time rate of change of momentum.
4. The total momentum of an isolated system of particles is an additive vector constant.

By the middle of the nineteenth century, the first of a number of crises for Newtonian mechanics was becoming apparent. The rapidly developing theories of electricity and magnetism were unified by James Clerk Maxwell (1831–1879), but his equations did not agree with Galilean concepts of how velocities should be added. Intrinsic to Maxwell's theory was a natural, universal, velocity scale consistent with the fixed value of the velocity of light in a vacuum. In free space, his electromagnetic fields are solutions to the wave equation

$$\left(\vec{\nabla}^2 - \frac{1}{c^2}\frac{\partial^2}{\partial t^2}\right)\begin{bmatrix}\vec{E}(\vec{x},t)\\ \vec{B}(\vec{x},t)\end{bmatrix} = 0, \tag{11.2}$$

where c is the speed of light.[1] From his equations, Maxwell correctly deduced that light was electromagnetic in character. In his 1864 presentation to the Royal Society, Maxwell stated:

> The agreement of the results seems to show that light and magnetism are affections of the same substance, and that light is an electromagnetic disturbance propagated through the field according to electromagnetic laws.   [Maxwell 1865]

This hypothesis was fully validated in 1888 with the production and detection of electromagnetic ultra-high-frequency radio waves in the laboratory by Heinrich Rudolf Hertz (1857–1894). By 1895, Guglielmo Marconi (1874–1937) had successfully demonstrated the long-range propagation of radio waves.

The observed isotropy of the light spectrum from the fixed stars argues against any gross universal variation of the speed of light. It was believed at that time, however, that light waves needed an underlying medium called the *luminiferous aether* (light-bearing aether) in which to propagate. Maxwell's equations with a special value for the speed of light could be made to comply with Galilean relativity if there was a preferred frame in which the aether was at rest. The assumption that light waves propagate in a medium was put

---

[1]The speed of light is currently considered to be an exact number, with the meter redefined in terms of it. According to the 17th General Conference on Weights and Measures (Conférence générale des poids et measures, or CGPM) in 1983, which maintains the SI system of units, "the meter is the length of the path travelled by light in vacuum during a time interval of 1/299792458 of a second."

to a stringent test by the Michelson-Morley experiment[2] in 1887 [Michelson 1887]. The Michelson-Morley experiment sought to determine the direction of the aether with respect to the proper motion of the Earth as measured relative to the fixed stars. Instead, the results indicated that the speed of light in vacuum is a constant independent of the Earth's motion. Either the aether is dragged along with the Earth, or there was no aether.

The aether concept was becoming increasing unrealistic for numerous reasons. To account for the polarization of light, light had to be a transverse wave, which required a solid rather than a fluid medium. The aether particles, moreover, had to be exceeding light, yet the medium had to be extremely rigid to get the high oscillation frequencies observed. As Newton himself had pointed out in his earlier rejection of light as a wave phenomenon, this medium also had to be nearly frictionless to avoid slowing down the heavenly bodies along their appointed paths.

It is now known that the Galilean velocity transformation is wrong at high speeds. Hendrik A. Lorentz (1853–1928), whose research greatly extended James Clerk Maxwell's theory of electricity and of light, developed the Lorentz transform[3] [Lorentz 1899; 1904], which paved the way for the Special Theory of Relativity. Maxwell's equations are form-invariant under Lorentz transforms, which are the correct transformation laws for velocities approaching light speed. The Lorentz transform is orthogonal, in the general sense of preserving the inner product on a space of indefinite signature, but mixes space and time degrees of freedom. One consequence is that there is a maximum limiting velocity for material objects in any frame of reference. Another consequence is that time is not a scalar. The new transformations introduced exotic effects such as length contraction and time dilation; however, it was unclear at first how to interpret these effects.

The year after the Lorentz velocity transform was republished by Lorentz in 1904,[4] Albert Einstein (1879–1955), like Newton before him, had his *annus mirabilis* (miraculous year) in which he published three seminal papers on

---

[2] Albert Abraham Michelson (1852–1931) and Edward Williams Morley (1838–1923).

[3] Joseph Larmor (1857–1942) actually published the Lorentz Transformations in 1897 [Larmor 1897; 1898] two years before Lorentz's first publication in 1899. He verified the FitzGerald-Lorentz contraction effect, and made the first apparent reference to time dilation. Larmor believed that the source of electric charge was a particle moving in the aether, which by 1897 he was already was referring to as the electron. He later became an opponent to the theories of special and general relativity, insisting on the absolute nature of time. Larmor precession is named after him.

[4] Lorentz was already aware of the Lorentz transformation as early as 1895. The first apparent mention of the Lorentz transform was by Woldemar Voigt (1850–1919) in 1887 and went entirely unnoticed at the time. It is not clear of how aware Einstein

the photoelectric effect [Einstein 1905a], Brownian motion [Einstein 1905b], and Special Relativity [Einstein 1905c], proving, in effect, that light comes in quantized packets, that atoms really exist, and that spacetime is four-dimensional in character.

Einstein based his special theory on two postulates:

1. The speed of light in vacuum is a constant in all inertia frames.
2. The laws of physics are the same in all inertia frames.

The critical postulate was that the speed of light in vacuum is a universal constant relating time to distance and thus independent of the frame of reference. The second postulate states the laws of physics must be invariant under Lorentz transformations in all "special" (inertial) frames. Maxwell's equations were invariant under Lorentz transformations, and Newtonian particle mechanics would have to be modified accordingly. Einstein's theory of General Relativity, published in 1915 [Einstein 1915], would latter remove the limitation of the special inertial frame constraint.

By making the Lorentz transforms an intrinsic property of space and time, rather than some mechanical effect of the aether, Einstein effectively reaffirmed the Galilean concept of the relative character of physical law. Galilean relativity was reinterpreted as the limiting case of a more general law of transformation, namely the Lorentz Law, and remained valid in the limit of velocities small compared to the speed of light. Not long after the special theory was announced, Hermann Minkowski (1864–1909) showed that the Lorentz transforms follow as a natural consequence from assuming that spacetime forms a single four-dimensional continuum, where space and time degrees of freedom differ in their relative metric signature [Lorentz 1952].

Einstein's publication of the photoelectric effect earlier in 1905, showing that light has quantized corpuscle properties, also hastened the demise of the aether. Particles, unlike waves, do not require a medium to propagate. Today, the concept of the aether has nearly universally been discarded, a victim of Occam's razor. It has been supplanted by the concept of the physical field. The physical field, such as the electromagnetic field, for example, is considered to be an intrinsic property of space and time, existing everywhere, requiring no underlying medium for its existence, support, or propagation.

If the aether has been discredited, however, in another sense, it has never truly gone away. The idealization of spacetime as a passive field of view for the action observed on the world stage is simplistic. General relativity, for example, envisions space and time as active participants in the interaction

_____

was of the work of others. Lorentz and Poincaré, at least, also made significant contributions to the development of the Theory of Special Relativity.

of particles and their fields, instructed by and instructing the behavior of the matter that they contain. Furthermore, due to quantum fluctuations, the vacuum state of matter, so-called "empty space," is neither empty nor simple. Ephemeral bubbles of matter and antimatter pairs, spontaneously arising and subsiding, contribute to the stress-energy tensor of the universe, and to the tension of the four-dimensional manifold in which matter exists and which physicists refer to as spacetime.

In the century since Einstein's work, special relativity has become an essential feature of human technology. Global Positioning Satellite (GPS) systems and Magnetic Resonance Imaging (MRI) probes, to give two examples, could work as predicted only if the Lorentz transformation laws were valid.

11.2 MINKOWSKI SPACE

For dimensional reasons, let $x^0 = ct$ denote the fourth (time-like) component of a four-dimensional spacetime. The convention used for dummy indices is that Greek indices start with an offset of zero and Latin indices start with an offset of one. Minkowski spacetime can be represented as either $GA(3, 1)$ or as $GA(1, 3)$. The $GA(3, 1)$ solution will be adopted here, since it contains $GA(3, 0)$ as a proper subgroup, and thus all previous three-space results will carry over without change. The $GA(3, 1)$ solution (space-like positive signatures) is commonly used in astrophysics. It is also the sign convention used by Misner, Thorne, and Wheeler [Misner 1973], whose results are referenced here. The $GA(1, 3)$ solution (time-like positive signatures) has many adherents and is often used in high-energy physics. There are some subtle differences in sign between expressions written in the two conventions, so caveat emptor. An infinitesimal time-like displacement in $GA(3, 1)$ has the overall signature

$$dx^2 = -c^2 d\tau^2 = g_{\mu\nu} dx^\mu dx^\nu = -c^2 dt^2 + dx^2 + dy^2 + dz^2 < 0, \quad (11.3)$$

where $d\tau = \sqrt{-dx^2/c^2}$ defines an element of *proper time* for time-like vectors.

11.2.1 Four-Velocity

Boldface notation is not commonly used for four-vectors and is avoided here by use of geometric algebra notation wherever possible. Boldface will continue to be used for dyadic equations, however, to maintain a consistent notation. Three-vector subcomponents of four vectors are represented by an over-arrow embellishment. One can characterize the trajectory of a time-like particle in terms of some scalar parameter τ

$$x(\tau) = \begin{bmatrix} x^0(\tau) \\ \vec{x}(\tau) \end{bmatrix} = \begin{bmatrix} x^0(\tau) \\ x^1(\tau) \\ x^2(\tau) \\ x^3(\tau) \end{bmatrix}, \tag{11.4}$$

where $x(\tau)$ defines the world line of the particle's trajectory. The *proper time* τ is a convenient uniform parameterization of the path of a massive point particle given by the path integral

$$\tau = \int_0^\tau d\tau = \int_0^\tau \sqrt{dt^2 - d\vec{x}^2/c^2} = \int_0^\tau dt\sqrt{1 - \vec{v}^2/c^2}. \tag{11.5}$$

It provides a measure for an internal clock, denoting the elapsed time in the particle's instantaneous rest frame. Every time-like particle maintains its own proper time for as long as it exists, but does it provide a reliable measure of physical and/or psychological time? Experience says yes, as will be shown.

Since proper time is a scalar, the derivative of displacement with respect to the proper time is a four-vector. Define the *four-velocity* u as the four-vector

$$u^u = \frac{dx^u}{d\tau} = \frac{dx^u}{dt}\frac{dt}{d\tau} = \gamma\frac{dx^u}{dt}. \tag{11.6}$$

Decomposing the four-velocity into its time and space components gives

$$u = \frac{dx}{d\tau} = \gamma\frac{dx}{dt} = \gamma\begin{bmatrix} c \\ \vec{v} \end{bmatrix} = \gamma c\begin{bmatrix} 1 \\ \vec{\beta} \end{bmatrix}, \tag{11.7}$$

where γ defines a relativistic scale parameter given by

$$\frac{dt}{d\tau} = \gamma. \tag{11.8}$$

The norm of the four-velocity of a time-like particle is a constant by definition (see Equation 11.3)

$$g_{uv}u^u u^v = \frac{g_{uv}dx^u dx^v}{d\tau^2} = -c^2. \tag{11.9}$$

Let the three-vector $\vec{\beta} = \vec{v}/c$ denote the *three-velocity* of the particle in units of the speed of light. Then there is a maximum limiting velocity for time-like particles given by the speed of light in vacuum. Given

$$u^2 = -c^2 = -c^2\gamma^2(1 - \beta^2), \tag{11.10}$$

the normalization constraint

$$\gamma = 1/\sqrt{1 - \beta^2} \tag{11.11}$$

restricts the magnitude of the three-velocity to $\beta < 1$.

Particle propagation obeys the rules of causal behavior. Only time-like massive particles and massless wave quanta can freely propagate. Time-like particles have a four-momentum given by

$$p = mu \tag{11.12}$$

with a negative metric signature

$$p^2 = -m^2 c^2 \leq 0 \tag{11.13}$$

and a frame-dependent, space-time split given by

$$p^\mu = \begin{bmatrix} p^0 \\ \vec{p} \end{bmatrix} = \begin{bmatrix} E/c \\ \vec{p} \end{bmatrix} = m\gamma c \begin{bmatrix} 1 \\ \vec{\beta} \end{bmatrix}, \tag{11.14}$$

where m is the invariant scalar mass of the particle. The norm of the four-momentum is an invariant under Lorentz transforms. The fourth (time-like) component of the momentum of a free particle can be associated with its total energy

$$E = p^0 c = -p_0 c = m\gamma c^2 \geq mc^2. \tag{11.15}$$

In the rest frame, a particle's internal energy is related to its invariant mass by $E_0 = mc^2$. The velocity of a massive particle with finite energy is strictly less than the speed of light in vacuum. The velocity of such a particle is given by

$$\vec{\beta} = \frac{\vec{v}}{c} = \frac{\vec{p}}{E/c}. \tag{11.16}$$

As a particle approaches light speed, both its momentum and energy increase but the velocity change becomes imperceptible, although the energy change can be substantial. Figure 11.1 illustrates the energy transfer T to a free particle at initially at rest, measured in units of $mc^2 = 1$. The total particle energy is $E = T + mc^2$, where mc^2 is the rest-energy of the particle. The particle's velocity in units of the speed of light is $\vec{\beta} = \vec{v}/c$. As the energy transfer becomes relativistic, the particle's speed approaches the limiting value of c. The particle's three-momentum is $\vec{p} = E\vec{\beta}/c$ where $|\vec{p}| \rightarrow E/c$ at high energies. The energy transfer T can be interpreted as the gain in kinetic energy of the particle.

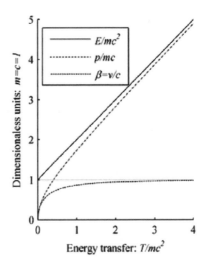

FIGURE 11.1 Energy transfer T to a free particle at initially at rest measured in units of $m = c = 1$. The total particle energy is $E = T + mc^2$. The particle's velocity in units of the speed of light is $\vec{\beta} = \vec{v}/c$. As the energy transfer becomes relativistic, the particle's speed approaches the limiting value of c. The particle's three-momentum is $\vec{p} = E\vec{\beta}/c$ where $|\vec{p}| \rightarrow E/c$ at high energies.

Massless *light-cone* particles are a distinct class unto themselves. They have a null four-momentum. In this case, the invariant measure of displacement is identically zero, so the motion of a light-cone particle cannot be parameterized by its own proper time. For null displacements, one has

$$d\tau^2 = 0 = dt^2 - d\vec{x}^2/c^2. \tag{11.17}$$

Dividing by the laboratory time interval dt^2, the three-velocity of a null-vector has a constant magnitude in all frames

$$\left| \frac{d\vec{x}}{dt} \right|_{p^2=0} = c. \tag{11.18}$$

Therefore, for massless particles, the speed $v = \beta c \rightarrow c$ is a constant

$$v|_{p^2=0} = \frac{|\vec{p}|c^2}{E} = c. \tag{11.19}$$

Light, in the vacuum, has the same velocity in all inertial frames because the four-momentum $p^2 = 0$ is invariant under Lorentz transforms. This explains the null result found in the Michelson-Morley experiment.

11.2.2 Light Cone

The norm of the four-momentum in orthonormal coordinates given by

$$p^2 = p_\mu p^\mu = |\vec{p}|^2 - |p^0|^2. \qquad (11.20)$$

The Lorentz spin group $Spin^+(3, 1)$ is the group of continuous maps of vectors into vectors, leaving the origin fixed and preserving all inner products. Both the parity of the space and the direction of time are unchanged under a spin transformation.

But, if space and time are allowed to mix, what happens to the concept of causality? What, for example, prevents a generalized rotation in spacetime taking a forward-moving particle into one that is moving backwards in time? The answer lies in the signature of the metric. The allowed continuous transformations of spacetime must preserve the length of four-vectors, therefore space-like vectors, with positive norm, and time-like vectors, with negative norm, form different invariant classes of objects. Null-vectors separate the two classes and form a third distinct class. Consider an event that occurs at some point in spacetime. Call this instance "the present". Draw a light cone given by the separatrix equation

$$\Delta x^2 = \Delta \vec{x}^2 - c^2 \Delta t^2 = 0. \qquad (11.21)$$

A Lorentz transform preserves the signature of the norm. The light cone therefore separates time-like from space-like displacements. Physical particles with non-zero rest mass are time-like in character. Furthermore, proper Lorentz transformations are continuous. Since the determinant of the transformation would have to change discontinuously, via a time reversal flip, there is no way to move in a continuous manner through the origin from a time-like vector moving forward in time to a time-like vector moving backwards in time, as seen from the light cone drawing shown in Figure 11.2. A particle with a time-like four-momentum $\beta < 1$ and positive energy signature can only propagate forward in time into the future. It can casually affect other events happening in its future. Events that happened in the time-like past, by reciprocity, can affect the particle's present. This frame-independent division based on the signature of scalar length preserves causality.

11.2.3 Paradox of Simultaneity

Events having a space-like separation (labeled "else when" in the diagram of Figure 11.2) cannot causally influence each other. In some frame of reference, two space-like separated events may be synchronized, occurring at the same time in that frame. This space-like synchronization, however, is subject to reordering, depending on the observer. Because time mixes with the space

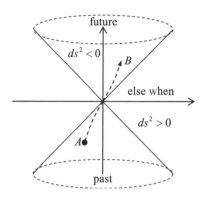

FIGURE 11.2 Relative to the present, past, future, and else when are invariant partitions of spacetime separated by the light cone of null vectors. The present is an instance in spacetime denoted by the vertex of the light cone. The dashed line shows the world line of a particle intercepting the present, while moving from event *A* in the past to event *B* in the future in a time-like manner.

coordinates, only point-like events (events happening at the same spacetime point) can be said to be truly simultaneous in a local Lorentz invariant way. One consequence is that action-at-a-distance (space-like interactions at the same time) cannot be allowed, since its time order is not Lorentz invariant; therefore, non-local interactions must be mediated by the time-like propagation of intermediate fields.

Since the measure of time is frame dependent, how does one determine which of two space-separated events occurred first, such as two runners crossing a finish line. Alternatively, one could ask how does one decide whether they finished *simultaneously* or not? Simultaneity for spacetime separate events is closely related to nature of observation. Observation requires a signal which has a maximum speed of propagation given by the speed of light. For a stationary observer located midway between two runners as they cross a finish line, one can strobe the runners with a flash of light as they cross the finish line. This is illustrated in Figure 11.3. Light emitted at a time $t_1 = -l/c$ returns at a time $t_2 = +l/c$ for two equally separated events. Therefore, the events are observed to be simultaneous, since the synchronized signals return simultaneously from the viewpoint of a stationary observer's frame of reference.

Simultaneity, however, is in the eye of the beholder. A second, moving, observer would observe something different. Consider light emitted simultaneously by a second observer at the same spacetime instance as the light emitted by the first observer. The second observer is moving towards one of the runners. Reflected light from the runner that the observer is approaching

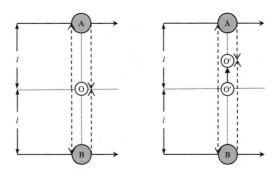

FIGURE 11.3 **Simultaneous measurements are frame-dependent. They depend on the propagation of synchronized signals. A fixed observer O midway between two runners uses a strobe of light and sees two runners A and B cross the finish line of a race simultaneously. A second moving observer O′ sees runner A, which it is approaching, cross the line first. By using a Lorentz transform, the two measurements can be reconciled with each other.**

comes back sooner than light bouncing from the other runner, since it covers a shorter return distance. Therefore, event A in Figure 11.3, where the runner A crosses the finish line, occurs earlier (from the point of view of the second observer) than event B, where the second runner is observed to cross the finish line. By using a Lorentz transform to compare results, the observations of the two observers can be reconciled with each other. Differences in observation are more apparent than real.

11.2.4 Poincaré Group

The Poincaré group, developed by Jules Henri Poincaré (1854–1912), is a ten parameter group that replaces the Galilean group and unifies relativistic particle dynamics with electromagnetism. The Poincaré group consists of the group of proper homogeneous Lorentz transforms that leave the length of a four-vector invariant, plus the set of homogeneous translations of the spacetime origin. Define a point $P(t,x,y,z)$ in a four-dimensional Minkowski space. Then, a displacement $\Delta P = P - O$ with respect to a fixed origin point can be transformed into another inertial frame by the Poincaré transformation

$$\Delta P' = S_L \Delta P \bar{S}_L + \Delta X_0, \tag{11.22}$$

where the Lorentz spin group S_L consists of all proper continuous mappings of 4-vectors into 4-vectors that preserve the invariant length of vectors and leave the origin fixed. This defines an similarity transformation that is orthogonal under the geometric adjoint $\bar{S}_L = S_L^{-1}$. The Lorentz spin group is a

6 parameter group consisting of the three-dimensional subgroup of space rotations and the 3-possible directions of velocity boosts. The displacements ΔX_0 are composed of the four possible translations of points with respect to the origin in space and time. Both the Galilean and Poincaré groups are ten parameter groups, but the structure of the Poincaré group is simpler.

11.3   GEOMETRIC STRUCTURE OF SPACETIME

In the Cartesian representation, a point in Minkowski space is given by

$$P(x) = e_\mu x^\mu. \tag{11.23}$$

where the constant pseudo-Euclidean vector basis is defined as

$$e_\mu = \frac{\partial}{\partial x^\mu} P(x^0, x^1, x^2, x^3) \tag{11.24}$$

A vector $V = V^a e_a$ is a rank-one tensor transforming like an infinitesimal displacement of the coordinates. For four spacetime dimensions, an infinitesimal vector displacement is given by

$$dx = e_\mu dx^\mu \tag{11.25}$$

The infinitesimal displacements dx^μ represent the components of a contravariant vector (or vector). The reciprocal basis is defined as

$$e^\mu = \nabla x^\mu, \tag{11.26}$$

where

$$\nabla = e^\mu \frac{\partial}{\partial x^\mu}. \tag{11.27}$$

The gradient operator ∇ defines the transformation properties of a covariant vector (or 1-form). For example, the four-momentum of a particle in a pseudo-Euclidean basis is given by

$$p_\mu = \frac{\partial L}{\partial \dot{x}^\mu}, \tag{11.28}$$

which transforms like a covariant vector if L is a scalar Lagrangian.

The Clifford Algebra for $GA(3, 1)$ has the pseudo-Euclidean metric

$$\{e_\mu, e_\nu\} = 2g_{\mu\nu} = 2\text{dia}(-1, 1, 1, 1). \tag{11.29}$$

It has a minimal, four-dimensional, real matrix representation of the form

$$
e_0 = \begin{bmatrix} 0 & \sigma_3 \\ \sigma_3 & 0 \end{bmatrix}, \quad
e_1 = \begin{bmatrix} \tau_1 & 0 \\ 0 & \tau_1 \end{bmatrix}, \quad
e_2 = \begin{bmatrix} \tau_2 & 0 \\ 0 & \tau_2 \end{bmatrix}, \quad
e_3 = \begin{bmatrix} 0 & \sigma_3 \\ -\sigma_3 & 0 \end{bmatrix},
\tag{11.30}
$$

where the two-dimensional submatrices $(\tau_1, \tau_2, \sigma_3)$ can be represented as

$$
\tau_1 = \begin{bmatrix} 1 & 0 \\ 0 & -1 \end{bmatrix}, \quad
\tau_2 = \begin{bmatrix} 0 & 1 \\ 1 & 0 \end{bmatrix}, \quad
\sigma_3 = \tau_1 \tau_2 = \begin{bmatrix} 0 & 1 \\ -1 & 0 \end{bmatrix}.
\tag{11.31}
$$

If the signature of the norm-squared (i.e., the product of a vector with its geometric dual $\bar{V}V = g_{\mu\nu}V^\mu V^\nu$) of a vector is positive, the vector is said to be space-like; if it negative, it is time-like; otherwise, it is a null or *light-cone* vector. Note, that the vector basis, although self-adjoint under reversion, is not Hermitian, since $(e_0)^2 = -1$.

11.3.1 Orthogonal Similarity Transformations

A similarity transformation is said to be orthogonal under the operation of reversion if

$$
S^{-1} = \bar{S},
\tag{11.32}
$$

so that

$$
S\bar{S} = 1.
\tag{11.33}
$$

The Lorentz group of spin transformations form the members of the special orthogonal similarity group $Spin^+(3,1)$. The plus superscript denotes that the time orientation of time-like vectors are preserved by the transformations as well as the inner product. Improper transformations, which cannot be evolved in a continuous manner from the identity element, are explicitly excluded. The group has six distinct generators, consisting of three rotational elements and three velocity boosts. The vector basis, by Grassmann extension, generates a bivector basis, which splits into two pseudo three-vectors

$$
\sigma^{ij} = e^i \wedge e^j = \varepsilon^{ijk}\hat{B}_k, \quad
\sigma^{k0} = e^k \wedge e^0 = \hat{E}^k, \quad
(\hat{E}^k)^2 = -(\hat{B}_k)^2 = 1,
\tag{11.34}
$$

where $\hat{E}^j$ $(\hat{B}_k)$ transform like a vector (axial-vector) basis under three-space inversion.

11.3.2 Geometric Algebra of Minkowski Space

The pseudoscalar element of four-dimensional spacetime is a totally antisymmetric rank-four tensor operator is given by

$$e_5 = e_0 e_1 e_2 e_3 = \begin{bmatrix} 0 & 1 \\ -1 & 0 \end{bmatrix}. \tag{11.35}$$

The four-dimensional pseudoscalar element of the enlarged algebra e_5 is anti-Hermitian but is self-adjoint in the geometric sense of reversion

$$e_5^2 = -1, \quad \bar{e}_5 = -e_5^\dagger = e_5. \tag{11.36}$$

The four-dimensional pseudoscalar element anticommutes with the four-vector basis and defines the improper similarity transformation of spacetime-inversion

$$\{e_5, e_\alpha\} = 0, \quad e_5 e_\alpha e_5^{-1} = -e_\alpha; \quad \alpha = 0, 1, 2, 3. \tag{11.37}$$

The notion used here for the pseudoscalar element e_5 is a compromise based on using the common high energy notation for the Dirac algebra. In geometric algebras, the symbol i is commonly used for the pseudoscalar element of a space of arbitrary dimension, but that is a greatly overloaded symbol, and can be confused with the imaginary i in complex algebras. In Dirac matrix notation, the most commonly used notation for the pseudoscalar element is e_5, possibly because some authors count four-vectors from 1 to 4 rather than from 0 to 3. But the pseudoscalar element is neither a line element in a five-dimensional space, nor is it a five-dimensional pseudoscalar, so the label e_5 is potentially confusing.

Table 11.1 summarizes the irreducible geometric structure of $GA(3, 1)$.

TABLE 11.1   Classification of the irreducible geometric structure of $GA(3,1)$

Geometric Class	Tensor Basis	Linearly Independent Elements
Scalar	$\Gamma^{\langle 0 \rangle} = 1$	1 identity element: 1
Vector	$\Gamma_\mu^{\langle 1 \rangle} = e_\mu$	4 vector elements: e_μ
Surface	$\Gamma_{\mu\nu}^{\langle 2 \rangle} = e_\mu \wedge e_\nu = \sigma_{\mu\nu}$	6 bivectors: $\sigma_{\mu\nu}$, $(\mu < \nu)$
Three-volume	$\Gamma_{\mu\nu\omega}^{\langle 3 \rangle} = e_\mu \wedge e_\nu \wedge e_\omega = \varepsilon_{\mu\nu\omega\lambda} e_5 e^\lambda$	4 pseudovector elements: $e_5 e^\lambda$
Four-volume	$\Gamma_{\mu\nu\omega\lambda}^{\langle 4 \rangle} = e_\mu \wedge e_\nu \wedge e_\omega \wedge e_\lambda = \varepsilon_{\mu\nu\omega\lambda} e_5$	1 pseudoscalar element: e_5

In high-energy physics applications, it is common to see the alternative $GA(1,3)$ solution, which has the inverted metric signature

$$\{\gamma_\mu, \gamma_\nu\} = 2g_{\mu\nu} = 2\text{dia}(1, -1, -1, -1). \tag{11.38}$$

The Clifford algebra of $GA(1,3)$ can be represented by the *Dirac matrices*, which have the minimal, two-dimensional, quaternion representation[5]

$$\gamma_0 = \begin{bmatrix} 1 & 0 \\ 1 & -1 \end{bmatrix}, \quad \gamma_i = \begin{bmatrix} 0 & \sigma_i \\ \sigma_i & 0 \end{bmatrix}, \quad \gamma_5 = \begin{bmatrix} 0 & 1 \\ -1 & 0 \end{bmatrix}. \tag{11.39}$$

11.3.3 Discrete Symmetries

Minkowski space is defined here as the pseudo-Euclidean space $GA(3,1)$ with three space and one time dimensions. $GA(3) \subset GA(3,1)$ is the subspace of $GA(3,1)$ consisting of those elements that commute with the three-dimensional pseudoscalar $i_3 = e_1e_2e_3$. The algebra is completed by adding a fourth, pseudo-Euclidean, dimension e_0 with signature $e_0^2 = -1$ to the 3-dimensional geometric algebra of Euclidean space. The additional degree of freedom defines a parity (space-inversion) operator

$$P = e_0, \quad P^{-1} = \bar{e}^0, \quad P^2 = -1, \tag{11.40}$$

Note that the parity similarity transformation in this algebra represents an improper, anti-unitary transformation.

$$PP^{-1} = -P\bar{P} = 1 \tag{11.41}$$

Applying the parity transformation to a Euclidean four-vector basis gives

$$Pe_iP^{-1} = -e_i, \quad Pe_0P^{-1} = e_0. \tag{11.42}$$

The parity operation inverts space, but not time. Note that P anticommutes with the 3-dimensional pseudoscalar element $i = i_3 = e_1e_2e_3$, therefore $Pi_3P^{-1} = -i_3$, thus changing the sign of the element of three-volume in $GA(3)$.

Time is considered a scalar in non-relativistic mechanics, but as the fourth component of a vector in relativistic mechanics. *Time reversal* is the effect of changing the direction of the flow of time. Dissipative systems, such as those described by the diffusion equation, have stable solutions in the forward time

---

[5]*Dirac matrix algebra* can also be thought of as the complex extension of $GA(3,1)$, with the Dirac gamma matrices defined as $\gamma_a = ie_a$, producing the complex Clifford algebra $Cl(1,3,\mathbb{C})$.

direction only. In $GA(3, 1)$, time reversal is given by the discrete similarity transformation

$$Te_0T^{-1} = -e_0, \quad Te_iT^{-1} = e_i, \tag{11.43}$$

resulting in a change of the time basis vector, while leaving the space basis unchanged. T has an explicit representation in the algebra given by

$$T = e_1e_2e_3 = e^0e_5 = i_3. \tag{11.44}$$

For the present choice of metric, time reversal is unitary under the tensor adjoint operator

$$T^{-1} = \bar{T} = e_5e^0 = \bar{i}_3 = -i_3. \tag{11.45}$$

Four-dimensional *spacetime inversion* is given by combining parity and time reversal

$$PT = e_0e_1e_2e_3 = e_5, \quad (PT)^{-1} = -\overline{PT} = -e_5. \tag{11.46}$$

Spacetime inversion changes the overall signs of irreducible geometric tensors of odd rank.

It is useful to redefine the concept of the Hermitian or matrix adjoint in terms of the geometric adjoint defined by reversion. This discrete transformation adds the effect of time-reversal to the geometric adjoint

$$e_\alpha^\dagger = T^{-1}\bar{e}_\alpha T, \quad \bar{e}_\alpha = Te_\alpha^\dagger T^{-1}, \tag{11.47}$$

where the Hermitian adjoint basis $e_\alpha^\dagger$ satisfies the positive definite property

$$e_\alpha^\dagger e_\alpha = 1, \quad \text{(nsc).} \tag{11.48}$$

The Hermitian adjoint defines a positive-definite measure for geometric tensors Ψ

$$\langle \Psi^\dagger \Psi \rangle = \langle e^0e_5\bar{\Psi}e_5e^0\Psi \rangle > 0 \text{ unless } \Psi \equiv 0. \tag{11.49}$$

The Hermitian adjoint, however, is not a tensor invariant. It represents a single component of a rank-2 tensor. The geometric adjoint is a more fundamental concept than the Hermitian adjoint. The former is a fundamental property of all inner product metric spaces, while the Hermitian adjoint is restricted to Hilbert spaces. The Hermitian adjoint is an attempt to convert the pseudo-Euclidean metric of spacetime back into a positive-definite four-dimensional Euclidean one. The Hermitian adjoint, however, plays an important role in function theory.

11.3.4 Lie Algebra of the Lorentz Spin Group

The Lie algebra of continuous transformations, formed from all bivectors $\sigma_{\mu\nu} = e_\mu \wedge e_\nu$ in $GA(3,1)$, generates the *Lorentz spin group*, $Spin^+(3,1)$, in four spacetime dimensions. $Spin^+(3,1)$ is the covering group for the *homogeneous proper Lorentz group* of special orthogonal transformations $SO^+(3,1)$. Elements of the Lorentz spin group satisfy the conditions

$$S_L \bar{S}_L = 1, \quad [e_5, S_L] = 0, \quad \det(S_L) = +1. \tag{11.50}$$

The generators of the Lorentz group are the bivector elements created by taking the exterior product of vector displacement elements

$$\sigma_{\mu\nu} = e_\mu \wedge e_\nu = \frac{1}{2}[e_\mu, e_\nu], \tag{11.51}$$

where the bivectors are antiadjoint under reversion

$$\overline{\sigma_{\mu\nu}} = \overline{e_\mu \wedge e_\nu} = e_\nu \wedge e_\mu = \sigma_{\nu\mu} = -\sigma_{\mu\nu}. \tag{11.52}$$

A frame-dependent space-time split can be made of an arbitrary rank-2 geometric tensor $F = \langle F \rangle_2$ into components having define parity with respect to the time basis,

$$F = F_E + F_B = F_{\mu\nu}\sigma^{\mu\nu}/2,$$
$$F_E = \frac{1}{2}(F - PFP^{-1}) = (F_{i0} - F_{0i})\sigma^{i0}/2, \tag{11.53}$$
$$F_B = \frac{1}{2}(F + PFP^{-1}) = (F_{ij})\sigma^{ij}/2,$$

where $F_E(F_B)$ behaves as a three-dimensional vector (axial-vector) under the subgroups of space rotations plus parity. The generator of space rotations is given by the $Spin(3)$ subgroup

$$\sigma_{ij} = e_i \wedge e_j = \varepsilon_{ijk}\sigma_k \quad i,j,k = 1,2,3, \quad (\sigma_{ij})^2 = -1, \tag{11.54}$$

while the generators of velocity boosts are given by

$$\sigma_{0i} = e_0 \wedge e_i = -e_5\sigma_i \quad i = 1,2,3, \quad (\sigma_{0i})^2 = +1. \tag{11.55}$$

For nonorthogonal basis vectors, this generalizes to the self-duality relationship

$$\sigma_{\alpha\beta} = \|g\|\varepsilon_{\alpha\beta\gamma\delta}e_5\sigma^{\gamma\delta}. \tag{11.56}$$

The six-element spin basis obeys the Lie algebra rules

$$[\sigma_i{}^k, \sigma_k{}^j] = 2\sigma_i{}^j, \quad [\sigma_0{}^k, \sigma_k{}^i] = 2\sigma_0{}^i, \quad [\sigma_i{}^0, \sigma_0{}^j] = 2\sigma_i{}^j \quad \text{(nsc)}. \tag{11.57}$$

The Lorentz spin group $Spin(3, 1)$ contains the three-dimensional rotation group $Spin(3)$ as a proper subgroup.

11.3.5 Infinitesimal Transformations

An infinitesimal Lorentz transform has the general form of

$$S(\sigma^{\mu\nu}, \delta\lambda) = e^{-\delta\lambda\sigma^{\mu\nu}/2} \approx 1 - \delta\lambda\sigma^{\mu\nu}/2. \qquad (11.58)$$

Expanding the transform applied to a vector gives

$$V'(\sigma^{\mu\nu}, \delta\lambda) = S(\sigma^{\mu\nu}, \delta\lambda)V\bar{S}(\sigma^{\mu\nu}, \delta\lambda) \approx V - \frac{\delta\lambda}{2}[\sigma^{\mu\nu}, V] + O(\delta\lambda^2). \quad (11.59)$$

This leaves the length of the vector invariant to order $\delta\lambda$, leading to the differential form

$$\frac{\delta V(\sigma^{\mu\nu}, \lambda)}{\delta\lambda} = -\frac{1}{2}[\sigma^{\mu\nu}, V]. \qquad (11.60)$$

A finite Lorentz spinor can be composed from a product of infinitesimal transformations using the well-known identity

$$S(\sigma^{\mu\nu}, \lambda) = e^{-\lambda\sigma_{\mu\nu}/2} = \lim_{N\to\infty} \left(1 - \frac{\lambda}{N}\sigma^{\mu\nu}/2\right)^N. \qquad (11.61)$$

Likewise, a finite Lorentz vector transformation can be built up out of infinitesimal transformations by iteration of Equation 11.60 in a Lie-Taylor series

$$V'(\sigma, \lambda) = V(0) + \left(\frac{-\lambda}{2}\right)[\sigma, V]|_0 + \frac{1}{2!}\left(\frac{-\lambda}{2}\right)^2[\sigma, [\sigma, V]]|_0 + \qquad (11.62)$$

11.4 PROPER LORENTZ TRANSFORMATIONS

The Lorentz spin group $S_L \in Spin^+(3, 1)$ is the subgroup of orthogonal maps $S_L\bar{S}_L = 1$ of four-dimensional Minkowski space onto itself, which can be generated continuously from the identity element and which leaves the metric invariant. The transformation maps vectors unto vectors while leaving the origin unchanged

$$V \to V' = S_L V\bar{S}_L. \qquad (11.63)$$

Vectors components transform by covariance

$$V'^{\mu} = \langle e^{\mu}V' \rangle = \langle e^{\mu}S_L e_{\nu}\bar{S}_L \rangle V^{\nu} = A^{\mu}{}_{\nu}V^{\nu}. \tag{11.64}$$

The dyadic form of this vector transformation is given by the proper homogenous Lorentz group of orthogonal transformations $A \in SO^+(3,1)$, which operate on ket vectors from the left-hand-side and on bra vectors from the right-hand-side. $Spin^+(3,1)$ is a double-valued cover for the restricted Lorentz group $SO^+(3,1)$ that preserves the space and time signatures of four vectors as well as their inner products. A vector transformation can be written in dyadic or bra-ket form as

$$\mathbf{V}' = \mathbf{A} \cdot \mathbf{V} \leftrightarrow |V'\rangle = A|V\rangle, \tag{11.65}$$

and has components

$$V'^{\mu} = \mathbf{e}^{\mu} \cdot \mathbf{V}' = \mathbf{e}^{\mu} \cdot \mathbf{A} \cdot \mathbf{V} = A^{\mu}{}_{\nu}V^{\nu}, \tag{11.66}$$

where

$$\mathbf{A} = A^{\mu}{}_{\nu}\mathbf{e}_{\mu}\mathbf{e}^{\nu}. \tag{11.67}$$

Comparing Equations (11.64) and (11.67) elements of the two groups are related by

$$A^{\mu}{}_{\nu} = \mathbf{e}^{\mu} \cdot \mathbf{A} \cdot \mathbf{e}_{\nu} = \langle e^{\mu}S_L e_{\nu}\bar{S}_L \rangle. \tag{11.68}$$

11.4.1 Rotations

A space rotation through an angle θ about a surface element $\sigma_i{}^{j} = e_i \wedge e^j$ is given by

$$S(\sigma_i{}^{j}, \theta) = e^{\sigma_i{}^{j}\theta/2} = \cos(\theta/2) + \sigma_i{}^{j}\sin(\theta/2). \tag{11.69}$$

For example, a counterclockwise rotation of a vector about the $\hat{3}$-space axis (illustrated in Figure 11.4) given by

$$S(\sigma_2^1, \theta) = e^{\sigma_2^1\theta/2} \tag{11.70}$$

has the effect

$$V' = V'^{\mu}e_{\mu} = S_R V \bar{S}_R = (e^{\sigma_2^1\theta/2}e_{\nu}e^{-\sigma_2^1\theta/2})V^{\nu}. \tag{11.71}$$

Evaluating components of $\mathbf{A} = A^{\mu}{}_{\nu}\mathbf{e}_{\mu}\mathbf{e}^{\nu}$ gives the contravariant transformation

$$A^{\mu}{}_{\nu}(\sigma_2^1, \theta) = \langle e^{\mu}e^{\sigma_2^1\theta/2}e_{\nu}e^{-\sigma_2^1\theta/2} \rangle, \tag{11.72}$$

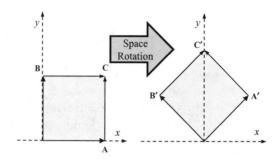

FIGURE 11.4 The lengths and relative orientations of space displacements are invariant under Lorentz space rotations, and the element of area $dx \wedge dy$ is preserved.

which has the orthogonal rotation matrix representation

$$\{A^{\mu}{}_{\nu}(\sigma_2^1, \theta)\} = \begin{bmatrix} 1 & 0 & 0 & 0 \\ 0 & \cos\theta & -\sin\theta & 0 \\ 0 & \sin\theta & \cos\theta & 0 \\ 0 & 0 & 0 & 1 \end{bmatrix}. \tag{11.73}$$

11.4.2 Velocity Boosts

A Lorentz velocity transform or velocity boost is a pseudorotation about a spacetime surface element $\sigma_0{}^i = e_0 \wedge e^i$ given by

$$S(\sigma_0{}^i, \lambda) = e^{\sigma_0{}^i\lambda/2} = \cosh(\lambda/2) + \sigma_0{}^i \sinh(\lambda/2). \tag{11.74}$$

Therefore, a velocity boost along the $\hat{3}$-space axis (illustrated in Figure 11.5) has the effect

$$V' = V'^u e_u = S_V V \bar{S}_V = (e^{\sigma_0^3\lambda/2} e_v e^{-\sigma_0^3\lambda/2}) V^v. \tag{11.75}$$

Evaluating

$$A^{\mu}_{\nu}(\sigma_0^3, \lambda) = \langle e^{\mu} e^{\sigma_0^3\lambda/2} e_v e^{-\sigma_0^3\lambda/2} \rangle \tag{11.76}$$

gives the contravariant transformation

$$\begin{bmatrix} V'^0 \\ V'^1 \\ V'^2 \\ V'^3 \end{bmatrix} = \begin{bmatrix} \cosh\lambda & 0 & 0 & \sinh\lambda \\ 0 & 1 & 0 & 0 \\ 0 & 0 & 1 & 0 \\ \sinh\lambda & 0 & 0 & \cosh\lambda \end{bmatrix} \begin{bmatrix} V^0 \\ V^1 \\ V^2 \\ V^3 \end{bmatrix}. \tag{11.77}$$

Substituting

$$\cosh\lambda = \gamma, \quad \sinh\lambda = \gamma(v/c) = \gamma\beta, \tag{11.78}$$

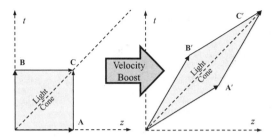

FIGURE 11.5   In a Lorentz velocity transformation, points (C) on the light cone remain on the light cone. The light cone separates space-like displacements (A) from time-like displacements (B). The element of area $dt \wedge dz$ is preserved under the transformation.

simplifies the matrix algebra, giving the matrix representation

$$\{A^{\mu}{}_{\nu}(\sigma_0^3, \beta)\} = \begin{bmatrix} \gamma & 0 & 0 & \gamma\beta \\ 0 & 1 & 0 & 0 \\ 0 & 0 & 1 & 0 \\ \gamma\beta & 0 & 0 & \gamma \end{bmatrix}, \tag{11.79}$$

Here, $\beta = \tanh(\lambda)$ denotes the three-velocity (in dimensionless units) imparted to a stationary particle by the transformation, and $\gamma = (1 - \beta^2)^{-1/2} \geq 1$ is the time dilation factor. This Lorentz velocity transformation about the $\hat{3}$-space axis can be generalized to be about any space axis and leads to the coordinate transformations

$$\Delta t' = \gamma \left(\Delta t + \frac{\vec{v} \cdot \Delta\vec{x}_{\parallel}}{c^2} \right),$$
$$\Delta\vec{x}'_{\perp} = \Delta\vec{x}_{\perp}, \tag{11.80}$$
$$\Delta\vec{x}'_{\parallel} = \gamma(\Delta\vec{x}_{\parallel} + \vec{v}\Delta t).$$

The set of proper Lorentz transforms (including space rotations) form a group. Any element in this group can be decomposed into the product of a rotation about some space axis plus a velocity boost along some space axis.

11.4.3 Composition of Velocities

Consider the transformation of the four-velocity of a particle under a Lorentz velocity boost, decomposed into velocity components parallel and perpendicular to the direction of the boost

$$\mathbf{u}' = \mathbf{A}_V(\sigma_0^{\parallel}, \beta_{\text{rel}}) \cdot \mathbf{u}, \tag{11.81}$$

The component equations are

$$\begin{bmatrix} \gamma' \\ \gamma'\vec{\beta}'_\parallel \\ \gamma'\vec{\beta}'_\perp \end{bmatrix} = \begin{bmatrix} \gamma_{rel} & \gamma_{rel}\vec{\beta}_{rel} & 0 \\ \gamma_{rel}\vec{\beta}_{rel} & \gamma_{rel} & 0 \\ 0 & 0 & 1 \end{bmatrix} \begin{bmatrix} \gamma \\ \gamma\vec{\beta}_\parallel \\ \gamma\vec{\beta}_\perp \end{bmatrix}, \tag{11.82}$$

Solving for the three-velocities gives

$$\beta'_\parallel = \frac{\beta_{rel} + \beta_\parallel}{1 + \beta_{rel}\beta_\parallel},$$

$$\vec{\beta}'_\perp = \frac{\vec{\beta}_\perp/\gamma_{rel}}{1 + \beta_{rel}\beta_\parallel}, \tag{11.83}$$

or

$$\vec{\beta}' = \frac{1}{1 + \vec{\beta}_{rel}\cdot\vec{\beta}}(\vec{\beta}_{rel} + \vec{\beta}_\parallel + \vec{\beta}_\perp/\gamma_{rel}). \tag{11.84}$$

The transverse components are suppressed by a factor of $1/\gamma_{rel}$ relative to the longitudinal components. This results in the enhanced relativistic shrinking of the *transverse divergence* of a beam of particles when Lorentz-boosted along their mean direction of motion

$$\tan\theta' = \frac{\beta'_\perp}{\beta'_\parallel} = \frac{1}{\gamma_{rel}}\frac{\beta\sin\theta}{\beta_{rel} + \beta\cos\theta} = \frac{1}{\gamma_{rel}}\tan\theta'_{NR}, \tag{11.85}$$

where θ'_{NR} is the expected shrinking of the angular divergence in the non-relativistic limit. In the small velocity limit, Equation (11.84) reduces to the Galilean velocity transformation

$$\vec{\beta}' = \vec{\beta}_{rel} + \vec{\beta}. \tag{11.86}$$

11.4.4 Time Dilation

Particles have well established mean lifetimes that can be measured with respect to their proper times in the particle frame. In its instantaneous rest frame, the point moves only with respect to its proper time

$$\Delta x = x(\tau) - x(0) = c\Delta\tau e_0. \tag{11.87}$$

Let τ denote the elapsed proper time from particle creation to annihilation. Applying a uniform Lorentz boost along the 3-axis, a particle displacement in the moving frame becomes

$$\Delta\mathbf{x}' = \begin{bmatrix} c\Delta t' \\ \Delta z' \end{bmatrix} = \mathbf{A}\cdot\Delta\mathbf{x} = \begin{bmatrix} \gamma & \gamma\beta \\ \gamma\beta & \gamma \end{bmatrix}\begin{bmatrix} c\Delta\tau \\ 0 \end{bmatrix}. \tag{11.88}$$

Therefore, the lifetime of a particle as observed by a time measurement in a uniform moving frame is given by

$$\Delta t' = \gamma \Delta \tau = \frac{\Delta \tau}{\sqrt{1 - \beta^2}} \qquad (11.89)$$

In moving frames, time becomes dilated. That is, particles appear to live longer when time is measured in a frame in which the particle is moving.

11.4.5 Space Contraction

In addition to time dilation, there is also a foreshortening effect, called *space contraction* that causes an extended distribution to appear to shrink when moving at relativistic speeds. Usually the concept of space contraction is presented in terms of its effect on a uniform measuring stick, where the velocity transformation is made along the length of the stick. The direction can be taken to be the $\hat{x}$-axis without loss of generality. In the rest frame of the stick its ends are at x_1 and x_2 and the stick has a length of $l_0 = x_2 - x_1$. In the moving frame, the separation of the ends must be measured at one and the same time t' in the moving frame (i.e., simultaneously). The equations for the end point coordinates become

$$x_1 = \gamma(x_1' - vt'), \quad x_2 = \gamma(x_2' - vt'). \qquad (11.90)$$

The length of the measuring stick in the rest frame is

$$\Delta l_0 = x_2 - x_1 = \gamma(x_2' - x_1') = \gamma \Delta l', \qquad (11.91)$$

with the result that the stick appears shorter in the moving frame

$$\Delta l' = \frac{1}{\gamma} \Delta l_0. \qquad (11.92)$$

In other words, the measurement of length is characterized by simultaneity in the moving frame, $\Delta t' = 0$, which can be combined with the first equation of Equation Block (11.80) to find the relation between the lengths in the two coordinate systems for fixed transverse position

$$\Delta \vec{x}_{\parallel}' = \gamma \left(1 - \frac{\vec{v}\vec{v}}{c^2} \right) \cdot \Delta \vec{x}_{\parallel},$$
$$\Delta \vec{x}_{\parallel}' = \frac{1}{\gamma} \Delta \vec{x}_{\parallel} \quad \text{when } \Delta \vec{x}_{\perp} = 0. \qquad (11.93)$$

Yet another way to see the effect is to make use of the fact that a Lorentz transformation leaves an element of four-volume unchanged, therefore,

$$dt'dx' = \begin{vmatrix} \dfrac{\partial t'}{\partial t} & \dfrac{\partial t'}{\partial x} \\ \dfrac{\partial x'}{\partial t} & \dfrac{\partial x'}{\partial x} \end{vmatrix} dtdx = \begin{vmatrix} \gamma & \gamma\beta/c \\ \gamma\beta c & \gamma \end{vmatrix} dtdx = dtdx. \tag{11.94}$$

$$\left.\frac{dx'}{dx}\right|_{t'} = \left.\frac{dt}{dt'}\right|_{x} = \frac{1}{\gamma}. \tag{11.95}$$

A measuring stick of uniform cross section oriented along the axis of motion appears shorter when viewed by an observer moving toward the stick. The apparent length of the stick is contracted by a multiplicative scale factor of $1/\gamma$. The particle density of the stick must therefore be increased by the same factor γ, since the total number of atoms in the stick is a scalar invariant. The atoms of the moving stick, of course, are not at all inconvenienced by this apparent compression, which is, after all, only an artifact of the frame of reference of the observer. By the principle of special relativity, there can be no physical effect on the stick due to its absolute motion in space. Only relative motions are physically significant.

Figure 11.6 shows the density profile of a spherical distribution of charge as viewed in the distribution's rest frame, on the left-hand side of the figure. The same distribution viewed in the frame of an approaching observer moving at a sizable fraction of light speed is shown on the right-hand side of the figure. The pancake effect is a consequence of making a Lorentz transformation to relativistic speeds. There is no physical difference between the two

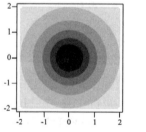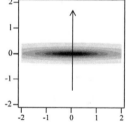

FIGURE 11.6 Space contraction: Equal density contours for a spherical distribution of charge is shown on the left as measured in the distribution's rest frame. On the right, the same distribution is shown as it appears to a moving observer. This pancake effect is referred to as space-contraction. The effect is due to the transformation laws mixing space and time degrees of freedom and are more apparent than real. All scalar measurements, such as the charge integral, are the same for both observers.

distributions. If the moving observer transfers his measurements into the rest frame of the charge distribution, he would recover the spherical shape seen in the rest frame portrait shown to the left.

Measurement implicitly involves the process of extracting scalar results. All scalar quantities are left invariant by Lorentz transforms. For example, the integrated charge measured by a moving observer is the same as that obtained in the stationary rest frame. Vector and tensor components are always measured relative to a choice of coordinate frame, in effect making a scalar projection of the tensor components into that basis. If a common frame of comparison can be agreed to, all observers should obtain the same measureable results, within the limits of experimental error.

11.4.6 Twin Paradox

The *twin paradox* is a thought experiment illustrating *time dilation*. The paradox postulates two twins born on Earth. Twin *A* remains at home while Twin *B* makes a round trip to Alpha Centauri, presumably by a fast rocket. If they both live to meet again (a doubtful hypothesis with current technology), the twin that stayed on Earth would have aged the most. One does not need to know the details of the trajectory to deduce which twin has aged the most. One needs only to recognize that the twin on Earth remains in a (nearly) inertial frame while the twin with a ticket to Alpha Centauri must have undergone significant acceleration somewhere on his journey. The reference frames are therefore not at all symmetric. Physical quantities in special relativity are well defined only when formulated with respect to an inertial frame of reference. One can plot the displacement of the moving twin with respect to an inertial Earth frame; the path might look something like Figure 11.7. There are two synchronous events, which one can use to compare the clocks of the two twins: the event at takeoff (departure from Earth) and again on landing (return to Earth). If one divides the path into infinitesimal time slices, one finds by numerical integration

$$T_A - T_B = \sum_{i=1...n} \left(\Delta t_i - \sqrt{(\Delta t_i)^2 - (\Delta \vec{r}_i)^2} \right) \geq 0, \qquad (11.96)$$

where all coordinates $(\Delta t_i, \Delta \vec{r}_i)$ are expressed in the inertial frame of the Earth. Due to the sign of the metric, the elapsed proper time measured along the curved path for Twin *B* is always shorter or equal to the proper time interval of Twin *A* remaining in the Earth's rest frame.

But how does one know that the measure of physiological time is equivalent to the proper time in the rest frame of an object? Aging, after all, could be influenced by stress due to acceleration, change of heart rate, etc.

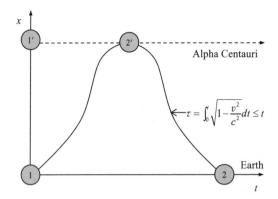

FIGURE 11.7 The twin paradox: the path length between two events depends on the path taken. The path of the twin in the accelerating frame is curved when measured with respect to the inertial frame of the twin that stays at home. The integral of the time is therefore shorter by the rules of the metric.

The twin paradox is probably not the best example for illustrating what has been measured and determined to be correct from repeated observations the decays of atomic particles. Namely, the statistical decay distributions of atomic particles are invariant when plotted with respect to their proper time. The measured distributions are observed to dilate when measured in a moving frame.

11.5   RELATIVISTIC PARTICLE DYNAMICS

The development of relativistic mechanics in Minkowski space parallels the development of Newtonian mechanics in Euclidean space. One can postulate that matter is composed of particles with invariant mass, having well-defined trajectories given by the movement of a point in spacetime with four-velocity $u = dx/d\tau$. Particles have an additive property called the four-momentum $p = mu$, the product of mass times four-velocity, which is conserved in the absence of an external applied force. The fourth component of the momentum vector in Minkowski space can be interpreted as the particle's energy. The rate of change of the four-momentum with respect to proper time is called the four-force $f = dp/d\tau$.

One important complication is that, unlike the Greek concept of stable and unchanging atoms, the elementary particles of high energy physics can be created or annihilated. Such properties are beyond the realm of Newtonian particle dynamics, since the transformation of particles requires detailed knowledge of the internal dynamics. The stability of normal particulate matter is preserved by the existence a number of conserved currents, resulting in charge, lepton and baryon number conservation. The free propagation of particles occurs for only certain discrete invariant masses. Particles can be transformed into other particles in atomic collisions and decay processes provided all the conservation rules are observed. Quantum field theories attempt to predict the transition rates for elementary processes. Fortunately, particle transitions are very short range in character, (typically on the scale of nuclear dimensions) and can often be handled by use of the *impulse approximation*.

11.5.1 Four-Force

The rate of change of the four-momentum with respect to a particle's proper time is a four-vector that can be referred to as the relativistic four-force

$$f = \frac{dp}{d\tau} = m\frac{du}{d\tau} = ma, \tag{11.97}$$

where $a = du/d\tau$ is the four-acceleration. Since $u^2 = -c^2$, this leads to a constraint on the four-force causing the scalar product of the four-force with the four-velocity to vanish

$$u \cdot f = u \cdot \frac{dp}{d\tau} = \frac{m}{2}\frac{du^2}{d\tau} = 0, \tag{11.98}$$

$$u^\alpha f_\alpha \equiv mu^\alpha a_\alpha = 0. \tag{11.99}$$

From Equation (11.99), since the four-velocity and four-momentum are time-like vectors, the four-force and four-acceleration are space-like vectors. Equation (11.99) leads to the constraint condition

$$u^0 f_0 = -u^i f_i \rightarrow f_0 = -\frac{\vec{v}}{c} \cdot \vec{f}. \tag{11.100}$$

The four-momentum can be separated into time-like and space like components. Its space-time split is

$$p^\mu = \begin{bmatrix} E/c \\ \vec{p} \end{bmatrix} = m\gamma \begin{bmatrix} c \\ \vec{v} \end{bmatrix}. \tag{11.101}$$

To compare with Newton, define a non-relativistic force $f_N = f/\gamma$, then the time rate of change of four-momentum is given by

$$f_{N\mu} = f_\mu/\gamma = \frac{dp_\mu}{dt} = \begin{bmatrix} -\dfrac{d}{dt}\left(\dfrac{E}{c}\right) \\ \dfrac{d}{dt}\vec{p} \end{bmatrix} = \begin{bmatrix} -\dfrac{\vec{v}}{c} \cdot \vec{f}_N \\ \vec{f}_N \end{bmatrix}. \tag{11.102}$$

The conservation of four-momentum in the absence of force results in energy and momentum conservation. Introducing a relativistic mass $m_R = \gamma m$, and a relativistic kinetic energy $T = E - mc^2$ gives

$$
\begin{aligned}
\vec{f}_N &= \frac{d(\vec{p})}{dt} = \frac{d(m_R \vec{v})}{dt}, \\
f_N^0 c &= \frac{dE}{dt} = \frac{dT}{dt} = \vec{v} \cdot \vec{f}_N,
\end{aligned}
\tag{11.103}
$$

For small velocities, T can be expanded in a power series, yielding the non-relativistic kinetic energy as its leading contribution in the small velocity limit

$$
T = \left(\frac{1}{\sqrt{1 - \vec{v}^2/c^2}} - 1 \right) mc^2 = \frac{1}{2} m\vec{v}^2 + O(\beta^4).
\tag{11.104}
$$

11.5.2 Electromagnetic Field Tensor

The electromagnetic field tensor is defined as the antisymmetric rank-2 tensor $F = \langle F \rangle_2$ given by

$$
F = F_{\mu\nu} \frac{\sigma^{\mu\nu}}{2!} = \nabla \wedge A,
\tag{11.105}
$$

with components in a pseudo-Euclidean basis

$$
F_{\mu\nu} = \langle \overline{e_\mu e_\nu} F \rangle =
\begin{bmatrix}
0 & -E_1/c & -E_2/c & -E_3/c \\
E_1/c & 0 & B^3 & -B^2 \\
E_2/c & -B^3 & 0 & B^1 \\
E_3/c & +B^2 & -B^1 & 0
\end{bmatrix},
\tag{11.106}
$$

where $F_{i0} = E_i$ and $F_{ij} = \varepsilon_{ijk} B^k$. In dyadic form,

$$
\mathbf{F} = F^\mu{}_\nu \mathbf{e}_\mu \mathbf{e}^\nu
\tag{11.107}
$$

giving

$$
F^\mu{}_\nu = \mathbf{e}^\mu \cdot \mathbf{F} \cdot \mathbf{e}_\nu =
\begin{bmatrix}
0 & E_1/c & E_2/c & E_3/c \\
E_1/c & 0 & B^3 & -B^2 \\
E_2/c & -B^3 & 0 & B^1 \\
E_3/c & +B^2 & -B^1 & 0
\end{bmatrix},
\tag{11.108}
$$

which has the structure of a generator of an infinitesimal Lorentz transform.

The electromagnetic field can be derived from a four-vector potential

$$F = \nabla \wedge A = (\partial_\mu A_\nu - \partial_\nu A_\mu)\frac{\sigma^{\mu\nu}}{2!} = F_{\mu\nu}\frac{\sigma^{\mu\nu}}{2!}, \tag{11.109}$$

$$F_{\mu\nu} = \partial_\mu A_\nu - \partial_\nu A_\mu,$$

where the four-vector potential has the following space-time split

$$A^\mu = \begin{bmatrix} \Phi/c \\ \vec{A} \end{bmatrix} \tag{11.110}$$

yielding

$$\vec{E} = F_{io}e^i = -\vec{\nabla}\Phi - \frac{\partial\vec{A}}{\partial t}, \tag{11.111}$$

$$\vec{B} = \vec{\nabla} \times \vec{A}.$$

This field is a solution to Maxwell's equations of motion

$$\nabla F = -\mu_0 J, \tag{11.112}$$

where

$$J^\mu = \begin{bmatrix} \rho c \\ \vec{J} \end{bmatrix} \tag{11.113}$$

is a four-current, having a vanishing divergence

$$\nabla \cdot J = \vec{\nabla} \cdot \vec{J} + \frac{\partial\rho}{\partial t} = 0. \tag{11.114}$$

The charge integral (in the flat Minkowski basis) is

$$q_e = \int \sqrt{\|g\|}\rho dx dy dz = \int \rho dx dy dz \tag{11.115}$$

This charge is conserved, provided that the current distribution is localized (i.e. it goes to zero faster than $1/r^2$ as $r \to \infty$, where $r = |\vec{x}|$)

$$\frac{dq_e}{dt} = \int_{\text{all space}} dV \frac{\partial}{\partial t}(\rho) = -\int dV \vec{\nabla} \cdot (\vec{J}) = -\lim_{r\to\infty} \oint r^2 d\Omega(\vec{J} \cdot \hat{r}). \tag{11.116}$$

The geometric algebra equation (11.112) separates into vector and axial-vector equations

$$\nabla \cdot F = -\mu_0 J, \tag{11.117}$$

$$\nabla \wedge F = 0,$$

which, in covariant component form, becomes

$$D_v F^{\mu\nu} = \mu_0 J^{\mu},$$
$$\varepsilon^{\mu\alpha\beta\gamma} D_\alpha F_{\beta\gamma} = \varepsilon^{\mu\alpha\beta\gamma} A_{\gamma,\beta\alpha} \equiv 0. \tag{11.118}$$

Separating the space and time parts of the four-vectors gives the standard form of Maxwell's equations

$$\vec{\nabla} \cdot \vec{E} = \rho/\varepsilon_0, \quad \vec{\nabla} \times \vec{B} = \frac{1}{c^2} \frac{\partial \vec{E}}{\partial t} + \mu_0 \vec{J},$$
$$\vec{\nabla} \cdot \vec{B} = 0, \quad \vec{\nabla} \times \vec{E} = -\frac{\partial \vec{B}}{\partial t}. \tag{11.119}$$

where $c^2 = 1/\varepsilon_0 \mu_0$ and $\mu_0 = 4\pi \times 10^7$ in MKSA units. In the absence of sources, and using the Lorentz gauge ($\nabla \cdot A = 0$), Maxwell's Equations reduce to the vector wave equation

$$\nabla F = \left(\vec{\nabla}^2 - \frac{1}{c^2} \frac{\partial^2}{\partial t^2} \right) A(\vec{x}, t) = 0. \tag{11.120}$$

11.5.3 Lorentz Force

The relativistic form of the *Lorentz force equation* is

$$f_u = m_i g_{uv} \frac{du^v}{d\tau} = q_e F_{uv} u^v, \tag{11.121}$$

This can be written in geometric notation as

$$\frac{du}{d\tau} = \frac{q_e}{m} F \cdot u = \frac{q_e}{2m} [F, u]. \tag{11.122}$$

where q_e is the particle charge. The electromagnetic field tensor is the generator of a local Lorentz transformation. To lowest order in dt, the evolution of the phase space is given by the coupled set of difference equations

$$du = \left[\frac{q_e}{2m} F, u \right] d\tau. \tag{11.123}$$
$$dx = u d\tau. \tag{11.124}$$

The first of above two equations generates an infinitesimal Lorentz transformation in time.

$$u' = u + du = e^{eFd\tau/2m} u e^{-eFd\tau/2m}.$$

By iteration, the time evolution of the four-velocity is therefore given by a continuous Lorentz transformation.

$$u(\tau) = S(\tau)u(0)\bar{S}(\tau). \tag{11.125}$$

The constraint condition given by Equation (11.100) is automatically satisfied since

$$u^\mu f_\mu = q_e u^\mu u^\nu F^{(2)}_{\mu\nu} \equiv 0 \tag{11.126}$$

To compare Equation (11.121) with the non-relativistic form of the force law, separate the space and time parts of the force equation using Equation (11.102) to get

$$\frac{dT}{dt} = \frac{dE}{dt} = \frac{d(m\gamma c^2)}{dt} = q_e \vec{v} \cdot \vec{E},$$
$$\frac{d\vec{p}}{dt} = \frac{d(m\gamma \vec{v})}{dt} = q_e(\vec{E} + \vec{v} \times \vec{B}). \tag{11.127}$$

The first of the two equations gives the time rate of change of the particle's relativistic kinetic energy. The second equation is the Lorentz force equation, corrected for relativistic effects.

11.5.4 Light Quanta

Einstein was able to explain the photoelectric effect by the assumption that light comes in quantitized particles with energy related to its angular frequency by $E = \hbar\omega$ ($\omega = 2\pi\nu$ where ν is the frequency). For massless photons, the four-momentum is quantized by

$$p = \frac{E}{c}\begin{bmatrix} 1 \\ \hat{\beta} \end{bmatrix} = \frac{\hbar\omega}{c}\begin{bmatrix} 1 \\ \hat{\beta} \end{bmatrix}. \tag{11.128}$$

where $\hat{\beta}$ is a unit vector in the direction of propagation. The wavelength of the photon is given by

$$c = \lambda\omega = \lambda\nu. \tag{11.129}$$

Since the speed of light is a constant, the *relativistic Doppler Effect* for the frequency of light from a stationary source as observed in a moving frame is given by

$$\mathbf{p}' = \mathbf{A}(\sigma_0^\parallel, \beta) \cdot \mathbf{p} \tag{11.130}$$

$$\hbar\omega' \begin{bmatrix} 1 \\ \hat{\beta}' \cos\theta' \\ \hat{\beta}' \sin\theta' \end{bmatrix} = \begin{bmatrix} \gamma & \gamma\beta & 0 \\ \gamma\beta & \gamma & 0 \\ 0 & 0 & 1 \end{bmatrix} = \hbar\omega \begin{bmatrix} 1 \\ \hat{\beta} \cos\theta \\ \hat{\beta} \sin\theta \end{bmatrix}. \tag{11.131}$$

giving a frequency shift

$$\nu' = \nu\gamma(1 + \beta\cos\theta). \tag{11.132}$$

When the observer is approaching the source ($\theta = 0$), the light is "blue-shifted" to higher frequency

$$\nu_{\text{obs}} = \nu_{\text{source}}\sqrt{\frac{1+\beta}{1-\beta}}, \tag{11.133}$$

and when it is receding from the source ($\theta = \pi$), it is "red-shifted" to lower frequency

$$\nu_{\text{obs}} = \nu_{\text{source}}\sqrt{\frac{1-\beta}{1+\beta}}. \tag{11.134}$$

The result is symmetric. It does not matter whether the source is moving or the observer is moving in an absolute sense. Only the relative motion of the two frames enters into the dynamics.

11.6 IMPULSE APPROXIMATION

Generalizing from Newton and Einstein, free matter comes in discrete particles which have well-defined energy and momentum. These are combined into a single vector called the four-momentum. Momentum is conserved in the absence of external force. The particle energy is given by the time component of the momentum

$$p^0 = E/c = \hbar\omega/c \tag{11.135}$$

Particles have well defined internal energy in their own rest frame given by

$$\sqrt{-p^2c^2} = mc^2. \tag{11.136}$$

Stable, freely propagating particles occur only for certain discrete values of mass. The four-momentum of a particle is given by its mass times its four-velocity where the quantum particle has a wave length and frequency satisfying the relationship

$$p = mu = mc\gamma \begin{bmatrix} 1 \\ \vec{v}/c \end{bmatrix} = \hbar \begin{bmatrix} \omega/c \\ \lambda \vec{v}/c \end{bmatrix}, \qquad (11.137)$$

$$\lambda \omega = \lambda \nu = v. \qquad (11.138)$$

In high energy physics applications, these kinematic relationships are often simplified by solving using the dimensionaless units

$$\hbar = c = 1, \qquad (11.139)$$

with energy given in units of *electron Volts* [6] and length in Fermis,[7] giving the conversion factor

$$\hbar c = 1 = 197.3269631(49) \text{ MeV} \cdot \text{fm}, \qquad (11.140)$$

where the 2006 CODATA recommended values are quoted [Mohr 2007].

Using the impulse approximation, quantum particles are stable except at the subatomic scale where they can collide with, or decay to, other particles in a manner that conserves energy and momentum of the reaction vertex

$$s_\mu = \sum_{i=1,n} p_u^{(i)} = \sum_{j=1,n'} p_u'^{(j)}, \qquad (11.141)$$

where s denotes the total four-momentum in the initial channel (unprimed variables) and final channel (primed variables).

11.6.1 Relativistic Rocket Problem

The acceleration of a relativistic rocket differs from the nonrelativistic case in that mass is no longer conserved. However, the problem is solved in essentially the same manner, using the impulse approximation and assuming that total four-momentum is conserved by the reaction in the absence of external force. Assume that the rocket is propelled by a reaction engine that converts mass into an impulse. The exhaust gases have a velocity $\vec{v}_{rel}$ relative to the instantaneous rest frame of the rocket. At a given instance in time, the system consists solely of the rocket, which has a four-momentum in the instantaneous rest frame given by

$$p_R|_{rest} = m_R u_R = m_R \begin{bmatrix} c \\ 0 \end{bmatrix}. \qquad (11.142)$$

---

[6]2006 CODATA recommended value is $1\text{eV} = 1.602\,176\,487(40) \times 10^{-19}$ J [Mohr 2007].

[7]1 Fermi = 1 fm = 10^{-15} m.

Since the four-acceleration must be orthogonal to the four-velocity, it is easy to show that $d\gamma_R|_{rest} = 0$. An instance latter, the momentum of the rocket has changed and this change in momentum is balanced by the momentum carried off by an element of fuel

$$dp_R = -dp_F \qquad (11.143)$$

The fuel has an element of mass dm_F and a velocity relative to the rocket $-\vec{v}_{rel}$, giving the rest frame equations

$$\begin{bmatrix} c \\ 0 \end{bmatrix} dm_R + m_R \begin{bmatrix} 0 \\ d(\gamma_R \vec{v}_R)|_{rest} \end{bmatrix} = -dm_F \gamma_{rel} \begin{bmatrix} c \\ -\vec{v}_{rel} \end{bmatrix}, \qquad (11.144)$$

This yields the constraint condition $\gamma_{rel} dm_F = -dm_R$. Note that the mass is not conserved in the reaction, since the reaction is inelastic. Eliminating the mass of the fuel gives the acceleration in the instantaneous rest frame

$$m_R d\vec{u}_R = -\vec{v}_{rel} dm, \qquad (11.145)$$

with four-acceleration

$$\frac{du_R}{d\tau}\Big|_{rest} = \begin{bmatrix} 0 \\ \frac{d\vec{u}_R}{d\tau}\Big|_{rest} \end{bmatrix} = \begin{bmatrix} 0 \\ -\vec{v}_{rel} \dfrac{dm_R}{m_R} \dfrac{dm_R}{d\tau} \end{bmatrix} \qquad (11.146)$$

Now make a Lorentz boost of the four-acceleration to the laboratory frame of the rocket. For simplicity, assume that the rocket velocity is in the direction of the thrust

$$\begin{aligned} du_R &= \begin{bmatrix} d(c\gamma_R) \\ d(\gamma_R \vec{v}_R) \end{bmatrix} = A(v_r) \begin{bmatrix} 0 \\ -\vec{v}_{rel} dm_R/m_R \end{bmatrix} \\ &= \begin{bmatrix} \gamma_R & \gamma_R v_R/c \\ \gamma_R v_R/c & \gamma_R \end{bmatrix} \begin{bmatrix} 0 \\ -\vec{v}_{rel} dm_R/m_R \end{bmatrix}. \end{aligned} \qquad (11.147)$$

Solving gives the relativistic rocket equation for the special case of linear motion in the direction of the thrust gives

$$m_R \frac{d\vec{v}_R}{dt} = -\vec{v}_{rel} \frac{dm_R}{dt}(1 - v_R^2/c^2). \qquad (11.148)$$

The term on the right is the thrust on the rocket which is positive since dm_R is negative. At small velocities, this equation is consistent with the nonrelativistic result found in Chapter 1, Section 3.9. Because of relativistic corrections, the velocity of the rocket cannot exceed the speed of light.

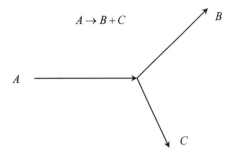

$A \rightarrow B + C$

A

B

C

FIGURE 11.8   Particle A decays into two particles B and C.

11.6.2 Particle Decay

Consider the decay of a particle A into two particles B and C as shown in Figure 11.8. Conservation of 4-momentum at the decay vertex requires

$$p_A = p_B + p_C. \tag{11.149}$$

The decay vertex is allowed in free space if all conserved quantum numbers are preserved and if the entry and exit channel particles are on their mass shells. This last constraint prevents a photon from spontaneously decaying into a particle and antiparticle pair. Some additional participant in the reaction is needed to keep the entry and exit particles on the mass shell.

Suppose the masses of all three particles are known. Then the particle C can be eliminated from the kinematics by evaluating its invariant mass

$$(p_C)^2 = (p_A)^2 - (p_B)^2 - 2p_A \cdot p_B. \tag{11.150}$$

Expanding the dot product gives

$$-p_A \cdot p_B = m_A m_B c^2 \frac{(1 - \beta_A \beta_B \cos\theta_{AB})}{\sqrt{1 - \beta_A^2}\sqrt{1 - \beta_B^2}} = \frac{1}{2}(m_C^2 + m_B^2 - m_A^2)c^2. \tag{11.151}$$

This is an equation with three parameters $(\beta_A, \beta_B, \theta_{AB})$ which can be solved for the third parameter if the other two are known.

Consider, for example, the kinematics of decay of a pion into a muon and a neutrino. The decay vertex conserves four-momentum

$$p_\pi = p_\mu + p_v. \tag{11.152}$$

Assume that the energy of the pion is known. One wants to solve for the energy of the neutrino in terms of its decay angle. One begins by eliminating the muon four-momentum from the equation by calculating its invariant mass

$$p_\mu^2 = (p_\pi - p_v)^2, \quad -m_\mu^2 c^2 = -m_\pi^2 c^2 - 2p_\pi \cdot p_v, \tag{11.153}$$

where the neutrino is assumed to be massless. Expanding the dot product

$$2p_\pi \cdot p_v = 2\frac{E_\pi E_v}{c^2}(-1 + \beta_\pi \cos\theta_v) = (m_\mu^2 - m_\pi^2)c^2 \qquad (11.154)$$

gives a formula for the neutrino energy in terms of the pion velocity and its scattering angle relative to the pion

$$E_v(\beta_\pi, \cos\theta_v) = \frac{(m_\pi^2 - m_\mu^2)c^2\sqrt{1 - \beta_\pi^2}}{2m_\pi(1 - \beta_\pi \cos\theta_v)}. \qquad (11.155)$$

There is no restriction on the neutrino angle in the laboratory frame since it is always in the allowed domain

$$1 - \beta_\pi \cos\theta_v > 0. \qquad (11.156)$$

Once the neutrino kinematics has been determined, the pion's four-momentum is given by energy-momentum conservation

$$p_\mu = p_\pi - p_v. \qquad (11.157)$$

The survival fractions of elementary particles and hadrons follow an exponential decay distribution

$$N(\tau) = N_0 e^{-\tau/\tau_e} \qquad (11.158)$$

with a well known mean $(1/e)$ lifetime τ_e when measured relative to their rest frame. A charged pion of mass $m_\pi = 139.6$ MeV/c^2 for example, decays via the electroweak interaction into a muon of mass $m_\mu = 105.7$ MeV/c^2 and a neutrino that is effectively massless. The pion has a mean life of $\tau_e = 2.6 \times 10^{-8}$ s in its rest frame, but when moving at high speeds with respect to the laboratory frame can be made to live much longer. The time dilatation effect increases the observed mean lifetime by a factor of γ compared to the rest frame. For example, a pion of total energy of $E_\pi = \gamma m_\pi c^2$ would have γ times the mean lifetime in the laboratory frame than that observed in its rest frame.

11.6.3 Particle Collisions

As an example of particle collisions, consider the elastic scattering of electrons and protons

$$e + p \rightarrow e' + p'. \qquad (11.159)$$

The reaction can be conceptualized by the impulse scattering diagram show in Figure 11.9. This reaction was studied by Robert Hofstadter (1915–1990) to

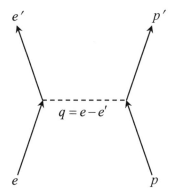

FIGURE 11.9 **Elastic scattering of electrons and protons. In the impulse approximation, the reaction is mediated by a single space-like virtual exchange photon.**

find the *root-mean-squared* (rms) charge radius of the proton [Hofstadter 1955]. The measured size yields a proton rms radius of about 0.7 fm. Hofstadter's experiment effectively demonstrated that the proton was not "elementary" in the sense that it must have internal structure. There can be no such thing as a perfectly rigid body in special relativity. An impulse traveling at light speed will cause the front surface of an extended object to begin to deform before the signal can reach the back surface. Therefore, a proton of finite size must have constituent parts or "partons" which are identified as constituent quarks. This internal structure reveals itself in the observed inelastic scattering baryon excitation spectrum.

This elastic scattering reaction is allowed. In the center of momentum frame, shown in Figure 11.10, the only effect of the reaction is to change incoming momenta into outgoing momenta and while rotating the scattering angle. There is no energy transfer in the center of momentum frame. Denote the total four-momentum by

$$s = \sum_i p_{(i)} = p_e + p_p = p'_e + p'_p. \tag{11.160}$$

The velocity of center of momentum frame is given by the velocity of the total four-momentum s_μ in the laboratory frame

$$\vec{v}_{cm} = \frac{\vec{p}_s}{E_s}c^2 = \frac{\vec{p}_e + \vec{p}_p}{E_e + E_p}c^2. \tag{11.161}$$

In the center of momentum frame, when the two particles collide, incoming momenta are replaced by outgoing momenta and the direction is rotated. The momentum transferred to the proton is twice the incoming momentum

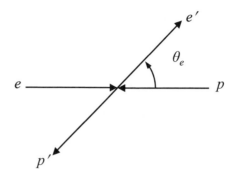

FIGURE 11.10   The elastic scattering of electrons and protons is illustrated in the center of momentum frame. There is no energy transfer in this frame. Incoming momenta re replaced by outgoing momenta and the scattering direction can be rotated. The momentum transferred to the proton is twice the incoming momentum of the electron. In the impulse approximation, the exchange is mediated by a single space-like virtual photon.

of the electron. Therefore, the exchange is mediated by a space-like virtual photon. Virtual exchange particles, unlike freely propagating physical particles need not be on the mass shell.

Now consider a relativistic electron scattering from a fixed proton target. Usually a cryogenic liquid hydrogen target is used. Denote the four-momentum transfer to the proton by q

$$q = p_e - p'_e = p'_p - p_p. \tag{11.162}$$

In the laboratory frame

$$p_e = \begin{bmatrix} E_e/c \\ \vec{p}_e \end{bmatrix}, \quad p_p = \begin{bmatrix} m_p c \\ 0 \end{bmatrix}, \quad q = \begin{bmatrix} \hbar\omega/c \\ \vec{q} \end{bmatrix} = p_e - p'_e, \tag{11.163}$$

where $\hbar\omega = E_e - E'_e$ is the energy transfer to the proton. Eliminating the final proton from the kinematics by evaluating its invariant mass gives

$$p'^2_p = (p_p + q)^2 = p^2_p + 2p_p \cdot q + q^2, \tag{11.164}$$

$$-m^2_p c^2 = -m^2_p c^2 - 2(m_p \hbar\omega) + q^2, \tag{11.165}$$

giving a space-like value for the four-momentum transfer

$$q^2 = 2m_p \hbar\omega = 2m_p(E_e - E'_e). \tag{11.166}$$

To complete the solution for the election kinematics, the extreme relativistic limit (ERL) is often used in high energy electron accelerators. In the

ERL, the mass of the electron is considered to be negligible

$$q^2 = (p_e - p'_e) \approx -2p_e \cdot p'_e \approx \frac{2E_e E'}{c^2}(1 - \cos\theta_e). \quad \text{(ERL)} \qquad (11.167)$$

This yields the kinematic relation

$$2E_e E'(1 - \cos\theta_e) = 2m_p c^2(E_e - E'_e) \qquad (11.168)$$

which can be solved for the scattered electron energy, giving

$$E'_e = \frac{E_e}{1 + (E_e/m_p c^2)(1 - \cos\theta_e)} = \frac{E_e}{1 + (2E_e/m_p c^2)\sin^2(\theta_e/2)}, \quad \text{(ERL)}$$
$$(11.169)$$

11.7   RELATIVISTIC LAGRANGIANS

To generate a relativistic Lagrangian, one can try to develop an action principle replacing the time parameter t with the scalar proper time τ_i. However, one immediately confronts several problems:

−   Every particle maintains its own proper time,
−   The relativistic field can be decomposed into real and virtual particles,
−   Particles can be created and annihilated, and
−   Particles propagate freely in space only if they are on their mass shell.

The Newtonian ansatz of ignoring the internal structure of particles fails when particles can be created or destroyed. Moreover, every particle maintains its own proper time which is defined from the moment of its creation to the moment of its destruction. This makes a relativistic particle Lagrangian treatment difficult.

The approach used here to develop relativistic single particle Lagrangians by treating the invariant mass as a constraint condition on the action. The resulting equations are valid in the limit that radiation field effects can be safely ignored or empirically modeled. Often, Monte Carlo simulations, based on the impulse approximation and cross-section lookup tables, can be used to generate statistical distributions of multiple-scattering radiation effects. When none these approximations are suitable, a field theoretical solution is necessary.

The free particle equations of motion can be derived from an action which for each particle introduces, up to a total derivative, a factor of the form

$$S = \int_1^2 p_\mu dq^\mu = \int_1^2 p_\mu \frac{dq^\mu}{d\lambda} d\lambda, \qquad (11.170)$$

where λ is some uniform parameterization of the motion, and where the momenta are not completely independent but subject to the scalar "Hamiltonian" mass-constraint

$$H_s(p) = mc^2 - c\sqrt{-m^2u^2} = mc^2 - c\sqrt{-p^2} = 0. \tag{11.171}$$

The following alternate, but equivalent, form for this constraint is often used

$$H'_s(p, x) = \frac{m^2}{2}(u^2(p, x) + c^2) = 0, \tag{11.172}$$

where the useful, but optional, convention of setting constraint degrees of freedom to zero is continued.

Defining the undetermined multiplier as Λ, the variational principle gives

$$\delta S = \int_1^2 \frac{dq^u}{d\lambda}\delta p_u + p_u\delta\frac{dq^u}{d\lambda} - \Lambda\frac{\partial H_s}{\partial q^u}\delta q^u - \Lambda\frac{\partial H_s}{\partial p_u}\delta p_u d\lambda = 0, \tag{11.173}$$

$$\delta S = \left(\int_1^2 \frac{dq^u}{d\lambda}\delta p_u - \frac{dp_u}{d\lambda}\delta q^u - \Lambda\frac{\partial H_s}{\partial q^u}\delta q^u - \Lambda\frac{\partial H_s}{\partial p_u}\delta p_u\right)d\lambda + \delta p_u q^u|_1^2. \tag{11.174}$$

The equations of motion are

$$\frac{dq^u}{d\lambda} = \Lambda\frac{\partial H_s}{\partial p_u}, \quad \frac{dp_u}{d\lambda} = -\Lambda\frac{\partial H_s}{\partial q^u}, \tag{11.175}$$

$$\frac{dq^u}{d\tau}\frac{d\tau}{d\lambda} = \Lambda\frac{\partial H_s}{\partial p_u}, \quad \frac{dp_u}{d\tau}\frac{d\tau}{d\lambda} = -\Lambda\frac{\partial H_s}{\partial q^u}, \tag{11.176}$$

where the constant term in H_s does not contribute to the motion. The Lagrange multiplier Λ can be determined by applying the constraint conditions, for example, in the equation

$$\frac{dq^u}{d\tau}\frac{d\tau}{d\lambda} = \Lambda\frac{p^u}{m}, \tag{11.177}$$

evaluating the norm of both sides gives

$$\frac{d\tau}{d\lambda} = \Lambda \tag{11.178}$$

The equations of motion therefore reduce to the relativistic form of Hamilton's equations of motion in free space

$$
\begin{aligned}
\frac{dq^u}{d\tau} &= \frac{\partial H_s(q,p)}{\partial p_u} = \frac{p^u}{m}, \\
\frac{dp_u}{d\tau} &= -\frac{\partial H_s(q,p)}{\partial q^u} = 0.
\end{aligned}
\tag{11.179}
$$

The scalar field H_s plays the role of a scalar Hamiltonian which is constant along the physical trajectory of a particle since the Hamiltonian does not explicitly depend on the proper time τ. This constant has no physical significance (other than that given by the value of the speed of light) and is therefore set to zero for convenience.

Therefore, for every particle with definite mass, one can shortcut the undetermined multiplier method by simply defining a scalar Lagrangian of the form

$$
L = p_u \dot{q}^u - H_s(p,q) + \frac{dF}{d\tau}
\tag{11.180}
$$

and apply variational principle

$$
\delta \int_1^2 \left(p_u \dot{q}^u - H_s(p,q) + \frac{dF}{d\tau} \right) d\tau = 0,
\tag{11.181}
$$

where the scalar H_s replaces the Hamiltonian H in the invariant form of the Lagrangian. This leads to the relativistic form of Hamilton's equations of motion given in Equation 11.179.

For a noninteracting "free" particle, the four-momentum is a constant of the motion. The evolution of the world line with respect to proper time is given by

$$
x(\tau) = x(0) + \frac{p}{m}\tau.
\tag{11.182}
$$

One can, of course, directly eliminate the constraint in the Lagrangian, by substituting a relativistic Hamiltonian that satisfies the constraint conditions for the fourth-component of the four-momentum, using $p_o c = H(\vec{p}, x)$ to get

$$
S = \int_1^2 \left(p_i \frac{dq^i}{dt} - H \right) dt = \int_1^2 \left(p_i \frac{dq^i}{dt} - \sqrt{(mc^2)^2 + (\vec{p}c)^2} \right) dt.
\tag{11.183}
$$

This gives the standard (noncovariant) Hamilton equations of motion, where the Newtonian nonrelativistic Hamiltonian H_N of a free particle has been replaced by the relativistic one H, which includes the rest energy of the particle. For a free particle, the Hamiltonian can be written as

$$E = H = p_0 c = \sqrt{(mc^2)^2 + (\vec{p}c)^2} = mc^2 + T. \tag{11.184}$$

In the absence of force, the relativistic Hamiltonian is due to the rest mass contribution mc^2 plus a contribution from relativistic kinetic energy T of the free particle. The nonrelativistic Hamiltonian formalism ignores the constant rest energy term.

11.7.1 Gauge Invariant Couplings

The free particle action is indeterminate up to a total derivative of the coordinates

$$S = \int_1^2 \left(p_u \frac{dq^u}{d\tau} - H_s + \frac{dF(q)}{d\tau} \right) d\tau \tag{11.185}$$

This implies that the momenta are indeterminate up to a total gradient of the coordinates

$$S = \int_1^2 (p_u - \lambda \nabla_u G(q)) dq^u - \int_1^2 \left(H_s + \frac{dF}{d\tau} - \lambda \frac{dG}{d\tau} \right) d\tau, \tag{11.186}$$

where λ is a scale factor. Note that

$$\int_1^2 \nabla_u G(q) \frac{dq^u}{d\tau} d\tau = \int_1^2 \frac{dG}{d\tau} d\tau = \int_1^2 dG, \tag{11.187}$$

and therefore the result is invariant under the combined set of gauge transformations

$$\begin{aligned} p_u &\to p_u - \lambda \nabla_u G(q), \\ dF &\to dF - \lambda dG(q), \\ H_s &\to H_s \end{aligned} \tag{11.188}$$

The electromagnetic field is the gauge field associated with this invariance. To generate the Lorentz force law, replace λ by the charge and introduce a four-vector potential

$$A(x) = \begin{bmatrix} \Phi(x)/c \\ \vec{A}(x) \end{bmatrix} \tag{11.189}$$

into the scalar Hamiltonian constraint

$$H_s(p, x) = mc^2 - c\sqrt{-(p - q_e A(x))^2} = 0, \qquad (11.190)$$

where the action is $S = \int_1^2 \left((p_u - q_e A_u)\frac{dq^u}{d\tau} - H_s + \frac{dF(q)}{d\tau}\right)d\tau$, and the relativistic Lagrangian is now invariant under the combined gauge transformation

$$\begin{aligned}
p_u &\to p_u - q_e \nabla_u G, \\
A_u &\to A_u - \nabla_u G, \\
dF &\to dF, \\
H_s &\to H_s.
\end{aligned} \qquad (11.191)$$

Note that the four-velocity is a gauge invariant construct

$$u = \frac{p - q_e A}{m}. \qquad (11.192)$$

The Euler-Lagrange equations of motion are given by

$$\frac{dx^\mu}{d\tau} = \frac{\partial H_s}{\partial p_\mu}, \quad \frac{dp_\mu}{d\tau} = -\frac{\partial H_s}{\partial x^\mu}, \qquad (11.193)$$

with solution

$$\frac{dx^\mu}{d\tau} = \frac{p^\mu - q_e A^\mu}{m} = u^\mu, \qquad (11.194)$$

$$\frac{dp_\mu}{d\tau} = q_e \frac{p^\nu - q_e A^\nu}{m}\frac{\partial A_\nu}{\partial x^\mu} = q_e u^\nu \frac{\partial A_\nu}{\partial x^\mu}. \qquad (11.195)$$

As in the nonrelativistic case, symmetries of the scalar Hamiltonian lead to constants of the motion. If the four-vector potential is time-independent, one gets a conserved energy

$$p^0 c = E^0 = m\gamma c^2 + q_e \Phi = mc^2 + T + q_e \Phi(\vec{r}). \qquad (11.196)$$

Likewise, if the scalar Hamiltonian is cyclic in the azimuthal angle, one gets a conserved angular momentum

$$p_\phi = m\gamma \frac{r^2 d\phi}{dt} + q_e A_\phi. \qquad (11.197)$$

Finally, eliminating the canonical momentum yields the standard relativistic form of the Lorentz equation for a massive charged particle moving in an electromagnetic field

$$f_u = g_{uv} m \frac{du^v}{d\tau} = q_e u^v F_{uv}, \qquad (11.198)$$

where the antisymmetric electromagnetic field tensor is defined as

$$F_{uv} = A_{v,\mu} - A_{\mu,v}. \tag{11.199}$$

If the electric field vanishes, and the magnetic field is time-independent, the equations of motion for a particle simplify to

$$\frac{dT}{dt} = 0 \rightarrow |\vec{v}| = \text{constant},$$

$$\frac{d\vec{v}}{dt} = -\left(\frac{q_e \vec{B}}{m_R}\right) \times \vec{v} = \vec{\omega} \times \vec{v}, \tag{11.200}$$

where $\omega = |q_e B|/m_R$ is the angular frequency of precession. The particle precesses in the magnetic field with an instantaneous radius of gyration given by

$$r_g = \frac{|p_\perp|}{|q_e B|}. \tag{11.201}$$

11.7.2 Stress-Energy Tensor of the Electromagnetic Field

Associated with the electromagnetic field is a totally symmetric second-rank *stress-energy tensor* given by

$$T_{\text{EM}}{}^{\mu v} = \frac{1}{\mu_0}\left(-F^\mu{}_\alpha F^{\alpha v} + \frac{1}{4}\delta^{\mu v}F^\beta{}_\alpha F^\alpha{}_\beta\right). \tag{11.202}$$

The stress-energy tensor defines a four-momentum density $T_{\text{EM}}{}^{\mu 0}$ with integrated four-momentum

$$p_{\text{EM}}{}^\mu = \frac{1}{c}\int d^3 x \sqrt{\|g\|} T^{\mu 0} \tag{11.203}$$

where

$$T^{00} = U = \frac{\varepsilon_0 \vec{E}^2}{2} + \frac{\vec{B}^2}{2\mu_0},$$

$$(T^{i0})e_i = \frac{\vec{S}}{c} = \frac{1}{\mu_0 c}\vec{E} \times \vec{B}, \tag{11.204}$$

and

$$\varepsilon_0 \mu_0 = 1/c^2. \tag{11.205}$$

The *energy density* of the electromagnetic field is given by

$$U = T^{00}, \tag{11.206}$$

while its linear *momentum density* T^{i0}/c is related to the *Poynting vector*[8] $\vec{S} = \vec{E} \times \vec{B}/\mu_0$ by

$$S^i/c^2 = T^{i0}/c. \tag{11.207}$$

The Poynting vector describes the energy flux of an electromagnetic field. Its magnitude is the power per unit area flowing through a surface normal to its direction. The time-averaged value of the Poynting vector divided by the speed of light is the *radiation pressure* exerted by an electromagnetic wave on the surface of a target.

It is left as an exercise to show that divergence of the stress-energy tensor is given by

$$D_\nu T_{\text{EM}}{}^{\mu\nu} = -F^\mu{}_\alpha J^\alpha. \tag{11.208}$$

If only electromagnetic forces act on the particles, and the current density condenses into localized charge singularities on a macroscopic scale, the current factors into a sum of discrete localized particle contributions[9]

$$J^\alpha(\vec{x}, t) = \sum_i q_{e(i)} u_{(i)}^\alpha \delta^3(\vec{x} - \vec{x}_i(t))/\gamma_{(i)}. \tag{11.209}$$

The time rate of change of the field's four-momentum is given by

$$\frac{dp_{\text{EM}}{}^\mu}{dt} = \int d^3x \frac{\partial T^{\mu 0}}{\partial t} = \int d^3x T^{\mu\nu}_{,\nu} + \text{surface integral}, \tag{11.210}$$

$$\frac{dp_{\text{EM}}{}^\mu}{dt} = -\int d^3x \sum_i q_{e(i)} F_{\text{EM}}{}^\mu{}_\alpha u_{(i)}^\alpha \frac{\delta(\vec{r} - \vec{r}_{(i)})}{\gamma_{(i)}} = -\sum_i \frac{dp_{(i)}{}^u}{dt}. \tag{11.211}$$

Therefore, the sum over the field and particle momenta are conserved

$$\frac{d}{dt}\left(p_{\text{EM}}{}^\mu + \sum_i p_{(i)}{}^\mu\right) = 0. \tag{11.212}$$

Even if the matter distribution is not discrete, and there are a number of gauge fields, one can define a total stress-energy tensor such that the sum of the matter and gauge field contributions is conserved

$$T^{uv}_{,v} = T^{uv}_{\text{matter},v} + T^{uv}_{\text{gauge},v} = 0, \tag{11.213}$$

---

[8] Named after John Henry Poynting (1852–1914).
[9] For a single point particle, the relationship between current density and location is given by $J_i^\alpha(x) = \int q_e u_i^\alpha \delta^4(x - x_i(\tau))d\tau$.

guaranteeing the existence of a conserved total four-momentum vector

$$p^\mu = \int d^3x \sqrt{\|g\|} T^{\mu 0}. \tag{11.214}$$

This separation of the world into distinct and separate particle and field components is overly simplistic, however. Both particle and fields share particle-like and wave-like characteristics. Fields are generated by particles, but the very existence of particles is revealed only by their field couplings to other particles. The asymptotic value of the gauge fields at large distances are constrained by the quantum numbers of the particle content. In a self consistent approach, including the self-coupling a particle to its field, the particle represents the localized, short-range part of the solution and its self-consistent field represents its long-range asymptotic part. Moreover, particle creation and annihilation mechanisms allow particle and field degrees of freedom to interact, exchanging energy and momentum. From this viewpoint, it is possible and even probable that particles and fields are only different manifestations of a single underlying physical reality; one that exhibits both a short-range localized quantized packet structure and a long-range asymptotic field behavior.

The attempt to reconcile the particle and wave nature of matter with relativity would eventually lead to the development of many-particle relativistic field theory. The standard model of particle physics,[10] for example, is just one attempt to build a field theoretical structure based on the observed gauge symmetries of nature. Although highly successful thus far, it is viewed as being too ad hoc in character to be a fundamental theory of everything. The continuing quest to develop a unified theory of particles and their fields represents the present holy grail of theoretical physics.

11.7.3 *Scalar Gravity in Flat Spacetime

Newton's theory of gravitation introduces a gravitational force on a particle labeled (i) due to a scalar potential Φ given by

$$m\frac{d\vec{v}_{(i)}}{dt} = -m\sum_{j\neq i}\vec{\nabla}\Phi_{(j)}(x - x_{(j)})|_{x=x_{(i)}}, \tag{11.215}$$

where the field source is due to a positive definite scalar mass density ρ summed over individual particle contributions

$$\vec{\nabla}^2\Phi_{(i)}(x - x_{(i)}) = 4\pi G\rho_{(i)}(x - x_{(i)}). \tag{11.216}$$

---

[10] For a simple introduction see [Griffiths 1987].

The nonrelativistic solution for a static point mass source is

$$m\Phi(\vec{x}, \vec{x}') = -mMG/|\vec{x} - \vec{x}'|. \tag{11.217}$$

As usual, the self-coupling terms have been omitted in Equation 11.215. Note that neither of Equations 11.215 or 11.216 are Lorentz invariant. To make the second equation Lorentz invariant, the minimal substitution would be to replace Poisson's equation for gravitation with the scalar wave equation

$$\nabla^2\Phi = \left(\vec{\nabla}^2 - \frac{1}{c^2}\frac{\partial^2}{\partial t^2}\right)\Phi = 4\pi G\rho, \tag{11.218}$$

where particle id suffixes have been suppressed. Next, one needs to replace the nonrelativistic particle Lagrangian with a relativistic one. However, there is no unique prescription for how to do this. Nevertheless, it is still interesting to study how a simple invariant Lagrangian might be constructed.

In order to upgrade Newtonian scalar gravity to a relativistic theory in flat space, one simple substitution proposed by Misner, Thorne, and Wheeler is to introduce a positive definite scale factor in the free particle action [Misner 1973, 178]

$$S = -mc\int e^{G\Phi/c^2}\sqrt{-u^2}d\tau, \tag{11.219}$$

where the gravitational potential satisfies the boundary conditions

$$\lim_{|\vec{r}|\to\infty}\Phi(r) = 0. \tag{11.220}$$

In the limit $G \to 0$, the solutions are geodesics in flat space, leading to the free particle equations of motion

$$\frac{du}{d\tau} = 0. \tag{11.221}$$

With the gravitational coupling turned on, the Euler-Lagrange equations of motion become

$$m\frac{du}{d\tau} = (u \cdot u\nabla - uu \cdot \nabla)\frac{mG}{c^2}\Phi, \tag{11.222}$$

which can be written in dyadic form as

$$\frac{d(m\mathbf{u})}{d\tau} = \frac{mG}{c^2}(\mathbf{u} \cdot \mathbf{u} - \mathbf{u}\,\mathbf{u}) \cdot \nabla\Phi. \tag{11.223}$$

Clearly, the constraint on the four-force is satisfied, since it is easy to show that

$$u \cdot \frac{dmu}{d\tau} = 0, \qquad (11.224)$$

Therefore, the square of the four-velocity $u^2 = -c^2$ can be considered as a first integral of the motion.

Substituting the four-momentum given by

$$p = me^{G\Phi/c^2}u, \qquad (11.225)$$

the equations of motion can be reduced to the canonical form

$$\frac{dp}{d\tau} = -e^{G\Phi/c^2}mG\nabla\Phi,$$
$$\frac{du}{d\tau} = e^{-G\Phi/c^2}\frac{p}{m}. \qquad (11.226)$$

If the field is static (time-independent), one gets a conserved energy

$$E = e^{G\Phi/c^2}m\gamma c^2. \qquad (11.227)$$

which gives the constraint condition

$$\gamma = \frac{E}{mc^2}e^{-G\Phi/c^2}. \qquad (11.228)$$

The orbit is bound if the total energy is less than the free particle rest energy

$$\frac{E}{mc^2} = e^{G\Phi/c^2}\gamma < 1. \qquad (11.229)$$

If, in addition, the field is spherically symmetric, $\Phi = \Phi(|\vec{x}|)$, the three-dimensional motion is restricted to the two-dimensional plane defined by $\vec{x} \wedge \vec{u}$, and the angular momentum will be conserved

$$p_\phi = e^{G\Phi/c^2}mu_\phi = r^2\frac{d\phi}{dt}\frac{E}{c^2} = \ell\frac{E}{mc^2}. \qquad (11.230)$$

where ℓ is the nonrelativistic value of the angular momentum and $r = |\vec{x}|$. This implies that the areal velocity is conserved just as in the non-relativistic case

$$\frac{dA}{dt} = \frac{1}{2}r^2\frac{d\phi}{dt} = \frac{1}{2}\frac{\ell}{m} = \frac{1}{2}\frac{p_\phi c^2}{E}. \qquad (11.231)$$

In plane polar coordinates, the Lagrangian is

$$L(u, r) = -mce^{G\Phi(r)/c^2}\sqrt{u_0^2 - \left(\frac{dr}{d\tau}\right)^2 - r^2\left(\frac{d\phi}{d\tau}\right)^2} \tag{11.232}$$

The solutions yield a conserved scalar Hamiltonian, which, however, identically vanishes

$$H_s = p_\mu u^\mu - L = me^{G\Phi/c^2}u^2 + mce^{G\Phi/c^2}\sqrt{-u^2} \equiv 0. \tag{11.233}$$

The radial motion is non-cyclic, but a first integral can be found using $u^2 = -c^2$. Letting $\gamma^2(c^2 - \vec{v}^2) = c^2$ gives

$$\left(\frac{dr}{dt}\right)^2 + r^2\left(\frac{d\phi}{dt}\right)^2 = c^2\left(1 - \frac{1}{\gamma^2}\right) = c^2\left(1 - \frac{m^2c^4}{E^2}e^{2G\Phi/c^2}\right) \tag{11.234}$$

The orbit equation is found by eliminating the time ($\dot{r} = \dot{\phi}dr/d\phi$) and replacing $r^2\dot{\phi} = \ell/m$, giving

$$\left(\left(\frac{1}{r^2}\frac{dr}{d\phi}\right)^2 + \frac{1}{r^2}\right)\frac{\ell^2}{m^2} = c^2\left(1 - \frac{m^2c^4}{E^2}e^{2G\Phi/c^2}\right). \tag{11.235}$$

Expanding to lowest order in the coupling constant, and letting $u = 1/r$ gives the nonrelativistic result previously found in Chapter 5

$$\left(\frac{du}{d\phi}\right)^2 + u^2 = \frac{m^2}{\ell^2}\frac{2}{m}(E_B - V'). \tag{11.236}$$

Where E_B is an effective binding energy that can be set equal to the non-relativistic energy and V' is the nonrelativistic potential energy with a small additional energy dependence

$$\frac{2E_B}{mc^2} = \left(1 - \frac{m^2c^4}{E^2}\right), \tag{11.237}$$

$$V' = \left(\frac{m^2c^4}{E^2}\right)mG\Phi = \left(1 - \frac{2E_B}{mc^2}\right)mG\Phi, \tag{11.238}$$

where $E_B < 0$ for bound orbits. Substituting the point source potential $V' = -k'/r$ reproduces the non-relativistic equation for an ellipse. A more careful analysis of the relativistic result shows that the ellipse precesses. Expanding to next order in the coupling constant gives

$$V' = \frac{mc^2}{2} \left(\frac{m^2 c^4}{E^2} \right) (e^{-2ha/r} - 1) = mc^2 \left(1 - \frac{2E_B}{mc^2} \right) \left(-\frac{\eta a}{r} + \left(\frac{\eta a}{r} \right)^2 - \cdots \right),$$

(11.239)

where the dimensionless η is

$$\eta = MG/c^2 a,$$

(11.240)

the mass of the gravitational source is M, and the radial distance is expanded in units of the mean separation distance $a = (r_p + r_a)/2$, where r_p and r_a are defined as the perihelion and aphelion distances from the center of force, respectively, and $e = (r_a - r_p)/2a$ defines the eccentricity of the bound orbit. The binding energy correction for the orbit of Mercury is miniscule, as is the leading order correction to the orbit

$$\frac{-2E_B}{mc^2} \approx \eta = 2.56 \times 10^{-8}.$$

(11.241)

In the case of Mercury, the resulting advance of the perihelion at this level of approximation is 14.4 arc-seconds per century, about one-third of the 42 arc-seconds per century that are needed to agree with the experimental discrepancy. See Chapter 5 for the details of this analysis.

Misner, Thorne, and Wheeler [Misner 1973, pp. 177–186] examine several relativistic extensions to Newtonian gravitation in flat spacetime. They reject scalar and vector gravitation couplings as not being able to account for the bending of light in a gravitational field. A tensor coupling does allow light to bend, but to get self-consistent results, in their opinion, leads inexorably to General Relativity.

11.7.4 *Brief Introduction to General Relativity

General Relativity is outgrowth of the attempt to make the Poincaré transformations local, so that all reference frames can be treated on an equal footing. Like electromagnetic gauge invariance, the theory is based on exploiting a fundamental symmetry of nature, with the gravitational field interpreted as the physical field that guarantees the translational and Lorentz invariance of spacetime Einstein based his theory on two fundamental principles:

- *The Principle of General Relativity:* The laws of physics cannot depend on the choice of reference frame used to express them, that is, there are no special frames of reference.

– *The Equivalence Principle:* There is no way to distinguish between the effects a uniform gravitational field and a uniform acceleration of a closed system.

The second statement is the crucial one. In an accelerating frame, light appears to follow a curved trajectory. Since photons are massless, to connect gravity to acceleration then requires that the field coupling be to the stress-energy tensor of a particle and not to its scalar mass.

According to John Wheeler [Wheeler 1990; Misner 1973], the main features of general relativity can be summarized as follows:

1. Spacetime is not a rigid arena in which events take place.
2. Its form and structure is influenced by the matter and energy content of the universe.
3. Matter and energy inform spacetime on how to curve.
4. Spacetime instructs matter on how to move.

Modern geometric approaches to gravitation theory are based on developing a gauge theory for the coupling, where the gravitational field is the generator of the local relativistic Poincaré group of Lorentz transforms (rotations and velocity boosts) and translations (continuous deformations). This is accomplished by coupling the curvature of spacetime to the stress-energy tensor of the matter fields.

In Einstein's theory, space and time represent a four-dimensional manifold. The link between geometry and gravitation is through the Riemann curvature tensor previously defined in Chapter 2. The Ricci tensor $R^{\alpha\beta}$ is a symmetric contraction of the Riemann tensor. The Ricci scalar R is a contraction of the Ricci tensor. The important combination

$$G^{\alpha\beta} = R^{\alpha\beta} - \frac{1}{2}g^{\alpha\beta}R \qquad (11.242)$$

is called the *Einstein tensor.* This tensor has a vanishing divergence

$$G^{\alpha\beta}_{,\beta} = 0, \qquad (11.243)$$

which gives rise to a conserved vector quantity, identified by Einstein as the total four-momentum of the universe. Einstein's tensor is connected in his *General Theory of Relativity* to the *stress-energy tensor* $T^{\alpha\beta}$ of other gauge fields and matter by the equation

$$G^{\alpha\beta} = \kappa T^{\alpha\beta}. \qquad (11.244)$$

where κ is a constant of proportionality. Einstein's gravitational coupling reduces to Newtonian gravitation in the weak field limit and for non-relativistic velocities.

The three classic tests of General Relativity are

- *The perihelion shifts of mercury,* whereby the orbit of Mercury is seen to deviate from Newtonian orbits.
- *The deflection of starlight,* whereby light passing close to a star has its path altered. This is also known as *gravitational lensing.*
- *The gravitational redshift,* whereby a clock in a gravitational potential well appears to run slower than a stationary clock distant from the source of gravity.

Einstein's theory of General Relativity explained the observed perihelion shift of Mercury and correctly predicted the outcomes of other two tests. Einstein's theory, however, is but one of many competing models for the gravitational interaction. Lasenby, Doran, and Gull, for example, have developed a gauge theory for gravity (GTG) in flat spacetime, suitable for use with spinor equations [Lasenby 1998; Doran 2003]. Hestenes has shown that, in the absence of spin torsion, GTG is consistent with General Relativity [Hestenes 2007]. A complete summary of the various approaches including quantum gravity would be beyond the scope of this text. For a detailed introduction to General Relativity see *Gravitation,* by Misner, Thorne, and Wheeler [Misner 1973].

11.8 SUMMARY

Invariance of Maxwell's equations under the Lorentz transformation can be explained if one assumes that spacetime is a four-dimensional continuum with metric

$$dx^2 = -c^2 d\tau^2 = -c^2 dt^2 + dx^2 + dy^2 + dz^2 = g_{\mu\nu} dx^\mu dx^\nu. \quad (11.245)$$

The proper time of a time-like particle is given by

$$\tau = \int_0^1 \sqrt{-dx^2} = \int_0^1 \sqrt{1 - \vec{\beta}^2} dt, \quad (11.246)$$

where

$$\vec{\beta} = \frac{\vec{v}}{c} = \frac{1}{c} \frac{d\vec{x}}{dt} \quad (11.247)$$

is the three-velocity in units of the speed of light and

$$\gamma = \frac{dt}{d\tau} = \frac{1}{\sqrt{1 - \vec{\beta}^2}} \tag{11.248}$$

is a time dilation factor.

Freely propagating massive particles are time-like in nature with a four-momentum given by

$$p = m\frac{dx}{d\tau} = mu = mc\gamma \begin{bmatrix} 1 \\ \vec{\beta} \end{bmatrix}, \tag{11.249}$$

where m is the particle mass and u is its four-velocity. Time-like particles with positive energy are required to propagate forward in time. The square of the four-momentum is an invariant

$$p^2 = -mc^2. \tag{11.250}$$

Photons are light cone particles with zero mass that propagate at the speed of light.

The differential of the four-momentum with respect to proper time is the four-force

$$f = \frac{dp}{dt} \tag{11.251}$$

which must satisfy the constraint

$$u \cdot f = mu \cdot \frac{du}{d\tau} = 0 \tag{11.252}$$

The four-acceleration is therefore a space-like four-vector. This constraint is automatically met for the electromagnetic Lorentz force law

$$m\frac{du}{dt} = \left[\frac{q_e}{2}F, u\right] = q_e F \cdot u, \tag{11.253}$$

where F is the electromagnetic field tensor with electric and magnetic field components

$$F^u{}_v = \begin{bmatrix} 0 & E_1/c & E_2/c & E_3/c \\ E_1/c & 0 & B^3 & -B^2 \\ E_2/c & -B^3 & 0 & B^1 \\ E_3/c & +B^2 & -B^1 & 0 \end{bmatrix}, \tag{11.254}$$

The electromagnetic field is a solution to Maxwell's equations

$$\nabla F = -\frac{J}{\mu_0} \tag{11.255}$$

where J is a conserved current. This field can be derived from a vector potential

$$F = \nabla \wedge A, \tag{11.256}$$

with components

$$A^\mu = \begin{bmatrix} \Phi/c \\ \vec{A} \end{bmatrix} \tag{11.257}$$

The Lorentz force on a particle can be derived from a relativistic action

$$S = \int_1^2 \left(p_u \frac{dq^u}{d\tau} - H_s + \frac{dF}{d\tau} \right) d\tau, \tag{11.258}$$

$$H_s(p, q) = mc^2 - c\sqrt{-(p - q_e A)^2} = 0.$$

The inner product of four vectors is invariant under the Lorentz transformations. A general Lorentz transformation can be written as a product of space rotations and velocity boosts, where a velocity boost in the z-direction takes the form

$$A^\mu_{\ v}(\sigma_0^3, \beta) = \begin{bmatrix} \gamma & 0 & 0 & \gamma\beta \\ 0 & 1 & 0 & 0 \\ 0 & 0 & 1 & 0 \\ \gamma\beta & 0 & 0 & \gamma \end{bmatrix}. \tag{11.259}$$

All other Lorentz boosts can be obtained by rotation.

Particle creation and annihilation at a reaction vertex is required to conserve the total four-momentum of the reaction channel

$$\sum_{i=1,n} p_u^{(i)} \bigg|_{in} = \sum_{j=1,n'} p_u'^{(j)} \bigg|_{out}. \tag{11.260}$$

The stress energy tensor of the electromagnetic field is given by

$$T_{EM}{}^\mu_{\ v} = \frac{1}{\mu_0 c} \left(-F^\mu_{\ \alpha} F^\alpha_{\ v} + \frac{1}{4} \delta^\mu_{\ v} F^\beta_{\ \alpha} F^\alpha_{\ \beta} \right). \tag{11.261}$$

This stress-energy tensor is conserved in the absence of charge currents

$$T_{EM}{}^{\mu v}_{\ \ ,v} = -F^\mu_{\ \alpha} J^\alpha. \tag{11.262}$$

In the presence of matter, one can define a total stress-energy tensor consisting of gauge field and matter contributions such that the total four-momentum of the interacting system is conserved

$$T^{\mu\nu}_{,\nu} = 0, \tag{11.263}$$

$$p^{\mu}|_{total} = \int \sqrt{\|g\|}d^3x T^{\mu 0} = p^{\mu}|_{gauge} + p^{\mu}|_{matter}. \tag{11.264}$$

11.9   EXERCISES

1. Show that the relativistic wave equation

    $$\left(\vec{\nabla}^2 - \frac{1}{c^2}\frac{\partial^2}{\partial t^2}\right)\psi(\vec{x}, t) = 0$$

    is invariant under a Lorentz velocity boost, but not a Galilean velocity boost.

2. Derive the Lorentz law for addition of velocities given by

    $$\vec{\beta}' = \frac{1}{1 + \vec{\beta}_{rel} \cdot \vec{\beta}}(\vec{\beta}_{rel} + \vec{\beta}_{\|} + \vec{\beta}_{\perp}/\gamma_{rel}).$$

3. A meter stick is oriented making an angle θ with respect to the $\hat{z}$-axis in its rest frame. If an observer approaches the meter stick along the $\hat{z}$-axis with a velocity βc what is the length of the stick as seen from the vantage point of the moving observer? What angle does it make with respect to the $\hat{z}$-axis.

4. Show that a particle with definite mass m moving in free space cannot spontaneously absorb or emit a single massless photon.

5. Show that an isolated photon cannot spontaneously decay to an electron positron pair in free space (Hint: work in the center of mass frame of the electron-positron pair).

6. Show that a massless photon cannot spontaneously split into two massless photons unless the two photons are moving collinearly in the direction of the initial photon.

7. An astronaut travels to and from Alpha Centauri, which is 4.3 light years away from Earth at a speed of 0.5 c in both directions. How much has the astronaut aged? Compare this to the aging of a twin that remains home on Earth.

8. A star is receding from the Earth with a velocity of 50 km/s as measured by the shift of the hydrogen alpha line ($\lambda = 656.3$ nm). By how much and in what direction does the hydrogen alpha line shift?

9. Show that

$$D_v T^{\mu v}_{EM} = -q_e F^{\mu v} J_v,$$

where $T^{\mu v}_{EM}$ is the stress-energy of the electromagnetic field given by

$$T^{\mu v}_{EM} = \frac{1}{\mu_0 c} \left(-F^\mu{}_\alpha F^{\alpha v} + \frac{1}{4} \delta^{\mu v} F^\beta{}_\alpha F^\alpha{}_\beta \right).$$

10. What is the minimum particle energy required for a beam of protons of momentum p_a with mass m_p to produce antiprotons $\bar{p}$ of the same mass from a stationary proton target with four-momentum p_b via the reaction

$$p_a + p_b \rightarrow p_c + p_d + p_e + \bar{p}.$$

11. A beam of electrons with four-momentum e produced in an accelerator scatters elastically from a stationary proton target via the reaction

$$e + p \rightarrow e' + p'.$$

Find the center of mass velocity for the reaction, and the energies and momenta of the particles in the center of mass. Make a Lorentz transform back to the laboratory frame to find a relationship between the laboratory and center of mass scattering angles and the energies of the scattered proton.

12. A charged pion of mass $m_\pi = 139.6\,\mathrm{MeV}/c^2$ decays into a muon of mass $m_\mu = 105.7\,\mathrm{MeV}/c^2$ and a neutrino that is effectively massless. The pion has a mean-life of $T = 2.6 \times 10^{-8}$s in its rest frame. If the pion is created with a total laboratory energy $E_\pi = 500\,\mathrm{MeV}$, find its mean-life in the laboratory frame. How far would the pion travel in this time? If the detected muon has 75% of the pion's total energy find the laboratory scattering angle of the emitted neutrino.

13. A muon is produced in a high energy accelerator with a laboratory velocity $v = 0.998$ c. Look up the mean life time of a muon in the Particle Data Tables [Yao 2006] and determine the mean distance in meters that it will travel in the laboratory frame before decaying. What is its laboratory energy in GeV?

14. A neutral pion decays in flight into 2 photons. If the two photons are emitted with the same energy in the laboratory frame, find a relationship between the pion's laboratory energy to the opening angle of the decay vertex.

15. In Compton scattering, a photon can scatter elastically from an electron via the reaction

$$\gamma + e \rightarrow \gamma' + e'.$$

Assume a photon with energy E in the laboratory frame scatters elastically from an electron of mass m_e at rest. Show that the energy of the scattered photon is related to its scattering angle by

$$E' = E\left(1 + \frac{2E}{m_e c^2}\sin^2\frac{\theta_\gamma'}{2}\right)^{-1}.$$

Find the energy and direction of the scattered electron.

16. Show that for a system of particles, the center of momentum frame has a velocity

$$\vec{v}_{cm} = \frac{\vec{s}}{s_0}c = \frac{\sum_i \vec{p}_i}{\sum_i E_i}c^2.$$

Find the energy of the system in the center of momentum frame and relate it to the invariant mass of the system.

17. Show that the solution for the velocity of a rocket of initial mass m_o starting from rest and accelerating in a constant direction is given by

$$\beta(m) = \frac{v}{c} = \frac{1 - (m/m_0)^{2\beta_{rel}}}{1 + (m/m_0)^{2\beta_{rel}}}.$$

Plot $\beta(m)$ as a function of m/m_0 for $\beta_{rel} = [0.001, 0.01, 0.1, 1]$. What fraction of the mass has to be converted to reach a final velocity of 0.9 c in each case?

18. Using the relativistic rocket equation, a space voyager travels to Alpha Centauri at a distance of 4.3 light years from Earth with a constant acceleration of $g = 9.8 \text{ m/s}^2$ in the instantaneous rest frame of the rocket for the first half of the trip. The rocket then decelerates at $-g$ in the instantaneous rest frame for the second half of the trip.

   a. What is the velocity of the spaceship in units of the speed of light at the half-way point?

   b. At the half-way point, how much time has elapsed (in units of Earth years) in the frame of the spaceship? How much time would have elapsed in the inertial frame of the Earth? Use numerical integration if you need to.

   c. Assume the relative exhaust velocity is constant and the spaceship has lost 95% of its original mass by the half-way point. What relative exhaust velocity would be required? Use the result from Exercise 11.17 (that is, ignore gravitational field effects).

d. What is the total proper time needed to reach Alpha Centauri?

e. What fraction of the spaceship's initial mass remains?

19. Verify that the Lorentz force law has the relativistic form

$$f_u = g_{uv} m \frac{du^v}{d\tau} = q_e u^v F_{uv}.$$

20. Show that the following Lorentz-invariant Lagrangian results in the Lorentz force law

$$L(u, x) = \frac{m(u^2 - c^2)}{2} + q_e u \cdot A.$$

21. Show that the Lagrangian of Exercise 11.20 can be rewritten in the Hamiltonian form

$$L = p \cdot u - \left(\frac{mc^2}{2} + \frac{(p - q_e A)^2}{2m} \right) = p \cdot u - H_s.$$

22. Find the velocity and position as a function of time of a particle of mass m and charge q_e accelerating from rest in a uniform, constant, electric field $\vec{E}_0$. Show that the velocity approaches the limiting value of c.

23. Show that, for a velocity boost $v = \beta c$ along the $\hat{x}$-direction, the electric and magnetic fields transform as

$$
\begin{aligned}
E'_x &= E_x & B'_x &= B_x \\
E'_y &= \gamma(E_y - \beta c B_z), & B'_y &= \gamma(B_y + \beta E_z/c), \\
E'_z &= \gamma(E_z + \beta c B_y), & B'_z &= \gamma(B_z - \beta E_y/c).
\end{aligned}
$$

Since the electric and magnetic fields transform into each other, they have no separate independent existence. Hint: The problem can be solved using either geometric algebra or dyadic notation, where

$$F' = SF\bar{S}$$

and

$$\mathbf{F'} = \mathbf{A} \cdot \mathbf{F} \cdot \bar{\mathbf{A}}$$

give the respective transformation rules, where S is an element of the spin group and $\mathbf{A}$ is an element of the special orthogonal group, respectively.

Chapter 12

WAVES, PARTICLES, AND FIELDS

In this chapter, the Euler-Lagrange equations of motion for continuous classical fields are derived from an action principle. Noether's theorem is generalized, relating the continuous symmetries of a system of fields to the existence of conserved currents. Elementary solutions to linear and non-linear scalar wave equations in one space dimension are examined. The Sine-Gordon Equation yields particle-like soliton solutions with a quantized topological charge. A brief overview of the electromagnetic wave equation is presented.

12.1 INTRODUCTION TO THE CLASSICAL THEORY OF FIELDS

The course of study of the dynamics of physical systems bifurcates at this point. Theoretical physicists will typically begin a study of relativistic field theory, while applied physicists and engineers have a number of other choices available to them. The paths to discovery are many and varied, while sharing unexpected interconnections.

A proper course of study of continuum mechanics requires a textbook of its own to begin to do it justice. However, having come so far, it would be an oversight to fail to indicate at least the common starting point for such studies. Rigorous study of the continuum begins with extending Hamilton's Action Principle. One defines a Lagrangian field density which is used to generate the equations of motion for the continuous physical system of interest. The symmetries of this Lagrangian density gives rise to the existence of conserved currents. Gauge couplings associated with these symmetries provide a means of defining the long-range field interactions. The analogy to the methods previously used to study particle systems is both straightforward and transparent.

The linear wave equation provides a core example illustrating the basic methodology. Because of wave-particle duality, one cannot begin to understand particles without understanding waves. Given the vast scope of the endeavor, only the simplest of examples can be covered in this introductory chapter. Although the general formalism is indicated, the provided examples will be limited to studying variants of the wave equation in one space dimension.

In Chapter 6, the nonrelativistic wave equation was shown to arise naturally from taking the continuum limit of a many-particle, periodic, lattice structure. This fits in naturally with Newton's scheme of trying to construct matter from point-like, particle, constituents. The converse is also true. Given a continuous system, it can be discretized. *Discretization* is the process of transforming continuous models and equations into discrete counterparts. It represents the initial step toward making them suitable for numerical evaluation. In order to be digitally processed, quantization is also essential. *Quantization* is the procedure of constraining something from a continuous set of values (such as the real numbers) to a discrete set (such as the integers). For example, a Fourier transform of a continuous wave in some region of space, can be replaced by a discrete Fourier transform quantized on some uniform space interval. The truncated, discrete, solution will be a good approximation for wavelengths long compared to the fundamental wavelength given by the discrete interval.

Going beyond the linear wave equation is challenging. The superposition principle, critical to the study of simple classical and quantum systems, no longer applies. There are few theorems providing guidance for finding self-consistent solutions to coupled systems of interacting fields. Once again, a single simple system will be examined in detail to illustrate the types of novel phenomena one might hope to see. The nonlinear Sine-Gordon equation is a relativistic equation with a non-linear self-coupling interaction. The self-coupling leads to localized solutions moving at slower than light speeds. The solutions are remarkable stable under perturbations. In fact, they exhibit particle-like *soliton* behavior. The single soliton solutions can be combined to get multiple soliton solutions, characterized by a quantized topological charge.

The quantum revolution led by Plank, Einstein, Bohr, Heisenberg, and others, showed that particles and fields share many properties in common. Waves carry energy and momentum, and they interact via the transfer of four-momentum by localized exchange quanta. Particles, like fields, manifest diffraction and interference scattering patterns. Because particles can be created and annihilated, it is no longer possible to ignore the internal structure of elementary particles at the relativistic and quantum levels. (Special relativity and causality requires that extended objects must have internal structure.) This structure must be treated as continuous fields in order to preserve local Lorentz invariance. Developments in classical field theory over the last half century have provided new tools for exploring and new ways of thinking about quantum systems.

It should be noted that the partition of matter into localized particles and long-range interaction fields is, in itself, a simplification, providing a convenient means of differentiating the diverse manifestations of a single underlying reality. Newton made tremendous strides in the analysis of the motion of particles by decoupling the two. His gravitational field worked, via action-at-a-distance, to mediate the motion of two massive particles relative to each other, thereby explaining the orbital motion of the planets of the solar system about the sun.

In calculating forces, the internal coupling of a particle to itself was deliberately, and according to Einstein, necessarily, ignored by Newton. For stable systems uniformly moving in free space, this self-coupling must internally cancel. Attempts to study the self-coupling of the electron to its self-generated field in Quantum Electrodynamics result in infinities in the perturbative calculations. Thus the self-coupling problem, which presumably is what leads to particle formation, is enormously complicated, and it cannot be said with any confidence that it is a tractable problem. The Sine-Gordon equation provides a simple example of how topological requirements can lead to a quantized classical system. The chapter ends with a review of the electromagnetic wave equation.

In this brief overview of classical field theory and the quantum limit, the treatment of continuous material systems such as the properties of compressible and incompressible fluids and elastic media will be regrettably deferred. Thermodynamics, which plays a major role in the analysis of dissipative systems, is not needed for the simple wave structures that will be examined in this chapter.

Nevertheless, bulk media can be treated using means similar to that used in deriving the simple wave equation. The dynamics of such systems are fascinating in their own right and are the subject of much current research. From interstellar star formation to the flow of water from a faucet, the beauty and complexity of nature's dynamical processes cannot be denied.

12.1.1 Euler-Lagrange Equations of Motion

The equations of motion for a set of continuous fields can be derived by a straightforward generalization of Hamilton's Principle. Continuous fields $\phi = \{\phi^{s;i}\}$ are defined on an outer product space where the index s denotes spacetime degrees of freedom and the index i denotes additional internal "isospin" degrees of freedom. The Lagrangian is replaced by a volume integral over a Lagrangian density which is a scalar contraction over all spacetime and internal degrees of freedom $\mathcal{L} = \langle L \rangle$

$$L = \int d^3x \sqrt{\|g\|} \mathcal{L}(\phi, \partial_x\phi, \partial_y\phi, \partial_z\phi, \partial_t\phi, x, y, z, t). \tag{12.1}$$

The Lagrangian density, in turn, is a functional of the continuous fields, their first-order partial derivatives with respect to space and time, and the space and time coordinates. It is important to realize that the space coordinates are no longer observables, but are merely parameters, like the time, and are not varied. The physical observables are the fields $\phi^{s;i}$, which may be mixed space and isospin tensors (or spinors) of arbitrary rank. These are the quantities that will be allowed to vary by the variation principle.

The action is defined as a four-dimensional integral over a region $\mathcal{R}$ of space and time. The variation principle minimizes the action over an interval of four-volume

$$\delta S = \frac{1}{c} \int_{\mathcal{R}} d^4x \sqrt{\|g\|} \delta\mathcal{L}(\phi^{s;i}, \partial_\mu\phi^{s;i}, x^\mu) = 0. \tag{12.2}$$

Here, the relativistic convention of measuring time in units of length is followed by defining a spacetime tuple $x^\mu = (ct, \vec{x})$, even though no assumption has yet been made as to whether the Lagrangian is invariant under Poincaré or Galilean transformations. Direct variation of the previous action gives the standard form of the Euler Lagrange equations of motion

$$\partial_u \frac{\partial\mathcal{L}}{\partial(\partial_\mu\phi^{s;i})}\Big|_\phi = \frac{\partial\mathcal{L}}{\partial\phi^{s;i}}\Big|_{\partial\phi}. \tag{12.3}$$

Note that the Euler-Lagrange equations of motion are formally the same, independent of whether one considers spacetime to consist of a single continuum, or as separate space and time continua. In the nonrelativistic case, the space and time components of the metric tensor simply split. Let

$$\partial_\mu = \begin{bmatrix} \dfrac{1}{c}\dfrac{\partial}{\partial t}, & \mu = 0 \\[2mm] \dfrac{\partial}{\partial x^j}, & \mu = j = 1, 2, 3 \end{bmatrix}. \tag{12.4}$$

Then, in the nonrelativistic case, Equation (12.3) separates in to separate time and space contributions

$$\frac{\partial}{\partial t}\frac{\partial\mathcal{L}}{\partial(\partial\phi^{s;i}/\partial t)} + \frac{\partial}{\partial x^j}\frac{\partial\mathcal{L}}{\partial(\partial\phi^{s;i}/\partial x^j)} = \frac{\partial\mathcal{L}}{\partial\phi^{s;i}}. \tag{12.5}$$

Note that the factor of c, introduced for the convenience of using the same units for time and space, cancels.

Since the fields may be components of a tensor, the ordinary derivatives $\partial_\mu \phi^{s:i}$ are not covariant tensors in general, and only gauge-independent covariant quantities can properly be considered as candidates for physical observables. However, since the Lagrangian itself is a scalar quantity, the partial derivatives only enter in covariant combinations of the space (or space-time) differentials. This suggests making the substitution $\partial_\mu \phi^{s:i} \rightarrow D_\mu \phi^{s:i}$ where $\phi^{s:i}$ denotes the components of a tensor of arbitrary rank and D_μ is the covariant derivative defined in Chapter 2.

The modified action principle which now involves only covariant observables is now

$$\delta S = \frac{1}{c} \int_{\mathcal{R}} d^4x \sqrt{\|g\|} \delta \mathcal{L}(\phi^{s:i}, D_\mu \phi^{s:i}, x^\mu) = 0. \tag{12.6}$$

The δ-variations have the following important properties:

1. Every covariant field is varied independent of every other tensor field.
2. The space and time parameters are not varied

$$\delta x^\mu \equiv 0. \tag{12.7}$$

3. The variations of the fields vanish at the enclosing three-dimensional boundary $\partial \mathcal{R}$ of the region

$$\delta \phi|_{\partial \mathcal{R}} = 0. \tag{12.8}$$

Before continuing, it is useful to define the four auxiliary "conjugate" current fields associated with a given field

$$\pi^\mu_{s:i} = \frac{\partial \mathcal{L}}{\partial(D_\mu \phi^{s:i})}\bigg|_\phi. \tag{12.9}$$

Sometimes the time-like component of the conjugate fields π^0 is singled out for special attention, but in the present treatment all four partial derivatives will be considered as having equal standing. The δ-variation of the Lagrangian density gives

$$\begin{aligned}
\delta \mathcal{L} &= \frac{\partial \mathcal{L}}{\partial \phi^{s:i}}\bigg|_{D\phi} \delta \phi^{s:i} + \frac{\partial \mathcal{L}}{\partial(D_\mu \phi^{s:i})}\bigg|_\phi \delta(D_\mu \phi^{s:i}) \\
&= \frac{\partial \mathcal{L}}{\partial \phi^{s:i}}\bigg|_{D\phi} \delta \phi^{s:i} + \pi^\mu_{s:i}\big|_\phi \delta(D_\mu \phi^{s:i}).
\end{aligned} \tag{12.10}$$

It is necessary to prove that virtual variation is interchangeable with covariant differentiation. This requires defining the δ variation more precisely. All well-behaved, differentiable and small, variations about the actual solution are allowed. It is useful to define the varied fields in terms of a small scalar expansion parameter α

$$\phi^{s:i}(x, \alpha) = \phi^{s:i}(x, 0) + \alpha \eta^{s:i}(x), \tag{12.11}$$

where $\phi(x, 0)$ is the actual solution to the system being studied, $\eta(x)$ is any continuous and differentiable function of the spacetime coordinates, and α is a scalar parameter. In the limit $\alpha \to 0$ (no variation) one recovers the desired solution

$$\lim_{\alpha \to 0} \phi^{s:i}(x, \alpha) = \phi^{s:i}(x, 0). \tag{12.12}$$

With this clarification, it is straightforward to show that variations and partial derivatives can be interchanged. Using Cartesian coordinates, so that $D_\mu = \partial_\mu$, it is trivial to show, using Equation (12.11), that

$$\partial_\mu(\delta\phi^{s:i}) = \delta(\partial_\mu\phi^{s:i}) = \partial_\mu\eta^{s:i}\delta\alpha. \tag{12.13}$$

This "pass-though" feature hold true as well for a general tensor basis using covariant differentiation, since in a variation of $D_\mu\phi^{s:i}$, the Christoffel symbols are not varied (or, more precisely, they are independently varied from the alpha variations) because the metric tensor is considered an independent field. Therefore, in general

$$D_\mu(\delta\phi^{s:i})|_{g^{\alpha\beta}} = \delta(D_\mu\phi^{s:i})|_{g^{\alpha\beta}} = D_\mu\eta^{s:i}\delta\alpha. \tag{12.14}$$

For Equation (12.10) one can now substitute

$$\delta\mathcal{L} = \frac{\partial\mathcal{L}}{\partial\phi^{s:i}}\delta\phi^{s:i} + \pi^\mu{}_{s:i}D_\mu(\delta\phi^{s:i}). \tag{12.15}$$

Recollect that covariant differentiation satisfies Leibnitz's rule

$$D_\mu(AB) = D_\mu(A)B + AD_\mu(B). \tag{12.16}$$

Returning to the varied action given by Equations (12.6) and (12.10), one can now integrate it by parts in a covariant manner

$$\delta S = \frac{1}{c}\int_{\mathcal{R}} d^4x\sqrt{\|g\|}\delta\mathcal{L} = \int_{\mathcal{R}} d^4x\sqrt{\|g\|}\left(\frac{\partial\mathcal{L}}{\partial\phi^{s:i}}\delta\phi^{s:i} + \pi^\mu_{s:i}D_\mu(\delta\phi^{s:i})\right)$$

$$= \int_{\mathcal{R}} d^4x\sqrt{\|g\|}\left(\left(\frac{\partial\mathcal{L}}{\partial\phi^{s:i}} - (D_\mu\pi^\mu_{s:i})\right)\delta\phi^{s:i}\right) + \int_{\mathcal{R}} d^4x\sqrt{\|g\|}D_\mu(\pi^\mu_{s:i}\delta\phi^{s:i})$$

$$= \int_{\mathcal{R}} d^4x\sqrt{\|g\|}\left(\frac{\partial\mathcal{L}}{\partial\phi^{s:i}} - (D_\mu\pi^\mu_{s:i})\right)\delta\phi^{s:i} + \oint_{\partial\mathcal{R}} \sqrt{\|g\|}(d^3x)_\mu\pi^\mu_{s:i}\delta\phi^{s:i}. \tag{12.17}$$

where the four-dimensional divergence theorem[1] is used to get the three-dimensional surface integral in the final step. This surface integral vanishes identically, since the variations are required to vanish at the boundary of the region.

The variation now takes the form

$$\delta S = \int_{\mathcal{R}} d^4 x \sqrt{\|g\|} \left(\left. \frac{\partial \mathcal{L}}{\partial \phi^{s:i}} \right|_{D\phi} - D_\mu \pi_{s:i}^\mu \right) \delta \phi^{s:i} = 0, \qquad (12.18)$$

where the quantity

$$\frac{\delta \mathcal{L}}{\delta \phi^{s:i}} = \left(\left. \frac{\partial \mathcal{L}}{\partial \phi^{s:i}} \right|_{D\phi} - D_\mu \pi_{s:i}^\mu \right) \qquad (12.19)$$

defines the *functional derivative* of the Lagrangian with respect to a field $\phi^{s:i}$.

Since every field is independently varied, by linear independence, the only way to satisfy Equation (12.18) is if the coefficients of all of the variations individually vanish, giving the explicitly covariant form of the Euler-Lagrange equations of motion

$$D_\mu \pi_{s:i}^\mu = D_\mu \left. \frac{\partial \mathcal{L}}{\partial (D_\mu \phi^{s:i})} \right|_\phi = \left. \frac{\partial \mathcal{L}}{\partial \phi^{s:i}} \right|_{D\phi}, \qquad (12.20)$$

where it is convenient to make use of the auxiliary fields defined in Equation 12.9.

The usual derivation of the Euler-Lagrange equation given by Equation 12.3 involves using $(\partial_\mu \phi, \phi)$ as independent variables so the results are not explicitly tensor invariant, but can be made so by noting[2]

$$\left. \frac{\partial \mathcal{L}}{\partial \phi^{s:i}} \right|_{\partial \phi} - \partial_u \left. \frac{\partial \mathcal{L}}{\partial (\partial_\mu \phi^{s:i})} \right|_\phi = \left. \frac{\partial \mathcal{L}}{\partial \phi^{s:i}} \right|_{D\phi} - D_u \left. \frac{\partial \mathcal{L}}{\partial (D_\mu \phi^{s:i})} \right|_\phi. \qquad (12.21)$$

Therefore, the two variational methods are equivalent and lead to the same equations of motion. This result also follows immediately from the replacement rule, a consequence of the equivalence principle [Misner 1973: 387]. The *replacement rule* states that the laws of physics, written in component form, change on passage from flat spacetime [and Euclidean coordinates] to

---

[1] $\int_{\mathcal{R}} d^4 x \sqrt{\|g\|} D_\mu V^\mu = \int_{\mathcal{R}} d^4 x \partial_\mu (\sqrt{\|g\|} V^\mu) = \oint_{\partial \mathcal{R}} \sqrt{\|g\|} d^3 x_\mu V^\mu$.
[2] The proof is trivial in flat spacetime. Simply work in the Cartesian representation, the affine connection then vanishes, and Equation 12.21 is identically satisfied.

curved spacetime [and generalized coordinates] by a mere replacement of partial derivatives by covariant derivatives

$$\partial_\mu \to D_\mu. \tag{12.22}$$

Suppressing indices, the equations can be expressed in the following simple form, which is reminiscent of Lagrange's equations of motion for particle systems

$$D_\mu \pi^\mu = \frac{\partial \mathcal{L}}{\partial \phi}\bigg|_{D\phi}, \tag{12.23}$$

where

$$\pi^\mu = \frac{\partial \mathcal{L}}{\partial D_\mu \phi}\bigg|_\phi. \tag{12.24}$$

12.1.2 Legendre Transformation of the Field Equations

The conjugate set of fields (π^μ, ϕ) play a role in linearizing the differential equations of motion, similar to the role played by the conjugate pairs (p, q) in the Hamiltonian formulation of particle mechanics. They can be used to generate a set of coupled first-order differential equations to be numerically integrated. Typically, the equations of motion are, at most, quadratic in the derivatives. One wants to put the equations into canonical form. Making a Legendre transform, one gets

$$\mathcal{L}(\pi^\mu, D_\mu\phi, \phi) = \langle \pi^\mu D_\mu\phi \rangle - \mathcal{H}(\pi^\mu, \phi), \tag{12.25}$$

where all indices (space and internal) are contracted over. $\mathcal{H} = \langle \mathcal{H} \rangle$ is a scalar field contracted over space and internal degrees of freedom that is independent of any derivatives. It can be written in the form

$$\mathcal{H}(\pi^\mu, \phi) = \langle \pi^\mu D_\mu\phi - \mathcal{L} \rangle. \tag{12.26}$$

The derivative contributions to $\mathcal{H}$ can be eliminated by inversion of Equation 12.9. The covariant variation of Equation 12.26 results in the canonical form of the Euler-Lagrange equations of motion

$$D_\mu \pi^\mu_{s:i} = -\frac{\partial \mathcal{H}}{\partial \phi^{s:i}}\bigg|_\pi, \qquad D_\mu \phi^{s:i} = \frac{\partial \mathcal{H}}{\partial \pi^\mu_{s:i}}\bigg|_\phi. \tag{12.27}$$

Suppressing indices, these equations become

$$D_\mu \pi^\mu = -\frac{\partial \mathcal{H}}{\partial \phi}\bigg|_\pi, \qquad D_\mu \phi = \frac{\partial \mathcal{H}}{\partial \pi^\mu}\bigg|_\phi, \tag{12.28}$$

which is suggestive of the skew-symmetric structure of Hamilton's equations of motion.

12.1.3 Noether's Theorem

Noether's theorem for particles states that there is a conserved charge associated with any continuous symmetry of a particle Lagrangian. When applied to the field equations derived from Lagrangian densities, Noether's theorem for continuous fields states that any continuous symmetry of a Lagrangian field density implies the existence of a corresponding conserved current density. Consider a Lagrangian of the form $\mathcal{L}(\phi, D_\mu \phi)$, which depends only on fields and covariant derivatives. There is no explicit dependence on the spacetime coordinates. Suppose also that the Lagrangian is invariant under a continuous variation of the fields with an expansion parameter ε. Consider first the case where there is no explicit dependence of the spacetime coordinates on ε. The variation of the action results in a variation of the Lagrangian that leaves the Lagrangian itself unchanged

$$\delta S = \int d^4 x \sqrt{\|g\|} \delta \mathcal{L}, \tag{12.29}$$

$$\mathcal{L}(x, 0) \rightarrow \mathcal{L}(x, \varepsilon) = \mathcal{L}(x, 0), \tag{12.30}$$

$$\phi^{s:i}(x, 0) \rightarrow \phi^{s:i}(x, \varepsilon) = \phi^{s:i}(x, 0) + \varepsilon G^{s:i}, \tag{12.31}$$

where connection coefficients $G^{s:i}$ generate the transformation. To maintain Lorentz invariance, the transformation is restricted to intermixing only space coordinates to other space coordinates and internal coordinates to internal coordinates. The variation of the Lagrangian with respect to the parameter ε gives (suppressing contracted indices)

$$\frac{\delta \mathcal{L}}{\delta \varepsilon} = \frac{\partial \mathcal{L}}{\partial \phi} \frac{\delta \phi}{\delta \varepsilon} + \frac{\partial \mathcal{L}}{\partial D_\mu \phi} \frac{\delta (D_\mu \phi)}{\delta \varepsilon}, \tag{12.32}$$

since the Lagrangian does not depend explicitly on the parameter ε. Using the Euler-Lagrange Equations 12.23 and 12.24, the preceding expression can be rewritten as

$$\frac{\delta \mathcal{L}}{\delta \varepsilon} = (D_\mu \pi^\mu) \frac{\delta \phi}{\delta \varepsilon} + \pi^\mu \frac{\delta (D_\mu \phi)}{\delta \varepsilon}. \tag{12.33}$$

Variation and differentiation pass through each other, therefore

$$\frac{\delta \mathcal{L}}{\delta \varepsilon} = (D_\mu \pi^\mu) \frac{\delta \phi}{\delta \varepsilon} + \pi^\mu \frac{D_\mu (\delta \phi)}{\delta \varepsilon}, \tag{12.34}$$

resulting in

$$\frac{\delta \mathcal{L}}{\delta \varepsilon} = D_\mu \left(\pi^\mu \frac{\delta \phi}{\delta \varepsilon} \right). \tag{12.35}$$

Therefore, assuming the Lagrangian is invariant under the transformation, such that $\delta \mathcal{L}|_{\varepsilon=0} = 0$, one gets a conserved current density

$$D_\mu \left(\pi^\mu \frac{\delta \phi}{\delta \varepsilon} \right)_{\varepsilon=0} = D_\mu J_\varepsilon^\mu |_{\varepsilon=0} = 0. \tag{12.36}$$

where current density is defined as

$$J_\varepsilon^\mu = \left(\pi^\mu \frac{\delta \phi}{\delta \varepsilon} \right) = \frac{\partial \mathcal{L}}{\partial D_\mu \phi} \frac{\delta \phi}{\delta \varepsilon}. \tag{12.37}$$

12.1.4 Stress-Energy Tensor

Consider now the more complicated case where the transformation results in a translation of the spacetime coordinates

$$x^\mu \to x'^\mu = x^\mu + \varepsilon^\mu. \tag{12.38}$$

The fields transform as

$$\phi(x) \to \phi(x + \varepsilon) \approx \phi(x) + \frac{\partial \phi}{\partial x^\mu} \bigg|_{\varepsilon=0} \frac{\delta x^\mu}{\delta \varepsilon^\nu} \delta \varepsilon^\nu$$
$$= \phi(x) + \delta \varepsilon \cdot \nabla \phi. \tag{12.39}$$

Partial derivatives can be replaced with covariant differentiation when applied to tensor invariants, therefore the functional derivative can be written as

$$\frac{\delta \phi}{\delta \varepsilon^\nu} = g_\nu^\mu (D_\mu \phi|_{\varepsilon=0}) \tag{12.40}$$

Appling this to tensor field components gives

$$\phi^{s:i}(x + \varepsilon) - \phi^{s:i}(x) \approx +D_\mu \phi^{s:i}|_{\varepsilon=0} g_\nu^\mu \delta \varepsilon^\nu. \tag{12.41}$$

This must be true for all fields and is therefore true for the Lagrangian itself. Combining this result with Equation (12.35) gives

$$\frac{\delta \mathcal{L}}{\delta \varepsilon^\nu} = (D_\mu \mathcal{L}) g_\nu^\mu = D_\mu \left(\pi^\mu \frac{\delta \phi}{\delta \varepsilon^\nu} \right). \tag{12.42}$$

or

$$D_\mu \left(\left(\pi^\mu \frac{\delta\phi}{\delta x^\nu} \right) - g^\mu_\nu \mathcal{L} \right) = 0. \tag{12.43}$$

This gives rise to a divergentless second-rank tensor called the *stress-energy tensor*

$$
\begin{aligned}
T^\mu{}_\nu &= \left(\pi^\mu \frac{\delta\phi}{\delta\varepsilon^\nu} \right) - g^\mu_\nu \mathcal{L} \\
&= \frac{\partial\mathcal{L}}{\partial(D_\mu\phi)} D_\nu\phi - g^\mu_\nu \mathcal{L}.
\end{aligned} \tag{12.44}
$$

The volume integral of the stress energy tensor gives rise to a conserved four-vector that one can attempt to identify as the conserved four-momentum of the field

$$p_\nu = \int \sqrt{\|g\|} T^0{}_\nu d^3x, \tag{12.45}$$

When dealing with gauge interactions, care must be taken, however, to assure that the results thus obtained are gauge invariant.

12.1.5 *Spin Angular Momentum Tensor

In the absence of external fields that explicitly depending on the spacetime coordinates, the relativistic Lagrangian must be globally invariant under an overall Lorentz transformation of all the fields. The generalized current associated with this invariance can be identified as a spin angular momentum density. Consider two types of spacetime fields, consisting of tensor fields ϕ and spinor fields ψ. Under a global, but infinitesimal, Lorentz transformation of the form

$$S_L = e^{-\varepsilon^{\alpha\beta}\sigma_{\mu\nu}/2} \approx 1 - \varepsilon^{\alpha\beta}\sigma_{\alpha\beta}/2, \tag{12.46}$$

tensor fields transform by similarity

$$\phi' = S_L \phi \bar{S}_L = \phi - \frac{1}{2}[\varepsilon^{\mu\nu}\sigma_{\mu\nu}, \phi]. \tag{12.47}$$

Tensor fields therefore contribute a term to the spin angular momentum density given by

$$J^\mu_{\alpha\beta}\Big|_{\text{tensor}} = -\frac{1}{2}\left\langle \frac{\partial\mathcal{L}}{\partial D_\mu\phi}[\sigma_{\mu\nu}, \phi] \right\rangle, \tag{12.48}$$

where the sum and scalar projections over all component degrees of freedom is implied.

Spinor fields, on the other hand, come in dual conjugate pairs $(\bar{\psi}, \psi)$ which transform as

$$\psi' = S_L \psi, \quad \bar{\psi}' = \bar{\psi} \bar{S}_L, \tag{12.49}$$

Such that bilinear product of the two fields $\psi \bar{\psi}$ forms a tensor *density matrix*. These spinor fields contribute a term to the total spin angular momentum current density given by

$$J^\mu_{\alpha\beta}\Big|_{\text{spinor}} = -\frac{1}{2} \left\langle \frac{\partial \mathcal{L}}{\partial D_\mu \psi} \sigma_{\mu\nu} \psi - \bar{\psi} \sigma_{\mu\nu} \frac{\partial \mathcal{L}}{\partial D_\mu \bar{\psi}} \right\rangle. \tag{12.50}$$

The total spin angular momentum density is a sum over tensor and spinor contributions

$$J^\mu_{\alpha\beta}\Big|_{\text{total}} = J^\mu_{\alpha\beta}\Big|_{\text{tensor}} + J^\mu_{\alpha\beta}\Big|_{\text{spinor}}. \tag{12.51}$$

12.2   NON RELATIVISTIC WAVE EQUATION

A common phenomena in continuous medium is wave motion. Waves are carriers of energy and momentum. Waves allow a disturbance in part of a medium to propagate rapidly to other parts of a medium, while the bulk of the medium remains nearly stationary. The mechanism works by nearest neighbor interactions. In the continuum limit, the nearest neighbor interaction is replaced by a derivative coupling. When a hammer is tapped against a long pipe, the individual molecules of which the pipe is composed barely move. Nevertheless, the acoustical disturbance generated is rapidly transmitted down the pipe, moving at the speed of sound for the given medium.

Consider the Lagrangian density of the vibrating string, which was derived in Chapter 6 as the continuum limit of a linear oscillator lattice structure

$$\mathcal{L} = \frac{\mu}{2} \left[-\left(\frac{\partial y}{\partial t}\right)^2 + v^2 \left(\frac{\partial y}{\partial x}\right)^2 \right]. \tag{12.52}$$

The velocity of propagation $v = \sqrt{T/\mu}$ is a function of the string's linear mass density μ and tension T. Assume a constant density and velocity. Evaluating the Euler-Lagrange equation for this Lagrangian density yields the wave equation

$$\left(\frac{1}{v^2} \frac{\partial^2 y}{\partial t^2} - \frac{\partial^2 y}{\partial x^2} \right) = 0. \tag{12.53}$$

This non-relativistic wave equation should be interpreted as the small velocity limit of a relativistic invariant equation, valid in the rest frame of the clamped string. One possible relativistic extension is to introduce the four-velocity of the laboratory frame of the stretched string $u_{\text{lab}} = (c, \vec{0})$, and use it to construct the invariant form

$$\left(\nabla^2 - \left(\frac{\kappa^2}{c^2} \right) (u_{\text{lab}} \cdot \nabla)^2 \right) y = 0, \tag{12.54}$$

where $\kappa^2 = 1 + c^2/v^2$. The actual relativistic tensor form chosen depends on the assumed dynamical model.

12.2.1 Standing Wave Modes of the Vibrating String

Assume that the non-relativistic string has length L and is clamped to zero amplitude at both ends. Then the normal standing wave modes for the vibrating string can be found by separation of variables

$$y(x, t) = y_n(x)e^{-i\omega_n t}. \tag{12.55}$$

The eigenfunctions $y_n(x)$ are solutions to the eigenvalue equation

$$\frac{d^2 y_n(x)}{dx^2} = -\frac{\omega_n^2}{v^2} y_n(x), \tag{12.56}$$

subject to Direchlet boundary conditions on the end points of the string

$$y_n(0) = y_n(L) = 0. \tag{12.57}$$

The eigenfunctions provide a complete orthogonal expansion basis

$$\int_0^L y_n y_m dx = \delta_{nm}, \tag{12.58}$$

having normalized solutions

$$y_n(x) = \sqrt{\frac{2}{L}} \sin\left(n\pi \frac{x}{L} \right) \tag{12.59}$$

where the eigenfrequencies satisfy the dispersion relation

$$\omega_n = vk_n = vn\pi/L. \tag{12.60}$$

The complete solution to the wave equation is a superposition of all possible normal modes. One can use Euler's relation $e^{\pm i\omega t} = \cos \omega t \pm i \sin \omega t$ to find the real-valued solutions. The general solution results in a Fourier Sine Series for the normal modes

$$y(x, t) = \sum_{n=1}^{\infty} \sqrt{\frac{2}{L}} \sin\left(n\pi \frac{x}{L}\right)(a_n \cos(\omega_n t) + b_n \sin(\omega t)). \qquad (12.61)$$

which can be inverted to get the series coefficients in terms of the initial conditions

$$a_n = \int_0^L \sqrt{\frac{2}{L}} \sin\left(n\pi \frac{x}{L}\right) y(x, t)\Big|_{t=0}, \qquad (12.62)$$

$$b_n = \frac{1}{\omega_n} \int_0^L \sqrt{\frac{2}{L}} \sin\left(n\pi \frac{x}{L}\right) \frac{\partial y(x, t)}{\partial t}\Big|_{t=0}, \qquad (12.63)$$

There are two sets of coefficients to be determined, since the initial value problem for the wave equation requires the specification of both the function and its first time derivative at some initial time. This is referred to as *Cauchy boundary conditions*.[3]

Consider, for example, a guitar string of length L. The string has a tension T and linear mass density μ resulting in a wave velocity $v^2 = T/\mu$. The string's response to being plucked is a linear superposition over all its allowed normal modes of oscillation. Suppose the string is plucked in its exact center at time $t = 0$. The effect of the plucking results in a triangular displacement of the string with an amplitude h at the center at the moment the string is released, as shown in Figure 12.1. The initial Cauchy boundary conditions are

$$y(x, t)\Big|_{t=0} = \begin{cases} 2hx/L & 0 \le x \le L/2 \\ 2h - 2hx/L & L/2 \le x \le L \end{cases}, \qquad (12.64)$$

$$\frac{\partial y}{\partial t}(x, t)\Big|_{t=0} = 0. \qquad (12.65)$$

The initial conditions require that the time behavior be an even function of time (a cosine series in time). Only those modes that are symmetric about the midpoint contribute. Therefore, only odd normal modes contribute, since

---

[3]Cauchy boundary conditions are named for mathematician Augustin Louis Cauchy (1789–1857) who rigorously formulated the theorems of calculus and contributed a number of important proofs in complex and real analysis.

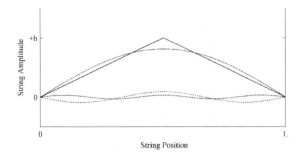

FIGURE 12.1 The plucked guitar string is displaced from equilibrium and released. The result is an initial triangular wave form that can be expanded as a sum over all contributing normal modes (relative contributions from the first four non-zero modes are shown).

these are the modes that are symmetric about the midpoint, giving

$$y(x,t) = \sum_{n=1,3,\ldots}^{\infty} a_n \sin \frac{n\pi x}{L} \cos \frac{n\pi v}{L} t, \tag{12.66}$$

At time $t = 0$, the solution must satisfy

$$y(x,0) = \sum_{n=1,3,\ldots}^{\infty} a_n \sin \frac{n\pi x}{L}, \tag{12.67}$$

with solution $a_n = (-1)^{(n-1)/2} \frac{8h}{(n\pi)^2}$ for odd n, or

$$y(x,t) = 8h \sum_{n=1,3,\ldots}^{\infty} \frac{(-1)^{(n-1)/2}}{(n\pi)^2} \sin \frac{n\pi x}{L} \cos \frac{n\pi v}{L} t, \tag{12.68}$$

The energy density of the wave can be calculated from the stress-energy tensor (Equation 12.44) giving

$$T^{00} = \frac{\mu}{2} \left(\left(\frac{\partial y}{\partial t} \right)^2 + v^2 \left(\frac{\partial y}{\partial x} \right)^2 \right). \tag{12.69}$$

The energy integral over the length of the string is a sum of the individual energies of the normal modes, which falls off as n^{-2}.

$$E = mv^2 \sum_{\text{odd}} n \left(\frac{4h}{n\pi l} \right)^2, \quad m = \mu L. \tag{12.70}$$

Figure 12.2 shows the integrated energy contained in each of the normal modes of oscillation.

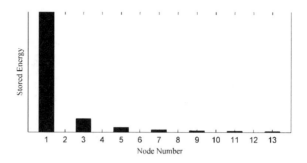

FIGURE 12.2   The energy stored in the normal modes of the plucked string falls off as n^{-2}.

12.2.2 Localized Wave Packets

If the string is very long, its ends can often be ignored and the string taken to be infinite. In this case the discrete normal modes of oscillation are replaced by a continuous number of frequencies. Mathematically, the Fourier series is replaced by a Fourier integral. A localized wave packet (with amplitude going to zero at $x \to \pm\infty$) can be expressed as an integral over all wave numbers, summed over right and left travelling wave packets

$$y(x,t) = \phi_+(x - vt) + \phi_-(x + vt)$$
$$= \frac{1}{\sqrt{2\pi}} \int_{-\infty}^{\infty} (g_+(\omega)e^{ik(x-vt)} + g_-(\omega)e^{ik(x+vt)})dk, \qquad (12.71)$$

where $k^2 v^2 = \omega^2$. The Fourier integral is an invertible transformation satisfying the normalization condition

$$\int_{-\infty}^{\infty} e^{ik(x-x')}dx' = 2\pi \delta(x - x'). \qquad (12.72)$$

This yields the right and left traveling wave-number amplitudes

$$g_\pm(k) = \frac{1}{\sqrt{2\pi}} \int_{-\infty}^{\infty} \phi_\pm(x' \mp vt)e^{-ik(x' \mp vt)}dx'. \qquad (12.73)$$

12.2.3 Transmission and Reflection of Waves at a Boundary

When a wave hits a boundary between two media, it is partially reflected and partially transmitted depending on the properties of the boundary interface. For a travelling wave on a string, a sudden change of the string density at the interface between two fused strings with a common tension results in a change of the wave velocity $v_{1,2} = \sqrt{T/\mu_{1,2}}$. The reflected and transmitted waves can be calculated by assuming that the wave amplitude and its first derivative are continuous at the interface. Consider an incident traveling wave $f_{+(in)}(x - vt)$ moving from left to right approaching an interface at $x = 0$. Originally, the wave is localized on the left of the interface, but after scattering from the boundary transmitted and reflected wave are observed.

The amplitudes of transmission and reflection are frequency dependent. Therefore, it is best to analyze this problem by considering infinite plane waves of fixed frequency. Let the incoming wave have unit magnitude, then the boundary conditions require

$$e^{i(k_1 x - \omega t)} + Re^{i(-k_1 x - \omega t)}\Big|_{x=0-\varepsilon} = Te^{i(k_2 x - \omega t)}\Big|_{x=0+\varepsilon} \qquad (12.74)$$

and

$$ik_1 e^{i(k_1 x - \omega t)} - ik_1 Re^{i(-k_1 x - \omega t)}\Big|_{x=0-\varepsilon} = ik_2 Te^{i(k_2 x - \omega t)}\Big|_{x=0+\varepsilon} \qquad (12.75)$$

where the limit $\varepsilon \to 0$ is taken and

$$k_1 = \omega/v_1,$$
$$k_2 = \omega/v_2.$$

The reflection and transmission coefficients can be complex in general. Eliminating the time dependence and solving for them gives

$$1 + R = T,$$
$$1 - R = \frac{v_1}{v_2}T, \qquad (12.76)$$

or

$$T = \frac{2}{1 + \frac{v_1}{v_2}}, \qquad (12.77)$$

$$R = \frac{\frac{v_1}{v_2} - 1}{1 + \frac{v_1}{v_2}}, \qquad (12.78)$$

The transmitted wave always has the same phase as the incident wave, but the phase of the reflected wave depends on the relative wave velocities in the two media. When a wave moves towards a medium where its wave velocity is higher than the first ($v_2 > v_1$), the phase of the reflected wave is inverted, while if the velocity is smaller in the second medium ($v_2 < v_1$), the refracted wave has the same phase as the initial wave.

12.2.4 Wave Equation in Three Space Dimensions

The wave equation for a complex field ψ in three-space dimensions is similar to the solution for a vibrating string. Consider a field required to vanish on the boundary ∂V of the three-dimensional volume V (Dirichlet[4] boundary conditions). Inside the boundary, the field satisfies the wave equation

$$\left(\frac{1}{v^2} \frac{\partial^2}{\partial t^2} - \vec{\nabla}^2 \right) \psi(\vec{x}, t) = 0. \tag{12.79}$$

This second order partial differential equation for ψ is formally real, so the same method of analysis can be used for real or complex fields, with the proviso that functions defined over the field of real numbers involve the projection of the real part of a complex solution. The complex version of the field equations, $\psi \in \mathbb{C}$, can be derived by variation of the Lagrangian density

$$\mathcal{L} = -\frac{1}{v^2} \frac{\partial \psi^*}{\partial t} \frac{\partial \psi}{\partial t} + \psi^* \overleftarrow{\nabla} \cdot \vec{\nabla} \psi, \tag{12.80}$$

where complex conjugate field ψ^* is the dual to the field ψ. Treating ψ and ψ^* as independent fields, they can be varied separately giving

$$\frac{1}{v^2} \frac{\partial^2 \psi}{\partial^2 t} - \vec{\nabla}^2 \psi = 0, \tag{12.81}$$

and

$$\frac{\partial^2 \psi^*}{\partial^2 t} \frac{1}{v^2} - \psi^* \overleftarrow{\nabla}^2 = 0, \tag{12.82}$$

where the second equation is recognized as the complex-conjugate reversion of the first.

Any function of the form $\psi(x \pm vt)$ gives rise to a dispersionless traveling wave solution to the wave equation. The wave equation, in general, has both longitudinal and transverse modes of oscillation. The two transverse modes give rise to linearly and circularly polarized waves.

The standing-wave normal mode solutions are found by separation of variables

$$\psi_n(\vec{x}, t) = \psi_n(\vec{x}) e^{i\omega_n t}. \tag{12.83}$$

---

[4]Dirichlet boundary conditions are named for mathematician Peter Direchlet (1805–1859) who determined the conditions under which Fourier series converge and who developed many other elegant proofs.

where the time-independent normal modes are solution to the *Helmholtz equation*,[5]

$$\vec{\nabla}^2 \psi_n(\vec{x}) = -k_n^2 \psi_n(\vec{x}) = -\frac{\omega_n^2}{v^2} \psi_n(\vec{x}) \qquad (12.84)$$

subject to Dirichlet boundary conditions on a closed boundary surface

$$\psi_n(\vec{x})|_{\partial V} = 0. \qquad (12.85)$$

The boundary conditions, when defined on a finite closed region of space, are only satisfied for certain discrete values of the eigenfrequencies.

The normal modes can be required to form an orthonormal expansion basis in three-space

$$\int_V d^3x \, \psi_n^*(\vec{x}) \psi_m(\vec{x}) = \delta_{mn}. \qquad (12.86)$$

The solution to the Helmholtz equation can be found by separation of variables, and, in that regard, the techniques used are similar to those needed to solve Hamilton's characteristic equation. A fuller explanation of techniques needed to solve the wave equation in multiple space dimensions go beyond the intent of this chapter, but most advanced students should be aware of them from their previous study of quantum mechanics and electromagnetic fields.

12.3 NON-LINEAR SINE-GORDON EQUATION

An example of an interesting nonlinear equation that has been well-analyzed is the *Sine-Gordon equation*. It is one of a class of nonlinear equations that are analytically solvable. The Sine-Gordon equation[6] represents a variant of the Klein-Gordon equation with the constant mass term replaced by a non-linear potential with multiple periodic minima. The equation is often expressed in unspecified reduced dimensionless units, which is acceptable,

---

[5] Hermann Ludwig Ferdinand von Helmholtz (1821–1894) was a philosopher, physician, and physicist. As a physicist, he was known for his work in electrodynamics and thermodynamics. In physiology, he is known for his studies of the eye and theories of vision. In philosophy, he made important contributions to the theory of empiricism, the relationship between the laws of perception and the laws of nature, the science of aesthetics, and for his ideas on the civilizing power of science.

[6] A pun on the name Klein-Gordon.

since it is a theoretical construct, used only for pedagogical purposes. This equation gives rise to soliton solutions and to a quantized topological charge.

A *soliton* is a self-reinforcing solitary wave that maintains its shape as it propagates. They were first described by railroad engineer John Scott Russell (1808–1882) who observed a solitary wave in the Union Canal near Edinburgh, Scotland. Russell later reproduced the effect in a wave tank, and called it the "wave of translation." Solitons arise from a delicate balance of dispersive and nonlinear effects. Drazin and Johnson [Drazin 1989] describe solitons as solutions of nonlinear differential equations which represent localized waves of permanent form that can interact strongly with other solitons, but emerge from the collision unchanged apart from a phase shift. Topological solitons, or *topological defects*, represent solutions that that are stable against decay to a trivial solution due to topological constraints. These arise due to boundary conditions, where the boundary has a non-trivial *homotopy group*, which is preserved by the differential equations of motion. Since there are no continuous transformations that map different *homotopy classes* into each other, the solutions maintain their integrity even in the presence of strong interactions.

The Sine-Gordon equation can be written as

$$\Box \phi = k^2 \sin \phi. \tag{12.87}$$

It can be derived from a Lagrangian density of the form

$$\mathcal{L} = \rho \left(\frac{1}{2} g^{\mu\nu} \partial_\mu \phi \partial_\nu \phi + k^2 (1 - \cos \phi) \right), \tag{12.88}$$

where ρ is a normalization constant. The equation is Lorentz invariant under velocity boosts and rotations. Given a solution, a continuous family of related Lorentz-transformed solutions can be generated.

In one-space and one-time dimension, the Sine-Gordon equation can be conceived as the continuum limit of a linear array of pendula coupled to their nearest neighbors by springs, as illustrated in Figure 12.3.[7] The individual pendula rotate and/or oscillate as a wave propagates down the array. The amplitude of the wave is a cyclical variable. The constant zero-energy vacuum state therefore has multiple minima for $\phi = 0$, modulo 2π (see Figure 12.4).

The linear systems of equations studied previously all satisfy the *superposition principle*. Any two solutions can be scaled by constant multiplicative factors and added to get a third solution. A large number of theoretical tools exist to handle such systems. Linear systems have their limitations, however.

---

[7]See [José 1998, 553–556] for an explicit derivation of the Sine-Gordon equation as the continuum limit of linear array of spring-coupled pendula.

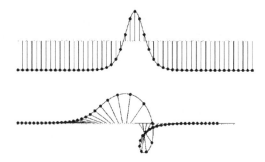

FIGURE 12.3   The Sine-Gordon equation is the continuous limit of a linear array of spring-coupled pendula in a uniform gravitational field. The amplitude of the wave is related to the rotation angles of the pendula. Solitons are solitary waves of fixed magnitude and shape introducing kinks in the lattice structure with integral winding number.

For example, there is no way to quantize the norm of a linear system except by fiat. Ultimately, real physical systems are non-linear, except in special limits, and may exhibit behavior that defies linear analysis techniques.

The study of non-linear dynamics, however, poses new challenges. There are few theorems predicting when, and under what circumstances, one can expect solutions to exist. The topology and stability of such equations is not well understood at present. The equations may have regions of chaotic behavior and non-perturbative solutions. Moreover, the coupling of nonlinear fields can indirectly give rise to non-local effects.

12.3.1 Solitons

The Sine-Gordon equation is solvable using the *inverse scattering transform* method [Dodd 1982]. Single soliton solutions can be found by assuming the

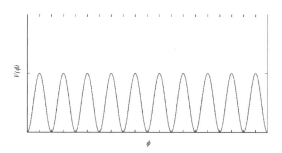

FIGURE 12.4   The Sine-Gordon equation has a vacuum energy state with multiple minima separated by a factor of 2π. Note that, depending on the local energy density, both librational and rotational modes are available to the pendula.

existence of traveling wave solutions with velocity v of the form $\phi(k(x - vt))$. With this ansatz, the Sine-Gordon equation becomes

$$\gamma^2 \frac{d^2\phi}{du^2} = \sin\phi \qquad (12.89)$$

where $u = k(x - vt)$ and

$$\gamma = 1/\sqrt{1 - v^2/c^2}. \qquad (12.90)$$

is a relativistic Lorentz contraction factor.

This leads the one-soliton or single-kink solutions

$$\phi_{\pm} = 4\arctan\exp\left(\pm k\gamma(x - vt)\right), \qquad (12.91)$$

Due to non-linearity, the solutions have a unique normalization. The solutions represent a traveling twist in the array localized at $x = vt$, where the amplitude function ϕ is rotated by a net factor of 2π. The positive root of Equation (12.91) gives a clockwise rotation and is referred to as a soliton or *kink* solution. The phase amplitude approaches 0 (modulo 2π) at $(x - ct) \to -\infty$ and increases by a single twist of 2π at $(x - ct) \to +\infty$. The negative root corresponds to a counterclockwise rotation and is referred to as an antisoliton or *antikink* solution. Figure 12.5 shows traveling wave kink and antikink solutions moving in the laboratory frame.

Analytic forms of two-soliton scattering solutions can be found by a transformation developed by A.V Bäcklund in 1873, which allows the use of known solutions to generate new solutions. An explanation of the Bäcklund transform can be found in [Jose 1998, 601–603]. The two soliton kink–kink scattering solution in the center of momentum frame is

$$\phi_{2s} = 4\arctan\left(\beta\sinh\left(\gamma kx\right)\operatorname{sech}\left(\gamma\beta kct\right)\right). \qquad (12.92)$$

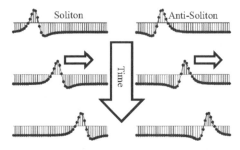

FIGURE 12.5 Traveling solitary soliton (left-hand side) and antisoliton (right-hand side) waves moving from left to right across the page.

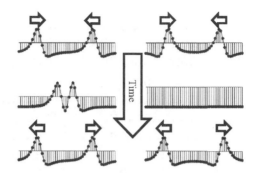

FIGURE 12.6   Scattering of two solitons (left-hand side) and a soliton-anti-soliton pair (right-hand side) shown in the center of momentum frame. In both cases the solitons emerge from the collision intact.

Note that this solution cannot be obtained by simple linear addition of the single-kink solutions. When the two solutions are far apart, they asymptotically approach the single kink limits; when they are close, they interact nonlinearly. The kink–antikink scattering solution in the center of momentum frame is

$$\phi_{s\bar{s}} = 4 \arctan\left(\beta^{-1} \operatorname{sech}(\gamma kx) \sinh(\gamma\beta kct)\right). \qquad (12.93)$$

When two kink solutions collide, they emerge from the scattering region unchanged as shown in Figure 12.6. When solitons collide, they emerged unscathed from the collision. This robustness of the structure under collisions is one of the criteria required for a solitary wave to be considered as a soliton.

Multiple kink solutions exist. A number of other interesting coherent solutions can be found, including a *breather mode* which represents a bound state of a kink-anti-kink solution. The equation for a stationary breather is given by

$$\phi_b = 4 \arctan\left((\beta\gamma)^{-1} \operatorname{sech}(\gamma^{-1}kx) \sin(\beta kct)\right) \qquad (12.94)$$

Figure 12.7 shows a stationary breather mode. Although the central pendula in this construct oscillates about the origin, there is no net twist, nor any net propagation of twist.

12.3.2 Topological Charge and Winding Number

Since the amplitude ϕ of the Sine-Gordon equation can be interpreted as a rotation angle for a lattice of pendula, the single-soliton solutions can be interpreted as an overall rotation of the lattice structure by a factor of 2π.

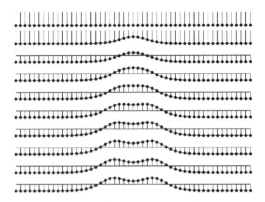

FIGURE 12.7   A stationary breather mode of oscillation: There is no net twist nor any propagation of net twist.

The total *winding number* of a solution is defined as the net number of complete twists in a solution integrated over all space

$$N = \frac{1}{2\pi}(\phi(x, +\infty) - \phi(x, -\infty)) = \frac{1}{2\pi}\int_{-\infty}^{\infty}\frac{\partial\phi}{\partial x}dx = \frac{1}{2\pi}\int_{-\infty}^{\infty}\frac{\partial\mathcal{L}}{\partial(\partial\phi/\partial x)}dx$$
$$(12.95)$$

If one assumes that there is no sources at infinity, then the limits of the amplitudes at infinity should approach a vacuum value of 0, modulo 2π. Thus, the winding number should be an integer. The net twist of a system is a topological invariant. Because of the finiteness of the speed of light, there is no way for a localized twist to propagate to infinity in a finite time; therefore the net twist is conserved.

12.4   ELECTROMAGNETIC WAVES

Maxwell equations in free space can be derived from a standard Lagrangian of the form

$$\mathcal{L} = \frac{1}{8\mu_0}F \cdot F = -\frac{1}{4\mu_0}F_{\mu\nu}F^{\mu\nu},$$
$$(12.96)$$

where the Maxwell bivector $F = \nabla \wedge A$ is given by

$$F_{\mu\nu} = \partial_\mu A_\nu - \partial_\nu A_\mu.$$
$$(12.97)$$

The field can be further expanded in terms of the electric field $\vec{E}$ and magnetic induction field $\vec{B}$ components

$$F_{\mu\nu} = \begin{bmatrix} 0 & -E_1/c & -E_2/c & -E_3/c \\ E_1/c & 0 & B^3 & -B^2 \\ E_2/c & -B^3 & 0 & B^1 \\ E_3/c & +B^2 & -B^1 & 0 \end{bmatrix}. \tag{12.98}$$

The Euler-Lagrange equations of motion in free space reduce to

$$D_\mu F^{\mu\nu} = 0, \tag{12.99}$$

which can also be written as

$$\nabla F = 0 \tag{12.100}$$

and further expanded to give Maxwell's equations in free space

$$\nabla \cdot \vec{E} = 0, \quad \nabla \cdot \vec{B} = 0,$$
$$\vec{\nabla} \times \vec{E} = -\frac{\partial \vec{B}}{\partial t}, \quad \vec{\nabla} \times \vec{B} = \frac{1}{c^2}\frac{\partial \vec{E}}{\partial t}. \tag{12.101}$$

Multiplying Equation 12.100 through from the left by ∇ results in the relativistic wave equation acting on the electromagnetic field tensor

$$\Box F = \nabla^2 F = \left(\vec{\nabla}^2 - \frac{1}{c^2}\frac{\partial}{\partial t^2} \right) F = 0. \tag{12.102}$$

This implies that the electric and magnetic fields in free space are individual solutions to the wave equation with velocity c. One can try substituting oscillatory plane wave solutions of form

$$\vec{E} = \vec{E}_0 e^{i(\vec{k}\cdot\vec{x}-\omega t)},$$
$$\vec{B} = \vec{B}_0 e^{i(\vec{k}\cdot\vec{x}-\omega t)}, \tag{12.103}$$

into Maxwell's free-space equations (Equation 12.101) giving

$$\vec{k} \cdot \vec{E}_0 = 0, \quad \vec{k} \cdot \vec{B}_0 = 0,$$
$$\vec{k} \times \vec{E}_0 = \omega \vec{B}_0, \quad \vec{k} \times \vec{B}_0 = -\frac{\omega}{c^2}\vec{E}_0. \tag{12.104}$$

with the result that the electromagnetic wave is a vector wave, polarized in a plane perpendicular to its direction of motion $\hat{k}$. The electric and magnetic fields are perpendicular to each other and differ in magnitude by a factor of c

$$\vec{E}_0 \cdot \vec{B}_0 = 0, \tag{12.105}$$

$$|\vec{E}_0| = c|\vec{B}_0|. \tag{12.106}$$

The dispersion relation between wave number and angular frequency is

$$\vec{k}^2 c^2 = \omega^2. \tag{12.107}$$

The polarization of the electromagnetic field is given by the orientation direction of the electric field. Both plane-polarized and circularly polarized transverse wave solutions are allowed.

The stress-energy tensor associated with the electromagnetic field can be calculated using the prescription of Equation 12.44. However, this does not lead to a gauge invariant result in this case, yielding instead

$$\left(\frac{\partial \mathcal{L}}{\partial (D_\mu \phi)} D_\nu \phi - g_\nu^\mu \mathcal{L} \right) \rightarrow \frac{-1}{\mu_0} \left(F^{\mu\lambda} D_\nu A_\lambda - \frac{g_\nu^\mu}{4} F^\alpha{}_\beta F^\beta{}_\alpha \right). \tag{12.108}$$

The problem with the derivation lies in the fact that the term $D_\nu A_\lambda$ is not gauge invariant. This difficulty is usually patched up by adding an additional term that restores gauge invariance, but the difficulty could have been avoided entirely, if only variations with respect to the gauge-invariant combination $(F_{\mu\nu} = D_\mu A_\nu - D_\nu A_\mu)$ were allowed to begin with. The correct gauge invariant result is

$$T^\mu{}_\nu = \frac{1}{\mu_0} \left(-F^{\mu\lambda} F_{\nu\lambda} + \frac{g_\nu^\mu}{4} F^{\alpha\beta} F_{\alpha\beta} \right). \tag{12.109}$$

Taking the divergence of this term and using Maxwell's equations gives

$$
\begin{aligned}
D_\mu T^\mu{}_\nu &= \frac{1}{\mu_0} \left(-(D_\mu F^{\mu\lambda}) F_{\nu\lambda} - F^{\mu\lambda}(D_\mu F_{\nu\lambda}) + \frac{1}{2}(D_\nu F_{\alpha\beta}) F^{\alpha\beta} \right) \\
&= \frac{1}{\mu_0} \left(-(D_\mu F^{\mu\lambda}) F_{\nu\lambda} - \frac{1}{2} F^{\mu\lambda}(D_\mu F_{\nu\lambda} + D_\lambda F_{\mu\nu} + D_\nu F_{\lambda\mu}) \right) \\
&= F_{\nu\lambda} J^\lambda, \tag{12.110}
\end{aligned}
$$

which vanishes in free space.

The electromagnetic field is known to come in quantized units of $\hbar = h/2\pi$, where h is Planck's Constant. The photon's spin angular momentum is also quantitized in units of $\hbar$. The quanta of the electromagnetic field are called photons. Other particles have similar quantum features with fermions being quantized in units of $\hbar/2$, and bosons, in general, in units of $\hbar$. The relationships between frequency $\nu = \omega/2\pi$, energy E, wave number

$k = 2\pi/\lambda$, momentum p, and dimensionaless velocity $\beta = v/c$, for quantum systems, are given by

$$E = h\nu = \hbar\omega, \quad p = h/\lambda = \hbar k,$$
$$\beta = pc/E = f\lambda/c = \omega/ck. \tag{12.111}$$

12.4.1 Interaction with Matter

The electromagnetic field interacts with matter through a gauge coupling. The simplest gauge algebra is the Abelian gauge of $Spin(2) = U(1)$, whose generator is often represented by the imaginary i and whose coupling constant e is the fundamental electric unit of charge. The covariant gauge derivative becomes

$$\mathcal{D}_\mu = D_\mu - i\frac{e}{\hbar}A_\mu. \tag{12.112}$$

One can then define spinor fields ψ coupled via the covariant spinor gauge derivative. The covariant gauge derivative $\mathcal{D}_\mu\psi$ is invariant under the combined set of local phase transformations

$$\psi' = e^{ieG/\hbar}\psi, \quad \bar{\psi}' = \bar{\psi}e^{-ieG/\hbar}, \tag{12.113}$$
$$A' = A + \nabla G. \tag{12.114}$$

The total Lagrangian of charged matter coupled to the electromagnetic field is a sum over the electromagnetic and matter Lagrangians. The interaction potential can be explicitly separated out, giving

$$\mathcal{L}_{\text{Total}} = \mathcal{L}_{\text{EM}} + \mathcal{L}_{\text{M}} = \frac{1}{8\mu_0}F \cdot F + (A \cdot J + \mathcal{L}'_{\text{M}}), \tag{12.115}$$

$$\mathcal{L}_{\text{EM}} = \frac{1}{8\mu_0}F \cdot F = \frac{1}{4\mu_0}F^{\mu\nu}F_{\nu\mu}, \tag{12.116}$$

$$J^\mu = \frac{\partial \mathcal{L}_{\text{M}}}{\partial A_\mu}. \tag{12.117}$$

The electromagnetic field Lagrangian is clearly gauge-invariant, as must be the sum of the remaining contributions to the Lagrangian. The electromagnetic field is required to be a solution to the Euler-Lagrange equations of motion, giving the expected result

$$\nabla F = -\mu_0 J.$$

The vector and axial vector components of this equation are

$$\nabla \cdot F = -\mu_0 J \rightarrow F^{\nu\mu}_{,\nu} = -\mu_0 J^\mu, \tag{12.118}$$
$$\nabla \wedge F = 0. \rightarrow \varepsilon_{\alpha\beta\gamma\delta}F^{\alpha\beta\gamma} = 0. \tag{12.119}$$

where the vanishing of the axial-vector component in the second equation is merely a reflection of the fact that the field was derived from a vector potential. Since the scalar projection of an antisymmetric tensor vanishes, the electromagnetic current is divergentless as can be seen from

$$\langle \nabla^2 F \rangle = 0 = -\mu_0 \langle \nabla J \rangle = -\mu_0 \nabla \cdot J.$$

This leads to a conserved charge

$$q_e = \frac{1}{c} \int_{\text{all space}} d^3x \sqrt{\|g\|} J^0.$$

12.4.2 Quantization of Charge

One of the great mysteries of nature is that particles come in quantized units of charge in terms of some natural unit of charge e

$$q_e = Ne. \tag{12.120}$$

This conserved charge can be related to a dimensionless *fine structure constant*, which is defined identically in all sets of units as

$$\alpha = \frac{e^2}{4\pi \varepsilon_0 \hbar c} = \frac{\mu_0 e^2 c^2}{4\pi \hbar c}. \tag{12.121}$$

The 2006 CODATA recommended value for α is given by [Mohr 2007]

$$\alpha^{-1} = 137.035999679(94) \tag{12.122}$$

Why charge is quantitized and why the fine structure constant should take on this precise value remains a mystery, as Richard Feynman, who developed the theory of *Quantum Electrodynamics* (QED), would note:

> "It has been a mystery ever since it was discovered more than fifty years ago, and all good theoretical physicists put this number up on their wall and worry about it. Immediately you would like to know where this number for a coupling comes from: is it related to π or perhaps to the base of natural logarithms? Nobody knows. It's one of the greatest damn mysteries of physics: a magic number that comes to us with no understanding by man."   [Feynman 1985, p. 129]

There is no a priori reason that α must be an absolute constant, and some have argued that it may have evolved over cosmological time scales to its present day value. The most precise experimental results to date, using gravitationally red-shifted absorption spectra from the Very Large Telescope [Chaud 2004], have put stringent limits of the fractional change of α over the last 10–12 billion years at

$$\delta\alpha/\alpha = (-0.6 \pm 0.6) \times 10^{-6}. \tag{12.123}$$

This does not mean that the fine structure constant could not have changed dramatically during the early inflationary period of the universe.

To avoid the fine tuning of nature, some of those that see the fine structure constant as an absolute property of spacetime argue that this may be so for deep topological reasons. An analogy can be made to the Sine-Gordon equation, where the existence of quantized solitons is due to the conservation of a topological winding number.

It should be noted that, within the context of perturbative QED, the charge of an electron (and thus the fine structure constant) is not directly calculable. The parameters that enter into the field equations are unknowns called the *bare charge* and *bare mass*. Infinities, due to radiative corrections, that occur in the perturbation expansion at every level beyond tree level are fixed by a *renormalization* of the bare charge and bare mass of the electron to their measured values, with added radiative corrections. Since α is proportional to e^2, it can be considered to be the square of an effective charge, as screened by the vacuum polarization and viewed from an infinite distance. Why this result should be the same for all systems up to an integer multiplicative factor is beyond simple explanation. In scattering experiments, the fine structure constant develops a running value that depends on the four-momentum transfer at the scattering vertex. The best single value for α currently comes measurements of the anomalous magnetic moment of the electron [Mohr 2007].

12.5   *GRAVITATIONAL FIELD THEORY

The Einstein-Hilbert[8] action [Hilbert 1915] for the gravitational field is

$$S = \int \left(\frac{1}{\kappa}R + \mathcal{L}_{\mathrm{M}} \right) \sqrt{\|g\|} d^4x, \tag{12.124}$$

$$\kappa = \frac{16\pi G}{c^4}, \tag{12.125}$$

where $\|g\| = \mathrm{abs}(\det(g_{\alpha\beta}))$, $\mathcal{L}_{\mathrm{M}}$ is the Lagrangian for matter and gauge fields, R is the Ricci scalar which is the contraction of the Ricci tensor. The Ricci

---

[8]David Hilbert (1862–1943), mathematician and physicist, was responsible for Hilbert Spaces, the axiomatization of geometry, and important work in functional analysis. He deserves a share of the credit for the discovery of General Relativity, although he never claimed one. He and Einstein had collaborated closely in the months leading up to Einstein's final breakthrough in 1915. Hibert was the first to derive the action that bears his name.

tensor in turn is a symmetric contraction of the Riemann curvature tensor. The chain of contractions is

$$R = R^\alpha{}_\alpha$$
$$R^\alpha{}_\beta = g^{\gamma\delta}R^\alpha{}_{\gamma\beta\delta} \qquad (12.126)$$
$$R^\alpha{}_{\gamma\beta\delta} = \partial_\beta\Gamma^\alpha{}_{\delta\gamma} - \partial_\delta\Gamma^\alpha{}_{\beta\gamma} + \Gamma^\alpha{}_{\beta\varepsilon}\Gamma^\varepsilon{}_{\delta\gamma} - \Gamma^\alpha{}_{\delta\varepsilon}\Gamma^\varepsilon{}_{\beta\gamma}.$$

The affine connections for a Riemannian basis $\Gamma^\gamma_{\alpha\beta}$ are functions of the metric tensor and its derivatives

$$\Gamma^\gamma_{\alpha\beta} = \langle\lambda^\gamma\partial_\alpha\lambda_\beta\rangle = \frac{1}{2}g^{\delta\gamma}(\partial_\alpha g_{\delta\beta} + \partial_\beta g_{\delta\alpha} - \partial_\delta g_{\alpha\beta}). \qquad (12.127)$$

Therefore, the variation of the preceding action can be expected to be complicated. However, this variation reduces to two intermediate results that are all that is required to short-circuit the derivation: (1) the functional variation of the determinant of the metric tensor and (2) the functional variation of the Ricci scalar. If the results for the variations of these quantities are accepted without proof, the remaining steps in the proof quickly fall into place. One takes the functional derivative of the action to get

$$\left(\sqrt{\|g\|}\frac{\delta\left(\frac{1}{\kappa}R + \mathcal{L}_M\right)}{\delta g^{\alpha\beta}} + \frac{\delta\sqrt{\|g\|}}{\delta g^{\alpha\beta}}\left(\frac{1}{\kappa}R + \mathcal{L}_M\right)\right)\delta g^{\alpha\beta} = 0. \qquad (12.128)$$

Without going to the going into the complexities of calculating the functional derivative for the metric field, the results are

$$\frac{\delta\sqrt{\|g\|}}{\delta g^{\alpha\beta}} = -\frac{1}{2}\sqrt{\|g\|}g_{\alpha\beta}, \qquad (12.129)$$

and

$$\frac{\delta R}{\delta g^{\alpha\beta}} = R_{\alpha\beta}. \qquad (12.130)$$

The variation of the action gives

$$\left(\frac{1}{\kappa}\left(R_{\alpha\beta} - \frac{1}{2}g_{\alpha\beta}R\right) + \left(\frac{\delta\mathcal{L}_M}{\delta g^{\alpha\beta}} - \frac{1}{2}g_{\alpha\beta}\mathcal{L}_M\right)\right)\delta g^{\alpha\beta} = 0, \qquad (12.131)$$

which results in the following formula for the Einstein tensor $G_{\alpha\beta}$

$$G_{\alpha\beta} = 2R_{\alpha\beta} - g_{\alpha\beta}R = \kappa T_{\alpha\beta}, \qquad (12.132)$$

where the gravitational stress-energy tensor of the material system is defined as

$$T_{\alpha\beta} = g_{\alpha\beta}\mathcal{L}_{\mathrm{M}} - 2\frac{\delta\mathcal{L}_{\mathrm{M}}}{\delta g^{\alpha\beta}}\bigg|_{f,Df}. \tag{12.133}$$

The Einstein tensor vanishes in the absence of matter and spacetime becomes flat. Note that the stress-energy tensor for the electromagnetic field evaluated using the preceding prescription is identical with the one defined by requiring translational invariance of the space given by Equation 12.109. This correspondence does not necessarily have to be true in general. Lagrangian theories are often constructed to make this be the case, but this is not always possible. In theories involving gravitational torsion, for example, there will an antisymmetric component to the stress-energy tensor defined by translational invariance that is not present in Equation 12.133.

In his original formulation of general relativity, Einstein, a proponent of the steady-state universe, included a *cosmological constant* to prevent gravitational contraction.[9] The modified Lagrangian is

$$S = \int \left(\frac{1}{\kappa}(R + 2\Lambda) + \mathcal{L}_{\mathrm{M}}\right)\sqrt{\|g\|}d^4x, \tag{12.134}$$

leading to the field equations

$$2R_{\alpha\beta} - g_{\alpha\beta}R + \Lambda g_{\alpha\beta} = \kappa T_{\alpha\beta}. \tag{12.135}$$

In 1929, Edwin Hubble (1889–1953) demonstrated that the universe is expanding. After the discovery of the Big Bang, Einstein would refer to his introduction of a cosmological constant as the biggest blunder of his career. The cosmological constant has the effect of giving an energy density to the vacuum. If the energy density is positive, the associated negative pressure would accelerate the expansion of the universe. In a surprising development, measurements of redshifts from distant galaxies in the late 1990's indicated that the expansion of the universe is, in fact, accelerating [Riess 1998; Perlmutter 1999]. The effect is similar to assigning a nonzero value of Einstein's cosmological constant and is currently attributed to the existence of a mysterious *dark energy* or *quintessence* whose nature not fully understood at the present time.

---

[9]This approach proved itself to be fundamentally flawed when the steady-state universe turned to be unstable under perturbations.

12.6 SUMMARY

Given a scalar Lagrangian density, written as a function over fields $\phi^{s;i}$, their partial spacetime derivatives $\partial \phi^{s;i}$, and Cartesian coordinates x

$$\mathcal{L} = \langle \mathcal{L}(\phi, \partial \phi, x) \rangle \tag{12.136}$$

the equations of motion can be found by variation of the resulting action,

$$\delta S = \frac{1}{c} \int_{\mathcal{R}} d^4 x \sqrt{\|g\|} \delta \mathcal{L} = 0. \tag{12.137}$$

resulting in the Euler-Lagrange equations of motion

$$\partial_u \frac{\partial \mathcal{L}}{\partial (\partial_\mu \phi^{s;i})}\bigg|_\phi = \frac{\partial \mathcal{L}}{\partial \phi^{s;i}}\bigg|_{\partial \phi}. \tag{12.138}$$

In Equation (12.136), the angle brackets denote contraction over all spacetime s and internal indices i. When using curvilinear coordinates, the resulting equations of motion are not explicitly obtained in a tensor covariant manner, however this can be remedied by use of the replacement rule. Replacing the partial derivatives in Equation (12.138) calculated using Cartesian coordinates by their covariant equivalents $\partial_\mu \to D_\mu$ gives the covariant equations of motion

$$D_\mu \pi^\mu_{s;i} = \frac{\partial \mathcal{L}}{\partial \phi^{s;i}}\bigg|_{D\phi}, \quad \pi^\mu_{s;i} = \frac{\partial \mathcal{L}}{\partial (D_\mu \phi^{s;i})}\bigg|_\phi. \tag{12.139}$$

where $\pi^\mu_{s;i}$ denote the conjugate four-currents associated with the fields $\phi^{s;i}$. Suppressing field indices, the Euler-Lagrange equations can be written in the compact form

$$D_\mu \frac{\partial \mathcal{L}}{\partial (D_\mu \phi)} = \frac{\partial \mathcal{L}}{\partial \phi}. \tag{12.140}$$

If there are gauge fields present, the covariant derivative can similarly be replaced by the covariant gauge derivative $D_\mu \to \mathcal{D}_\mu$. Ideally, variation should be carried out in terms of explicitly gauge invariant quantities.

The equations of motion can be expressed as first order partial differential equations by making an appropriate Legendre transformation. Defining a scalar density $\mathcal{H} = \langle \mathcal{H} \rangle$, such that

$$\mathcal{H}(\pi^\mu, \phi) = \langle \pi^\mu D_\mu \phi - \mathcal{L} \rangle, \tag{12.141}$$

gives rise to the couple set of first order differential equations of motion

$$D_\mu \pi^\mu = -\left.\frac{\partial \mathcal{H}}{\partial \phi}\right|_\pi, \quad D_\mu \phi = \left.\frac{\partial \mathcal{H}}{\partial \pi^\mu}\right|_\phi. \tag{12.142}$$

Noether's theorem states that the current associated with a continuous symmetry of the Lagrangian is conserved

$$\frac{\delta \mathcal{L}}{\delta \varepsilon} = D_\mu \left(\pi^\mu \frac{\delta \phi}{\delta \varepsilon} \right) = D_\mu J_\varepsilon^\mu,$$
$$\left.\frac{\delta \mathcal{L}}{\delta \varepsilon}\right|_{\varepsilon=0} = 0 \rightarrow D_\mu J_\varepsilon^\mu \big|_{\varepsilon=0} = 0. \tag{12.143}$$

Translational invariance of the Lagrangian leads to the conservation of the stress-energy tensor

$$T^\mu{}_\nu = \frac{\partial \mathcal{L}}{\partial (D_\mu \phi)} D_\nu \phi - g^\mu_\nu \mathcal{L}. \tag{12.144}$$

The Lagrangian for a string clamped at both ends, and free to vibrate in a transverse direction, is

$$\mathcal{L} = \frac{\mu}{2} \left[v^2 \left(\frac{\partial y}{\partial x} \right)^2 - \left(\frac{\partial y}{\partial t} \right)^2 \right]. \tag{12.145}$$

which gives rise to the wave equation in one space dimension

$$\left(\frac{1}{v^2} \frac{\partial^2}{\partial t^2} - \frac{\partial^2}{\partial x^2} \right) y(x,t) = 0. \tag{12.146}$$

The three-dimensional secular wave equation is

$$\left(\frac{1}{v^2} \frac{\partial^2}{\partial t^2} - \vec{\nabla}^2 \right) \psi(\vec{x},t) = 0. \tag{12.147}$$

Both transverse and longitudinal vibrations are allowed, usually with different wave velocities. The wave equation can be solved by a number of standard techniques, including separation of variables and Green's function/propagator methods.

The Sine-Gordon equation

$$\Box \phi = k^2 \sin \phi. \tag{12.148}$$

is an example of a non-linear, relativistic wave equation having quantized soliton solutions

$$\phi_\pm = 4 \arctan \exp\left(\pm k\gamma(x - vt) \right), \tag{12.149}$$

Due to nonlinearity, the superposition principle cannot be used to generate multiple soliton states, but the Bäcklund transform can be used to generate new solutions from existing known solutions.

Maxwell's electromagnetic field equations in free space give rise to the tensor wave equation

$$\Box F = \nabla^2 F = \left(\vec{\nabla}^2 - \frac{1}{c^2} \frac{\partial}{\partial t^2} \right) F = 0. \tag{12.150}$$

The electromagnetic field is quantized in units of Planck's Constant. The relationships between frequency $v = \omega/2\pi$, energy E, wave number $k = 2\pi/\lambda$, momentum p, and dimensionaless velocity $\beta = v/c$, for quantum systems, are given by

$$E = hv = \hbar\omega, \quad p = h/\lambda = \hbar k,$$
$$\beta = pc/E = f\lambda/c = \omega/ck. \tag{12.151}$$

In Einstein's theory of General Relativity, the Gravitational Field Tensor $G_{\alpha\beta}$ is related to the Ricci Curvature Tensor and to the stress-energy tensor of the matter fields via the equation

$$G_{\alpha\beta} = 2R_{\alpha\beta} - g_{\alpha\beta}R = \kappa T_{\alpha\beta}. \tag{12.152}$$

12.7   EXERCISES

1. Derive the wave equation for the vibrating string from the Lagrangian density

$$\mathcal{L} = \frac{\mu}{2}\left[-\left(\frac{\partial y}{\partial t}\right)^2 + v^2 \left(\frac{\partial y}{\partial x}\right)^2 \right].$$

2. Derive the three dimensional wave equation from the Lagrangian density

$$\mathcal{L} = \bar{\psi}\overset{\leftarrow}{\vec{\nabla}} \cdot \vec{\nabla}\psi - \frac{1}{v^2}\frac{\partial\bar{\psi}}{\partial t}\frac{\partial\psi}{\partial t}.$$

3. Derive the Schrödinger equation from the Lagrangian density

$$\mathcal{L} = \left\langle -\bar{\psi} i\hbar \frac{\partial \psi}{\partial t} + \frac{1}{2m} \bar{\psi} \overleftarrow{\wp} \cdot \overrightarrow{\wp} \psi \right\rangle, \quad \overrightarrow{\wp} = \frac{\hbar}{i}\vec{\nabla}, \overleftarrow{\wp} = -\frac{\hbar}{i}\overleftarrow{\nabla},$$

4. Derive the Klein-Gordon equation from the Lagrangian density

$$\mathcal{L} = \frac{\bar{\psi} \overleftarrow{\wp}^{\mu} \overrightarrow{\wp}_{\mu} \psi + \bar{\psi}\, m^2 c^2\, \psi}{2m}, \quad \overrightarrow{\wp}_{\mu} = \frac{\hbar}{i}\frac{\partial}{\partial x^{\mu}}, \overleftarrow{\wp}_{\mu} = -\frac{\hbar}{i}\frac{\partial}{\partial x^{\mu}}.$$

5. Find the Fourier wave number coefficients required for a Gaussian wave packet to be transmitted without dispersion in the positive $\hat{x}$ direction with a wave velocity of v. The Gaussian packet has a shape and amplitude given by

$$f_+(x - x_0 - vt) = \frac{A}{\sqrt{2\pi}} e^{-(x-x_0-vt)^2/2\sigma^2}.$$

Show that σ is the root mean square radius of the distribution.

6. Show that if two Lagrangians differ by a total divergence $D_\alpha F^\alpha(\phi, x)$ they satisfy the same Euler-Lagrange equations of motion.

7. Consider a solution to the one-dimensional wave equation having the Fourier coefficients

$$\tilde{f}_\pm(k) = \frac{B}{\sqrt{2\pi}} e^{-\sigma^2(k-k_0)^2/2}.$$

Find the solution for the amplitude $f(x, t)$.

8. A one dimensional string of length L is under tension and is clamped at both ends. Assume a small damping term b such that the string is a solution to the linear damped wave equation.

$$\left(\frac{\partial^2}{\partial x^2} - \frac{1}{v^2}\frac{\partial^2}{\partial t^2} \right) y(x, t) = b\frac{\partial}{\partial t}y(x, t).$$

find the normal modes of motion by separation of variables into product solutions of the form $y(x, t) = y(x)e^{i\Omega t}$. ($\Omega$ is complex in general). Find the oscillation frequencies of the normal modes and the rate at which they are damped.

9. An infinitely long string has a tension T and a linear mass density μ. Assume that its initial transverse displacement at $t = 0$ is given by $y(x, 0) = f(x)$ and its initial velocity distribution is $\partial y/\partial t|_{t=0} = g(x)$. Find the amplitude $y(x, t)$ of the wave for all latter times $t > 0$.

10. By analogy to the one-dimensional wave equation solutions for scattering from a potential barrier, find the transmitted and reflected wave coefficients for a Schrödinger plane wave of positive energy E and mass incident from the left on a potential barrier. The potential function is given by the step function

$$V = \begin{cases} 0 & x < 0 \\ V_0 & x \geq 0 \end{cases}.$$

Note that the dispersion relation for the Schrödinger equation differs from that of the wave equation, and one gets oscillatory transmitted waves only if $E > V_0$. What happens if $E < V_0$?

11. Show explicitly that the following functions are solutions to the Sine-Gordon equation
   a. Single-soliton solution: $\phi_\pm = 4 \arctan \exp (\pm k\gamma (x - vt))$,
   b. Two-soliton solution: $\phi_{2s} = 4\arctan(\beta \sinh (\gamma kx) \operatorname{sech} (\gamma \beta kct))$,
   c. Stationary-breather solution: $\phi_b = 4 \arctan ((\beta \gamma)^{-1} \operatorname{sech} (\gamma^{-1} kx) \sin (\beta kct))$.

12. Find the stress-energy tensor for the Sine-Gordon field given by the Lagrangian density of $\mathcal{L} = \rho(-\frac{1}{2}g^{\mu\nu}\partial_\mu \phi \partial_\nu \phi + k^2(\cos \phi - 1))$. Show that the energy density is localized for the one-dimensional single-kink solution $\phi = 4 \arctan \exp (k\gamma (x - vt))$, by plotting the energy density at a given time $t = 0$.

13. Use gauge invariance under phase rotations to derive a conserved current density for the Klein-Gordon equation. See Exercise 12.4 for the Lagrangian of the Klein-Gordon equation.

14. Evaluate the stress-energy tensor for the Klein-Gordon equation given by the Lagrangian density of Exercise 12.4.

15. Evaluate the stress-energy tensor for the Schrödinger equation given by the Lagrangian density of Exercise 12.3.

16. Find the conditions that must be satisfied for the following circularly polarized electric fields $\vec{E}_\pm$ moving in the $\hat{z}$ direction to be solutions to Maxwell's equations in free space

$$\vec{E}_\pm = E_0(e_x \cos k(z - ct) \mp e_y \sin k(z - ct)).$$

INTERNATIONAL SYSTEM OF UNITS AND CONVERSION FACTORS

Appendix **A**

D etail information on SI units can be found in NIST Special Publication 330, with a focus on history [Taylor 2001], and NIST Special Publication 811, 1995 Edition [Taylor 1995], with a focus on usage and unit conversions. The following summary and associated tables for the International System of Units (SI) has been extracted from *The NIST Reference on Constants, Units, and Uncertainties*.[1] The SI system is based on seven base units that are independently defined. These are listed in Table A.1

For ease of understanding and convenience, 22 SI derived units have been given special names and symbols, as shown in Table A.3.

The 20 SI prefixes used to form decimal multiples and submultiples of SI units are given in Table A.4. Note that the kilogram is the only SI unit with a prefix as part of its name and symbol. Because multiple prefixes may not be used, in the case of the kilogram the prefix names of Table A.4 are

TABLE A.1   SI base units

Base quantity	Name	Symbol
length	meter	m
mass	kilogram	kg
time	second	s
electric current	ampere	A
thermodynamic temperature	kelvin	K
amount of substance	mole	mol
luminous intensity	candela	cd

---

[1] Available from *http://physics.nist.gov/cuu/Units/units.html.*

TABLE A.2   Examples of SI derived units

Derived quantity	Name	Symbol
area	square meter	m^2
volume	cubic meter	m^3
speed, velocity	meters per second	m/s
acceleration	meters per second squared	m/s^2
wave number	reciprocal meter	m^{-1}
mass density	kilogram per cubic meter	kg/m^3
specific volume	cubic meter per kilogram	m^3/kg
current density	ampere per square meter	A/m^2
magnetic field strength	ampere per meter	A/m
concentration	mole per cubic meter	mol/m^3
luminance	candela per square meter	cd/m^2
mass fraction	kilogram per kilogram	$kg/kg = 1$

used with the unit name "gram" and the prefix symbols are used with the unit symbol "g." With this exception, any SI prefix may be used with any SI unit, including the degree Celsius and its symbol °C.

Certain units are not part of the International System of Units, that is, they are outside the SI, but are important and widely used. Consistent with the recommendations of the International Committee for Weights and Measures (CIPM, *Comité International des Poids et Mesures*), the units in this category that are accepted for use with the SI are given in Table A.5. Note the following definitions:

1. The electronvolt is the kinetic energy acquired by an electron passing through a potential difference of 1 V in vacuum. The value must be obtained by experiment, and is therefore not known exactly.
2. The unified atomic mass unit is equal to $\frac{1}{12}$ of the mass of an unbound atom of the nuclide ^{12}C, at rest and in its ground state. The value must be obtained by experiment, and is therefore not known exactly.
3. The astronomical unit is a unit of length. Its value is such that, when used to describe the motion of bodies in the solar system, the heliocentric gravitation constant is $(0.017\ 202\ 098\ 95)^2$ $ua^3 \cdot d^{-2}$. The value must be obtained by experiment, and is therefore not known exactly.

TABLE A.3   SI derived units with special names and symbols

Quantity	Name	Sym	SI units	SI base units
plane angle		rad*	—	$m \cdot m^{-1} = 1$
solid angle		sr*	—	$m^2 \cdot m^{-2} = 1$
frequency		Hz	—	s^{-1}
force		N	$Nm \cdot kg \cdot s^{-2}$	
pressure, stress	pascal	Pa	N/m^2	$m^{-1} \cdot kg \cdot s^{-2}$
energy, work, heat	joule	J	$N \cdot m$	$m^2 \cdot kg \cdot s^{-2}$
power, radiant flux	watt	W	J/s	$m^2 \cdot kg \cdot s^{-3}$
electric charge	coulomb	C	—	$s \cdot A$
electric potential, emf	Volt	V	W/A	$m^2 \cdot kg \cdot s^{-3} \cdot A^{-1}$
capacitance	farad	F	C/V	$m^{-2} \cdot kg^{-1} \cdot s^4 \cdot A^2$
electric resistance	ohm	Ω	V/A	$m^2 \cdot kg \cdot s^{-3} \cdot A^{-2}$
electric conductance	siemens	S	A/V	$m^{-2} \cdot kg^{-1} \cdot s^3 \cdot A^2$
magnetic flux	weber	Wb	$V \cdot s$	$m^2 \cdot kg \cdot s^{-2} \cdot A^{-1}$
magnetic flux density	tesla	T	Wb/m^2	$kg \cdot s^{-2} \cdot A^{-1}$
inductance	henry	H	Wb/A	$m^2 \cdot kg \cdot s^{-2} \cdot A^{-2}$
Celsius temperature	°Celsius	°C	—	K
luminous flux	lumen	lm	$cd \cdot sr$	$m^2 \cdot m^{-2} \cdot cd$
illuminance	lux	lx	lm/m^2	$m^2 \cdot m^{-4} \cdot cd$
activity (of isotope)	becquerel	Bq	—	s^{-1}
absorbed dose, kerma	gray	Gy	J/kg	$m^2 \cdot s^{-2}$
dose equivalent	sieverts	Sv	J/kg	$m^2 \cdot s^{-2}$
catalytic activity	katal	kat	—	$s^{-1} \cdot mol$

*The radian and steradian may be used advantageously in expressions for derived units to distinguish between quantities of a different nature but of the same dimension. In practice, the symbols rad and sr are used where appropriate, but the derived unit "1" is generally omitted.

A.1   CGS Units

Although SI units have been widely adopted for engineering and educational use, much of the basic research literature continues to use the older centimeter-gram-second (CGS) system of units. The mechanical CGS units are fairly standardized, but there are a number of different conventions in use for electromagnetic quantities. Table A.6 lists the most common CGS units. See [Jackson 1999: Appendix on Units and Dimensions] for more detailed information on converting electromagnetic units.

TABLE A.4   SI prefixes

Factor	Name	Symbol	Factor	Name	Symbol
10^{24}	yotta	Y	10^{-1}	deci	d
10^{21}	zetta	Z	10^{-2}	centi	c
10^{18}	exa	E	10^{-3}	milli	m
10^{15}	peta	P	10^{-6}	micro	μ
10^{12}	tera	T	10^{-9}	nano	n
10^{9}	giga	G	10^{-12}	pico	p
10^{6}	mega	M	10^{-15}	femto	f
10^{3}	kilo	k	10^{-18}	atto	a
10^{2}	hecto	h	10^{-21}	zepto	z
10^{1}	deka	da	10^{-24}	yocto	y

A.2 CONVERSION FACTORS BETWEEN VARIOUS ELECTROMAGNETIC SYSTEMS OF UNITS

Maxwell's equations have different coefficients depending on the system of units employed. Further confusing the issue, many engineers commonly mix systems of units. The defining relationship for Maxwell's equation is given in

TABLE A.5   Units outside the SI that are accepted for use with the SI

Name	Symbol	Value in SI units
minute (time)	min	$1\,\mathrm{min} = 60$ s
hour	h	$1\,\mathrm{h} = 60\,\mathrm{min} = 3600$ s
day	d	$1\,\mathrm{d} = 24\,\mathrm{h} = 86\,400$ s
degree (angle)	°	$1° = (\pi/180)$ rad
minute (angle)	′	$1′ = (1/60)° = (\pi/10\,800)$ rad
second (angle)	″	$1″ = (1/60)′ = (\pi/648\,000)$ rad
liter	L	$1\,\mathrm{L} = 1\,\mathrm{dm}^3 = 10^{-3}\,\mathrm{m}^3$
metric ton	t	$1\,\mathrm{t} = 10^3$ kg
neper	Np	$1\,\mathrm{Np} = 1$
bel	B	$1\,\mathrm{B} = (1/2)\ln 10$ Np
electronvolt	eV	$1\,\mathrm{eV} = 1.602\,18 \times 10^{-19}$ J
unified atomic mass unit	u	$1\,\mathrm{u} = 1.660\,54 \times 10^{-27}$ kg
astronomical unit	ua (AU)	$1\,\mathrm{ua} = 1.495\,98 \times 10^{11}$ m

TABLE A.6 Common CGS units (None of these units, other than the gram, centimeter and second, are approved for use with SI units)

Dimension	Unit	Definition	SI conversion factor
charge	esu	1 esu = 1 statC = 1 Fr	3.33564×10^{-10} C
capacitance	—	1 cm	1.113×10^{-12} F
electric current	—	1 esu/s	3.33564×10^{-10} C/s
electric potential	statvolt	1 statV = 1 erg/esu	299.792458 V
electric field	—	1 statV/cm = 1 dyn/esu	2.99792458×10^4 V/m
inductance	—	1 s^2/cm	8.988×10^{11} H
length	centimeter	1 cm	10^{-2} m
magnetic field H	oersted	1 Oe	$1000/(4\pi)$ A/m
magnetic flux	maxwell	1 M = 1 G $\cdot$ cm^2	10^{-8} Wb
magnetic induction B	gauss	1 G = 1 M/cm^2	10^{-4} T
mass	gram	1 g	10^{-3} kg
time	second	1 s	1 s
force	dyne	1 dyn = 1 g·cm/s^2	10^{-5} N
energy	erg	1 erg = 1 g $\cdot$ cm^2/s^2	10^{-7} J
power	erg per second	1 erg/s = 1 g $\cdot$ cm^2/s^3	10^{-7} W
pressure	barye	1 Ba = 1 dyn/cm^2	10^{-1} Pa
resistance	—	1 s/cm	$8.988 \times 10^{11}\Omega$
resistivity	—	1 s	$8.988 \times 10^9 \Omega \cdot$ m
viscosity	poise	1 P = 1 g/(cm $\cdot$ s)	10^{-1} Pa·s
wave number	kayser	1/cm	100/m

SI units in terms of the permittivity of free space (the electric constant ε_0) and the magnetic permeability of free space (the magnetic constant μ_0). These quantities are not independent, but are related by

$$\varepsilon_0 \mu_0 = 1/c^2. \tag{A.1}$$

In terms of these parameters, Maxwell's equations in free space can be written in SI units as

$$\nabla \cdot \mathbf{E} = \rho/\varepsilon_0, \quad \nabla \times \mathbf{B} = \frac{1}{c^2}\frac{\partial \mathbf{E}}{\partial t} + \mu_0 \mathbf{J},$$
$$\nabla \cdot \mathbf{B} = 0, \quad \nabla \times \mathbf{E} = -\frac{\partial \mathbf{B}}{\partial t}, \tag{A.2}$$

where $\mathbf{E}$ is the electric field strength $\mathbf{B}$ and is the magnetic induction. In addition, the macroscopic electric displacement field $\mathbf{D}$ and magnetic field

TABLE A.7   Values of electromagnetic constants in various systems of units [Jackson 1999: Appendix on Units and Dimensions]

Units	μ_0	ε_0	λ_0
SI	$4\pi \times 10^{-7}$	$\dfrac{10^7}{4\pi c^2}$	1
Electrostatic (esu)	c^{-2}	1	4π
Electromagnetic (emu)	1	c^{-2}	4π
Gaussian	1	1	4π
Heaviside-Lorentz	1	1	1

strength **H** have different normalization conventions in the various systems of units. In all common systems of units, one can define these fields as

$$\mathbf{D} = \varepsilon_0 \mathbf{E} + \lambda_0 \mathbf{P}, \tag{A.3}$$

$$\mathbf{B} = \mu_0 (\mathbf{H} + \lambda_0 \mathbf{M}). \tag{A.4}$$

where **P** is the electric polarization of the medium and **M** is the magnetic magnetization of the medium. $\lambda_0 = 1$, if the Coulomb force law is given in rationalized units; and $\lambda_0 = 4\pi$, if the units are unrationalized; for SI units, $\lambda_0 = 1$. The Lorentz force law in SI units is

$$\mathbf{F} = q_e (\mathbf{E} + \mathbf{v} \times \mathbf{B}). \tag{A.5}$$

Table A.7 show the values of electromagnetic constants in various systems of units [Jackson 1999: Appendix on Units and Dimensions].

Table A.8 shows common electromagnetic equations expressed in various systems of units [Jackson 1999: Appendix on Units and Dimensions].

A.3   DIMENSIONLESS UNITS

Various systems of dimensionaless units are used in theoretical physics, quantum physics, and cosmology. Most common dimensionless units in quantum physics set $\hbar = c = 1$. The speed of light then sets the scale between time and length units, while Plank's constant $h = 2\pi\hbar$ sets the scale between energy and time units

$$c = 299792458 \text{ m} \cdot \text{s}^{-1} (\text{exact}), \tag{A.6}$$

$$\hbar = h/2\pi = 1.054571628(53) \times 10^{-34} \text{ J} \cdot \text{s}. \tag{A.7}$$

TABLE A.8   Common electromagnetic equations expressed in various systems of units [Jackson 1999: Appendix on Units and Dimensions]

Units	Maxwell's Equations		Polarizations	Force/charge
SI	$\nabla \cdot \mathbf{D} = \rho$	$\nabla \times \mathbf{E} = -\dfrac{\partial \mathbf{B}}{\partial t}$	$\mathbf{D} = \varepsilon_0 \mathbf{E} + \mathbf{P}$	$\mathbf{E} + \mathbf{v} \times \mathbf{B}$
	$\nabla \cdot \mathbf{B} = 0$	$\nabla \times \mathbf{H} = \dfrac{\partial \mathbf{D}}{\partial t} + \mathbf{J}$	$\mathbf{H} = \mathbf{B}/\mu_0 - \mathbf{M}$	
ESU	$\nabla \cdot \mathbf{D} = 4\pi\rho$	$\nabla \times \mathbf{E} = -\dfrac{\partial \mathbf{B}}{\partial t}$	$\mathbf{D} = \mathbf{E} + 4\pi\mathbf{P}$	$\mathbf{E} + \mathbf{v} \times \mathbf{B}$
	$\nabla \cdot \mathbf{B} = 0$	$\nabla \times \mathbf{H} = \dfrac{\partial \mathbf{D}}{\partial t} + 4\pi\mathbf{J}$	$\mathbf{H} = c^2\mathbf{B} - 4\pi\mathbf{M}$	
EMU	$\nabla \cdot \mathbf{D} = 4\pi\rho$	$\nabla \times \mathbf{E} = -\dfrac{\partial \mathbf{B}}{\partial t}$	$\mathbf{D} = c^{-2}\mathbf{E} + 4\pi\mathbf{P}$	$\mathbf{E} + \mathbf{v} \times \mathbf{B}$
	$\nabla \cdot \mathbf{B} = 0$	$\nabla \times \mathbf{H} = \dfrac{\partial \mathbf{D}}{\partial t} + 4\pi\mathbf{J}$	$\mathbf{H} = \mathbf{B} - 4\pi\mathbf{M}$	
Gaussian	$\nabla \cdot \mathbf{D} = 4\pi\rho$		$\mathbf{D} = \mathbf{E} + 4\pi\mathbf{P}$	$\mathbf{E} + \dfrac{\mathbf{v}}{c} \times \mathbf{B}$
	$\nabla \times \mathbf{E} = -\dfrac{1}{c}\left(\dfrac{\partial \mathbf{B}}{\partial t}\right)$		$\mathbf{H} = \mathbf{B} - 4\pi\mathbf{M}$	
	$\nabla \cdot \mathbf{B} = 0$	$\nabla \times \mathbf{H} = \dfrac{1}{c}\left(\dfrac{\partial \mathbf{D}}{\partial t} + 4\pi\mathbf{J}\right)$		
Heaviside-	$\nabla \cdot \mathbf{D} = \rho$	$\nabla \times \mathbf{E} = -\dfrac{1}{c}\dfrac{\partial \mathbf{B}}{\partial t}$	$\mathbf{D} = \mathbf{E} + \mathbf{P}$	$\mathbf{E} + \dfrac{\mathbf{v}}{c} \times \mathbf{B}$
Lorentz	$\nabla \cdot \mathbf{B} = 0$	$\nabla \times \mathbf{H} = \dfrac{1}{c}\left(\dfrac{\partial \mathbf{D}}{\partial t} + \mathbf{J}\right)$	$\mathbf{H} = \mathbf{B} - \mathbf{M}$	

Depending on the application, this leaves either a length scale or an energy scale that can also be made dimensionless. The relationship of length to energy scales is given by the conversion factor

$$\hbar c = 197.326968(17) \text{ MeV} \cdot \text{fm}. \tag{A.8}$$

The unit electric charge is dimensionless as well, with its relationship to the fine structure constant depending on whether the electromagnetic units are rationalized or not as summarized in Table A.9.

In cosmological and gravitational theoretical papers, *dimensionless Plank units*, shown in Table A.10, are often used with the Newtonian gravitational coupling constant set to one.

TABLE A.9   Fundamental unit of charge in high-energy dimensionless units, with $\hbar = c = 1$

Unit type	Charge	Electric constant
rationalized	$e^2 = \alpha$	$\varepsilon_0 = 1/4\pi$
unrationalized	$e^2/4\pi = \alpha$	$\varepsilon_0 = 1$

TABLE A.10   Dimensionless Plank units

Unit	Symbol
Speed of light in vacuum	$c = 1$
Gravitational constant	$G = 1$
Planck's constant	$\hbar = 1$
Coulomb force constant	$1/4\pi\varepsilon_0 = 1$
Boltzmann constant	$k = 1$

Planck units simplify theoretical formulas, but the base units are very different from the laboratory scale, as shown in Table A.11, which compares the values of Planck base units to SI units.

TABLE A.11   Values of Planck base units compared to SI units

Name	Expression	SI equivalent
Planck length	$l_P = \sqrt{\dfrac{\hbar G}{c^3}}$	1.61624×10^{-35} m
Planck mass	$m_P = \sqrt{\dfrac{\hbar c}{G}}$	2.17645×10^{-8} kg
Planck time	$t_P = \dfrac{l_P}{c} = \dfrac{\hbar}{m_P c^2} = \sqrt{\dfrac{\hbar G}{c^5}}$	5.39121×10^{-44} s
Planck charge	$q_P = \sqrt{\hbar c 4\pi \epsilon_0}$	$1.8755459 \times 10^{-18}$ C
Planck temperature	$T_P = \dfrac{m_P c^2}{k} = \sqrt{\dfrac{\hbar c^5}{G k^2}}$	1.41679×10^{32} K

PHYSICAL CONSTANTS AND SOLAR SYSTEM DATA

The following table (Table B.1) of common physical constants has been extracted from *The NIST Reference on Constants, Units, and Uncertainties.*[1] The values quoted are based on the 2006 CODATA recommended values; see [Mohr 2007; Yao 2006] for a more detailed listing of constants.

TABLE B.1 Common physical constants

Constant name	Symbol and Value (Error)
atomic mass constant	$m_u = m(^{12}C) / 12 = 1.660\,538\,782\,(83) \times 10^{-27}$ kg
Avogadro constant	$N_A = 6.02214179(30) \times 10^{23}$ mol^{-1}
Boltzmann constant	$k = 1.380\,6504(24) \times 10^{-23}$ J $\cdot$ K^{-1}
conductance quantum	$G_0 = 2e^2/h = 7.748\,091\,7004(53) \times 10^{-5}$ s
electric constant	$\epsilon_0 = 1/\mu_0 c^2 = 8.854\,187\,817\ldots \times 10^{-12}$ F $\cdot$ m^{-1} (exact)
electron mass	$m_e = 9.109\,382\,15(\,45\,) \times 10^{-31}$ kg
electron volt	$eV = 1.602\,176\,487(\,40\,) \times 10^{-19}$ J
elementary charge	$e = 1.602\,176\,487(\,40\,) \times 10^{-19}$ C
Faraday constant	$F = N_A e = 96485.3399(24)$ C $\cdot$ mol^{-1}
fine-structure constant	$\alpha = e^2/4\pi\epsilon_0\hbar c = 7.297\,352\,537\,6(50) \times 10^{-3}$
inverse fine-structure constant	$\alpha^{-1} = 4\pi\epsilon_0\hbar c/e^2 = 137.035\,999\,679$ (94)
magnetic constant	$\mu_0 = 4\pi \times 10^{-7}$ N $\cdot$ A^{-2} (exact)
magnetic flux quantum	$\Phi_0 = h/2e = 2.067\,833\,667$ (52) $\times 10^{-15}$ Wb
molar gas constant	$R = 8.314472(15)$ J $\cdot$ mol^{-1} $\cdot$ K^{-1}
gravitational constant	$G = 6.67428(67) \times 10^{-11}$ m^3 $\cdot$ kg^1 $\cdot$ s^2

---

[1] Available from *http://physics.nist.gov/cuu/Units/units.html.*

TABLE B.1   (*Continued*)

Constant name	Symbol and Value (Error)
Planck constant	$h = 6.62606896(33) \times 10^{-34} \, \text{J} \cdot \text{s}$
Planck constant/2π	$\hbar = h/2\pi = 1.054571628(53) \times 10^{-34} \, \text{J} \cdot \text{s}$
proton mass	$m_p = 1.672621637(83) \times 10^{-27} \, \text{kg}$
proton-electron mass ratio	$m_p/m_e = 1836.15267247(80)$
Rydberg constant	$R_\infty = \alpha^2 m_e c/2h = 10973731.568527(73) \, \text{m}^{-1}$
speed of light in vacuum	$c = 299792458 \, \text{m} \cdot \text{s}^{-1} \, (\text{exact})$
Stefan-Boltzmann constant	$\sigma = \pi^2 k^4/60\hbar^3 c^2 = 5.670400(40)10^{-8} \, \text{W/m}^2 \cdot \text{K}^4$

Table B.2 summarizes useful data for the Earth-moon-sun system compiled from a variety of sources. Table B.3 contains key planetary orbital and composition parameters for the major planets of the solar system, plus Pluto. Solar system data available from NASA.

TABLE B.2   Earth-Moon-Sun System Data

Parameter	Sun	Earth	Moon
Mass (kg)	1.989×10^{30}	5.975×10^{24}	7.349×10^{22}
Equatorial radius (km)	695 000	6 378.388	1 737.4
Mean density (gm/cm^3)	1.410	5.515	3.34
Distance from Sun or Earth (km)	—	149 600 000	384 400
Rotational period (days)	25–36*	0.997 22	27.321 66
Orbital period (days)	—	365.256	27.321 66
Mean orbital velocity (km/sec)	—	29.79	1.03
Orbital eccentricity	—	0.016 7	0.054 9
Tilt of axis (°)	—	23.45	1.542 4
Orbital inclination (degrees)	—	0.000	5.145 4
Equatorial escape velocity (km/sec)	618.02	11.18	2.38
Equatorial surface gravity (m/sec^2)	—	9.78	1.62
Mean surface temperature	6000°C	15°C	−153°C 107°C[†]
Standard Earth surface pressure (bar)	—	1.012 53	—
Standard Earth surface gravity (m/s^2)	—	9.806 65	—

*Surface rotation varies from 25 days at equator to 36 days at poles.
[†]night/day ranges.

TABLE B.3   Planetary Parameters and Orbital Data[2]

Planet	Distance AU	Period* yr	Inclin. °	Eccen. —	Mass* $M_\oplus$	Radius* $R_\oplus$	Rot. Period days|hr
Mercury	0.387	0.241	7	0.206	0.0552	0.382	59 d
Venus	0.723	0.615	3.39	0.007	0.814	0.949	243 d
Earth	1.00	1.00	0	0.017	1.00	1.00	24 hr
Mars	1.524	1.88	1.85	0.093	0.107	0.532	24.6 hr
Jupiter	5.203	11.86	1.3	0.048	318	11.2	9.8 hr
Saturn	9.539	29.46	2.49	0.056	95.2	9.46	10.2 hr
Uranus	19.18	84.01	0.77	0.047	14.5	4.01	15.5 hr
Neptune	30.06	164.8	1.77	0.009	17.1	3.88	15.8 hr
Pluto	39.53	247.7	17.15	0.248	0.00216	0.180	6.4 d

*Units relative to Earth.

[2]Source: *http://vathena.arc.nasa.gov/curric/space/planets/planorbi.html.*

Appendix C

GEOMETRIC ALGEBRAS IN N DIMENSIONS

The angle bracket notation $\langle\,\rangle_r$ will be used to denote the projection of a totally antisymmetric rank-r geometric tensor, which is also known as an r-blade. For the scalar projection the index 0 will be suppressed $\langle\,\rangle_0 = \langle\,\rangle$. Repeated exterior products of irreducible geometric tensors maximally extend the tensor rank of the product. The exterior product of two geometric tensors of ranks r and s, respectively, where $r, s \geq 1$, has rank $r + s$ denoted by

$$\langle A \rangle_r \wedge \langle B \rangle_s = \begin{cases} \langle C \rangle_{r+s}, & r + s \leq n \\ 0, & r + s > n \end{cases}, \tag{C.1}$$

where n is the dimension of the vector space. For example, the exterior product of a vector with an irreducible geometric tensor of arbitrary rank $m < n$ increases the tensor rank by one to $m + 1$. The operation of extension can be written in terms of the group product as

$$\langle V \rangle_1 \wedge \langle T \rangle_m = \frac{1}{2}(\langle V \rangle_1 \langle T \rangle_m + (-1)^m \langle T \rangle_m \langle V \rangle_1). \tag{C.2}$$

Likewise, the dot product of two geometric tensors will be defined to produce a maximally contracted tensor

$$\langle A \rangle_r \cdot \langle B \rangle_s = \langle C \rangle_{|r-s|}. \tag{C.3}$$

The dot product of a vector with an irreducible tensor of rank $m \geq 1$ reduces the rank of the product to $m - 1$

$$\langle V \rangle_1 \cdot \langle T \rangle_m = \frac{1}{2}(\langle V \rangle_1 \langle T \rangle_m - (-1)^m \langle T \rangle_m \langle V \rangle_1). \tag{C.4}$$

The elements of a geometric tensor can be grouped by the irreducible tensor ranks generated by taking all possible totally antisymmetric exterior

557

products of the basis vectors

$$\Gamma^{\langle 0 \rangle} = 1,$$
$$\Gamma^{\langle 1 \rangle}_a = \lambda_a,$$
$$\Gamma^{\langle 2 \rangle}_{a<b} = \lambda_a \wedge \lambda_b = \sigma_{ab}, \quad \forall a < b,$$
$$\Gamma^{\langle m \rangle}_{a<b<c\cdots} = \lambda_a \wedge \lambda_b \wedge \lambda_c \cdots, \quad \forall a < b < c \cdots,$$

(C.5)

where the indices have been ordered to prevent double counting. The sequence of extensions ends when the tensor rank m equals n, since there are only n linearly independent vectors in an n-dimensional space. The pseudoscalar element is

$$\Gamma^{\langle n \rangle}_{1<2\cdots<n} = \sqrt{\|g\|} i_n,$$

(C.6)

where

$$i_n = \sqrt{\|g\|}^{-1} \lambda_1 \wedge \lambda_2 \wedge \cdots \wedge \lambda_n$$

(C.7)

denotes the unit pseudoscalar ($i_n^2 = \pm 1$) in n dimensions and $\|g\|$ denotes the absolute value of the determinant of the metric tensor. In general, there are

$$\binom{n}{m} = n!/(n-m)!m!$$

(C.8)

linearly-independent tensor elements of rank m such that $0 \leq m \leq n$. There are 2^n independent tensor elements in the geometric algebra of n-dimensional space, since

$$\sum_m \binom{m}{n} = 2^n.$$

(C.9)

Every geometric tensor has a geometric adjoint tensor given by reversion of the order of vector products, such that a generalized tensor basis has an adjoint basis given by reversion

$$\bar{\Gamma}^{\langle r \rangle}_{abc} = \overline{\lambda_a \wedge \lambda_b \wedge \lambda_c} = \lambda_c \wedge \lambda_b \wedge \lambda_a,$$

(C.10)

where a rank-three tensor basis element is used for specificity. The adjoint basis projects out its dual in the reciprocal basis

$$\langle \bar{\Gamma}^{\langle r \rangle a' < b' < c' \cdots} \Gamma^{\langle r \rangle}_{a<b<c\cdots} \rangle = \delta^{a'}_a \delta^{b'}_b \delta^{c'}_c \cdots.$$

(C.11)

For a tensor of irreducible geometric rank, the signature of the dual under reversion is given by

$$\bar{\Gamma}^{\langle m \rangle} = \begin{cases} (-1)^{(m-1)!} \Gamma^{\langle m \rangle}, & m \geq 2, \\ \Gamma^{\langle m \rangle}, & m = 0, 1. \end{cases}$$

(C.12)

The extraction of an irreducible geometric tensor of rank $0 \leq r \leq n$ is given by the projection operators

$$
\begin{aligned}
\langle T \rangle_0 &= \langle T \rangle, \\
\langle T \rangle_1 &= \langle \bar{\lambda}^i T \rangle \lambda_i, \\
\langle T \rangle_2 &= \sum_{i<j} \langle \bar{\sigma}^{ij} T \rangle \sigma_{ij}, \\
\langle T \rangle_3 &= \sum_{i<j<k} \left\langle \overline{\lambda^i \wedge \lambda^j \wedge \lambda^k} T \right\rangle \lambda_i \wedge \lambda_j \wedge \lambda_j, \ldots \\
\langle T \rangle_n &= \langle \bar{\Gamma}^{\langle n \rangle 12 \ldots n} T \rangle \Gamma^{\langle n \rangle}_{1,2 \ldots n},
\end{aligned}
\tag{C.13}
$$

where by completeness

$$
T = \sum_{m=0}^{n} \langle T \rangle_m.
\tag{C.14}
$$

Let Γ_j be shorthand for one of the n^2 linearly-independent geometric tensor basis elements of the algebra. One can represent an arbitrary geometric tensor as an expansion over this basis

$$
T = T^i \Gamma_i.
\tag{C.15}
$$

The tensor basis operators Γ_j act as projection operators for any element of the geometry. Define the reciprocal tensor basis as

$$
\langle \bar{\Gamma}^i \Gamma_j \rangle = \delta^i_j, \quad \forall i,j = 1, \ldots, n^2.
\tag{C.16}
$$

Then, by projection,

$$
\langle \bar{\Gamma}^i T \rangle = T^i, \quad \langle \bar{\Gamma}_i T \rangle = T_i.
\tag{C.17}
$$

TABLE C.1   Minimum matrix representations of $GA(s,t)$ shown for $s+t \leq 4$. Note that $n = s+t$, $m = s-t$, and $\mathbb{F}(n)$ stands for an $n \times n$ matrix in the field structures $\mathbb{F} = \{\mathbb{R}, \mathbb{C}, \mathbb{H}, {}^2\mathbb{R}, {}^2\mathbb{C}, {}^2\mathbb{H}\}$

$n \backslash m$	-4	-3	-2	-1	0	1	2	3	4
0					$\mathbb{R}$				
1				$\mathbb{C}$		${}^2\mathbb{R}$			
2			$\mathbb{H}$		$\mathbb{R}(2)$		$\mathbb{R}(2)$		
3		${}^2\mathbb{H}$		$\mathbb{C}(2)$		${}^2\mathbb{R}(2)$		$\mathbb{C}(2)$	
4	$\mathbb{H}(2)$		$\mathbb{H}(2)$		$\mathbb{R}(4)$		$\mathbb{R}(4)$		$\mathbb{H}(2)$

Table C.1 shows the minimum matrix representations for $GA(s, t)$, for $s + t \leq 4$. Matrix algebras can be based on any of the associative division algebras, See [Lounesto 1997] for higher dimensional representations. When complex and quaternion fields are use to minimize the matrix representation, the fields acquire geometric significance. The imaginary and hyper-complex quaternion operators should not be interpreted as scalars, and, in fact are defined to have zero scalar projection.

REFERENCES

Abramowitz, M.; and I.A. Stegun, editors. 1965. *Handbook of Mathematical Functions: With Formulas, Graphs, and Mathematical Tables.* New York: Dover Publications (New Ed edition).

Adelberger, E.G.; B.R. Heckel; and A.E. Nelson. 2003. "Tests of the Gravitational Inverse-Square Law." *Annual Review of Nuclear and Particle Science* **53**: 77–121.

Arfken, George B. and Hans J. Weber. 2000. *Mathematical Methods for Physicists, 5th Edition.* San Diego: Harcourt/Academic Press.

Arnold, V. I.; A. Weinstein (Trans); and K. Vogtmann (Trans). 1989. *Mathematical Methods of Classical Mechanics, 2nd Edition.* New York: Springer Verlag.

Aspect, A.; P. Grangier; and G. Roger. 1982. "Experimental Realization of Einstein-Podolsky-Rosen-Bohm-Gedank En Experiment: A New Violation of Bell's Inequalities." *Phys. Rev. Lett..* **49**: 91–94.

Barrow-Green, J. 1997. "Poincaré and the Three-body Problem," *History of Mathematics,* V. 11. American Mathematical Society.

Bell, J. S. 1964. "On the Einstein-Podolsky-Rosen Paradox." *Physics* **1**: 195–200.

Bennett, G.W. et al. 2004. "Measurement of the Negative Muon Anomalous Magnetic Moment to 0.7 ppm." *Physical Review Letters* **92**: 1618102.

Beyer, W. H. 1987. *CRC Standard Mathematical Tables, 28th Edition.* Boca Raton. FL: CRC Press.

Bohm, David. 1951. *Quantum Theory.* New York: Prentice Hall. 1989 reprint. New York: Dover.

Bohm, D. 1952. "A Suggested Interpretation of the Quantum Theory in Terms of 'Hidden' Variables, I and II." *Physical Review* **85**: 166–193.

Bohm, D. 1953. "Proof that Probability Density Approaches $|\psi|^2$ in Causal Interpretation of Quantum Theory." *Physical Review* **89**: 458–466.

Bohr, Niels. 1913. "On the Constitution of Atoms and Molecules, Part 1." *Philosophical Magazine* **26**: 1–25; Part II. *Philosophical Magazine* **26**: 476–502. Part III. *Philosophical Magazine* **26**: 857–875.

Brown, H. N. et al. 2001. "Precise measurement of the positive muon anomalous magnetic moment." *Phys. Rev. Lett.* **86**: 2227–2231.

Cavendish, H. 1798. "Experiments to determine the Density of the Earth." *Philosophical Transactions of the Royal Society of London (part II)* **88**: 469–526.

Chenciner, A. and R. Montgomery. 2000. "A remarkable periodic solution of the three-body problem in the case of equal masses." *Annals of Mathematics* **152**: 881–901.

Clifford, W. 1878. "Applications of Grassmann's Extensive Algebra." *American Journal of Mathematics* **1**: 350–358.

Clotfelter, B. E. 1987. "The Cavendish experiment as Cavendish knew it." *American Journal of Physics,* **55**: 210–213.

Cohen, I. Bernard. 1955. "An Interview with Einstein." *Scientific American,* **193**: 68–73.

Cooper, William H. 2001. *Great Physicists: The Life and Times of Leading Physicists from Galileo to Hawking.* Oxford, England: Oxford University Press.

Cornish, Neil J. undated. *The Lagrange Points.* Unpublished. Available at *http://map.gsfc.nasa.gov/ContentMedia/lagrange.pdf.*

Chand H.; R. Srianand; P. Petitjean; and B. Aracil. 2004. "Probing the cosmological variation of the fine-structure constant: Results based on VLT-UVES sample." *Astron. Astrophys.* **417**: 853–871.

Dirac, P. A. M. 1928. "The Quantum Theory of the Electron II." *Proceedings of the Royal Society of London* **A117**: 610–624.

Dirac, P. A. M. 1930. "A Theory of Electrons and Protons." *Proceedings of the Royal Society of London* **A126**: 360–365.

Dirac, P. A. M. 1981. *The principles of quantum mechanics.* The International Series of Monographs on Physics, 4th, (rev.) ed., Vol. 27. Oxford, England: Clarendon Press.

Dodd, R. K.; J. C. Eilbeck; J. D. Gibbon; and H. C. Morris 1982. *Solitons and Nonlinear Wave Equations.* London. Academic Press.

Doran, C. J. L.; D. Hestenes; F. Sommen; and N. van Acker. 1993. "Lie Groups as Spin Groups." *J. Math. Phys.* **34**: 3642–3669.

Doran, Chris and A. N. Lasenby. 2003. *Geometric algebra for physicists.* Cambridge, New York: Cambridge University Press.

Drazin, P. G. and R. S. Johnson. 1989. *Solitons: an introduction.* Cambridge University Press.

Einstein, Albert. 1905a. "On a Heuristic Viewpoint Concerning the Production and Transformation of Light." *Annalen der Physik* **17**: 132–148.

Einstein, Albert 1905b, "On the Motion—Required by the Molecular Kinetic Theory of Heat—of Small Particles Suspended in a Stationary Liquid." *Annalen der Physik* **17**: 549–560.

Einstein, Albert. 1905c. "On the Electrodynamics of Moving Bodies." *Annalen der Physik* **17**: 891–921.

Einstein, Albert. 1915. "Die Feldgleichungen der Gravitation (The Field Equations of Gravitation)." *Koniglich Preussische Akademie der Wissenschaften*: 844–847.

Einstein, Albert. 1920. *Relativity. The Special and General Theory.* New York: Henry Holt. See also New York. Bartleby.com (2000), available online at *http://www.bartleby.com/173/.*

Einstein, Albert. 1931. Maxwell's Influence on the Evolution of the idea of Physical reality: On the one hundredth anniversary of Maxwell's birth, in *James Clerk Maxwell: A Commemoration Volume.* Cambridge, UK: Cambridge University Press, 1931. Paper included in [Einstein 1954, pp. 266–269].

Einstein, A.; B. Podolsky; and N. Rosen. 1935. "Can quantum-mechanical description of physical reality be considered complete?" *Phys. Rev.* **47**: 777–780.

Einstein, Albert. 1940. "The Fundaments of Theoretical Physics," originally published in *Science*, May 24, 1940. Paper included in [Einstein 1954, pp. 323–335].

Einstein, Albert. 1954. *Ideas and Opinions.* New York. Crown Publishers.

Einstein, Albert; Roger Penrose (Foreword); John Stachel (Editor). 2005 *Einstein's Miraculous Year: Five Papers That Changed the Face of Physics.* P: Princeton University Press (New Ed edition).

Erler, Jens; Michael J. Ramsey-Musolf. 2005. "Low Energy Tests of the Weak Interaction." *Prog. Part.Nucl. Phys.* **54**: 351–442.

Farrington, Benjamin. 1966. *Greek Science: Its Meaning to Us.* Harmondsworth: Penguin.

Fetter, Alexander L. and John Dirk Walecka. 1980. *Theoretical Mechanics for Particles and Continua.* New York: McGraw-Hill College.

Feynman, Richard P. 1985. *QED: The Strange Theory of Light and Matter.* Princeton: Princeton University Press, 1985.

Gabrielse, G.; D. Hanneke; T. Kinoshita; M. Nio; and B. Odom. 2006. "New Determination of the Fine Structure Constant from the Electron g Value and QED." *Phys. Rev. Lett.* **97**: 030802.

Galileo Galilei; Stillman Drake (Translator); Albert Einstein (Foreword). 1967. *Dialogue Concerning the Two Chief World Systems.* Berkeley, CA: University of California Press. (Initially published in 1632).

Gamov, George. 1961. *The Great Physicists from Galileo to Einstein.* New York: Dover Publications.

Gibbs, J. Willard; Henry Andrews Bumstead; and Ralph Gibbs Van Name. 1906. *Scientific Papers of J. Willard Gibbs.* London, New York and Bombay: Longmans, Green and Co.

Gibson, James J. 2007. *The fine structure constant, a 20th century mystery.* Queen Mary College, University of London. Available at http://www.maths.qmul.ac.uk/~jgg/page5.html, last accessed July 25, 2007.

Goldstein, Herbert; Charles P. Poole; and John L. Safko. 2002. *Classical Mechanics, 3rd Edition.* San Francisco: Addison Wesley.

Gomez, G.; W. S. Koon; M. W. Lo; J. E. Marsden; J. Masdemont; and S. D. Ross. 2004. "Connecting orbits and invariant manifolds in the spatial restricted three-body problem." *Nonlinearity* **17**: 1571–1606.

de Gossen M. A. 2001. *Principles of Newtonian and Quantum Mechanics: The Need for Planck's Constant, h.* London: Imperial College Press.

Grassmann. 1844. *Die Lineale Ausdehnungslehre - Ein neuer Zweig der Mathematik.* (The Linear Extension Theory - A new Branch of Mathematics) For an English translation see [Grassmann 1995].

Grassmann, Hermann. 1995. A new branch of mathematics: The "ausdehnungslehre" of 1844 and other works [Selections.]. Chicago. Open Court.

Griffiths, David J. 1987. *Introduction to Elementary Particles.* New York: Wiley.

Hamilton, W. R. 1834. "On a General Method in Dynamics." *Philosophical Transaction of the Royal Society* Part I 247–308, (1834); Part II 95–144, (1835).

Hamilton, W. R. 1844. "On a new Species of Imaginary Quantities connected with a theory of Quaternions." *Proceedings of the Royal Irish Academy* **2**: 424–434.

Hestenes, David. 1966. *Space-time algebra.* Documents on Modern Physics, Vol. 7. New York: Gordon and Breach.

Hestenes, David; and Garret Sobczyk. 1984. *Clifford Algebra to Geometric Calculus: A Unified Language for Mathematics and Physics.* Fundamental Theories of Physics. Dordrecht; Boston; Hingham, MA, U.S.A.: D. Reidel; Distributed in the U.S.A. and Canada by Kluwer Academic Publishers.

Hestenes, David. 2002. *New Foundations for Classical Mechanics.* Fundamental Theories of Physics, 2nd ed., Vol. 99. New York: Kluwer Academic Press.

Hestenes, D. 2007. *Spacetime Geometry with Geometric Calculus.* To be published in the Proceedings of the Seventh International Conference on Clifford Algebra.

Hagiwara, K.; et al. 2002. "2002 Review of Particle Physics." *Phys. Rev. D* **66**: 010001.

Hilbert, D. 1915. "Die Grundlagen der Physik, Konigl. Gesell. d. Wiss. Gottingen," *Nachr. Math.-Phys.* **Kl**: 395–407.

Hofstadter, R. and R. W. McAllister. 1955. "Electron scattering from the proton." *Phys. Rev.* **98**: 217–218.

Jackson, John David. 1999. *Classical Electrodynamics*, 3rd ed. New York: Wiley.

José, Jorge V. and Eugene J. Saletan. 1998. *Classical Dynamics: A Contemporary Approach.* Cambridge, UK: Cambridge University Press.

Lagrange, J. L., Auguste Claude Boissonnade (trans), and Victor N. Vagliente (trans). 1997. *Analytical mechanics.* Boston studies in the philosophy of science [Mécanique analytique]. Vol. 191. Dordrecht; Boston, Mass.: Kluwer Academic Publishers.

Lanczos, Cornelius. 1986. *The Variational Principles of Mechanics.* Dover Books on Physics and Chemistry, 4th ed. New York: Dover Publications.

Landau, L. D. and E. M. Lifshitz. 1976. *Mechanics*, 3rd edition. Oxford, England: Pergamon Press.

Larmor, Joseph. 1897. "On a dynamical theory of the electric and luminiferous medium." *Phil. Trans. Roy. Soc.* **190**: 205–300 (third and last in a series of papers with the same name).

Larmor, Joseph. 1898, "Note on the complete scheme of electrodynamic equations of a moving material medium, and electrostriction." *Proceedings of the Royal Society* **63**: 365–372.

Lasenby, A.; C. Doran; and S. Gull. 1998. "Gravity, gauge theories and geometric algebra." *Phil. Trans. R. Lond. A* **356**: 487–582.

Lorentz, H. A. 1899. "Simplified theory of electrical and optical phnomena in moving systems." *Proc. Acad. Science Amsterdam* **I**: 427–443.

Lorentz, H. A. 1904. "Electromagnetic phenomena in a system moving with any velocity smaller than that of light," in: *KNAW, Proceedings*, Amsterdam **6**, 1903–1904, 809–831.

Lorentz, H. A., Albert Einstein, Hermann Minkowski, and Hermann Weyl. 1952. *The Principle of Relativity: A Collection of Original Memoirs.* Dover.

Lorenz, E. N. 1963. "Deterministic nonperiodic flow." *J. Atmos. Sci.* **20**: 130–141.

Lounesto, P. 1997. *Clifford Algebras and Spinors.* Cambridge: Cambridge University Press.

Mandelbrot, Benoit. 1977. *Fractals: Form, Chance and Dimension.* San Francisco: W H Freeman and Co.

Maxwell, James Clerk. 1865. "A Dynamical Theory of the Electromagnetic Field." *Philosophical Transactions of the Royal Society of London* **155**: 459–512. This article accompanied a December 8, 1864 presentation by Maxwell to the Royal Society. See also [Maxwell 1996].

Maxwell, James C. and Thomas F. Torrance (editor). 1996. *A Dynamical Theory of the Electromagnetic Field.* Eugene OR: Wipf & Stock Publishers. This essay was originally printed by the Royal Society of London in 1864.

Michelson, A. A. and E. W. Morley. 1887. "On the Relative Motion of the Earth and the Luminiferous Ether." *American Journal of Science* **34** : 333–345. See also Philos. Mag. S.5, 24 (151): 449–463.

Misner, C. W.; K. S. Thorne; and J. A. Wheeler. 1973. *Gravitation.* San Francisco: W. H. Freeman.

Mohr, P. J. and B. N. Taylor. 2002. "CODATA recommended values of the fundamental physical constants: 2002." *Reviews of Modern Physics* **77**, 1–107 (2005). One can find the latest CODATA values online at [Mohr 2007].

Mohr P. J., B. N. Taylor, and D. B. Newell. 2007. The 2006 CODATA Recommended Values of the Fundamental Physical Constants (Web Version 5.0). This database was developed by J. Baker, M. Douma, and S. Kotochigova. Available: http://physics.nist.gov/constants [2007, July 14]. National Institute of Standards and Technology, Gaithersburg, MD 20899.

Morii, Masahiro. 2003. *Mechanics lecture notes* (Harvard University). Unpublished. Available at *http://huhepl.harvard.edu/~masahiro/phys151/lectures/*.

Moulton, Forest Ray. 1970. *An Introduction to Celestial Mechanics*. New York: Dover Publications.

Murdoch, Dugald. 1987. *Niels Bohr's Philosophy of Physics*. Cambridge, New York: Cambridge University Press.

Newton, Isaac. 1687. *Philosophiae Naturalis Principia Mathematica*. The Principia was first published in 1687, with the final edition in 1726 before his death considered the most authoritative. For English translations from the Latin see [Newton 1969, 1999].

Newton, Isaac. 1675. Letter to Robert Hooke, dated February 5, 1675.

Newton, Isaac; and Robert Thorp (trans). 1969. *The Mathematical Principles of Natural Philosophy*. London: Dawsons.

Newton, Isaac; I. Bernard Cohen (Trans); Anne Whitman (Trans). 1999. *The Principia: Mathematical Principles of Natural Philosophy*. Berkeley: University of California Press, 1st Edition. (A modern English translation of the *Principia*.)

Noether, Amalie Emmy. 1918. "Invariante Variationsprobleme." *Nachr. v. d. Ges. d. Wiss. zu Göttingen*, 235–257. For an English translation see [Noether 1971].

Noether, A. E. (author) and M. A. Tavel (Trans). 1971. "Invariant variational problems." *Transport Theory and Statistical Mechanics* **1**: 183–207. Available at *http://arxiv.org/PS_cache/physics/pdf/0503/0503066.pdf*.

Perlmutter, S. et al (The Supernova Cosmology Project). 1999. "Measurements of Omega and Lambda from 42 High-Redshift Supernovae." *Astrophys. J.* **517**: 565–586.

Pais, Abraham. 1991. *Niels Bohr's times : In Physics, Philosophy, and Polity*. Oxford, New York: Clarendon Press, Oxford University Press.

Poincaré, Henri. 2001. *The Value of Science: Essential Writings of Henri Poincare*. New York: Modern Library Science.

Riess, A. G. et al (High-z Supernova Search Team). 1998. "Observational Evidence from Supernovae for an Accelerating Universe and a Cosmological Constant." *Astron. J.* **116**: 1009–1038.

Sakurai, J. J. 1967. *Advanced Quantum Mechanics*. Addison-Wesley Series in Advanced Physics. Reading, Mass.: Addison-Wesley Pub. Co.

Sethna, James P. 1996. *Jupiter: The Three Body Problem*. Last updated: 1996. The software for this computer simulation of chaos is available at *http://pages.physics.cornell.edu/sethna/teaching/sss/jupiter/jupiter.htm*.

Szebehely, Victor. 1967. *Theory of Orbits*. New York: Academic Press.

Tabor, M. 1989. *Chaos and Integrability in Nonlinear Dynamics: An Introduction*. New York: Wiley.

Taylor, Barry N. 1995. Guide for the Use of the International System of Units (SI) NIST Special Publication 811, 1995 Edition. (Guide to the SI, with a focus on usage and unit conversions.)

Taylor, Barry N., Editor. 2001. *The International System of Units (SI)* NIST Special Publication 330, 2001 Edition. (Guide to the SI, with a focus on history.)

Thomas, L. H. 1927. "The kinematics of an electron with an axis." *Phil. Mag.* **3**: 1–22.

Thornton, Stephen T., and Jerry B. Marion. 2004. *Classical Dynamics of Particles and Systems*. 5th ed. Belmont, CA: Brooks/Cole.

Tong, David. 2005. *Lectures on Classical Dynamics* (University of Cambridge). Unpublished. Available at *http://www.damtp.cam.ac.uk/user/dt281/teaching.html*.

Yao W. M. et al. 2006. "Review of Particle Physics." *Journal of Physics G* **33**: 1. Available online at the Particle Data Group website: *http://pdg.lbl.gov/*.

Weinberg, S. 1972. *Gravitation and Cosmology: Principles and Applications of the General Theory of Relativity.* New York: Wiley.

Weihs, G.; T. Jennewein; C. Simon; H. Weinfurter; and A. Zeilinger. 1998. "Violation of Bell's Inequality under Strict Einstein Locality Conditions." *Phys. Rev. Lett.* **81**: 5039–5043.

Wheeler, John Archibald. 1990. *A Journey into Gravity and Spacetime.* Scientific American library series. Vol. 31. New York: Scientific American Library: Distributed by W. H. Freeman and Co.

Whittaker, E. T. 1988. *A Treatise on the Analytical Dynamics of Particles and Rigid Bodies: With an Introduction to the Problem of Three Bodies.* Cambridge Mathematical Library. 4th ed. Cambridge: Cambridge University Press.

Wollnik, Hermann. 1987. *Optics of Charged Particles.* London; New York: Academic Press.

Web Based Resources (links last updated in December 16, 2007)

WWW: DLMF. *Digital Library of Mathematical Functions.* (Under development, Expected to be available in book form and online sometime in 2008.) For current status see *http://dlmf.nist.gov/Contents/*.

WWW: History of Mathematics Archive. (Contains biographical materials on many of the physicists and mathematicians referred to in this textbook.) Available at *http://www-history.mcs.st-andrews.ac.uk/*.

WWW: Holistic Numerical Methods Institute. (The HNMI Web site at the University of South Florida contains numerical recipes code for use with Maple, Mathematica, Mathcad, and MATLAB.) Available at *http://numericalmethods.eng.usf.edu/index.html*.

WWW: Matlab Central. (Contains resources for MATLAB users.) Available at *http://www.mathworks.com/matlabcentral/*.

WWW: NASA. *World Wind software (Open Source).* (Arial maps of France require the French *Geoportail plug in*, available from the same site.) Available for download at *http://worldwind.arc.nasa.gov/*.

WWW: NIST. *The NIST Reference on Constants, Units, and Uncertainties.* Available from http://physics.nist.gov/cuu/Units/units.html.

WWW: Wikipedia. (Online encyclopedia containing a useful compendium of scientific, mathematical, and historical information.) Available at *http://www.wikipedia.org/*.

WWW: Wolfram MathWorld. (This Web site contains extensive mathematical resources.) Available at *http://mathworld.wolfram.com/*.

INDEX